SYSTEMS ANALYSIS &
DESIGN METHODS
SECOND EDITION

The Whitten-Bentley-Barlow System of Instruction

- Instructor's Guide with Transparency Masters
- Projects and Cases
- Test Items
- Computest II Computerized Testing Package

All of these supplements were written by Victor M. Barlow, Jeffrey L. Whitten, and Lonnie D. Bentley. When combined with the *Systems Analysis and Design Methods* text, they provide a complete system of instruction for teaching systems analysis and design. For more information see the *Preface* in this text or call Irwin College Publishing at 800-634-3963.

SYSTEMS ANALYSIS & DESIGN METHODS
SECOND EDITION

Jeffrey L. Whitten, MS, CDP
Associate Professor

Lonnie D. Bentley, MS, CDP
Associate Professor

Victor M. Barlow, MBA, CDP
Assistant Professor

All at Purdue University
West Lafayette

1989
Homewood, IL 60430 / Boston, MA 02116

© Richard D. Irwin, Inc. 1989

Senior Editor: Larry Alexander
Developmental Editor: Rebecca J. Johnson
Production: Stacey C. Sawyer, San Francisco
Text Design: Nancy Benedict
Cover Design: Michael Rogondino
Copyeditor: Elizabeth Judd
Illustrator: Pat Rogondino
Typesetter: Graphic Typesetting Service, Los Angeles
Episode Photos: Sara Hunsacker

Library of Congress Cataloging in Publication Data
Whitten, Jeffrey L.
    Systems analysis & design methods.

    Rev. ed. of: Systems analysis & design methods. 1st ed.
    1986.
    Includes bibliographies and index.
    1. System design.   2. System analysis.   I. Bentley, Lonnie D.   II. Barlow, Victor M.   III. Whitten, Jeffrey L. Systems   analysis   &   design   methods.   IV.   Title.
QA76.9.S88W48   1989   003   89-1915
ISBN 0-256-07493-3

Printed in the United States of America
1 2 3 4 5 6 7 8 9 0 - DO - 6 5 4 3 2 1 0 9

CONTENTS IN BRIEF

CONTENTS IN DETAIL

PART TWO   Systems Analysis Tools and Techniques   134

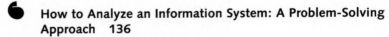

How to Analyze an Information System: A Problem-Solving Approach   136

PART THREE

Systems Design: Tools and Techniques 394

14 Designing Modern Computer Databases   470

15 Designing and Prototyping Computer Outputs and Controls   506

PART FOUR   Skills That Overlap Systems Analysis and Design Phases   704

▲ Project Management Tools and Techniques   706

▉ Fact-Finding Techniques   726

The Intended Audience for This Book

Systems Analysis and Design Methods, second edition, is intended to support a practical first course in computer information systems development. This course is normally taught at the sophomore, junior, or senior level in both two- and four-year colleges and trade schools. We recommend that students have taken an introductory data processing or computer concepts course and at least one programming course before using this book. The book can be used for any introductory systems development or software engineering course in either the DPMA, ACM, or independent computer science or information systems curriculum.

Why We Wrote This Book

Today's students are "consumer oriented"—they expect their money's worth for every course. They expect to walk away from a course with more than just a grade and the promise that someday they'll appreciate what they've learned. They prefer to practice (and, we hope, master) career-oriented skills and techniques. Like most instructors, we have addressed the consumer students by providing practical projects for learning and applying systems analysis and design methods. Our enthusiasm for teaching the subject was tempered by both student and instructor dissatisfaction with the available textbooks. We wrote the first edition to solve the following problems found in other books:

- **Many books are too conceptual.** While they reinforce concepts with examples, the examples are either too late or too few to be helpful when dealing with real projects.

- **Some books are more practical, but too mechanical.** They leave the student with the impression that systems analysis and design is only tools and techniques, so the students do not get the sense of the "people" side of systems work—working with users, resolving conflicts, and the like.

- **Many books still perpetuate the myth that classical, structured, and prototyping techniques are mutually exclusive.** Those practicing analysis and design know that this isn't true. All techniques have their relative advantages and disadvantages. Unfortunately, few books attempt to integrate the popular techniques in a complementary fashion.

- **Most books lack sufficient examples to demonstrate concepts and techniques.** This problem has been addressed by some of our competitors since our first edition: however, we still perceive a lack of **sufficient** examples in most books. Consequently, students find it difficult to apply the techniques **after** completing the course.

- **Most books do not offer a glimpse into the future directions of systems analysis and design.** Very few information systems professions are changing as quickly as analysis and design. Students need to be aware of the changes on the horizon. Our goal was to write a book that overcomes this problem. Additionally, we wanted to write a text that not only teaches the students but serves as a useful post-course reference guide.

Changes in the Second Edition

We have made numerous changes for the second edition. We have preserved all the features that you liked in the first edition, including the continued use of the pyramid model (improved!) to provide a visual framework for concepts and principles, the extensive use of examples and walkthroughs, the people-oriented running case study, and the Next Generation box feature. But we've also given you more. Some of the newer features are:

- **Explicit coverage of systems development methodologies for systems development, especially the so-called structured techniques.** This coverage incorporates some of the latest changes as introduced in Yourdon's modern structured analysis. Included is a complete chapter on structured methodologies (Chapter 5) that is reinforced in the subsequent techniques-oriented chapters.

- **Separate chapters for physical and logical data flow diagramming.** This edition substantially improves coverage of logical process modeling as a requirements analysis technique.

- **A complete chapter on data modeling using Chen's entity relationship technique.** Data modeling has become the method of choice in many systems analysis circles. Chapter 8 includes a visually supported approach to **normalization**.

- **A complete chapter on database design.** Chapter 14, with appropriate threads to the data modeling chapter, has evolved from Part Four, Module E in the first edition. In addition to introducing the database concept to students who are unfamiliar with the approach, the chapter demonstrates a simplified transition from the conceptual data model to each of the three alternative physical data models: relational, network, and hierarchical.

- **Prototyping is a running theme.** Prototyping is an important, complementary technique for students. Although it is nearly impossible to fully demonstrate prototyping outside of the context of a live demonstration, we felt it important to introduce the topic early (Chapter 5) and reinforce the importance of prototyping throughout relevant technique chapters in Parts Two and Three.

- **Computer-assisted systems engineering (CASE)** (also known as computer-aided **software** engineering) **is a running theme.** We adopted the **systems** interpretation since it is more consistent with industry **trends** and the systems orientation of this book. CASE is introduced in Chapter 5 and reinforced by the use of CASE diagrams and dictionary printouts throughout the text. Even if students don't have hands-on access to CASE technology, they should see the technology being used since they will likely encounter it in industry.

- **Minicases have been added to the problems and exercises at the end of most chapters.** The new minicases are an addition to the first edition's chapter-opening case and the running case study.

- **A new running "Analysts in Action" case study was written.** This new running case is from a setting that many students will be familiar with: a record and tape club.

Why We Think You Should Consider This Book

We are even more excited by this second edition than we were about the first edition. In the first edition, our publisher challenged you to "discover the difference." For the second edition, we think you'll find that we still offer you a different, better alternative to teaching a first course in systems analysis and design. We think you should consider adopting this text because:

- **This book covers more topics than other texts.** We cover several subjects that receive either little or no coverage in other texts. For example, we provide **entire** chapters on:
  —Systems development methodologies (separate from, but consistent with, the life cycle)
  —Data dictionaries and codes
  —User interface and terminal dialogue design
  —Physical data flow diagrams
  —Logical data flow diagrams
  —Data modeling with entity relationship diagrams
  —Database design (separate from file design)

- **We present both concepts and principles as tools and techniques.** The text provides a solid learning path from concepts to principles to techniques to tools.

- **Our book is practical and skills-oriented, but not mechanical.** We emphasize the role of the analyst, interpersonal skills, human engineering, and user interaction.

- **Our book is highly integrated in its use of classical and modern techniques.** In the first edition, we presented both the classical and the so-called structured techniques. This strength has been enhanced in the second edition by integrating the use of classical and structured techniques with modern techniques such as prototyping and CASE.

- **We have included many more examples than other systems texts.** Reviewers have consistently noted the abundance of practical examples in this text. It makes the book longer; however, students learn by doing, which must be preceded by seeing. They need moderately sized sample systems so that they can see the relationships between the various concepts, tools, and techniques.

- **This book has value beyond the classroom.** Students frequently discard texts because the texts don't contain enough substance to help them beyond the classroom requirements. No matter what percentage of our book is covered in class, it will remain a practical, example-loaded reference book, whose value extends far beyond the course in which it is used.

- **This book is up-to-date and points to the future of systems analysis and design.** We cover the latest tools and techniques. For example, the book teaches an integrated approach to data and process modeling. It teaches prototyping and CASE. It teaches structured analysis and design. At the same time, its Next Generation box features draw the students' attention to emerging and future applications, tools, and techniques.

- **The book offers flexibility of topic coverage.** Specialized skills that overlap systems analysis (Part Two) and Systems Design (Part Three) are presented as modules in a separate unit (Part Four). These skills include project management, interpersonal communications, fact-finding, and cost/benefit analysis. The instructor is encouraged to weave these modules into his or her course at whatever points are most appropriate.

- **Our in-text aids have been carefully developed to maximize student learning.** They include:
  —**Chapter-Opening Minicases.** Most chapters begin with a minicase and discussion questions to sensitize the reader to the chapter's issues. Some minicases demonstrate successes and others demonstrate problems or failures.

—**What Will You Learn in This Chapter?** After the minicase, each chapter begins with objectives that the student will achieve by reading and applying the chapter.

—**Running Case Study: Analysts in Action.** The case study features a typical project going through its system analysis and design phases. The running case study is presented in two ways:

- Eight episodes emphasize the interpersonal aspects of systems work, especially the interaction between analysts and users. Tools and techniques are demonstrated, but not taught, to motivate reading of the tools and techniques chapters that follow each episode.

- Within the tools and techniques chapters, the running case provides the examples used to demonstrate proper use of the tools and techniques.

—**Problems and Exercises.** Most problems and exercises have been designed to avoid having their effectiveness reduced by the buildup of files containing the correct answers.

—**Projects and Minicases.** New to the second edition. Most chapters include projects or case problems that require more thought and effort from the students than the problems and exercises.

—**Annotated References and Suggested Readings.** This is not a bibliography. The references are annotated with the authors' comments about their favorite books, papers, and articles.

- **We have written the book using a lively, conversational tone.** Today's students often lose interest in textbooks because they are written in a dry, factual tone. We are concerned when students perceive systems analysis and design to be a dry subject—it is such an exciting field! Therefore, we wrote this text in a conversational, "talk with you, not at you" tone. We hope the style doesn't offend anyone or patronize any specific audience. And we apologize if it does. Our hope, supported by the success of the first edition, is to hold the readers' interest over a long period of time.

We hope these reasons will convince you to consider this book. We also hope you'll discover it to be the book you've been waiting for.

How to Use This Book

Systems Analysis and Design Methods, second edition, is divided into four parts. The first three parts are generally sequential, although experience with the first edition suggests that instructors can easily omit chapters that they feel are less appropriate for their students and resequence chapters (especially within a part) to meet their preferences.

Part One, "Systems Analysis and Design: Concepts, Philosophies, and Trends," presents the information systems development situation and environment. This part introduces the analyst, information systems, the systems development life cycle, and the structured methodologies. A visual model, the **information systems pyramid**, is developed to organize and relate the concepts and principles. Part One can be covered quickly or in depth, depending on students' backgrounds, course schedule, and instructor's preferences.

Parts Two and Three make up the major subject matter of the text. They cover "Systems Analysis Tools and Techniques" and "Systems Design Tools and Techniques," respectively. We organized these chapters around the proven model for successful systems, the **systems development life cycle**. Each part begins with a chapter that expands on specific phases of the life cycle (originally surveyed in Part One, Chapter 4). Subsequent chapters develop specific skills using tools and techniques that the student can immediately apply to projects and problems.

Part Four is unique. Market research identified a number of subjects whose course "location" differs widely from one school to another. Examples included project management, fact-finding, interpersonal communications, and cost/benefit analyses. This variability was not surprising since these subjects overlap the various phases of the life cycle. We organized these subjects into self-contained modules in Part Four. Instructors may elect to cover none, some, or all the modules, cover them in part or in whole, cover them in any sequence, and introduce (or review) them at any time. Alternatively, the instructor could assign the modules for supplemental reading and not cover them in class.

Additional course design alternatives and textbook use guidelines may be found in the **Instructor's Guide** that accompanies our text.

A Word About Size

This is a large book. But it is not nearly as large as you might think. First, the conversational writing style contributes to greater size—but the readability is improved. Second, the large number of examples add pages, but they also enhance the student's level of understanding. Finally, in response to a frequently asked question, no, not even we cover the entire book in a single course.

Supplements

It has always been our purpose to provide adopters with a course, not just a textbook. To this end, we provide comprehensive supplements.

Instructor's Guide

This guide is not just your everyday combination of key terms, topical outlines, problem answers, and transparency masters! We've appreciated your positive response to the first edition's guide and have retained its unique features:

- **Course planning, design, scheduling, and control suggestions** for both quarters and semesters, as well as single- and multi-course plans of study. Planning aids include a template for a course syllabus.
- **Textbook conversion aids** to help instructors convert from their current text (and our first edition!) to the second edition of Whitten/Bentley/Barlow.
- **Lesson planning guidelines** that offer numerous options for using classroom time. Lessons are designed around student objectives. For each objective, alternative classroom approaches (lecture, lab, discussion, workshop, demonstration) are directed toward the objective. Evaluation mechanisms, other than those in the book and Projects and Cases supplement, are offered.
- **Additional references** for the instructor's preparation.
- **Guidelines** for integrating and using the **Projects and Cases** and **Excelerator** supplements.
- **Blank forms and charts** with duplication permission (contingent on adoption of the textbook or supplements).
- **Transparency masters** of (1) key graphics from the book, (2) adaptations of textbook graphics, and (3) new graphics.

Projects and Cases

Not just a case book! The second edition has preserved the best features of the first edition case book and added some new value.

Many first-edition adopters felt our **Build Your Own Case** option was the most innovative and valuable feature of the case book. This approach attempts to overcome the most common flaw of case studies—the difficulty of distilling the complexities of a realistic, practical situation to the English language. As a result, these "canned" case studies often contain inadvertent omissions and discrepancies that confuse the students and instructors.

Build Your Own Case is a controlled approach that allows students to build their own systems analysis and design project from their own work or life experience (which doesn't have to be in computers or information systems). The controls ensure each system reinforces textbook concepts and techniques as well as preventing the system size from growing uncontrollably.

Student interest in this approach has been positive. Many first-edition instructors found the variety of systems projects to be stimulating.

One alternative use of the **Build Your Own Case** approach allows the instructor to build a single common case study for each term's students to prevent build-up of "solutions" shared from previous semesters' students. In this scenario, the control features of **Build Your Own Case** serve the instructor.

The major addition to the Projects and Cases book for the second edition is the inclusion of two "canned" case studies. Despite the inevitable (and previously mentioned) problems with canned case studies, we now recognize that some instructors simply don't have enough free time or class time to create their own case studies or help students create their own case studies. Our canned case studies were developed from our **Build Your Own Case** approach to ensure consistency with the original case book.

No matter which approach you adopt, the case book provides term project milestones that are cross-referenced to the textbook. Each milestone gives the student a checklist for common errors and omissions.

CASE Tutorial

Systems Analysis and Design Methods, second edition, can be used in conjunction with our tutorial text, **Using Excelerator for Systems Analysis and Design**. This tutorial, currently compatible with versions 1.7 and 1.8 of **Excelerator**, walks the student through a series of lessons that demonstrate the use of CASE (computer-assisted systems engineering) for systems analysis and design. The **Instructor's Guide** provides cross-references for those who choose to use this tutorial.

At the time of this writing, adopting schools of **Systems Analysis and Design Methods**, second edition, are entitled to receive one free copy of **Excelerator** (not a limited or crippled version). Additional reduced-cost copies may be available through Index Technology's educational support programs.

Test Bank

A written test bank includes approximately 3000 items using the following formats: true/false, multiple choice, matching, and sentence completion. The test bank includes both correct answers and explanations or rationales for incorrect answers.

Acknowledgments

We are indebted to many individuals who have contributed to the development of the first and second editions of this textbook. First, we wish to thank those many individuals who reviewed the first edition of the text—we have tried to retain all of the features that you endorsed in that edition.

Second, we would like to thank the reviewers of the second edition. Their patience, constructive criticism, and suggestions were essential! As you might guess, reviewers' opinions varied. We have incorporated the majority opinions and the essence of the reviewers' advice into this second edition. We early look forward to continued guidance for the next edition! The second edition reviewers were:

Charles P. Bilbrey
James Madison University

Barbara B. Denison
Wright State University

Patricia J. Guinan
Boston University

Riki S. Kuchek
Orange Coast College

Ronald J. Norman
San Diego State University

Charles E. Paddock
University of Nevada at Las Vegas

June A. Parsons
Northern Michigan University

Jerry Sitek
Southern Illinois University
at Edwardsville

Craig W. Slinkman
University of Texas at
Arlington

Ahmed S. Zaki
College of William and Mary

We also include our thanks to our developmental editors for the first and second editions. Susan Solomon, your special talents, advice, and friendship have been as much a part of this book as our own thoughts! Liz Currie, much of the essence of the first edition was a direct result of your efforts. Larry Alexander and Rebecca Johnson, your patience and perseverance in this second edition was "above and beyond the call of duty." And to Stacey Sawyer, our production coordinator, who demonstrated superior attention to detail in incorporating the complicated elements of design in this book: we could not have done this without you! For the second edition, additional thanks is offered to Nancy Benedict, Pat Rogondino, and Gail Mandaville. Special thanks is given to Dorothy Jane Miller, our secretary, who provided encouragement, overtime, patience, and "freedom from interruption" at appropriate times during this project, and to Judith Vanderkay, Index Technology.

We also thank our students. You are truly special and make teaching a rewarding experience. This book is for you!

We hope we haven't forgotten anyone. And we assume full responsibility for any errors in this text. Any comments, suggestions, or improvements are welcome and appreciated. Write to us in care of Irwin Publishing, 1818 Ridge Road, Homewood, IL, 60430.

To those who used the first edition, thank you for your continued support. For new adopters, we think you'll really see a difference in this text. And until we unveil the third edition, enjoy!

Jeff Whitten
Lonnie Bentley
Vic Barlow

To my lovely wife, Debbie. Your love and encouragement have given new significance to my every accomplishment. Also, to my mother and father. You have always been a source of inspiration.—Jeff

To my wife and best friend, Cheryl. And to my children, Coty, Robert, and Heath. God blessed me with a wonderful family.—Lonnie

To my lovely wife, Linda. Your love and support made this possible. Also, to my parents, who always encouraged me to attain my goals.—Vic

To the students of the Computer Technology Department at Purdue University, West Lafayette, and Statewide Technology sites. May all your experiences be successful.—Jeff, Lonnie, and Vic

SYSTEMS ANALYSIS &
DESIGN METHODS
SECOND EDITION

PART ONE

Systems Analysis and Design: Concepts, Philosophies, and Trends

1
The Information Systems Analyst

2
The Business and Its Users as a System

3
Modern Information Systems

4
A Systems Development Life Cycle

This is a practical book about systems analysis and design methods. So why start with concepts, philosophies, and trends? Many concepts units seem only to fill space in textbooks. They have that "because I told you so!" feel to them. So why should you study this unit?

Systems analysis and design aren't mechanical activities. There are no magical secrets for success; no perfect tools, techniques, or methods. Programming computers may be a skill, but analyzing and designing systems are still very much an art. Fortunately, most of you have the natural ability and interest to succeed in this art. But you need to be trained to discover, refine, and extend your natural abilities. We start here with concepts, philosophies, and trends . . . the basics. If you understand these concepts and philosophies, you will be able to apply, with confidence, the practical tools and techniques we present. Furthermore, you will find yourself able to adapt to new situations and methods. In addition to concepts and philosophies, we will also discuss trends that will affect systems analysis and design.

Five chapters make up this unit. Chapter 1 introduces you to the **information systems analyst**, the professional most likely to practice the methods presented in this book. You'll also learn about the relationships among systems analysts, computer programmers, and end-users of computer applications. Finally, you learn how to prepare yourself for a career as an analyst.

Chapter 2 introduces the **business as a system** and examines more closely the information needs and perspectives of the **end-users**. A business and user perspective on systems analysis and design is a critical success factor for developing good information systems.

Chapter 3 focuses on the **information system** as the product developed by systems analysts for businesses and their users. We will describe information systems in terms of their capabilities, inputs, outputs, memory, and processing components.

Chapter 4 introduces *a* (not *the*) general process for developing systems. This is called a **systems development life cycle**. This life cycle will provide the framework for the rest of the book's discussion of tools, techniques, and methods.

5

The Structured
Methodologies

Chapter 5 introduces a framework for systems analysis and design methods. This framework, called the **structured methodologies**, is gradually changing systems analysis and design from an art to an engineering-like discipline. The actual structured tools will then be taught in Parts Two and Three of the book.

As we've stated, these five chapters teach concepts, philosophies, and trends; however, we think you'll soon see that the presentation is very practical and down-to-earth. Part One will give you a solid foundation on which to build your study of the tools and techniques presented in Parts Two, Three, and Four of this book.

CHAPTER ONE

The Information

Systems

Analyst

What Will You Learn in This Chapter?

This chapter introduces the systems analyst, the person who usually performs the tasks described in this book. You will know that you understand what a systems analyst is when you can

1. Define the systems analyst's role and responsibilities in a typical organization.
2. Define *systems analysis, systems design, systems implementation,* and *systems support*—the principal activities performed by the systems analyst.
3. Differentiate between a *systems analyst*, a *programmer/analyst*, and other common synonyms.
4. Differentiate between the types of work done by the systems analyst and the computer programmer.
5. Describe how the systems analyst fits into the information systems department.
6. Develop a plan of study for your education/training that will prepare you for a career as a systems analyst.

Who Should Read This Book?

Even if you're not planning to become a systems analyst, you can use this book! Whether you are preparing for a career as a computer professional or as a professional who will be an end-user of computers and computing services, systems analysis and design is an important subject to know and understand. Why? Because, regardless of your career choice, your job will probably bring you in contact with systems analysts. But just what is a systems analyst and what does one do? This book presents the answers to these questions.

Where Did the Systems Analyst Come From?

In the beginning, there was the computer—or so it seems! Truly, computers have become a way of life for today's high-tech society. Look around you. There are computers in your home appliances and new cars. Your bills are generated by computers. You receive junk mail because you are on everybody's computerized mailing list. You use computerized bank machines to get money from the bank at all hours. Do you have any idea how much information a person could obtain about you if he or she gave a computer your social security number? The computer revolution has changed all our lives. But you haven't seen anything yet!

Computer technology is improving at an astonishing rate. Mainframe and minicomputers get more powerful with each passing year. And what about microcomputers? When IBM announced its 80386-based PS/2 Model 80, it described its processing power as being equivalent to a single-user, Model 370 mainframe from the not-too-distant past. Today, we are seeing major systems designed for desktop microcomputers. Furthermore, we are now connecting mainframes, minis, and micros in more sophisticated ways to provide complex networks of support for computer applications.

Still, despite all of their current and future technological capabilities, computers owe their power and usefulness to people. In other words, computers are only tools that offer the opportunity to collect and store enormous volumes of data, process business transactions with great speed and accuracy, and provide timely and relevant information for management. Businesspeople define the applications and problems to be solved by the computer, and computer programmers and technicians apply the technology to well-defined applications and problems. Unfortunately, the potential of computers has not been fully or even adequately realized in most businesses. Business users do not completely understand the capabilities and limitations of modern computer technology. In fact, it is becoming increasingly difficult for computer professionals to keep up with new technological developments. Likewise, computer programmers and technicians frequently do not understand the business applications that they are trying to computerize. In short, a communications gap has developed between those who need the computer and those who understand the technology.

The systems analyst bridges the gap between business users and computer programmers/technicians. The information systems analyst evolved from the need to improve use of the computer resource for the information processing needs of business applications. Many of the failures and disappointments of business computing can be traced to the preoccupation of computer professionals with the computer and the failure of these professionals to recognize the problems and needs of the business system. All too frequently the computer solutions are designed without regard for the people who will use them.

Has the systems analyst truly bridged the communications gap between the business users and computer programmers/technicians? Not really! The need for the systems analyst was recognized long before we developed effective tools, techniques, and methodologies to build the bridge. Today, these methods exist! And new methods promise higher productivity and increased success for future systems analysts. This book is about some of those methods, current and future.

What Does a Systems Analyst Do?

Let's summarize the systems analyst's role and responsibilities:

> A **systems analyst** studies the problems and needs of an organization to determine how people, methods, and computer technology can best accomplish improvements for the business. When computer technology is used, the analyst is responsible for the efficient capture of data from its business source, the flow of that data to the computer, the processing and storage of that data by the computer, and the flow of useful and timely information back to business users.

Analysts do more than analyze and design systems. Analysts sell the services of information systems. They sell change. They exploit the latest technological innovations.

The title *systems analyst* is frequently misused in the industry. Organizations assign this title to anyone from a computer programmer to a sophisticated designer of computer applications. What is even more confusing is that, in many organizations, computer programmers actually perform some or all of the duties of a systems analyst. In other organizations systems analysts perform the duties of the computer programmer. So how can you identify a true systems analyst? It is best to regard the title with some skepticism until you see or hear a job description. A realistic job description is depicted in Figure 1.1.

There are several legitimate variations on the title *systems analyst*. The most common of these is *programmer/analyst*, whose job includes the responsibilities of both the computer programmer and the systems analyst. Other common variations on the title are listed in the margin.

**SYNONYMS FOR
SYSTEMS ANALYST**

Systems Designer

Systems Engineer

Systems Consultant

Management
Consultant

Operations Analyst

Information Analyst

Data Analyst

Business Analyst

Comparing Responsibilities of the Systems Analyst and Programmer

Essentially, the systems analyst performs systems analysis, systems design, systems implementation, and systems support for computer-based business applications. These terms are defined on page 9.

JOB DESCRIPTION

JOB TITLE: Systems Analyst   REPORTS TO: Development Center Manager

DESCRIPTION: Gathers and analyzes data for developing information systems. A systems analyst shall be responsible for studying the problems and needs set forth by this organization to determine how computer equipment, business procedures, and people can best solve these problems and accomplish improvements. Designs and specifies systems and methods for installing computer-based information systems and guides their installation. Makes formal presentations of findings, recommendations, and specifications in formal reports and in oral presentations.

RESPONSIBILITIES
1. Evaluates projects for feasibility.
2. Analyzes current business systems for problems and opportunities.
3. Defines requirements for improving or replacing systems.
4. Evaluates alternative solutions for feasibility.
5. Selects hardware and software products (subject to approval).
6. Designs system interfaces, flow, and procedures.
7. Supervises system implementation.

DUTIES
1. Estimates personnel requirements, budgets, and schedules for systems projects.
2. Develops and implements systems development plans according to CIS standards.
3. Performs interviews and other data gathering.
4. Documents and analyzes current system operations.
5. Formulates applications of current technology to business problems.
6. Educates user management on capabilities and use of current technology.
7. Evaluates technological possibilities for technical, operational, and economic feasibility.
8. Reviews and presents proposed systems solutions for approval.
9. Designs and tests system prototypes.
10. Designs file and database structures.
11. Designs user interfaces (input, output, and dialogue) to computer systems.
12. Designs data collection forms and techniques.
13. Designs systems security and controls.
14. Prepares specifications for applications programs.
15. Writes, tests, and integrates applications programs.
16. Supervises applications programming.
17. Develops and guides systems testing and conversion plans.   (*continued*)

FIGURE 1.1   Job Description for a Typical Systems Analyst   The job description for a systems analyst will vary from firm to firm. This description is representative of a systems analyst.

EXTERNAL CONTACTS
1. Users of computing services
2. User management
3. Technical support personnel
4. Data administration personnel
5. Systems and programming personnel
6. CIS operations personnel
7. Applications programmers
8. Computer hardware and software vendors
9. Other systems analysts (on project teams)

QUALIFYING EXPERIENCE
1. A bachelor's or master's degree in computing, business, statistics, or industrial engineering is required.
2. Programming experience is mandatory.
3. Training or experience in business functions is desirable.
4. Training in systems analysis, especially structured methods, is desirable.
5. Good communications skills—oral and written—are mandatory.

TRAINING REQUIREMENTS
1. Systems development standards
2. Database methods
3. Data communications methods
4. Structured systems development methods
5. Prototyping methods
6. Feasibility and cost-benefit methods

CAREER PATH

Promotion based on time in rank and performance. Job levels are:

1. Systems Analyst 1: 30% analysis and design, 70% programming and implementation.
2. Systems Analyst 2: 50% analysis and design, 50% programming.
3. Systems Analyst 3: 70% analysis and design, 30% programming (mostly prototyping).
4. Senior Analyst: 30% project management, 60% analysis and design, 10% programming (mostly prototyping).
5. Lead Analyst: 75% management of several project managers, 25% analysis and design.

Systems analysis   is the study of a current business system and its problems, the definition of business needs and requirements, and the evaluation of alternative solutions.

Systems design   is the general and detailed specification of a computer-based solution that was selected during systems analysis. Design specifications are typically sent to computer programmers for systems implementation.

Systems implementation   is placing the system into operation. Computer programs are written and tested, managers and users are trained to use the new system, and operations are converted to the new system.

Systems support   is the ongoing support of the system after it has been placed into operation. This includes program maintenance and system improvement.

This book assumes that you are familiar with the job of a computer programmer. With this in mind, we should compare the responsibilities and job characteristics of the systems analyst with those of the computer programmer. The jobs are certainly not similar.

For the computer programmer, we note the following characteristics (adapted from DeMarco, 1978). First, the programmer's responsibility rarely extends beyond the computer program itself. The scope of the programmer's world includes the computer, the operating system and utilities, and the programming languages in use—for example, COBOL or BASIC. Second, the programmer's work is very *precise;* that is, a program instruction is right or wrong. The program's logic is correct or incorrect. Finally, the programmer's job involves fewer interpersonal relationships. The programmer normally deals with other programmers and with the systems analysts who prepared the program specifications.

How do the analyst's responsibilities and job characteristics differ from those of a programmer? First, the systems analyst must deal with more than computer programs. The analyst is responsible for the computer equipment selected, the people who will use the system, the procedural aspects of the system (such as how data will be captured), and the files/databases of the system. Second, the work performed by the analyst cannot be considered precise. There are few right or wrong answers. System solutions are often compromise solutions. This can be frustrating for the student or inexperienced analyst, who often feels confused or insecure about not always being able to find *the* answer. Actually, the opportunities for creativity in finding *an* answer can be very satisfying. Finally, the interpersonal relationships are more numerous and complex. The analyst must be willing and able to deal with his or her business clients (often called *end-users,* or *users*), business management, programmers, information systems management, auditors, and information systems salespeople. All of these people have their own motivations, desires, and jargon.

Consequently, although the job of systems analyst is complex and demanding, it presents a fascinating and exciting challenge to many individuals. It offers high management visibility and opportunities for important decision making and creativity that may affect an entire organization. Furthermore, this job can offer these benefits relatively early in your career, compared to other jobs at the same level.

The Analyst's Position and Role Within Computer Information Systems

SYNONYMS FOR
INFORMATION
✓ *SYSTEMS*
DEPARTMENT

Information Systems (IS)

Management
Information Systems
(MIS)

Computer Information
Systems (CIS)

Management Systems

Data Systems

Information Resources

Information or Data
Services

Systems analysts normally work for the computer information systems (CIS) department in an organization. Common synonyms for that department are listed in the margin. Externally, CIS may report to one of several departments, although many organizations have learned that operating a CIS shop under any one application area results in conflicts with other areas that need computer support. In many organizations, the CIS manager has achieved vice president status or, more recently, the designation of *chief information officer (CIO)*.

The internal organization of CIS can be depicted by organization charts. Although the structure of the CIS function varies from company to company, the generic structure presented in Figure 1.2 is becoming common. It reflects changes that are currently being driven by trends in database, microcomputers, and data communications usage.

Although the names may vary, there are four main centers of activity: development center, information center, database center, and computer center. *Development center* is a relatively new term that encompasses systems analysts and programmers who develop and support key systems for the organization's users and management. *Information center,* another relatively new term, includes analysts and programmers (although they are not referred to as such) who help the organization's users and management develop and maintain their own systems, which are usually smaller and less strategic than the systems developed by the development center. The creation of information centers was largely motivated by the explosion of microcomputers in the offices of end-users and management. The *database center,* often called *data administration,* tracks and manages the enormous volume of data that is stored in the average organization. The *computer center* provides central computing resources and services such as data entry, operations, systems programming, and data communications (between the central computers, remote sites, and microcomputers).

Most analysts and programmers work in the development center. They are organized into temporary work teams called *projects*. These teams are formed and disbanded as projects come and go. As shown in Figure 1.2, the organization of a typical project team includes a project leader (often an experienced analyst), systems analysts, and programmers.

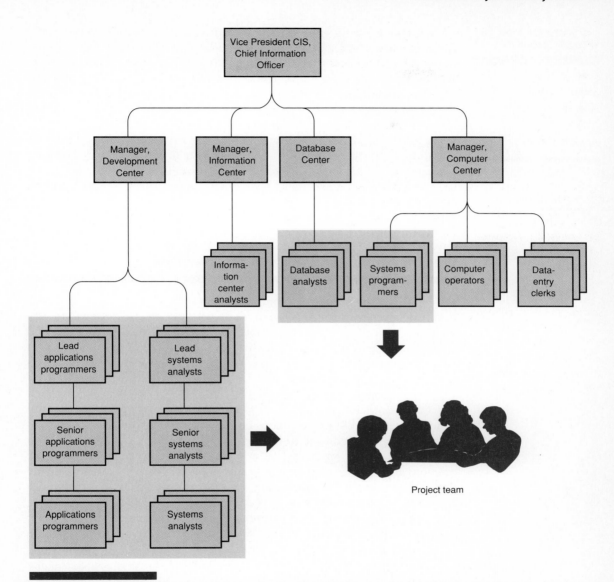

FIGURE 1.2 Organization of the Information Systems Function Every information systems shop develops its own unique structure; however, this structure is fairly typical. Notice how programmers and systems analysts are organized into project teams. These teams will be created and disbanded as projects are started and completed or canceled.

What is the analyst's role in the system project? The analyst may be viewed as a *facilitator.* The analyst acts as the interface among many different types of people and facilitates the development of computer applications through these people (see Figure 1.3). The analyst may well be the only individual who sees the big picture of the system! Within the systems analyst job category, various levels of responsibility may exist. They are listed in the margin, in order of decreasing responsibility.

Before we leave this topic, we should note that not all analysts work in CIS shops. Software companies require analysts to design generic software packages, those that will appeal to the most customers (which, in many cases, are CIS shops for organizations). Consulting companies also require analysts, although they usually call them consultants.

Preparing for a Career as a Systems Analyst

This book will not make you a competent systems analyst any more than your first programming book or course made you a competent computer programmer. You will, however, be able to immediately apply the skills and concepts you learn in this book to systems projects, although you may need some supervision. Furthermore, you will have a solid foundation on which to base additional systems training.

We have suggested that the analyst is the principal link between business users and computer programmers. What does it take to become a successful systems analyst? Most organizations consider computer programming experience to be a prerequisite to systems analysis and design experience. The reasoning is sound. An analyst must be able to develop technically feasible solutions and precise program specifications. You should not, however, assume that a good programmer will become a good analyst or that a bad programmer could not become a good analyst. There is no such correlation. Unfortunately, many organizations insist on promoting good programmers who become poor or mediocre systems analysts. Worse still, poor programmers are often passed over in the belief that they cannot become good analysts.

That brings us back to our original question, *what does it take to become a successful systems analyst?* One writer suggests the following:

I submit that systems analysts are people who communicate with management and users at the management/user level; document their experience; understand problems before proposing solutions; think before they speak; facilitate systems development, not originate it; are supportive of the organization in question and understand its goals and objectives; use good tools and approaches to help solve systems problems; and enjoy working with people (Wood, 1979).

FIGURE 1.3   People with Whom the Analyst Must Work   As facilitators of systems development, the analyst must work with many types of people, both technical and nontechnical. The joint efforts of these professionals, as coordinated by the analysts, will result in successful computer applications.

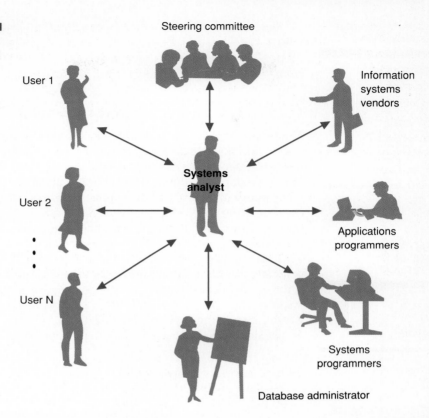

Steering committee

User 1

Information systems vendors

Systems analyst

User 2

Applications programmers

User N

Systems programmers

Database administrator

TECHNICAL TOPICS OF INTEREST TO WELL-INFORMED ANALYSTS ✓

Database Techniques

Data Communications

Artificial Intelligence

Microcomputers

Microcomputer Software

Graphics

Computer-Integrated Manufacturing

Fifth-Generation Computers

Computer Security

A tall order, no? It is often difficult to pinpoint those skills and attributes necessary to succeed. However, the following text describes those skills most frequently cited by practicing systems analysts as essential to their work.

Working Knowledge of Information Systems Techniques and Technology

The systems analyst has been accurately described as an "agent of change." End-users must be shown how new technologies can benefit their operations, so that the analyst must be cognizant both of existing technologies and techniques and of future trends. Initial knowledge can be acquired in college courses, professional development seminars/courses, and in-house, corporate training programs. Some of the technologies that you should be studying today are listed in the margin. The best way to keep up on what's happening is to develop a disciplined and organized habit of skimming and reading various trade periodicals on information systems. Examples of helpful trade publications can also be found in the margin on the next page. Most of these magazines should be available in your college or business library.

GENERAL CIS TRADE PUBLICATIONS

Computerworld
Datamation
Computer Decisions
MIS Week
Infosystems
InformationWEEK
EDP Analyzer

MICROCOMPUTER TRADE PUBLICATIONS

Byte
PC Week
InfoWorld

SPECIALIZED TRADE PUBLICATIONS

Data Communications
Telecommunications
Manufacturing Systems
Information Center
Office Systems

Additionally, you should consider joining a professional association for information systems specialists. Students and professionals alike are encouraged to join organizations such as the Data Processing Management Association (DPMA) and the Society for Information Management (SIM).

Computer Programming Experience and Expertise

Although we mentioned programming earlier, we can expand on this skill. Become proficient in at least one high-level programming language. For business applications, this language should probably be COBOL. For engineering or scientific applications, FORTRAN or Pascal may be the better choice. Some companies with critical real-time applications (for instance, computer-assisted manufacturing) may soon switch to Ada. Software houses—companies that build software products for sale to other companies—are increasingly using the language C, especially for microcomputer software packages.

Today's analyst should also become familiar with at least one fourth-generation language/applications generator. Examples are listed in the margin at the top of the next page. Most of these languages are built around database systems. These languages are used to support prototyping, a technique that we will discuss throughout this book. Language familiarity will help analysts design systems to better exploit the capabilities of new languages.

General Business Knowledge

In most instances, the systems analyst need not be an expert in a specific business application or function, such as accounting or production. However, analysts should be able to communicate with business experts to gain knowledge of problems and needs. Much of this knowledge will be acquired while on the job. Still, we strongly suggest that you include one course in each of the subjects listed in the margin at the bottom of the next page. Specializations such as accounting or production can also be very valuable in some instances.

Problem-Solving Skills

The systems analyst must have the ability to take a large business problem, break that problem down into its component parts, analyze the various aspects of the problem, and then assemble a system to solve the problem. The analyst must also learn to analyze problems in terms of causes and effects rather than in terms of simple remedies. Methodologies, such as *Structured Analysis,* have emerged for assisting the analyst in the problem-solving process. Being well organized is also part of developing good problem-solving skills.

Analysts must be able to creatively define alternative solutions to problems and needs. Creativity and insight are more likely to be gifts than skills, although they can certainly be developed to some degree. Perhaps the best inspiration

**FOURTH-
GENERATION
LANGUAGES**

FOCUS
IDEAL
NATURAL
ADS/O
RAMIS
SAS
MANTIS
RBASE
dBASE
ORACLE

BUSINESS SUBJECTS

Financial Accounting

Managerial or Cost
Accounting

Quantitative Methods
(for example,
statistics)

Marketing

Production and
Inventory Operations

Personnel Management

Business Finance

Organizational Behavior

Business Law and Ethics

for students and beginning analysts comes from Rear Admiral Grace Hopper, USN Retired and founder of the language COBOL. She suggests that "the most damaging phrase in the language is 'We've always done it that way.' " Always be willing to look beyond your first idea for other solutions.

Interpersonal Communications Skills

An analyst must be able to communicate effectively, both orally and in writing. The analyst should actively seek help or training in business writing, technical writing, interviewing, presentations, and listening. A good command of the English language is considered essential. These skills are learnable, but most of us must force ourselves to seek help and work hard to improve them. Communications skills are probably the single most important ingredient to success. This is because of the number and complexity of the interpersonal relationships previously discussed. Some of the modules in Part Four of this book survey important communications skills for the systems analyst.

Interpersonal Relations Skills

It has been suggested that analysts "need to exercise the boldness of Lady Godiva, the introspection of Sherlock Holmes, the methodology of Andrew Carnegie, and the down-home common sense of Will Rogers" (Lord and Steiner, 1978, p. 349). In other words, systems work is people-oriented and systems analysts must be extroverted or people-oriented. Interpersonal skills help us work effectively with people. Although these skills can be developed, some people simply do not possess an extroverted personality. The interpersonal nature of systems work is demonstrated in the eight-part case study, *Analysts in Action,* that appears at intervals throughout this book.

Interpersonal skills are also important because of the political nature of the systems analyst's job. The analyst's first responsibility is to the business, its management and workers. Individuals frequently have conflicting goals and needs. They have personality clashes. They fight turf battles over who should be responsible for what and who should have decision authority over what. The analyst should try to mediate such problems and achieve benefits for the business as a whole.

Another aspect of interpersonal relations is recognition of the analyst's role as an agent of change. The systems analyst is frequently as welcome as an IRS auditor! There is comfort in the status quo. An analyst should study the theory and techniques of affecting change. Persuasion is an art that can be learned! For starters, begin by studying sales techniques—after all, systems analysts sell change.

Finally, systems analysts work in teams, and so the ability to function as part of a team, to cooperate and compromise, is critical for the success of most projects.

THE NEXT GENERATION

Applications Programming Without Programmers: How Will the Analyst's Job Change?

For many years now, computer programs have been written in procedural languages such as COBOL, FORTRAN, BASIC, and, more recently, C and Ada. The term *procedural* refers to the programmer's use of instructions to describe, in considerable detail, exactly how to accomplish an intended task. But there is an alternative. Both information systems professionals and end-users are increasing their reliance on very high level, nonprocedural languages, often called *fourth-generation languages* or *4GLs*. (4GLs are used to prototype and implement new systems.) Instead of specifying exactly how to perform a task, nonprocedural languages simply specify *what* is to be accomplished. Nonprocedural languages use menus, question-and-answer combinations, and a simpler, English-like syntax to design and implement systems, update databases, generate reports, create graphs, and answer inquiries. At the risk of oversimplifying the result, "the computer does the rest." Examples of nonprocedural languages include FOCUS, RAMIS, NATURAL, IDEAL, SAS, RBASE, and dBASE. Many of these languages can now be used on microcomputers. And it is generally believed that end-users—those with little or no information systems background—will soon be using such languages to write many of their own computer programs and systems.

How will information systems jobs change? The change has already started. Many organizations now run information centers where analysts teach these tools to the users and provide advice to users as they develop their own systems. If you are a programmer, you may be wondering about job security. Fear not! Current 4GLs are still immature—occasionally awkward and sometimes limited in capability. Also, the poorer processing throughput of systems written in 4GLs frequently forces CIS to rewrite the programs in more traditional languages. Finally, complex requirements may not be as easy to implement in 4GLs. For the time being, we will still be needing programmers for many more years!

And what about the systems analyst? We will need analysts more than ever! 4GLs can greatly speed the design and implementation of current computer applications. However, the analysis of those applications is still necessary. Why? Because 4GLs, like any other tool, can be misused. End-users, like analysts, may create incorrect, incomplete, and inflexible solutions to their own problems. Additionally, other tasks await systems analysts in the nonprocedural language environment.

Most 4GLs are built around databases that must be carefully developed to fully exploit the language. Therefore, many systems analysts will become more involved in *database analysis* and *design*. Other analysts will become *techniques specialists,* who will help users determine whether or not specific problems are suited to 4GLs. And still other analysts will become experts and consultants for end-users, helping them learn and apply the nonprocedural languages to their problems. Finally, larger applications cross departmental boundaries and affect the entire organization. Analysts will be needed to develop and maintain those larger systems since the end-users of those systems tend not to think of the system as a whole. The roles will change, but the systems analyst will survive!

Does this environment sound too futuristic to believe? If so, read James Martin's book *An Information Systems Manifesto* (1984). The book describes a systems development environment that is very different from most of today's CIS shops. It's a real eye-opener!

Formal Systems Analysis and Design Skills

Most analysts could use formal training in systems analysis and design skills. These skills can be conveniently factored into three subsets: *concepts and principles, tools and techniques, and methodologies.*

When all else fails, the systems analyst who remembers the basic concepts and principles of systems work will still succeed. No tool, technique, process, or methodology is perfect in all situations! Concepts and principles will help you adapt to new and different situations and methods as they become available. We have purposefully emphasized applied concepts and principles in this book. This is not a mechanical, "monkey see, monkey do" book! We believe that if you carefully study the concepts presented in Part One, you will be better able to communicate with potential employers, business users, and computer programmers alike!

Not too long ago, it was thought that the systems analyst's only tools were paper, pencil, and flowchart template. Over the years, many tools and techniques have been developed to help the analyst build systems faster and better, and to document information systems as they are developed. This book covers the classical, the so-called *structured,* and the modern tools and techniques of the trade. You will learn how these tools can and should coexist and work together to build superior systems. Also, this book includes discussions and demonstrations of an important trend—the use of computer tools to automate and support systems analysis and design tasks.

Methodologies are specific strategies for applying specific tools and techniques in a disciplined manner to successfully develop systems. There are numerous popular methodologies, and each has its own supporters. We will discuss the most popular methodologies and categories later; however, we endorse no single, specific methodology. You will learn to use the tools and techniques of several methodologies. Consequently, we hope you will avoid the pitfall of blind devotion to any one methodology.

Good Old-Fashioned Experience

Plainly stated, there is no substitute for good old-fashioned experience in systems analysis and design. Experience in applying skills to an actual systems project is necessary for you to become a "good" systems analyst. Experience tends to improve an important quality of good systems analysts, *patience.* On a typical project there will be periods of time during which you feel you are not making visible progress. You are! But only experience will convince you. No textbook can provide that experience! Many schools require live projects as an experience-providing capstone to their systems analysis and design training.

Summary

The information systems analyst is a professional who studies business problems and needs to determine how people, methods, and computers can best accomplish improvements for the business. When computers are used, the systems analyst is responsible for the efficient capture of data, the flow of that data to the computer, the processing and storage of that data by the computer, and the flow of accurate and timely information back to business users.

A systems analyst performs many duties including systems analysis, systems design, and systems implementation. Because of the emphasis on systems, the systems analyst must frequently apply the wisdom and insight attributed to the experienced business executive. The systems analyst is usually part of the information systems function in the business. Within the CIS function, systems analysts are usually classified according to experience as lead analysts, senior analysts, analysts, and junior analysts. Systems analysts are organized into temporary teams, along with programmers, and are assigned to projects.

To properly prepare for a career as a successful systems analyst, you should develop or refine all of the following skills: your working knowledge of CIS techniques, technology, and programming; your knowledge or experience with typical business functions; your ability to analyze and solve problems; your ability to communicate with others; your ability to work well with many types of people; and formal systems analysis and design skills. This book focuses primarily on formal systems analysis and design skills including concepts and principles, tools and techniques, and methodologies. Finally, you need to seek out opportunities to gain experience with systems analysis and design methods.

Problems and Exercises

1. Explain why a noncomputer professional (for instance, engineer, business manager, accountant, and the like) needs to understand systems analysis and design.

2. What is the role of the systems analyst when developing a computer application? To whom is the systems analyst responsible?

3. Make an appointment to visit with a systems analyst or programmer/analyst in a local business. Try to obtain a job description from the analyst. Compare and contrast that job description with the job description provided in this chapter.

4. Using the definitions of *systems analysis, systems design, systems implementation,* and *systems support* provided in this chapter, write a letter to

your instructor that proposes the development of an improved personal financial management system (to plan and control your own finances). Tell your instructor what has to be done. Assume that your instructor knows nothing about computers or systems analysts. In other words, be careful with your use of new terms.

5. Differentiate between the four centers of activity in modern information systems: development, information, database, and computer.

6. Visit a local information systems department in your business community. Compare and contrast its organization with the generic organization described in this chapter. Are the four centers of activity present? What are they called? How is their organization different? How is it better? Do you see any disadvantages? Where do the systems analysts fit in?

7. Visit a local information systems department in your business community. How are its project teams formed? How are they organized? How does this compare and contrast with the generic structure described in this chapter?

8. Kathy Thomas has been asked to reclassify her systems analysts. Virtually all of her analysts perform systems analysis, systems design, and systems implementation (including computer programming). However, depending on their experience, the percentage of time in the three phases varies. Younger analysts do 80 percent of the programming. The most experienced analysts do 80 percent systems analysis, largely due to their greater understanding of the business and its users and management. How should Kathy reclassify her personnel?

9. Diversified Plastics, Inc., has adopted an unusual information systems organization. The Management Systems Group consists of systems analysts who perform only systems analysis and very general systems design. The Technical Systems Group consists of programmers who perform only detailed systems design and implementation. All analysts must come from a business, engineering, or management background, with no computer experience requirement. Programmers must come from a computing background. Transfers between the groups are discouraged and in most cases not allowed. What are the advantages and disadvantages of such an organization and its policies?

10. Based on the systems analyst's job characteristics and requirements described in this chapter, evaluate your own skills and personality traits. In what areas would you need to improve?

11. Federated Mortgages' corporate information officer (CIO) is facing a budget dilemma. End-users have been buying microcomputers at an alarming rate. In a sense, the business users of these microcomputers, who have little or no CIS background, are developing their own applications systems. Because of this, the CIO has been asked to justify the continued

growth of the budget, especially that of the programmer and systems analysis staff. There is even some feeling that the number of programmers and analysts should be reduced. How can the CIS manager justify the staff? Will the roles of the programmers and analysts change? If so, how? Can the users completely replace the programmers and analysts?

12. The students in an introductory programming course would like to know how systems analysis and design differs from computer programming. Specifically, they want to know how to choose between the two careers. Help them out by explaining the differences between the two and pointing out factors that might influence their decision.

13. Your library probably subscribes to at least one big-city newspaper. Additionally, your library, academic department, or instructor may subscribe to an information systems publication such as *Computerworld* or *MIS Week*. Study the job advertisements for systems analysts and programmer/analysts. What skills are being sought? What experience is required? How are these skills and experiences important to the role of the analyst as described in this chapter?

14. You need to hire two systems analysts. Explain to a personnel department recruiter the characteristics and background you seek in an experienced systems analyst.

15. Prepare a curriculum plan for your education as a systems analyst. If you are already working, prepare a statement that expresses your personal need for continuing education to become a systems analyst.

Projects and Minicases

1. A systems analyst applies new technologies to business and industrial problems. As a prerequisite to this "technology transfer," the analyst must keep abreast of the latest trends and techniques. The best way to accomplish this is to develop a disciplined reading program. This extended project will help you develop this program.
   a. Visit your local school, community, or business library. Make a list of all computing and information systems publications.
   b. Skim two or three issues of each publication to get a *feel* for their contents and orientation. Select five periodicals that you find most interesting and helpful. We recommend that you select five publications that will give you the breadth of microcomputers, mainframes, data communications, applications, and management issues.
   c. Set up a browsing schedule. This should consist of one or two hours a week that you will spend browsing the list of journals. You should try to maintain this schedule for 10 to 15 weeks. If you miss a day, make it up within one week.

d. Set up a journal to track your progress. Record the date, the journals browsed, the title or subject of the cover story or headlines, and the title of one other article that caught your eye.

e. Learn to browse. You won't have time to read. If you try, you will get discouraged and quit the program. Study the table of contents. Read only the first paragraph or two of each article along with any highlighted text in the article. Move on, no matter how interesting. Read the conclusion or last paragraph of the article. Note any article that you want to fully read after browsing your reading list.

f. After browsing each of your selected publications, select at least two articles to read thoroughly. The number you read is limited only by your interest and available time. Record these articles in your journal.

This project will show you how to keep up with a rapidly changing technological world without consuming excessive time and effort.

2. Who should the systems analyst work for? Rolland Industries is facing an information systems reorganization dilemma. Non-CIS management is pressing for a new structure whereby most systems analysts would directly report to their application user group (such as Accounting, Finance, Manufacturing, Personnel) as opposed to reporting to Information Systems management. Non-CIS management feels that, in the existing structure, systems analysts are too influenced to "change everything" for the sake of computing because they report to Information Systems. To ensure that systems meet Information Systems standards, a small contingent of analysts would remain in Information Systems as a Quality Assurance Group that has final signoff on all systems projects.

Information Systems management is resisting this change. They feel that if systems analysts become removed from CIS, they will become technologically "out-of-tune." They also feel that separating the systems analysts from one another will result in less sharing of ideas, and will subsequently reduce innovation. They also feel that data files and programs will be unnecessarily duplicated. Conflicts between analysts and programmers, who will remain in CIS, will likely increase. CIS also feels that users will hire new systems analysts without regard to programming and technical experience or familiarity with CIS's technical environment.

Systems analysts, themselves, are split on the issues. They see the benefits of users being more directly in control of their own systems destinies; however, they are concerned that users and user management will be less forgiving when faced with budget overruns and schedule delays that historically plague CIS. Analysts are also concerned that they will become more prone to technological obsolescence if they are physically relocated outside of Information Systems and its more technically oriented staff.

The decision will likely be made at a higher level than CIS. What do you think should be done?

Annotated References and Suggested Readings

ben-Aaron, Diana. "Amazing Grace Hopper: Computing's First Lady." *InformationWEEK,* no. 107, March 9, 1987. Grace Hopper is a champion for students, open-mindedness, and change.

DeMarco, Tom. *Structured Analysis and System Specification.* Englewood Cliffs, N.J.: Prentice-Hall, 1978.

Lord, Kenniston W., Jr., and James B. Steiner. *CDP Review Manual: A Data Processing Handbook.* New York: Van Nostrand Reinhold, 1978. A review manual for the Certificate in Data Processing examinations. Chapter 8, "Systems Analysis and Design," traces the history, functions, and responsibilities of the systems analyst.

Martin, James. *Application Development Without Programmers.* Englewood Cliffs, N.J.: Prentice-Hall, 1982. This book describes the trend toward use of nonprocedural languages and the effect they will have on systems analysts.

Martin, James. *An Information Systems Manifesto.* Englewood Cliffs, N.J.: Prentice-Hall, 1984. This book describes several important trends in information systems, technology, management, and systems development. It also describes how educators, students, and various computer professionals, including analysts, should prepare for this coming age.

Wood, Michael. "Systems Analyst Title Most Abused in Industry: Redefinition Imperative." *Computerworld,* April 30, 1979, pp. 24, 26. This article sums up our feelings about systems analysts and what it takes to be successful as analysts. It has become our battle cry for the need to train analysts to develop their interpersonal relations and communications skills.

SOUNDSTAGE

SoundStage Entertainment Club— A Preview of Your Demonstration Case

This is the story of Sandra Shepherd and Robert Martinez, systems analysts for SoundStage Entertainment Club.

EPISODE 1   Systems analysis and design is more than concepts, tools, techniques, and methods. It is people working with people. Although experience is the best teacher, you can learn a great deal by observing other systems analysts in action. Ms. Nancy Yan, Chief Information Officer (CIO) and Director of Information Systems Services (ISS) for SoundStage Entertainment Club, has kindly consented to let you watch two of her analysts on a typical project.

Sandra Shepherd, a senior systems analyst and project manager, has volunteered for this demonstration. She has successfully implemented several information systems for SoundStage and should be able to provide you with a valuable learning experience. Bob Martinez, Sandra's partner, is a new programmer/analyst at SoundStage. In fact, today is his first day! Bob has to go through orientation today, and Nancy has invited you to observe the orientation. It'll be a good way for you and Bob to get acquainted with SoundStage.

Welcome to SoundStage Entertainment Club

SoundStage Entertainment Club is one of the fastest growing record and tape clubs in America. All operations are located in Indianapolis, Indiana. We begin the preview by joining Bob in Nancy's office.

"Hi, Bob!" Nancy extended her hand. "It's great to have you aboard. My name's Nancy Yan, and I'm the Chief Information Officer and Director of Information Systems Services. I didn't get a chance to meet you when you interviewed last month. Why don't you tell me a little about yourself?"

"Well, I just graduated from college," replied Bob, somewhat nervously. "I received my bachelor's degree in Computer Information Systems from State University. I was also president of my local student chapter of the Data Processing Management Association and I hope to get involved in the Indianapolis professional chapter. My career goals are oriented toward applying the systems analysis and design skills I learned in college.

That's the main reason I accepted this job. It looked like I'd get a chance to do some analysis and design here —not just programming."

"That's why we hired you, Bob," Nancy replied. "You may be interested to know that we were especially impressed by your classroom experience with computer tools for systems analysis and design. We just bought a package called Excelerator from Index Technology Corporation. We want you to learn and use that package on your first project and report your experiences back to the management staff. You'll learn more about this technology from your partner, Sandra Shepherd, a senior systems analyst who is very familiar with the tools and techniques you learned in college."

"I learned Excelerator in my systems analysis courses. I'm going to enjoy using it on a real project. And I met Sandra during the interview."

Nancy continued, "Sandra will show you the ropes and help you learn about SoundStage and our

Episode 1, continued ▶

way of doing things. You'll meet with Sandra soon."

Nancy stood up and motioned Bob to the door. "Okay, let's take a tour of the building."

As they walked up to the second floor, Nancy continued her orientation. "The second floor houses several departments including Personnel, Building Services, Accounting, Marketing, and Budget. These outside wall offices belong to executive managers, staff, and assistants."

"I don't know how much you know or remember about Sound-Stage, so I'll give a quick overview. We used to be called SoundStage Record Club. It used to be a record subscription service. Customers join the club through advertisements and member referrals. The advertisements typically dangle a carrot such as *Choose any ten records for a penny and agree to buy ten more within two years at regular club prices.* I'm sure you've seen such offers."

"Yes," replied Bob.

Nancy continued, "Club members receive monthly promotions and catalogs that offer a record of the month. They must respond to the offer within a few weeks or that selection will automatically be shipped and billed to their account. Customers can also order alternative selections and special merchandise from the catalogs. After members fulfill their original subscription agreement, they are eligible for bonus coupons that may be redeemed for free merchandise from our catalogs."

"How many members are there?" asked Bob.

"About 14,750," replied Nancy.

"Of that total, about 8,000 accounts are active, having purchased merchandise in the last 12 months. We recently reorganized and changed our name to SoundStage Entertainment Club. Our new name reflects our changing product mix. We are selling more cassettes than anything else. Compact digital discs is a new venture that will probably overtake cassettes someday. We are also getting into video sales and audio/video equipment sales. There is even talk of selling home computer software. The range of entertainment media for the modern household has forced us to diversify."

They walked along the hallway by the executive offices. Nancy continued, "The office on your left belongs to our President and Chief Executive Officer (CEO), Steven Short."

As they moved past Steve's office, Nancy handed Bob a piece of paper (see Figure E1.1) and continued. "This is our organization chart, Bob. As you can see, SoundStage is divided into three main divisions including the Information Systems Services Division. The offices you walked through on your way to this office belong to the Administrative Services Division. They include offices for our accounting, budgeting, marketing, personnel, and building services managers and staff."

"Down on first floor we have the Operations Division, which handles day-to-day operations including customer services, purchasing, inventory control, warehousing, and shipping and receiving. I will soon take you through those facilities."

"In the basement, you'll find the Information Systems Services Division, where you'll have your office. Our mainframe computers are located down there along with my staff, which totals about 25 people."

They walked down to the first floor. Once again, Bob was confronted with a maze of offices. Nancy explained, "These are the offices for the Operations Division, including Purchasing and Inventory Control. They buy the merchandise we will resell to the customer."

"And this office area we are approaching is the Customer Services Division. Those clerks are processing orders, backorders, follow-ups, and other customer transactions. That's where your first project is going to be. Sandra will be taking you there to meet your end-users."

The mention of his first project got Bob excited. "Where to next?"

Nancy guided Bob through a pair of double doors. "This is the warehouse. This is where the action is!"

The warehouse was quite large. Bob was somewhat surprised by the size of the operation. Clerks were loading and unloading trucks, stocking shelves, and filling orders. As Bob and Nancy walked through the warehouse, Bob commented, "This must be a difficult operation to coordinate."

Nancy answered, "I wouldn't want to do it. We haven't done much in the way of information systems support for the warehouse. We are just starting to investigate that possibility. We do support Purchasing and Inventory

Episode 1, continued ▶

SOUNDSTAGE

FIGURE E1.1   SoundStage Organization Chart   This organization chart identifies the organization structure for SoundStage Entertainment Club. Note the relatively high status of the Information Systems Services. This reflects a modern, forward-thinking attitude toward the information systems function in the company.

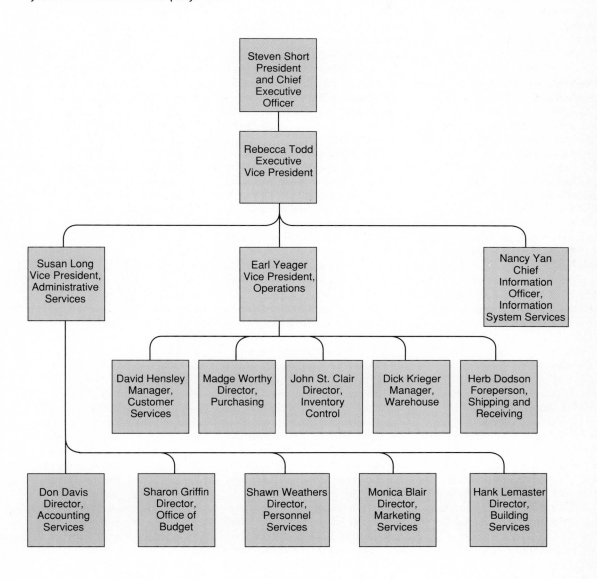

Control, but that is as close as we come to supporting the warehouse."

Nancy continued, "Well, let's go downstairs and tour our own Information Systems Services Division, and then I'll take you to Sandra."

The Information Systems Services Division at SoundStage

Bob and Nancy walked into the Computer Room, where Nancy removed another sheet of paper from her folder (see Figure E1.2) and handed it to Bob. "This is an organization chart for Information Systems Services, Bob. You might want to refer to it as we complete this part of the tour. I now report directly to Steve Short. That reflects the recent reorganization. I used to report to the vice president of the Administrative Services Division, but it caused some problems with prioritizing requests from the other divisions."

"We are entering the Computer Operations Center right now. You are looking at our IBM 3081 mainframe computer. We have an audiovisual self-study course that you will take to learn more about the IBM computer and its MVS operating system. For now, it should suffice to say that this machine supports most of our computer processing needs. It has several disk drives capable of storing 100 million bytes of data. We also have a full complement of tape drives, optical character readers, and printers. If you have any questions, just ask."

Bob replied, "I've never used this particular computer before. Is that going to be a problem?"

"No," answered Nancy. "You have a good, solid education in computers. No matter what machines you learn in school, the technology is constantly changing. You should be able to learn any new system fairly fast. Most people do. I don't know what your expectations were when you graduated, but I think you'll find that your education is only starting now that you are out of school. You'll catch on quickly. Trust me."

They moved into an adjacent office area. Nancy said, "And this is where you'll be working. It's called the Development Center. Different groups of systems analysts and programmers are located in different parts of the office. Each group supports a certain group of business users. You are assigned to the Customer Services Support Group."

Bob interrupted, "I see a couple of groups on the organization chart that I'm not familiar with: Technical Support and the Information Center. What are they?"

"Good question," Nancy said. "The Technical Support Group doesn't really serve users. It serves us. When you have a technical question or problem concerning the IBM computer system, that group will help you out. They're also the people who will be teaching you about that IBM computer system. The Information Center Group is a new function. They are helping all SoundStage users learn how to apply microcomputers to do some of their own work. They teach the users how to do word processing, spreadsheets, and

things like that. In other words, they teach the users how they can help themselves. That frees us up to do the really difficult projects. It's a novel concept for our operation!"

"I agree," responded Bob.

"This office is assigned to our Database Administrator. He manages our IDMS database management system. You'll work with his staff on any information systems projects that use IDMS. We are mandated by policy to implement all new systems using that database; therefore, I'm sure you'll soon be visiting his staff."

Bob and Nancy returned to the Development Center and walked into an office. Nancy continued, "And this is your office—I wonder where . . . Here she is!"

A woman entered the office. Nancy continued, "Sandra, I want you to meet Bob Martinez, your new partner. Bob, . . ."

Sandra interrupted, "Bob and I met during his interview, Nancy. Welcome, Bob. I'm glad you accepted our offer. I really wanted to work with you—so much so that I asked to be your partner."

Nancy replied, "Terrific! I didn't know you had already met. I don't know what you've told Bob about yourself, Sandra, but I'd like to do a little biographical sketch."

"Sandra has been with us for seven years. She was recently promoted to senior systems analyst because she has proven herself to be one of our most competent, progressive, and personable analysts. Sandra has a bachelor's degree in business. She had little formal computing education but

Episode 1, continued ▶

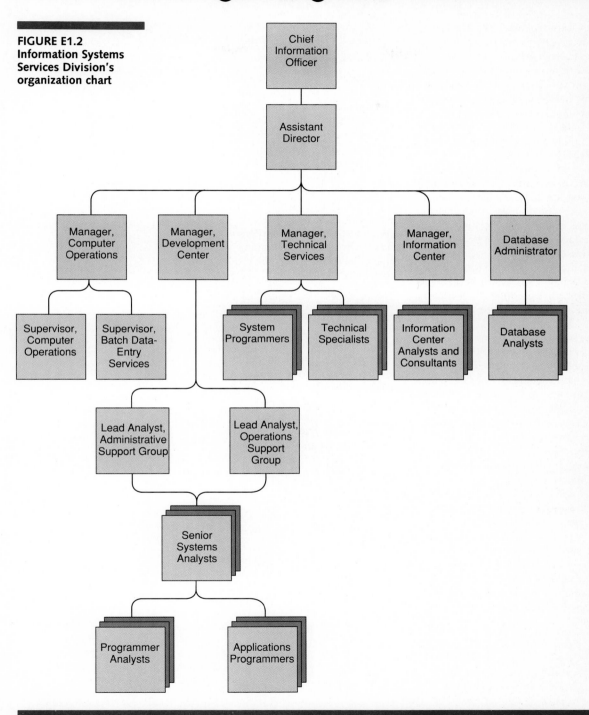

**FIGURE E1.2
Information Systems
Services Division's
organization chart**

Chief Information Officer

Assistant Director

Manager, Computer Operations

Manager, Development Center

Manager, Technical Services

Manager, Information Center

Database Administrator

Supervisor, Computer Operations

Supervisor, Batch Data-Entry Services

System Programmers

Technical Specialists

Information Center Analysts and Consultants

Database Analysts

Lead Analyst, Administrative Support Group

Lead Analyst, Operations Support Group

Senior Systems Analysts

Programmer Analysts

Applications Programmers

she has done well because she always seeks out opportunities to learn more through reading, seminars, and company training courses. Sandra's credentials also include recognition as a Certified Systems Professional or CSP."

Bob interrupted, "Is that anything like the Certificate in Data Processing or CDP?"

Sandra answered, "Yes, but the CSP is a relatively new certification program that recognizes systems professionals. I was required to pass examinations on various aspects of systems work and provide evidence of my systems experience. I am certified for three years, after which I must be recertified through evidence of continuing education or reexamination."

Nancy responded, "So you see, Bob, you'll be learning from one of our best people!" With that, Nancy started to leave the office. "Once again, Bob, welcome! I'll leave you with Sandra now. She'll help you get organized and start teaching you all the things you'll need to start learning. We'll sit down in a week or so to set some goals for your first six months. Bye!"

How to Use the Demonstration Case

You've just been introduced to a case study that will be continued throughout this book. It's important that you understand the purpose of the case study:

The purpose of the continuing case is to show you that tools and techniques alone do not make a systems analyst. Systems analysis and design

involves a commitment to work for and with a number of people.

When we started writing this book, we wanted to make sure that the chapters would teach you the important concepts, tools, and techniques. But we were also afraid that you might begin to believe that, if you knew those tools and techniques, you'd have all the knowledge necessary to be a systems analyst. Chapter 1 emphasized the importance of communications and interpersonal skills; however, this continuing case demonstrates the *people* side of the job.

The case study is divided into *Episodes* that represent various stages of a typical project. Although each episode will *introduce* new tools and techniques, the episodes are *not* intended to *teach* the tools and techniques. Instead, they introduce a new situation in which you will need to use a new tool or technique. The episodes show that need in terms of what Sandra and Bob must do to develop a new computer information system for SoundStage. In almost all cases, you will see Sandra and Bob working closely with their business users.

The chapters that follow an episode will teach you how to use the tools and techniques that were introduced in that episode. And those chapters will apply the tools and techniques to a SoundStage project.

A brief transition called *Where Do You Go from Here?* concludes each episode and introduces the chapters that follow it. We hope you'll find these demonstrations

interesting and informative. We think they'll help you place the subject of systems analysis and design into its most practical setting—people working with people!

CHAPTER TWO

The Business

and Its Users

as a System

Custom Shades and Drapes

Custom Shades and Drapes (CSD) is a large business specializing in customized window shades and draperies. Through the 1980s, CSD matured and began to experience its largest increase in business. The manual methods by which customer orders were processed were no longer effective, and these orders were frequently backlogged, lost, or incorrectly processed.

In the spring of 1988, Sales and Services Manager Martha Beck, who was aware of the need to improve the order-processing operation, decided to take action and submitted an Information Systems Request to Henry Mikovich, asking him to automate the current order-processing operations. Henry managed information systems as well as serving as systems analyst, programmer, computer operator, and microcomputer consultant.

After receiving the request, Henry spent four months developing a new computer information system for order entry and processing. He purchased one of IBM's latest microcomputers, the IBM PS/2 Model 70, on which to implement the new system. After implementing the new computer information system, he scheduled a meeting to discuss it with Martha. Overall, Martha was satisfied with the system he had developed. She was especially impressed with how easy the new system (particularly the microcomputer) was to learn and use—not at all what she had feared! Henry had obviously spent a lot of time making the new system friendly to her staff, most of whom had never used a computer. Martha felt that her order-processing staff would surely be excited when the new operations were installed.

"I'm impressed!" Martha said. "This should certainly solve a lot of our order-processing problems. There's just one small thing I'd like you to do for me. Here are a couple of samples of reports I sometimes prepare to help myself make some important decisions. I was hoping you could get the computer to generate them for me. This is the most essential one. Ms. Wellman (the company president) has informed me that our company is going to start becoming more aggressive in the mail order and TV–home buyer markets.

By 1991, we hope to generate 20 percent of total sales in these markets. This report summarizes the company's monthly sales according to different markets. We need to add the two new markets so that the report will help me track progress toward the sales goals. This other report gives me some idea of the volume of business we've handled during a given month. It summarizes the sales orders according to their type. I use this report to keep track of the volume and type of orders our department is handling."

Henry replied, "Okay, I see no problem with writing a couple of new programs. But one last question. What do you mean by type of order?"

"You know," answered Martha. "We classify orders according to type of product and whether the product is a regular stock item, modified regular stock item, or customized product."

Henry had a puzzled look on his face. "Gee, I didn't realize you could distinguish between different types of orders. Should the system have processed these different order types differently?"

"I don't really know," she replied. "My clerks will have to answer that question. Why don't you talk to my shift supervisor, Kevin Chen. He'll be very interested in knowing what we're up to."

Henry still seemed a little concerned when he responded, "That's a good idea. I talked to the order clerks. They didn't mention the order types. I'll call Mr. Chen. No, I think I'd better go talk with him right now."

Henry went straight to Kevin Chen's office to discuss the new plans for the order information system. Henry learned that the system he had created accommodated the bulk of all orders—those for regular stock items. Unfortunately, the processing of the other order types was slightly different. Kevin had been struggling to improve the productivity of his staff to efficiently handle the increasing number of custom product orders. He was very enthusiastic about the system and its potential for decreasing order-processing time, reducing customer backlogs, and improving order-processing accuracy.

As Kevin looked over a written description of the system, he said, "This system sounds terrific! I like the fact the computer will be doing a lot of the work. It should also reduce a lot of common mistakes like arithmetic and copying errors."

"Right!" exclaimed Henry. "You can be assured that I included a lot of error checking routines in the programs. They should catch any bad data they might key in."

"How about the credit check?" asked Kevin. "We've had a lot of inconsistencies in our credit checking of customer orders. Sometimes we find we've approved credit sales that exceed the customer's credit limit."

Henry proudly answered, "They won't be able to do that anymore. I . . ."

"Just a second!" interrupted Kevin. "Sometimes they should do it and other times they shouldn't. It all depends on the customer's credit history."

Once again, Henry looked concerned. He said, "This could be a problem. I've made some incorrect assumptions. I had no idea the credit limit should be extended in certain instances. You'd better explain to me how credit checks should be done. I have to make some modifications to one of my programs."

Kevin handed Henry a document. "This describes how my staff checks customer credit orders. If I can change the subject, do you suppose I could also send you a description of a few reports that I would like to have? I have some good ideas of reports that would be very beneficial. They'd help me a lot more than these two reports that your system currently produces."

Henry's anxiety was increasing. "But your order-entry clerks told me that those are the reports they're supposed to send you!"

"They are," answered Kevin. "But you should see what I have to do to get the information I need from them. It takes me several hours every week. Can't the computer help me out?"

"No, I can do it. I guess I just wish I had known. I'll just have to make the modifications and additions. But I don't think this system will be ready in two weeks. I'm going to get back to work now. I'll be expecting your notes concerning those reports you'd like."

Henry spent several weeks making the corrections and additions. Not only did he have to change computer programs; he also had to change the procedure manuals for the order-entry staff. He had meetings with Martha and Kevin to obtain their final approval before proceeding to install the new system. Both of them eagerly encouraged him to begin installing it. One month later the automated order-processing system was up and running.

Unfortunately, the new system didn't solve all the problems. Some of the order-entry clerks—those dealing with the not regular stock item orders—found the new system cumbersome and unacceptable. They insisted that several less common but still important processing steps had been omitted. Kevin's boss called to complain that the reports approved by Kevin were unacceptable to upper management. The accounts receivable manager had called to complain that credit checking policies were still not entirely consistent with corporate policy. Apparently Kevin's description was not complete. Henry didn't understand. He wondered why Kevin and Martha were so positive about the system and how other company personnel were so negative.

Discussion

1. Why wasn't Henry able to implement an acceptable automated order-processing system?

2. Whose fault is it that the automated order-processing operations proved less than completely acceptable?

3. What should Henry have done before he developed the initial automated order-processing system?

4. Why didn't Kevin recognize that the proposed automated order-processing system would not work?

What Will You Learn in This Chapter?

Systems analysts build information systems for businesses. This chapter discusses the beneficiaries of the information system—the business and its end-users. You will understand the business and end-user dimensions of information systems when you can

1. Identify end-users in an organization and characterize them as clerical and service staff, supervisory staff, middle-management and professional staff, or executive-management staff.
2. Define information system.
3. Describe how information systems serve end-users.
4. Using the systems concepts, explain the relationship between the business system, information systems, computer systems, and end-users.
5. Describe a business in terms of its purpose, goals, objectives, and policies.

One myth suggests that the first systems analyst appeared on the scene some 6,000 years ago, during the construction of the Egyptian pyramids. This self-made systems analyst, concerned about the inefficient methods used to construct the great monuments, offered the following suggestion to Khufu (or Cheops), builder of the Great Pyramid:

"O Noble Khufu, it's time we got organized. We've been pushing this rock through the desert in the wrong direction for seven years. What we need is this Pyramid Erection and Routing Technique." Rumor also has that he was flogged on the spot and never heard from again—at least not until the mid-20th century (Lord and Steiner, 1978, p. 349).

That first systems analyst was looking for a better way to do business—building a product (the pyramids). Today's computer-systems analyst is also looking for a better way to do business—building a new product, information systems to improve business productivity and decision making.

As a tribute to that first systems analyst, we will use a pyramid to illustrate information systems concepts. Each face of the pyramid represents a different dimension or point of view and raises issues that you, as an analyst, must consider when developing a system. The pyramid serves as a framework for understanding modern information systems. The different faces of the pyramid also demonstrate interrelationships between concepts. Thus, the pyramid model is a simple and effective tool to learn and understand the information system.

This chapter presents the end-user and business faces of the pyramid model. Chapter 3 presents the remaining faces.

Information Systems Pyramid

End-Users, the Beneficiaries of Information Systems

The term **knowledge worker** has been coined to describe those people whose jobs involve the creation, processing, and distribution of information. The livelihood of these workers depends on information and the decisions made from information. Today, more than 60 percent of the U.S. workforce is involved in the production, distribution, and use of information. The information services sector of the economy has grown to very great proportions.

Is this significant? You bet! If ever there were a watchword for today's economy, that watchword would be *productivity*. Although much attention is currently directed toward automating our factories with robots and computers, we should also be asking ourselves how we can improve the productivity of the knowledge workers. If and when systems analysts stop improving knowledge worker productivity, analysts will become expendable!

Knowledge workers are the **end-users** of the systems analyst's services. Consequently, we will use this term to describe knowledge workers from now on. End-users have varying information needs. We can classify end-users into four groups according to their level of responsibility and authority: clerical and service staff, supervisory staff, middle management and professional staff, and executive management.

Clerical and Service Staff

Clerical and service workers perform the day-to-day information activities in the organization. Sample activities include screening and filling orders, typing correspondence, and responding to customer inquiries. Examples of clerical and service staff are listed in the margin. Data is captured or created by these workers, many of whom perform manual labors in addition to their information roles. For example, the warehouse clerk who fills an order is both packing the products and recording valuable data about the transaction. To some degree, most nonmanagers will qualify as clerical and service staff. The order-entry clerks in the chapter minicase would be classified as clerical and service staff.

Most clerical and service workers either respond to information (for instance, filling an order) or capture the raw data needed to produce information (for instance, taking an order).

In addition to clerical and service workers, there are three generally recognized management levels. Management performs the planning, organizing, controlling, and decision-making activities in a business. The three levels of management differ in the time frame of their activities, ranging from day-to-day operations to extremely long-range planning.

✔ CLERICAL AND SERVICE STAFF

Secretaries
Office workers
Clerks
Tellers
Salespeople
Bookkeepers

Supervisory Staff

Supervisors, the lowest level of management, control day-to-day operations in the business. Examples of supervisory staff are cited in the margin. Most supervisory managers are concerned with no more than the current day's or the next week's operations. Thus, supervisors are users of day-to-day detailed and historical information about those activities performed by their subordinates. For example, an order-entry supervisor might need an order register that indicates the status of all orders recently processed. A production line supervisor might need the detailed production schedule for any given day. Supervisors also frequently prepare reports that summarize the activities performed by their subordinates. For instance, a sales manager may prepare a report that summarizes the sales generated by each sales representative. When studying the needs of supervisors, the systems analyst should seek to understand the day-to-day activities supervised and how they can be measured and improved. Can you identify anyone in the chapter minicase that can be classified as supervisory staff? If you guessed Kevin Chen, you are correct.

SUPERVISORY STAFF

Forepersons
Supervisors
Group leaders
Union stewards

MIDDLE MANAGEMENT; PROFESSIONAL STAFF

Operations Managers
Industrial Engineers
Accountants
Marketing Specialists
Engineers
Business Lawyers
Personnel Managers
Sales Managers

Middle Management and Professional Staff

Middle management is concerned with relatively short-term (tactical) planning, organizing, controlling, and decision making. They are not interested in the detailed, day-to-day operations that involve the supervisor. Instead, they are concerned with a longer time frame, perhaps a month or a quarter. Some of their functions include gathering operating information for higher levels of management, developing tactical strategies and plans that implement executive management's wishes, and designing products and services. Middle-level managers buffer executives from having to deal with the day-to-day operations of the business.

Professional staff consists largely of business and industrial specialists who perform highly skilled, often technical work. Examples of both middle management and professional staff are listed in the margin.

In the chapter minicase, Martha Beck is an example of middle management. That explains why Martha was unable to tell Henry how special and regular customer orders were processed. She was more concerned with how order-processing operations functioned as a whole than with the detailed step-by-step activities.

Executive Management

Executive management is responsible for the long-term (strategic) planning and control for the business. Executive managers frequently look a year or more into the past and future. They examine trends, establish long-range plans and policies for the business, and then evaluate how well the business

FIGURE 2.1   End-User
Dimension of the
Information System
When developing any
information system, the
systems analyst should
identify the end-users
and their responsibilities.

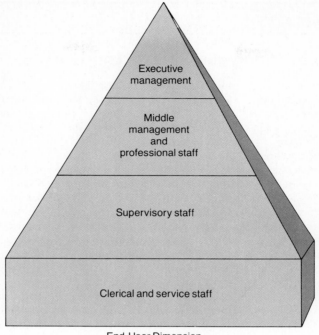

Executive
management

Middle
management
and
professional staff

Supervisory staff

Clerical and service staff

End-User Dimension

✓ **EXECUTIVE
MANAGEMENT**

Board of Directors

Presidents

Chief Executive Officers
  (CEOs)

Vice Presidents

Partners

Comptrollers

carries them out. They allocate the scarce resources of the business, including
land, materials, machinery, labor, and capital (money). Because they are concerned primarily with the overall condition of the business, executive managers usually want highly summarized information to support important decisions. Examples of executive managers are listed in the margin.

Figure 2.1 depicts the organization of the end-user groups as the first
dimension of our pyramid. We'll call this the *end-user dimension* of the information system. Note that it has the same familiar shape as organization charts.

The Importance of the End-User
to the Systems Analyst

Systems analysts must be responsive to the needs of their end-users. The
overriding philosophy of systems analysis and design must be that *the system
is for the end-user.* This is one lesson Henry Mikovich should have learned in
the minicase. Henry made the mistake of not identifying and working with
all of the end-users before developing the new order-processing system. As
a result, he had to make major additions and modifications to the automated
system to fulfill the requirements of Martha Beck and Kevin Chen. Even then,

THE NEXT GENERATION

Dealing with the Computer-Literate End-User

It used to be that information systems professionals could get away with almost any proposal. The end-users didn't understand enough about computing to seriously question technical recommendations, schedules, costs, and alternatives. But times are rapidly changing.

Few graduates from higher education are without computing background. Many (if not most) high school graduates have also been exposed to computers and programming. And even elementary school students are learning about computers. Much of this trend is fueled by the explosion of microcomputers.

Consider now, if you will, the probable computing backgrounds of those students currently going through the school system. What types of end-users are these people going to be?

These end-users will be less threatened by the pace of computer technology. They very well may be more receptive to technological change. They will probably be more participative than today's average end-user. Sounds like utopia at first glance, right? Don't be too hasty.

These same end-users are going to expect more from their systems analysts and consultants. They are going to be less

inclined to accept answers like, "We can't do that" or "That's not possible." They are going to expect much higher quality systems (fewer errors and greater adaptability) than today's end-user. Instead of just complaining about computer types, they'll try to do something about it. They will probably expect us to be more engineeringlike in our methods (greater rigor) and more businesslike in our attitude (proving economic feasibility). In other words, the next generation of end-users will be more demanding because they are more computer literate.

We are already seeing evidence of this trend, beyond the fact that the new generation has simply had more exposure to computers in school. End-users, through information centers, are building a greater and greater number of their own applications (see the Next Generation box feature for Chapter 1). Within the next decade, end-users will directly fulfill more than half of their own computing requests, although most of those requests will simply involve extracting data from existing files and databases.

As further evidence, look at the microcomputer industry. Hardware and software sales have been more influenced by the buying habits of end-users

than by computer professionals. In virtually all businesses, end-users are buying micros and software in great volumes.

The potential for superior analyst/end-user relationships and cooperation will likely result from this computer literacy revolution. Stated more simply, there will be less of a communications gap between technical and nontechnical workers.

Also, the analyst should become liberated from the more mundane work such as generating new reports against existing systems. Instead, the analyst will be called on to develop sophisticated databases and data networks on which the more sophisticated end-users can draw.

Finally, the applications will probably become much more interesting as end-users look for new ways to exploit technology to gain some competitive advantage in their marketplace. The fundamental business systems will already be done!

For some analysts it may prove intimidating. To our minds, it will be exciting!

the system was inadequate. Why? Because, not realizing that Martha and Kevin did not understand the detailed tasks performed by the order-entry clerks, Henry had failed to involve all the order-entry clerks when he was investigating needs.

The ultimate value of any information system is not determined by the analyst or any computing professional. These people only build the information systems. End-users are served by the information system, and only they can determine the system's worth!

In the chapter minicase, Martha Beck and Kevin Chen were impressed with the new automated order-processing system. On the other hand, the order-entry clerks were frustrated with the new system and thought it less than satisfactory. Why? Because, by their standards, the new system was unworkable.

As a footnote, you may be surprised to discover that the systems analyst is also an end-user. Do you have any idea where the analyst is in our hierarchy? If you guessed professional staff, you are correct!

Information Systems and the Business

Some people will tell you that an information system can't be defined, but you'll know one when you see it. Not being able to precisely define the product you are trying to build can make communicating with business end-users difficult. Indeed, if you can't define an information system, how can you justify its cost to management? Fortunately, classical systems concepts exist to help us define an information system. We can begin with the definition of a *system:*

> A **system** is a set or arrangement of interdependent things or components that are related, form a whole, and serve a common purpose.

There are two types of systems, natural and fabricated. The solar system and the human body are natural systems. They exist in nature. Fabricated systems, on the other hand, must be built by people. A manufacturing operation, an accounting system, and an information system are all examples of fabricated systems. The purpose of our information system is to collect, process, and exchange information among business workers. More specifically, an information system should support the day-to-day operations, management, and decision-making information needs of business workers. This leads us to a definition of *information system:*

> An **information system** is an arrangement of interdependent components that interact to support the operations, management, and decision-making information needs of a business.

Note that the computer has yet to play a role in our definition. Information systems exist in all organizations. Whenever end-users get together in an organization, they work out some sort of system to collect, process, and exchange information. Today, these systems are usually implemented without the aid of the computer. Although manual systems usually get the job done, they are often inefficient and error-prone. Why does computerization help? Because the computer amplifies the potential of the information system by increasing efficiency, reducing errors, and increasing effectiveness. The computer complements rather than replaces the end-users (most of the time). With this in mind, we suggest the following definition:

> An **information system** is an arrangement of interdependent human and machine components that interact to support the operational, managerial, and decision-making information needs of an organization.

Information systems that use the computer are sometimes called **computer information systems**. In this book, we use the term *information system* to describe both computerized and noncomputerized systems.

The business itself is also a system. Like other fabricated systems, the business system serves a purpose, profit, or service. The purpose of an information system should be compatible with the purpose of the business. What is the relationship between the information system, the business system, and end-users? Why is understanding the business system important to the analyst? These questions will be answered in this section.

System Concepts for Business and Information Systems

One of the difficult tasks of an analyst is overcoming size and complexity to identify the scope of an information system. System concepts help us deal with size and complexity by showing us the general characteristics that *all* systems have in common.

All systems, save perhaps the universe, have a **boundary** that separates the system from its **environment**. For all but the simplest systems, that boundary is somewhat artificial—you cannot see or touch it. The boundary of an information system is defined by the scope of the business activities to be supported by that information system. For example, a small information system may support only a few of the business activities in a single department— say, accounting. A somewhat larger information system could be built to support all of the activities of the accounting department. And an even larger information system could be built to support several business departments, including accounting.

This brings us to an important concept—the concept of systems and subsystems. Any system may contain numerous **subsystems**, each of which is a complete system within the larger system. For instance, in Figure 2.2, we

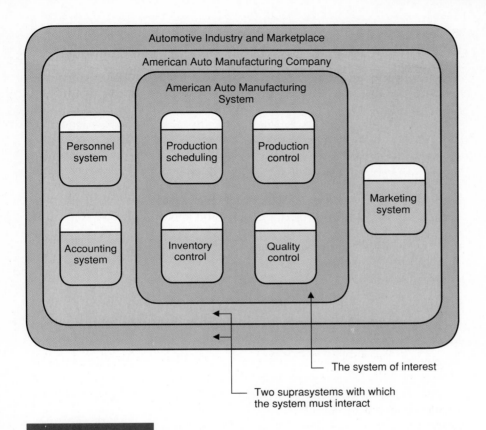

The system of interest

Two suprasystems with which
the system must interact

FIGURE 2.2   Environment, System, and Subsystem   Production
scheduling, production control, inventory control, and quality control are
subsystems of the American Auto Manufacturing Company (AAMC)
system. AAMC operates in a suprasystem of the automotive industry
along with other AAMC systems.

see depicted a manufacturing system that contains four subsystems: production scheduling, production control, inventory control, and quality control. These subsystems, in turn, may consist of other subsystems (for example, the inventory control system may consist of a shipping subsystem and a receiving subsystem). The concept also works in reverse. For example, the manufacturing system is a **suprasystem** of the four functional subsystems, and the business as a whole is a suprasystem of the manufacturing system.

Systems interact with their environment. That environment may include other systems in the same company. For instance, the manufacturing system in Figure 2.2 probably interacts with American Auto's personnel, accounting, and marketing systems. The environment might also include external inter-actions with customers, suppliers, the government, and so forth. For American Auto, the external interfaces are to the automotive industry and marketplace. Interactions take the form of inputs, events, outputs, and responses.

Many people have confused the terms *business system* and *information system*. They are not one and the same. Information systems—there are usually several—support the activities of the business system. Consequently, information systems can be viewed as subsystems of the business system.

It is important that the systems analyst consider the interrelationships that might exist between information systems. For example, in the chapter mini-case, Henry devoted his attention to the order-processing operations (sys-tem). He likely viewed the accounts receivable department or operations as outside the scope of his project. Yet the credit approval process conducted by the order-processing system was dependent on policies and procedures established by accounts receivable.

Information System: A Redefined Definition

Now that we've considered systems concepts, let's refine our definition of an information system:

> An **information system** is a subsystem of the business. Specifically, it is an arrangement of interdependent human and machine components that interact to support the operational, managerial, and decision-making information needs of the business's end-users.

Because the information system must support the business system, a business perspective and attitude are important to the analyst.

The Business Scope as It Applies to Information Systems

A business *system* can be characterized in terms of its purpose, goals, objec-tives, and policies. To fully understand the needs of the end-users in a business system, the analyst must understand the business itself.

The **purpose** of a company is a general statement of its reason for exis-tence. Purpose usually addresses who the company's customers are and what products or services the company provides its customers. The analyst should always be aware of strategic plans relative to the business's primary purpose. For example: Is the company planning to change or add to its customer base or products and services? Will such changes dramatically impact the infor-mation needs of clients? A good analyst keeps abreast of the strategic direction of the business (as it becomes known)!

You probably have some idea of what goals and objectives are, but they are terms that are commonly confused. **Goals** are very general statements of the degree to which the business purpose is to be realized. For example, does the business want to maximize profit, minimize costs, maximize market share, or maximize return on investments? These are all broadly stated goals. **Objectives** are specific targets that are directed toward goals. In order to maximize profits (a goal), we will likely establish many objectives such as "Decrease production costs by 10 percent by August 30th." One end-user's objective may be another's goal. The previous cost reduction objective may not be specific enough for a lower-level end-user who considers it a goal. One of this worker's objectives may be to reduce scrap and waste of materials at the lathes by 15 percent by August 30th. Objectives should specify the activity and the performance criteria. Most objectives are defined at the work-unit level. That is, each work unit, such as accounting or personnel, will have separately defined objectives.

Goals and objectives should both drive information systems development and mirror the business. When working in any business unit (for instance, accounts payable), the analyst should consider the purpose, goals, and objectives of that particular unit as well as those of the business as a whole.

Most organizations have adopted policies that must be followed while fulfilling objectives. A **policy** is a set of rules that govern the activities involved in achieving objectives. For instance, most companies have a credit policy for determining whether or not to accept or reject an order. The Custom Shades and Drapes Company in our minicase did. And Henry, like any other systems analyst, should recognize that policies may place restrictions on how a system can function. To ensure that the policies don't deter achievement of the goals and objectives that have been determined, policies must be set *after* goals and objectives.

We can now rotate our information system pyramid to show the business dimension. The faces in Figure 2.3 show the correlation between the business and end-user dimensions of an information system. The business purpose is established by the highest levels of executive management. Most goals and high-level objectives are developed by middle- to executive-level managers. Detailed objectives tend to be established by middle managers. Policies are set by middle- to supervisory-level managers. Clerical and service workers must perform operational activities in accordance with company policies.

The Business Importance to the Systems Analyst

The importance of thoroughly understanding the purpose, goals, objectives, and policies of an organization should be clear. The analyst should work closely enough with the different levels of end-users to understand the full business implications of any project. The value of any information system is ultimately a function of how well it serves the business.

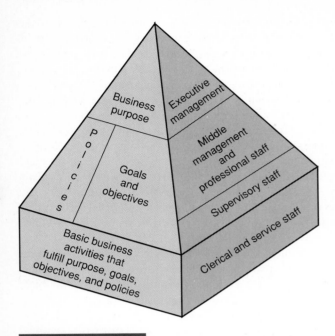

FIGURE 2.3   Business Dimension of the Information System   When building an information system the systems analyst should clearly understand or define the purpose, goals, objectives, and policies of the business and application. Note that the business purpose is established by executive management. All management levels participate in policy establishment. Similarly, all management levels establish goals and objectives for their subordinate levels. Finally and ultimately, clerical and service staff execute day-to-day business activities that fulfill the business purpose, policies, goals, and objectives.

Summary

An information system is a business subsystem. Specifically, it is an arrangement of interdependent human and machine components that interact to support the operational, managerial, and decision-making information needs of its business end-users.

The overriding theme of this chapter is that the information system is for the end-user. End-users are those individuals whose jobs involve the creation, processing, and distribution of information. Because these people represent the majority of today's workforce, much emphasis is being placed on improving their productivity by building more effective and efficient information systems. When building an information system, the systems analyst should address the problems and needs of all involved end-users, including clerical and service staff, supervisory staff, middle management and professional staff, and executive management.

The systems analyst must understand the relationship between the business system and the information system. The business represents a much larger system within which the information system functions. Along the same lines, the information system is a subsystem of the business. Because of the close relationship between business and information systems, the information system must support the business. The business can be characterized in terms of its purpose, goals, objectives, and policies. The effectiveness and value of an information system are dependent on its fulfillment of the business's purpose, goals, objectives, and policies.

End-users are an integral component of the information system. The computer system is a common but not essential subsystem of many information systems. Information systems that contain computer components are often called *computer information systems*. Additional components of the information system are introduced in Chapter 3.

Problems and Exercises

1. Define *information system*. Information systems are all around you. From your last job, in information systems or any other field, give an example of a completely manual information system. Describe how a computer might improve that information system.

2. Explain why systems analysts are instrumental in improving the productivity of their end-users.

3. For each of the following job titles, identify the worker as clerical or service staff, supervisory staff, middle management, professional staff, or executive management. Defend your answer.
   a. Receptionist
   b. Shop floor line supervisor
   c. Financial manager
   d. Assistant store manager
   e. Chief operating officer
   f. Manufacturing control manager
   g. Terminal operator/data entry
   h. Applications programmer
   i. Sales representative
   j. Programmer analyst
   k. Warehouse clerk who fills orders
   l. Stockholder
   m. Product engineer
   n. Consultant

4. Give three examples of knowledge workers from each of the following classifications: clerical and service staff, supervisory staff, middle man-

agement and professional staff, and executive management. Explain the job responsibilities of each example you provided and state why your example represents that particular classification of end-user. For each knowledge worker, describe a situation that would make that worker an end-user who might work with an analyst.

5. Obtain an organization chart from a local company or at your library. Classify each person and/or job position appearing on the chart according to the type of end-user represented, such as clerical and service staff. (*Alternative:* Substitute the organization charts appearing in the Analysts in Action episode that preceded this chapter.)

6. Consider a business by which you were, or are, employed in any capacity. Identify the knowledge workers at each level in the end-user hierarchy. (*Alternative:* Substitute your school for the business. Students are part of the clerical and service staff classification. Do you see why?)

7. Identify an information system with which you are somewhat familiar. Remember, information systems are everywhere, so you shouldn't have to look far. What are some of the possible subsystems of your business systems? What system is the information system a subsystem of? With what other information systems does your system interact?

8. Identify an information system with which you are somewhat familiar. What are some other systems that interface with the system? What are the implications of a systems analyst failing to identify those systems and how they interface with the information system you identified?

9. Obtain an organization chart from a local company or at your library. What are some of the subsystems you see on the organization chart? Explain how an organization chart is, by definition, a chart of systems and subsystems.

10. Select a typical business, and identify its purpose. What are some goals this business might establish? Identify some objectives that may direct the business toward reaching these goals. What are some of the policies that might exist within the business?

11. Make an appointment to discuss your curriculum with your advisor or an instructor. Consider your curriculum to be a system (it is!). What is the purpose of your major? What are the goals? Choose a specific course in your major. What are the objectives of that course? Are those objectives consistent with some of the goals of your curriculum? Finally, identify a few policies established in the course. Are those policies consistent with the course's goals and objectives?

Projects and Minicases

1. William Holmes, president of Holmes Tire & Supplies, was under the impression that computerization of his clerical and service functions would effectively reduce the size of that staff. Since installing computer support eight years ago, he has seen his clerical and service staff increase 10 percent, while business volume has remained steady. His friends in the local Chamber of Commerce have experienced the same phenomenon. Mr. Holmes is particularly puzzled by another curious trend. Although the use of computers did not decrease his need for clerical and service staff, it has reduced his dependence on middle-level management and professional staff. This seems exactly the opposite of his expectations; however, it is typical.

   Do background research into a couple of local information systems shops and some of their implemented computer-based systems. Were their clients' expectations similar to those of Mr. Holmes before completion of the system? How did the analysts deal with the expectations, if they did so at all? How would you have dealt with Mr. Holmes' expectations prior to beginning the systems project? Were the outcomes of the implemented system similar? Why do these outcomes happen?

2. A systems analyst for Demartini Office Supplies has uncovered a sensitive problem while working on a purchasing and inventory control project. The purchasing manager's job performance is based on minimizing purchase cost per unit on all products (all products are purchased direct from various manufacturers). She does this by taking advantage of large quantity discounts offered by suppliers; in other words, the suppliers offer lower per-unit prices for large order quantities. Meanwhile, the inventory manager's job performance is based on minimizing inventory carrying costs while avoiding stockouts (running out of stock). He wants to keep the quantities of each product to a minimum. These managers have conflicting goals and objectives that prevent the analyst from suggesting a system that at least one of the managers is certain not to accept. What should the analyst do?

3. For the Analysts in Action episode that preceded this chapter, brainstorm a business purpose, some goals, some objectives that might be directed to each goal, and some policies that might constrain the system.

Annotated References and Suggested Readings

Davis, Gordon B. "Knowing the Knowledge Workers: A Look at the People Who Work with Knowledge and the Technology That Will Make Them Better." *ICP Software Review* (Spring 1982), 70–75. This article provided our first exposure to the concept of knowledge workers as the new majority.

Davis, Gordon B., and Margrethe H. Olson. *Management Information Systems: Conceptual Foundations, Structure, and Development.* 2d ed. New York: McGraw-Hill, 1985. This is our all-time favorite book on information systems concepts and principles.

Lord, Kenniston W., Jr., and James B. Steiner. *CDP Review Manual: A Data Processing Handbook.* 2d ed. New York: Van Nostrand Reinhold, 1978.

Naisbitt, John. *Megatrends: Ten Directions Transforming Our Lives.* New York: Warner Books, 1982. A best-selling book that discusses, in considerable depth, the impact that technology has had and will have on our society and business.

Sprague, Ralph H., Jr. "A Framework for the Development of Decision Support Systems." *MIS Quarterly* 4 (December 1980), 1–26. This paper presents the forerunner of our information systems pyramid model.

CHAPTER THREE

Modern

Information

Systems

Starboard Boats, Inc.

Carol Stones, a recently hired systems analyst for Starboard Boats, is leaving the Information Systems manager's office after receiving her first systems project assignment. She has been assigned to join Mark Greenspan, a systems analyst who has been with Starboard for approximately four years, on an ongoing inventory control project. She is very excited. Her first three weeks were spent attending training workshops, studying the Information Systems Department standards, meeting with various members of the staff, learning about the hardware and software environment, and the like.

Carol walks into Starboard's computing library and training center, where she finds Mark poring over his systems documentation. She sits down at the same table and interrupts, "Hello, Mark. I guess you've heard that I've been assigned to work with you on your inventory control project?"

"Hi. Yes, I've been waiting for you. In fact, I personally requested that you be assigned to work with me on the inventory control project. In a nutshell, I've been working on this project for the last two months. I've studied the existing inventory control system, and I've defined our users' requirements for a new and improved system. Unfortunately, I've been stuck for the past week. I'm still trying to specify the computer solution. I've drawn a blank. That's why I requested that you be assigned to work with me. I thought two heads might be better than one."

"I'll try my best," Carol offered. "Actually, one of the projects I worked on at my last job dealt with automating that company's inventory control operations. They seemed quite pleased with that new system. Maybe my experiences can help. As far as I'm concerned, after three weeks of orientation, the sooner I get started the better."

"Well, let's get started. I guess I should provide you with a little background," he suggested. "Specifically, the project involves automating a manual purchasing system to replenish inventory. Currently, a warehouse clerk does a daily physical count of inventory and prepares a list of items that are low in stock."

Carol interrupted, "That seems very time consuming. I assume you're thinking about automating the inventory file and having a computer program generate the report?"

"Yes," Mark answered. "Right now it takes the clerk approximately three hours to do this inventory check. Anyway, the purchasing manager uses that list of items low in stock to initiate purchasing decisions. For each item listed on the report, the manager scans a notebook containing copies of purchase orders that were prepared and sent out in the previous week. If the item doesn't appear on any of the outstanding purchase orders, that item must be ordered. I'm convinced we should store this notebook in a computer file and have a program search the file for the outstanding purchase orders. The manager shouldn't have to spend time leafing through that notebook."

"I agree!" Carol said. "And don't forget the potential that file has for generating other useful reports for the end-users. In my previous job, we implemented an interesting inventory reporting system. For instance, you mentioned a report that identifies items low in stock. We also created a report that identified items that were both low in stock and currently on order. Our warehouse people found the second report reassuring because it gave them an idea when new stock might be received."

"I didn't think of that. Did you generate many other reports?" Mark queried.

"Sure! We must have generated 15 different reports off a file similar to your notebook. Different reports are intended for different audiences. Some reports were very detailed, and others summarized those details. Still others reported situations that required immediate or eventual attention. When we developed the system, we studied the responsibilities and activities of all levels of management. Then we brainstormed different types of reports that management might need. It really got them thinking about computer support, and they began recommending useful reports on their own initiative!"

Mark was beginning to feel more optimistic than he had felt over the past week. "I can already tell that you're going to contribute significantly to this project. I'm lucky you were available. Getting back to my manual system—after the items that need to be ordered are identified, the manager then checks through another notebook containing information about suppliers and their products to determine the ideal supplier from which to order the item. This is the tough part. I'm not sure I can automate this step. I know this notebook could also be stored in a computer file, but the data is volatile. Each day, I'd have to reproduce the equivalent of the notebook so that managers can make their decisions."

"Why?"

"Well, first of all," Mark replied, "a variety of information is needed to determine which supplier to order the part from. For example, the manager needs to know which suppliers carry the item, each supplier's price for that item and ability to meet our requested delivery date, and any discounts the suppliers might offer. But I can't justify printing that report every day."

"We were faced with the same problem, Mark, and we also decided a report was a waste of paper. Our people wanted to be able to get at the information for only the items they needed to order. So we gave them an on-line system. The workers use the system's inquiry function to retrieve needed information on demand. The program can provide the end-user with a menu of inquiries that the end-user can perform, such as inquiries to obtain supplier, product, or purchase order information."

Mark pondered this for a moment. "It would be a more difficult system to develop, but we could do it. In fact, I suppose I could also have the program provide the manager with the option of entering the final decision on supplier. But I still have a tricky problem to solve."

"What is it?" Carol asked.

"Well, the problem is that we don't want to generate a purchase order right when the supplier decision is made for the item. Let me explain the steps we currently go through. After making a decision for an item, the manager transfers information about that item and supplier to something called an order pad. The order pad is then given to a purchasing clerk, who generates the final purchase order. The purchasing clerk uses the order pad to generate a list that consolidates those items that are to be ordered from the same supplier. This way, we don't send multiple purchase orders to the same supplier. Once this list has been prepared, the clerk completes the purchase orders. Additional information to be written on the purchase orders, such as the supplier address or unit of measure for an item, would have to be obtained from the supplier notebook. If I put the supplier selection decision on-line, how do I continue consolidating items into single purchase orders for individual suppliers?"

"I see." Carol shook her head. "You're right, that is a problem, one we didn't have to face." Then suddenly an idea came to mind. "Listen. Why not store the supplier selection decisions in a pending purchase orders file? New items could be added to a pending purchase order as decisions are made. Then . . ."

"I get it!" Mark exclaimed. "Then, at the end of the day, the computer could print the final purchase orders. It looks like the new system will work. Of course, with all these new files, we have to make sure that there are mechanisms for capturing all the data used to create, modify, and delete records for files. The accuracy of reports and inquiries will be dependent on that data."

"You're absolutely right," interrupted Carol. "I think I can start helping you do that right away."

"You've been extremely helpful, Carol. Thanks to you, I feel pretty optimistic now. How about I buy you lunch and we'll get back to work afterward?"

Carol and Mark successfully implemented the automated inventory purchasing system. The new system proved to be more efficient and effective than the old system.

Discussion

1. Did you notice that Carol and Mark were primarily focusing on the procedures for preparing purchase orders? Why were they mainly discussing how purchase orders were generated? How will an understanding of those procedures help them generate new and useful reports for management?

2. What are some examples of reports that the new system might provide the purchasing clerks and managers? Where would the data for those reports come from?

3. Why was it important to Mark to define the mechanisms for capturing all the data used to update the automated files?

4. Who were the end-users mentioned in the inventory purchasing system? Did the new system serve each end-user? How?

What Will You Learn in This Chapter?

This chapter expands on the information systems, specifically systems functions and components. You will understand the functional capabilities and input-process-output components of an information system when you can

1. Explain the relationships between data, information, input, processing, and output, and give examples of each item.

2. Recognize, describe, and give examples of the following three types of information support that an information system can provide to end-users: transaction processing, management reporting, and decision support.

3. Recognize, describe, and give examples of the components of an information system, including data, information, end-users, methods and procedures, data storage, hardware, software, and internal controls.

4. Describe the importance of keeping up with information systems trends and explain several current trends in modern information systems.

In Chapter 2, you learned that an information system was a tool created by businesses to generate information for the business and its end-users. In this chapter, we look at the types of information that a modern information system can produce for its end-users.

Chapter 2 also defined an information system as an *arrangement of components*. So far, we have not identified those components. Because the systems analyst plays a major role in selecting and designing the components in information systems, this chapter will carefully define them.

We continue to unveil our information systems pyramid model in this chapter, rotating the pyramid to reveal the last two of its faces—the information systems functions and the input-process-output (IPO) components. We remind you that we'll use the pyramid throughout the book to note important issues and concerns as we study systems analysis and design methods.

Information Systems Functions

The purpose of an information system is to fulfill the information needs of its end-users. But what is information? Is it the same as data? What information support functions should be considered by the systems analyst? These questions are addressed in this section.

Data and Information

Most people use the terms *data* and *information* interchangeably. But data and information are not the same thing! The distinction is important when studying systems analysis and design. Throughout this book, we use a classic definition for each term:*

Data   are raw facts in isolation. Data describe the business. These isolated facts convey meaning but generally are not useful by themselves.

Information   is data that has been manipulated to be useful to someone. In other words, information must have value, or it is still data. Information tells people something they don't already know or confirms something that they suspect.

One person's information may be another person's data. For example, consider a report that lists all customer accounts. If customers call to find out their current balance due, a clerk can answer their questions by looking at the report. To the clerk, this report is information. But suppose a manager wants to know the total dollar amount of delinquent accounts. The manager would have to identify delinquent accounts on the report and sum their balances. Thus, the report (as it is) represents data to that manager. A report that listed those accounts and the summed balances would represent information to the manager.

What about the items-low-in-stock report that was mentioned in the chapter's minicase? Did the report represent data or information to the manager? When the report was produced manually it probably represented data. It's true that the report told the manager what items were low in stock, but it didn't tell the manager which items needed to be ordered. The manager had to check the purchase order notebook to determine whether or not the item was already on order. If the report had included only those items both low

*Throughout the book, we use *data* in both the singular and the plural, depending on its context. This use is becoming common.

CUSTOMER NUMBER
CUSTOMER NAME
CUSTOMER ADDRESS
CREDIT RATING
BALANCE DUE

in stock and not currently on order, then that report would have been information to the manager.

Data describes the fundamental components and events of a business system: assets, liabilities, accounts, products, materials, customers, suppliers, transactions (such as orders and invoices), and so forth. Each of these business components and transactions can be described by data attributes. For instance, a customer may be described by the data attributes in the margin.

An information system creates, collects, and stores data and processes that data into useful information. This is an important concept of systems analysis and design:

$$information = f(data, processing)$$

Information is a function of data and processing. This formula is as important to the systems analyst as $E = mc^2$ is to the physicist, and you will find most systems methods easier to apply if you commit it to memory.

The concepts of data and information will be useful as we study different information systems functions and capabilities. These support capabilities include transaction processing, management reporting, and decision support systems.

Transaction Processing

Transactions are business events such as orders, invoices, requisitions, and the like. Some organizations consider their **transaction-processing** systems to be separate from their information systems. But we've just learned differently! To produce information, we must first capture data. Business transactions represent 90 percent or more of all data that must be captured. If we capture that data, we can support most information needs. That explains why, in the chapter's minicase, Mark was so intent on being sure that he identified the mechanisms for capturing the data used to update files. If he failed to do so, the data used to prepare the various systems outputs would be either inaccurate or nonexistent.

INPUT TRANSACTIONS

Customer Orders
Accounting Vouchers
Course Registration
Time Cards
Airline Reservations
Payments
Charge Card Slips
Bank Deposit Slips

There are two types of transaction processing. The most common is the processing of *input transactions*. Examples of input transactions are listed in the margin. All of these transactions input new data into the information system. For example, the order data input may consist of an ORDER NUMBER, CUSTOMER NUMBER, ORDER DATE, PRODUCT NUMBER (or NUMBERS), and QUANTITY (or QUANTITIES). In this chapter's minicase, the input transaction that was being processed was the order.

Another type of transaction processing is the *output transaction*. Output transactions trigger responses from or confirm actions to those who eventually receive those transactions. Examples of output transactions are listed in the margin at the top of the next page. In each case, the output transaction confirms an action or triggers a response, such as confirming that you're

✓ **OUTPUT TRANSACTIONS**

Customer Invoices (Bills)

Course Schedules

Paychecks

Airline Reservation Confirmations

Airline Tickets

Payment Receipts

Sales Receipts

booked on a flight or acknowledging the receipt of your payment of a bill. In this chapter's minicase, the purchase order forms that were being generated and sent to suppliers represent an output transaction.

An example of transaction processing is illustrated in Figure 3.1, which portrays an inventory purchasing system similar to the one envisioned in the chapter's minicase. Do not concern yourself that the system is not complete. We have tried to keep it simple enough to demonstrate the concepts you've just learned. Transaction processing is usually performed by and for clerical and service end-users. To some extent, the clerical tasks performed by these workers can be replaced by automation, which is faster and more accurate. These workers can, in turn, assume more stimulating and productive assignments. The reduction of clerical work is frequently offset by an increase in data-entry tasks for the system. One trend in transaction processing is on-line support. On-line systems displace fewer workers because clerks must enter data on computer terminals that are on-line with the computer. Data is captured and processed immediately. Discretionary decisions can still be handled by people rather than machines. The computer provides the information, and the end-user makes the decision.

Despite the fact that transaction processing was the first generation of information systems support, it still provides exciting new applications. For instance, there are still older batch systems that need to be converted to on-line, database versions. Also, many companies are directly linking their customers and suppliers into order-processing systems to get a competitive advantage in their market. Telemarketing with computers is another fast-growing transaction-processing option.

Management Reporting

Management reporting is the natural extension of transaction processing. Data that was captured and stored during transaction processing can be used to produce information of value to end-users, especially management. The information is used to plan, monitor, and control business operations. Because the information is intended for the three levels of management (discussed in Chapter 2), the term **management information system** (or **MIS**) was coined to describe this type of application.

The term *MIS* was popular at one time, but it never really lived up to its advanced billing. The original MIS concept was the idea that a single, large-scale, integrated system could be developed to support all levels of management. Companies were unable to achieve this goal. The basic fault rested in the assumption that members of management could identify all of their information needs; this is simply impossible.

Fortunately, many information needs are definable, particularly at the middle-management levels. This is because many middle-management functions and decisions occur frequently enough that we are able to define rules, models,

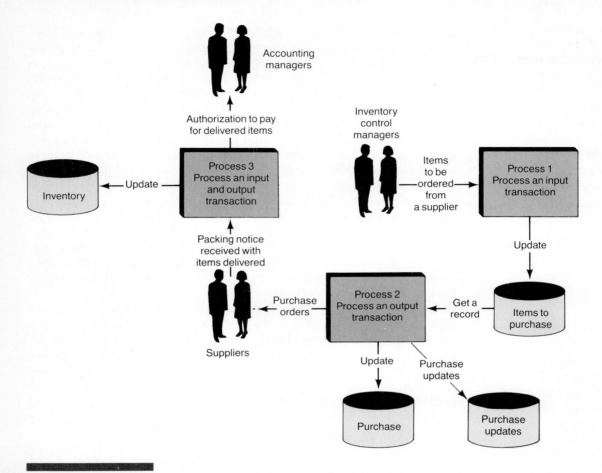

FIGURE 3.1   Transaction Processing   Two types of transaction processing are illustrated. In process 1, we see an input transaction being captured and processed. In process 2, we see an output transaction being generated. In process 3, we see an input transaction being processed *and* an output transaction being generated.

and information requirements for those activities. Examples include information for planning and control functions such as production scheduling, inventory control, and quality control. Other information requirements are regulatory, such as Equal Employment Opportunity Commission and Internal Revenue Service reports. It is these remnants of the MIS concept that we will refer to as *management reports*.

Management reporting systems typically produce four types of information: detailed, historical, summary, and exception reports.

Detailed reports present information with little or no filtering or restrictions. Examples include a detailed listing of all customer accounts or products

✓ **DETAILED REPORTS**

Customer Accounts
  Reports
Student Transcripts
Year-to-Date Capital
  Expenditures

✓ **HISTORICAL
  REPORTS**

Daily Orders Register
Daily Payments Register
Custmer Update
  Report
Deposit/Withdrawals
  Register

✓ **SUMMARY REPORTS**

Sales-by-Store Report
Sales Department
  Report
Salesperson
  Commissions Report
Income Statement
Balance Sheet
Cash Flow Statement

✓ **EXCEPTION REPORTS**

Outstanding Purchase
  Orders
Delinquent Accounts
  Report
Expired Subscriptions
  Report
Backorders Report

in inventory. Other examples are listed in the margin. These reports assist management planning and controlling by generating schedules and analysis.

Historical reports are similar to detailed reports, yet serve a different purpose. These reports provide information on all transactions processed and serve as an audit trail that confirms transaction processing and ensures that the data can be recaptured in case it gets lost somewhere in the system. Examples are listed in the margin.

Summary reports categorize information for managers who do not want to wade through details. The data is categorized and summarized to indicate trends and potential problems. Summary reports can be tabular (tables of numbers); however, graphics are rapidly gaining acceptance because they show trends at a glance. An income statement is a summary report of income and expenses for a business. Another example of a summary report is a sales analysis report that summarizes sales by salesperson, region, and product. Other examples are listed in the margin. Mark, in this chapter's minicase, might be asked to have the new system produce a report that summarizes the outstanding orders for different suppliers. Summary reports are very popular with modern managers.

Exception reports filter data before it is presented to the manager as information. Only exceptions to some condition or standard are reported. A classic example of an exception report is the items-low-in-stock report that was to be generated in the chapter's minicase. The information system reports only those materials for which the quantity on hand is less than is desired. This allows the purchasing manager or clerk to concentrate on only those materials that require immediate attention. Another exception report may tell a sales manager which salespersons are 10 percent or more below their sales quotas. Other examples are listed in the margin.

Figure 3.2 illustrates the management reporting capabilities of a typical system. The management reporting subsystem receives its data from the files maintained by the transaction-processing subsystem (Figure 3.1).

Management reporting will always be fertile ground for new and improved systems. End-users seem to have limitless imagination for new or better reports. And government control will continue to exert new reporting pressures on business.

Decision Support Systems

You have just learned that MIS never fully lived up to its expectations because managers cannot precisely define all their information needs. Sometimes managers don't know what information will help them until the need to make a decision arises. A decision support system (DSS) provides the end-user with decision-making information when a decision-making situation occurs. We will not argue whether DSS—as the third generation of information systems applications—is a new concept or an extension or realization of the original

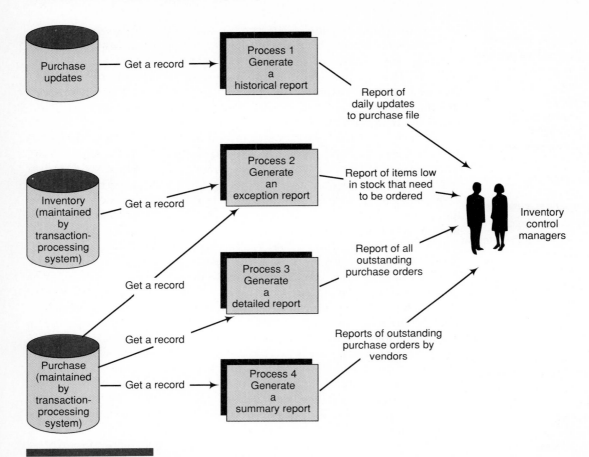

FIGURE 3.2 Management Reporting Management reporting systems typically produce four types of reports: detailed, historical, summary, and exception. These reports are produced from the data that was originally captured and stored in files or databases by the transaction-processing system. Notice that the exception report triggers a decision.

MIS concept. Instead, let's study decision making and determine how the information system can be made to support it.

There are three types of decisions: structured, semistructured, and unstructured. **Structured decisions** are those we can predict will happen. We can't always predict *when* they will happen, but we can predict *that* they will happen. Managers can usually define the information requirements to support a structured decision. For example, we can provide an on-line credit check to help a manager or clerk decide whether or not to let someone charge a sale at a local department store. We know those charge sales will occur. We can also define what information is needed to check a person's credit.

Unstructured decisions cannot be predicted. We don't know when the decision-making need will occur, and we also don't know the nature of such a decision. Can you think of an unstructured decision? If you could, it wouldn't be unstructured! Because the unstructured decision can't be predefined or predicted, you can't define what information will be necessary to assist the decision-making process. DSS is primarily intended to support the unstructured decision. Does this seem impossible?

The concept behind DSS is that the data for many unstructured decisions have already been captured by the transaction-processing and management reporting systems. Additional data may be available in national and international databanks around the world. It should be noted that managers may have to input additional data that isn't stored in any files. The objective of DSS is to subsequently make it easy for managers to get at needed data and manipulate that data when a decision must be made.

We have seen two types of decisions. They represent extremes, because many decisions are **semistructured**; that is, they can be predicted, but not all variables can be considered. How are decisions made? Ideally, a DSS should support the way decisions are made. A classic model of decision making used by managers is known as the Simon model. It suggests that most decision situations pass through three steps before the decision is made:

- need identification
- alternatives analysis
- choice

Need identification (called *intelligence* by Simon) is the process of identifying that there is a decision to be made. Decisions can usually be identified by examining and analyzing existing data, presumably from the transaction and management reporting subsystems. Specifically, decision support systems provide managers with powerful tools to look at data in new ways. These include easy-to-use query languages, ad hoc report generators, statistical analyzers, and graphics. Need identification is the simplest level of decision support and is designed into most new systems.

Alternatives analysis (called *design* by Simon but renamed here to avoid confusion with the same term used throughout this book) identifies and analyzes alternative courses of action in a decision situation. Decision support systems can provide managers with "what if" tools that can pretest the potential outcome of different decisions. The best example of such a tool is the *spreadsheet*, such as Lotus 1-2-3. Most of you have probably learned about spreadsheets in other courses or experiences. With them, you can test the financial implications of various budgeting or funding decisions before you make them. Now, just extend that concept one step further—extract the data from other systems and then play these "what if" games with the spreadsheet.

Another level of decision support for alternatives analysis is *simulation*. Simulators use existing business data and complex mathematical models to

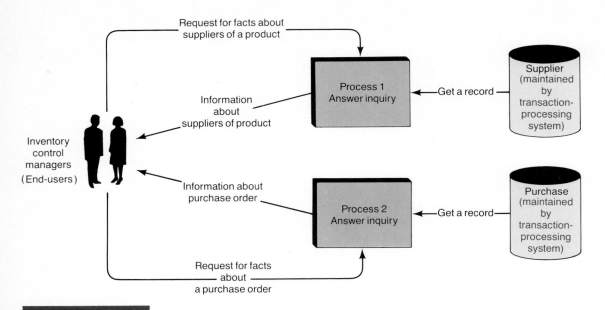

FIGURE 3.3   Decision Support System (DSS)   Decision support allows the end-user to obtain information needed to solve a problem. Data needed to support the decision is obtained from the files or databases created and maintained by the transaction-processing and management reporting systems.

project the implications of alternative decisions. To learn more about such tools, take a course in finite mathematics, linear programming, or operations research or management.

Choice (the Simon term) is only beginning to be exploited by DSS. **Expert systems** and **programmed decision systems** support the choice stage. They are primarily intended to support middle-management, professional-staff, and executive-management decision making. The key to these systems is advancement of the technology of **artificial intelligence (AI)**. AI and expert systems are systems that can learn and reason in a manner similar to humans. See the "Next Generation" feature on page 64 for more on expert systems.

Decision support (Figure 3.3) and expert systems are among the newest and most exciting applications for information systems.

Integrating Information Systems Functions

We now rotate our information system pyramid to reveal the *information systems functions* face (Figure 3.4). Note that transaction processing is the foundation on which the information system is built. Looking over to the end-user face, we also see that transaction processing primarily (there are exceptions) serves the clerical and service end-users in the business.

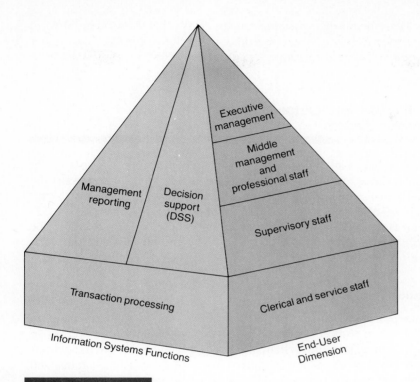

**FIGURE 3.4   Information Support Provided by the Information
System**   The systems analyst should design information systems that
provide complete information support to end-users, including transaction
processing, management reporting, and decision support (DSS).

Next, note that the management reporting and DSS primarily support all
three management levels of end-users: supervisors, middle managers and
professional staff, and executive managers. Think about it. *All* managers can
make use of appropriate management reports. These reports differ only in
their level of detail. Along the same lines, *all* managers make decisions—
some operational, some tactical, and some strategic. Management reporting
and DSS applications can be designed for managers at all levels of the
organization.

The systems analyst should become familiar with all of the potential support capabilities of the modern information system. The systems analyst is
responsible for educating the end-user about these support capabilities and
for choosing the best support for fulfilling their needs. The three levels of
functions described in this chapter are not separate systems. They can and
should be integrated, especially because they all share the fundamental data

that describes the business system. Many projects you will encounter involve extending the capabilities of a transaction-processing system to include better management reporting, decision support, or maybe even expert systems.

Input, Process, and Output Components of an Information System

Recall that an information system is a subsystem of the business—specifically, a person/machine arrangement of components that interact to support the operational, managerial, and decision-making information needs of end-users.

When we talk about arrangement of components, what components do we mean? How are these components arranged? What trends should we be aware of? The answers to these questions complete our introduction to the information system.

Information Systems Components

Most systems can be described in terms of inputs, outputs, and processes. In fact, the purpose of most systems is the processing of inputs into outputs. In the case of a manufacturing system, the inputs are raw materials, the outputs are the finished products, and the processors are the machines, tools, and people that transform raw materials into finished products. These inputs, outputs, and processors make up the arrangement of components that we call a manufacturing system.

Any system's components can be classified according to this input-process-output model. For an information system, recall our fundamental formula, **information = f(data, processing)**! Information is output. Data is input. Therefore, we could modify our formula to read

output = f(input, processing)

This is an appropriate way to look at any information system. Rotating our pyramid, we introduce the last face, the *input-process-output* (*IPO*) components for an information system (see Figure 3.5). Data is the input component, and information is the output component. The process that transforms the data into information may itself consist of and relate to the components listed in the margin. Let's examine each of these internal components of the information system.

End-Users The computer is important, but the people who use and work with the information system are the most vital and most overlooked aspect of the system. Information systems are for end-users! How often have you referred to your latest project as "my system" (or program)? Is the system

✓ INFORMATION
SYSTEM
COMPONENTS

End-Users
Methods and
   Procedures
Computer Software
Computer Hardware
Data Storage

FIGURE 3.5 The Input-Process-Output Components of the Information System When studying or building an information system, the systems analyst should consider all of the input, process, and output components of the information system illustrated here.

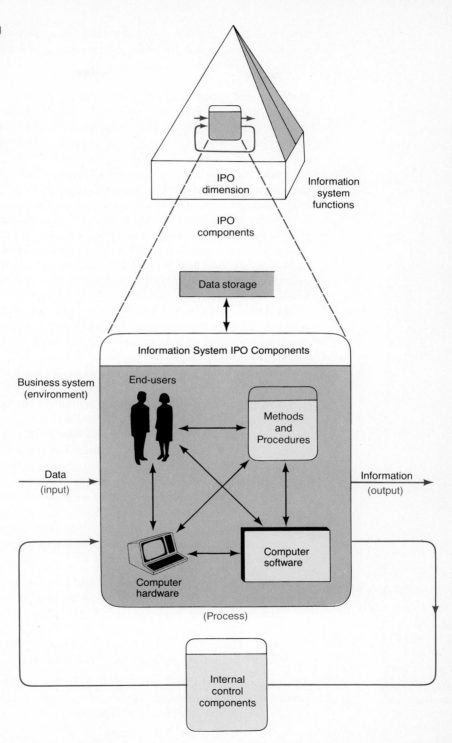

really yours? Hardly! It is important that a systems analyst not view end-users as simply those people for whom he or she builds information systems. The analyst must recognize that the end-user is an integral *part* of the system.

Methods and Procedures   Methods and procedures are also underemphasized in many information systems. Methods typically refer to information systems methods employed in computer information systems. Computer-based systems generally process data into information by one of three methods: batch, interactive, or real-time processing.

Under **batch processing** methods, data must be accumulated into batches and processed periodically, usually on a specific schedule. For example, we could collect orders into batches and process them once or twice each day.

On the other hand, using **interactive processing** (also called *on-line processing*) methods, data is processed either as it occurs or as it becomes available. For instance, we could use computer terminals to enter and process each order as soon as it is received. Batch and interactive processing are not mutually exclusive. You could collect data interactively and process it as a batch. This is referred to as **deferred processing**.

Real-time processing is of interest to applications that require very rapid response times. Examples include missile tracking and robotics. There are many implications of information systems methods, and we discuss them more fully in Chapter 18.

Procedures describe how end-users perform their jobs and make decisions. Procedures do not refer to computer programs. It is important to realize that any information systems method will affect the procedures used by its end-users. How will the data be collected and entered? How will the reports be distributed? What decisions and activities will the worker still be responsible for? What is the sequence of activities? These are all *procedural* questions to be addressed by systems analysts. Thus, business procedures are as important as computer procedures (programs)!

The key is *human engineering*. We are slowly learning that technological improvements must be counterbalanced with improved human elements. Human engineering is the design of anything that is easy to use, appreciated, and nonthreatening. A key factor in resistance to new systems is poorly designed procedures. Does this sound familiar?

> Before the computer system, all we salesmen had to do to get our expenses was to complete a form, get it authorized and present it to the cashier at virtually any time. Now we have to complete the form with all sorts of codes, once a month. We have to put the form in by the last Friday of the month. Then, we eventually get a cheque [check]—up to three weeks later. Some service those computers give [London, 1976, p. 41].

This situation is still not unusual. It indicates that analysts are frequently preoccupied with computer aspects of systems. Hence, they tend to poorly design the manual aspects of the system.

Computer Hardware   Information systems may include a variety of hardware components, including duplicating machines, calculators, and computer systems. In this discussion, we will concern ourselves mainly with the computers. Computer components include peripheral devices such as optical scanners (input), line printers (output), CRT terminals (input and output), magnetic disk drives (data storage), and central processing units (processor). We are assuming that you are familiar with these components, either from your first information systems course or through personal experience.

It's always frustrating to try to predict trends in the fast-changing world of computer technology. But certainly one trend that will continue is the proliferation of microcomputers. Another trend, supported by advanced telecommunications and networking technology, will be the continued emphasis on distributed information systems, in which processing power, in the form of mini- and microcomputers, will be located closer to the end-users. And still another trend will be improved connectivity, the ability of these distributed systems to interface and integrate with one another.

Computer Software   Computer programs have long been the responsibility of the computer programmer. There are many applications programs that support specific business functions. There are also systems programs that run the computer itself and allocate resources to the computer users. Programmers are responsible for the programs, and systems analysts are responsible for the program specifications. You should, however, be aware of two significant software trends: the use of software packages and applications development without programmers.

Software packages are computer programs that are purchased rather than built. The concept is "Why reinvent the wheel?" If a software package meets all or most of your needs, then programmers can be reassigned to projects for which software packages do not exist. Because programmers are a scarce commodity, this concept makes sense. Most packages require some modification before they operate properly. Many such packages are flexible and adaptable so they can be more easily customized. The systems analyst's job frequently requires the selection of a software package(s) for an information system.

Applications development without programmers was introduced in the "Next Generation" feature for Chapter 1. Many decision support needs are so spontaneous that information systems professionals cannot hope to respond quickly enough. Often, information systems cannot justify the analysis and programming cost of a one-time—or few-times—output requirement. But today we are seeing software tools (see margin for examples) that enable end-users to personally and rapidly generate their own reports from existing files and databases.

✓ SOFTWARE TOOLS 4(GLs)

Focus
RAMIS
dBASE IV
Nomad
On-Line English

Data Storage   Data storage is another fundamental component of the information system. The internal data of the information system is stored in files

Expert Systems and Knowledge Engineers

A new breed of systems analysts are being created. They are called *knowledge engineers*. Their job? To implement expert systems for their end-users.

Expert systems mimic the reasoning of experts in various fields. They use artificial intelligence technology to mimic reasoning to real problem solving and decision support. And it has become a $4 billion industry.

A more formal definition of an expert system is a computer-based information system that has been encoded with human knowledge and experience to achieve expert levels of problem solving.

Expert systems are finally finding their way out of the research laboratories into the business sector. The following examples are real:

- A food manufacturer uses an expert system to preserve the production expertise of experienced engineers who are nearing retirement.

- A major credit card broker uses an expert system to accelerate credit screening that requires data from multiple sites and databases.

- An auditing firm uses an expert system to expand its customer base by providing tax accrual accounting services, for which accountants are scarce and time is even more scarce.

Virtually any business can benefit from automation of internal or external expertise.

How do expert systems work? The technology of artificial intelligence separates expert systems into two distinct components, the inference engine and a knowledge base.

The **inference engine** is a software product that you buy from an expert systems vendor. It contains the problem-solving/decision-making software. Vendors usually refer to the software as a **shell**. The shell is an expert systems generator that eliminates much of the need to write programs in traditional AI languages like LISP and PROLOG. Examples of popular shells include GCLISP, EXPEROPS, INSIGHT, VP-EXPERT, and GURU.

The **knowledge base** is an extension of the database concept. It contains not only data, but also facts and rules for decision making (called *heuristics*). The knowledge base is created by or for the end-user using the shell. The inference engine executes the knowledge base to simulate the expertise.

You don't need millions of dollars to get into expert systems. Indeed, microcomputer expert systems technology can be acquired for less than $10,000 (including the microcomputer).

Also needed is a new type of systems analyst, commonly called a knowledge engineer. A knowledge engineer needs to understand the technology of artificial intelligence and expert systems. This person also needs to understand tools for concisely documenting and analyzing rule-based logic (complex combinations of conditions that result in different conclusions, actions, or recommendations). Even more importantly, a knowledge engineer needs to have superior communications and interpersonal skills to extract expertise from the experts. For instance, a knowledge engineer must sensitively probe expertise. It isn't enough to know that if A and B are true, then C must be true. The engineer needs to find out how the expert drew that conclusion and why.

Expert systems are one of the next generation of computer-based information systems applications. And better systems analysts are the ideal candidates to become the knowledge engineers who will develop these systems.

and databases. These data stores may be either manual (for example, a file cabinet) or automated (a computer disk). One significant trend in this area is the gradual replacement of traditional files with modern database technology. This requires some explanation.

The terms *file* and *database* are frequently used as synonyms. But the terms are not interchangeable, despite their frequent misuse in the software industry. Traditional systems were built using file technology. Each file was designed and structured for the specific programs that would use it. If a later program needed the file's data to be structured differently, then you had to write special programs to extract, sort, merge, and create new or temporary files for that purpose—clearly redundant data! Why not change the existing files? Because changes would force costly changes to many programs that use that file's existing structure. In other words, the data in traditional files is dependent on programs that use the data, and vice versa.

Today, database offers an alternative. The notion of *files* as they were just defined disappears. A **database** is an integrated collection of data, stored with a minimum of redundancy and structured such that multiple applications can share the data. Ideally, the data is structured independent of the programs that use it. This allows the structure to be changed without having to change the existing programs that use the data. Both database and file methods are taught in this book.

Data storage has become such a critical component of systems that systems analysts are turning to *data-centered* development approaches that emphasize understanding data *before* worrying about processing it. This book will address this approach in several chapters.

Internal Controls   Feedback and control is a systems concept and feature that must be added to any system to ensure proper operation. In Figure 3.6, you can see a feedback and control model for our information system. As you can see, the **internal control** component is an information subsystem in and of itself. It consists of end-users, methods and procedures, data storage, and possibly hardware and software. It reacts to information generated by the information system (which is data to the internal control subsystem). The internal control subsystem either directly implements change to the information system or notifies the end-user that the system is not functioning properly.

Controls ensure that input data is valid before it is processed, that file and database data are protected against unauthorized access and update, that recovery procedures are available for a system catastrophe, that output information is disseminated to proper end-users, and that other relevant conditions are met. Again, controls ensure that the system works properly! We will discuss controls throughout Part Three of the book.

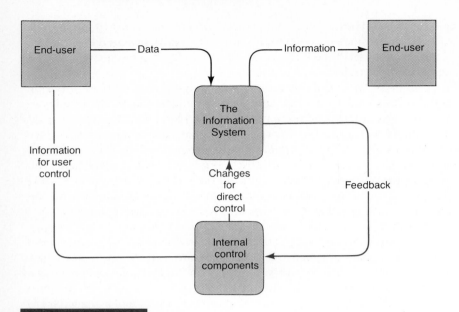

FIGURE 3.6 Internal Controls for an Information System Good information systems contain a subsystem of components (end-users, methods and procedures, data stores, hardware, and software) that make sure the system is properly operating. These components react to feedback from the information system and make or recommend changes for the information system.

The Importance of Input-Process-Output Components to the Systems Analyst

The implications of the IPO components of information systems are profound. Although it is easy to become caught up with the computer components in a system, the end-users, methods and procedures, manual data storage, and internal controls are equally important. Until you have designed all components, your task is incomplete. And if you choose not to define all components, the system will likely come back to haunt you. The analyst must ensure that the components, both manual and computerized, are coordinated to support the business.

Summary

In addition to the end-user and business dimensions of an information system, information support and input-process-output dimensions exist. To understand these dimensions, it is important to differentiate between data and information. Data are raw facts in isolation. Information is data that has been manipulated so that it is useful to someone. An important concept of systems analysis and design is the formula information = f(data, processing).

An information system is capable of providing end-users with three types of information. Transaction processing is the means by which most data is captured. There are two types of transactions, *input* and *output*. Management reports are another type of information. Management reports fulfill the defined information needs of managers. Four types of reports can be produced: *detailed*, *historical*, *summary*, and *exception* reports. Each report is produced from the data that were captured and stored during transaction processing. The third type of information supports decision making. Decision support systems (DSS) provide end-users with the ability to obtain needed information when a decision-making situation arises. Expert systems are extensions of DSS.

In addition to providing different types of support in information systems, an analyst must design and select the working components of the information system. The components that make an information system work can be classified according to an input-process-output (IPO) model. The input component of an information system is data. The output component of an information system is information. The process component that transforms the data into information consists of end-users, methods and procedures, data storage, computer equipment, and computer programs. Information systems also include the internal controls component. Internal controls ensure that the information system is operating properly. A systems analyst must consider each of these components when developing an information system.

Problems and Exercises

1. Identify each of the following as data or information. Explain why you made the classification you did.
   a. A report that identifies, for the purchasing manager, parts that are low in stock
   b. A customer's record in the Customer Master File
   c. A report your boss must modify to be able to present statistics to his boss
   d. Your monthly credit card invoice
   e. A report that identifies, for the inventory manager, parts low in stock
2. Give two examples of a report or document that might be considered information to one person but data to another. What would have to be done to transform the report into information for the second person?
3. Identify the type of information support provided for each of the following applications.

  a. A customer presents a deposit slip and cash to a bank teller.
  b. A teller gets a report from the cash register that summarizes the total cash and checks that should be in the drawer.
  c. Before cashing a customer's check, the teller checks the customer's account balance.
  d. The bank manager gets an end-of-day report that shows all tellers whose cash drawers don't balance with the cash register summary report.
  e. The system prints a report of all deposits and withdrawals for a given day.

4. Give an example of a transaction (input or output). What data describe that transaction? Now describe a management report that might include those data. Finally, describe a decision that might require access to those data.

5. What is the relationship between data, information, input, processing, and output? Give an example of this relationship.

6. Identify the transaction-processing, management-reporting, and decision-support functions that were discussed in the minicase at the beginning of this chapter.

7. Make an appointment to visit a systems analyst at a local information systems installation. Discuss one of the information systems projects the analyst has worked on. What were some of the transaction-processing, management reporting, and decision support provisions in the system?

8. Give an example—or better still, collect an example—of each of the following: a detailed, a historical, a summary, and an exception report. *Hint:* Examples of reports are everywhere. They are in your programming and information systems textbooks, the places where you work (CIS or otherwise), your mail, on your teacher's or secretary's desk, and in all sorts of other places.

9. Identify all of the information components discussed in the minicase at the beginning of this chapter.

10. Define a decision that you—or your boss or a friend—have recently made or are making. The decision can be either personal or business. Describe the decision-making process for that decision. What information or facts might help you make a better decision?

11. The end-user is the most important and overlooked component of an information system. Explain why.

12. Visit a systems analyst at a local information systems installation. Discuss one of the information systems projects that the analyst has worked on. Who were some of the end-users supported by the information system? What were some of the automated and manual processing methods and procedures of the system? What types of data storage were being sup-

plied? What computer equipment or hardware was being used? What were some of the applications programs? What type of internal controls were installed in the system?

13. Why is it important to keep up with information trends? Explain how you plan to keep abreast of information trends.

Projects and Minicases

1. Last year, John Freneau—the Comptroller at Hologram, Inc.—purchased an IBM PC and the spreadsheet Lotus 1-2-3. John learned how to use the product to support his own budgeting and cost control decision-making needs. He likes Lotus, but he's getting tired of reentering the same data into different spreadsheets. He wants to know why he can't store the data in a database on his microcomputer. Better still, John knows that much of the data currently resides on the Burroughs mainframe computer in Information Systems. Why not tap that database to minimize the amount of data he has to input personally? In terms of the information systems functions model, what is John requesting? Ignoring technical implications, are his expectations reasonable?

2. Liz Rakowsky, an Account Collections Manager for the bank card office of a large bank, has a problem. Each week she receives a listing of accounts that are past due. This report has grown from a listing of 250 accounts two years ago to 1,250 accounts today. She has to go through the report to identify those accounts that are seriously delinquent. A seriously delinquent account is identified by several different rules, each requiring her to examine one or more data fields for that customer. What used to be a half-day job has become a three-days-per-week job. Even after identifying seriously delinquent accounts, Liz cannot make a final credit decision (such as a stern phone call, cutting off credit, or turning the account over to a collection agency) without accessing a three-year history on the account. Additionally, she needs to report what percentage of all accounts are past due, delinquent, seriously delinquent, and uncollectable. The current report doesn't give her that information. What kind of report does she have— detail, summary, or exception? What kind of reports does she need? What kind of decision support aids would be useful?

3. Knowledgeable University plans to support course registration and scheduling on a computer. The following end-user community has been designated:
   a. Curriculum deputy—one per department, responsible for estimating demand by that department's own students for each course offered by the university. This person may revise demand estimates from time to time.

b. Schedule deputy—one per department, responsible for deciding which courses from that department will be offered, at what times, and by what teachers, and for determining what enrollment limits will be. These parameters may change during the registration period. This is the only person who can increase or decrease enrollment limits for a course, including adding or deleting a course to or from the schedule.
c. Schedule director—in charge of allocating classroom and lecture hall space and time to departments. Also prints the schedule of classes to show students what will be offered and when.
d. Students—submit course requests and revisions and receive schedules and fee statements.
e. Counselors—advise students and approve all course requests and revisions. Also help students resolve time conflicts where they have registered for two courses that meet at the same time.

This is the cast of characters; it may be revised or supplemented by your instructor to more closely match your school. For each end-user, brainstorm input and/or output transactions, management reports (especially summary and exception reports), and decision support inquiries.

Annotated References and Suggested Readings

London, Keith. *The People Side of Systems*. New York: McGraw-Hill, 1976. Includes numerous examples of what can happen when information systems are not designed for ease of use by people.

Naisbitt, John. *Megatrends: Ten Directions Transforming Our Lives*. New York: Warner Books, 1982.

Sprague, Ralph H. "A Framework for the Development of Decision Support Systems." *MIS Quarterly,* vol. 4, December 1980, pp. 1–26. This paper presented the forerunner of our pyramid model. We are especially indebted to it for our information systems functions face of that model.

Wetherbe, James. *Systems Analysis and Design: Traditional, Structured, and Advanced Concepts and Techniques*. 2d ed. St. Paul, Minn.: West, 1984. This book has an excellent systems concepts chapter that inspired us to look at what we called the IPO components of an information system. Although our IPO components model differs somewhat, the same fundamental systems concept is being applied.

SOUNDSTAGE

A New Information Systems Project Is Started

Sandra and Bob get an assignment: "How can we improve our order processing through improved information systems support?"

EPISODE 2   Sandra entered her office with the Monday morning "blahs." Things had been pretty slow since the new Accounts Receivable system had been placed into operation. Very few errors had been made and the end-users were happy. But she eagerly wished for a new project. She wouldn't have to wait long.

Debbie Lopez phoned at 8:45 A.M. Debbie was Administrative Assistant to David Hensley, Customer Services Manager for SoundStage.

"Good morning, Sandra! How was your weekend?"

"Just fine, Debbie. What can I do for you?"

"Can I come and see you for a moment?" asked Debbie.

"Sure," Sandra replied.

A few minutes later, Debbie entered Sandra's office. She gave Sandra a piece of paper (Figure E2.1). It was a project request form.

Debbie said, "This is a project request. And I wouldn't trust it to anyone but you. That Accounts Receivable system you created is a

godsend. Bill is so pleased with the system! Meanwhile, this one is just as important, probably more so. David handed this proposal to me last week. I think we ought to get some preliminary analysis quickly."

Sandra quickly skimmed the request.

Debbie continued, "Well, in a nutshell, we want to see what you can do to improve our Order Entry and Follow-Up system. As you know, our product mix is rapidly changing, especially with the addition of compact digital discs. We want to start a separate compact disc club, but first we need a system to support it. Since we need this new system, we thought that this would be an ideal time to reexamine support for the record and tape club also. There are some problems there. Also, we would like to implement some new interclub services for members who are either switching to disc or collecting both tapes and discs."

"This may be a stupid question," Sandra suggested, "but why don't you just hire more clerks?"

"I'd love to. But I can't! David won't approve the additional payroll. Besides, we're convinced that many problems are information and decision oriented."

Sandra responded, "Could you clarify that?"

"Well," said Debbie, "we experience delays because we can't validate customer data, credit status, part numbers, part availability, and the like quickly enough. Even if David hired more people, where would we put them? And sooner or later, sales would increase to a point where the added staff would again be inadequate. You know how it is. Increase sales! More! More! More!"

Sandra was already enthusiastic. "One thing is certain. If we redesign the system, we will be required to convert it to an on-line system. ISS is phasing out batch systems over the next five years. We would also have to use the new database system since we are directed to develop all new systems with that technology."

Debbie replied, "I'm all for it.

Episode 2, continued ▶

```
REQUEST FOR SYSTEM SERVICES                              FORM 100


SUBMITTED BY ___Susan Lopez_____   DATE ___July 25, 1989_____
DEPARTMENT ___Member Services Division (Mgr. David Hensley)_____
TYPE OF REQUEST      [xx] New System
                     [  ] System Redesign
                     [  ] System Modification


    Order transactions are expected to increase 25% over the next
    three years. Current order-processing and follow-up procedures
    cannot support this demand. The problem is inefficient transaction
    data capture, inadequate decision support, and lack of management
    information to support operations.


REQUEST FOR SERVICE TO ____Implement an improved member services___
    system with immediate emphasis on order entry and order followup.




ACTION (to be completed by Steering Committee)

    [  ] Request Approved      Assigned to _____

                               Start Date _____

    [  ] Request Delayed (backlogged) Until _____
    [  ] Request Rejected for Reason: _____
         _____
         _____
```

FIGURE E2.1   Project Request Form

I've seen some of the on-line systems ISS has already converted. I have some transfers from some of the departments that use those systems. They complain about our outdated methods. As far as database technology is concerned, I know nothing about it, but I assume that you wouldn't be using it if it weren't feasible. Speaking of technology, we'd like you to investigate how microcomputers might be used or integrated into whatever solution you come up with. We have some PCs; however, we aren't really using them for much more than word processing. Five thousand dollars is a lot to pay for glorified typewriters."

"Let me get to David on this, Debbie," said Sandra. "I think we have a priority project here but I need to do some preliminary analysis. This one seems sure to get through the Steering Committee." Sandra knew that the results of a preliminary investigation are always reviewed by the ISS Steering Committee to determine if the proposed project is worthwhile and of a high priority.

Debbie stood up. "Well, I just wanted to deliver the request personally. Thanks! I'll see you later, Sandra."

Sandra studied the Request for System Services form carefully. Then she phoned David, the Manager of Customer Services.

"David? Hi. Listen, I got your project proposal today. Can we meet over lunch to discuss this order-entry problem of yours? Say, about 11:30?"

"Great!" said David. "I've got an earful for you! See you at 11:30!"

Sandra invited Bob to join her for the luncheon meeting. She knew Bob was anxious to start a real project. And she wanted to give Bob a tour of the Member Services Division's offices. The meeting started on schedule.

"David, I'd like you to meet Bob Martinez, my new partner. Bob, this is David Hensley, the Manager of Member Services."

"I'm glad to meet you, David," said Bob. "How are you today?"

"Just fine, Bob. So, Sandra, you two got my project request?" David was openly enthusiastic.

They ate lunch and then directed the conversation to the proposed project. Sandra opened the questioning. "Tell me something about how you do order entry and approval today. I'm just trying to get a general feel for your situation. We'll study the system in detail later."

"Well," David responded, "I'll try to give you an overview. We process three kinds of orders: dated orders, priority orders, and merchandise orders for the record/tape club and forthcoming compact disc club. Dated orders are orders to be filled if a member doesn't return a dated order card to cancel or change the monthly 'title of the month' offer. Priority orders are for those cards that were returned. Merchandise orders are for any other type of merchandise we sell, including shirts, posters, computer software, videotapes, and so on. Video and computer software may soon merge into a third club.

"The orders are manually screened for accuracy and completeness. Some orders have to be transcribed to . . ."

"Transcribed?" asked Bill.

"Yes, copied to our standard order forms. Most orders are then sent to ISS to be processed. ISS performs credit checks but cannot reject orders officially. We do the follow-up letters to customers if they must pay on their account before we can release their order. Most orders pass the credit check and are split into multiple orders corresponding to merchandise ordered."

Sandra reentered the conversation. "What kind of problems initiated your project request, David?"

"First off, I purposefully simplified my description of order entry and approval. You need to spend time in Member Services to really appreciate the magnitude of my operation. But, to answer your question, I have a few problems to share. First, I've got a big problem coming! I had a meeting with Rebecca Todd (a Vice President) yesterday. I was told to expect an aggressive new marketing program over the next three years . . . TV, radio, newspapers, and magazines. New clubs, the first of which will focus on compact digital discs, will be created. Corporate management expects sales to increase at least 25 percent over the next three years. I can't handle that load with my existing staff. I have no place to put new staff. I'm convinced we need better computer information support. Meanwhile, my response time to orders is getting worse each month. Orders are getting more complicated. New A/R policies allow customers to carry balances forward on account. We didn't use to do that. You see, Accounts Receivable's new system has complicated my operation. By the way, I forgot to mention that we are taking over the subscription function that enrolls new members through advertisements and special orders."

"I'm beginning to see the magnitude of your problem, and you're right; we're going to have to spend some time in Member Services to fully understand. I'm a bit lost already. Can you explain what you think you need?" Sandra asked.   **Episode 2, continued ▶**

"Yes. I would like more significant support of order processing. I want to be able to better follow up on outstanding orders. I want some way to prioritize backorders. I'd also like to get some decent information that shows me where I can improve my operation. I don't really know *what* information. I'm kind of hoping that you may help me figure that out. At least help me harness my data so I can get at it when I want to." David's tone indicated some frustration with his current operation.

"This sounds interesting. I hope . . . ," said Bob.

David interrupted, "Oops. I'm sorry, Bob. But I almost forgot an important constraint. We want to simultaneously rework our Member Bonus Credits program and implement it with whatever system

you come up with. The new program is being documented right now under the name Preferred Member Program."

"Whew!" said Sandra. "Bob and I will need to gather some more facts. We should talk with you again very soon. We'd better get going, Bob. We have a review conference on the Accounts Receivable system in 15 minutes. Thanks for lunch, David. We'll get back to you soon!"

Bob and Sandra completed their preliminary study with one more in-depth interview with David. They presented their findings in a memorandum to the Steering Committee which is shown in Figure E2.2 on the next two pages. The Steering Committee approved the project.

Where Do We Go from Here?

Sandra and Bob now have a project. Where should they start? If you were assigned to write a proposal that outlines a step-by-step process to develop the new system, what would your steps be? Take a moment to jot down your ideas.

In Chapter 4 you will study a general process for developing an information system—Sandra and Bob are being asked to develop an Order Information System. The process you'll be studying is called a **systems development life cycle**. The steps are called *phases*. Each phase may consist of several tasks. Those tasks will be studied in later chapters. The systems development life cycle is the basis for all of the tools and techniques you will learn in the remainder of this book.

Episode 2, continued ▶

FIGURE E2.2

MEMORANDUM

DATE: July 28, 1989
TO: ISS Steering Committee
FROM: Sandra Shepherd, Systems Analyst
           Robert Martinez, Programmer/Analyst
SUBJECT: Feasibility Assessment for New Member Services Information System

We have just completed a preliminary investigation for a proposed member services information system for the Member Services Division. If a successful system can be developed (and we think that highly likely), then the system will provide substantially improved support for both the division and our customers.

Problem Statement

Member Services handles membership subscriptions and member orders. Subscription and order processing is, for the most part, a manual process. Users have developed a rudimentary database on a microcomputer system to help with some aspects of the system; however, there are many more clerks than there are micros. The primary benefit of the database is a simple daily printout distributed to the clerks. The following problems in the existing system have been identified:

1. A changing product mix (from records to cassettes and compact discs) has led to a decision to factor memberships into distinct clubs. The current system does not support this structure.
2. Directives to increase membership and orders through aggressive advertising have led to subscription and order-processing workloads that cannot be handled by the current system.
3. Response times to orders have doubled. In other words, orders that used to be filled in one week now require two weeks. This leads to member dissatisfaction and reduced cash flow (since members won't pay until they receive merchandise).
4. Management has approved a new "Preferred Member Program" that cannot be implemented using existing information.
5. Unpaid orders have increased to 4% (from 2% only two years ago). This has been attributed to a poor credit check interface with the accounts receivable system.
6. Backorders are not receiving priority. Some backorders go unfilled for three months even though inventory was received. New orders frequently deplete that new inventory before backorders are even checked.

Scope of the Project

The users and their mangement are requesting a new system that will:
1. Expedite the processing of subscriptions and orders through improved data capture methods and decision support.
2. Provide improved marketing analysis of promotional programs.
3. Provide improved mechanisms for follow-up on orders and backorders.
4. Factor memberships into distinct clubs based on members' preferred media (e.g., records, cassettes, videotapes, compact discs, etc.).
5. Reduce credit losses.

(continued)

The primary users of the new system would be the management and staff of the Member Services Division. The proposed system will likely interface with, and therefore affect, the following business units:

1. Purchasing
2. Warehouse
3. Accounts Receivable
4. Marketing

Constraints

The new system is needed in approximately one year. The new system cannot alter any existing file or database structures in the accounts receivable or inventory systems. Such alterations would impose extensive maintenance to the programs in those systems.

Preliminary Alternatives

Any final solution would be subject to an extensive identification and analysis of alternatives; however, a preliminary checklist of options has been prepared. These options include:

1. Design and implementation of a on-line, IBM mainframe system (which, by management decree, requires use of the IDMS database management system).
2. Purchase a suitable on-line membership and order-processing system from a vendor. Both mainframe and microcomputer products would be considered.
3. Develop an improved system around the existing, user-developed, dBASE III microdatabase system. This would likely involve an upgrade to dBASE IV and complete redesign of existing dBASE files to reflect proper relational database design.
4. Submit this project, to the steering committee that is investigating the decentralization of systems, as a candidate for the new IBM AS/400 minicomputer system.

If this project is approved, other alternatives may become apparent and, subsequently, analyzed for feasibility.

Recommendations

We highly recommend that this project be approved for systems analysis. The results of a detailed study, user requirements, and proposed technical solution would be submitted to the steering committee before proceeding to systems design and implementation.

The basis for our recommendation is that this system is essential to the business mission of SoundStage. Unless the member services system is dramatically improved and the existing problems are solved, it is likely that new directives to expand membership, orders, and services will fail. Indeed, it is also likely that the existing membership base and orders will decline unless the system is improved. We believe that you will concur that this system, because of its direct membership interface, should be a higher priority than most projects currently being considered.

Resource Requirements for Systems Analysis

The detailed study of the current system, definition of user requirements, and specification of a target solution will require approximately $14,950, budgeted as follows:

2.5 months for systems analysts	$9,500
1.0 months of user release time	3,550
0.5 secretarial support	1,000
overhead	900
	$14,950

The systems analysis report will include a budget for design, implementation, and support for the new system. If you have questions, our phone extensions are 355 and 356. You can also contact us through electronic mail using the addresses SANDRA-S or BOB-M.

CHAPTER FOUR

A Systems

Development

Life Cycle

Century Tool and Die, Inc.

Valerie and Larry had just sat down to discuss their current project, improving the recently implemented Accounts Receivable (A/R) information system. Larry is the Assistant A/R Manager at Century Tool and Die, Inc., a manufacturer of industrial tools and machines. Valerie is a systems analyst for the Information Systems Department. As they start to discuss the project, they are interrupted by Robert Washington, the Executive Vice President of Finance, and Gene Burnett, the A/R Manager.

Larry suddenly looked very nervous. And for good reason—he had suggested the new system. It had not turned out as promised. Gene's support had been lukewarm, at best. Mr. Washington initiated the conversation.

"We've got big problems. This new A/R system is a disaster. It has cost the company more than $625,000, not to mention lost customer goodwill and pending legal costs. I can't afford this when the Board of Directors is complaining about declining return on investment. I want some answers. What happened?"

Gene spoke first. "I was never really in favor of this project. Why did we need this new computer system?"

Larry replied defensively, "Look, we were experiencing cash flow problems on accounts. The existing system was too slow to identify delinquencies and incapable of efficiently following up on those accounts. I was told to solve the problem. A manual system would be inefficient and error-prone. Therefore, I suggested the improved computer-based system."

"I'm not against the computer," Gene replied. "I approved the original computer-based system. And I realized that a new system might be needed. It's just that Larry and Valerie decided to redesign the system without considering alternatives—just like that! To my mind, you should always analyze options. And let's suppose that a new system was our best option. Why did we have to build the system from scratch? There are good A/R software packages available for purchase."

Washington responded, "Gene has a point, Larry! Still, the system you proposed was defended as feasible. And yet it failed! Valerie, as lead analyst, you proposed the new system, correct?"

Valerie answered, "Yes, with Larry's help."

Washington continued, "And you wrote this feasibility report early in the project. Let's see. You proposed replacing the current batch A/R system with an on-line system using a database management package."

Valerie interrupted, "Strictly speaking, the database management package wasn't needed. We could have used existing VSAM files."

"The report says you needed it. I paid $15,000 to get you that package!" replied Washington.

"Bill, our database administrator, made that recommendation," answered Valerie. "The A/R system was to be the pilot database project."

Washington continued, "You also proposed using a network of microcomputers as a front end to the mainframe computer?"

"Yes," answered Valerie. "Larry felt that a mainframe computer-based system would take too long to design and implement. With microcomputers, we could just start writing the necessary transaction programs and then transfer the data to the database on the mainframe computer."

Washington interrupted, "It seems to me that some sort of design work should have been done no matter what sized computer you use . . . [brief pause] In any case, the bottom line in this report is that your projected benefits outweighed the lifetime costs. You projected a 22 percent annual return on investment. Where did you get that number?"

Valerie answered, "I met with Larry four times—about six hours total, I'd say—and Larry explained the problems, described the requirements, made suggestions, and then projected the costs, benefits, and rate of return."

Washington responded, "But that return hasn't been realized, has it? Why not?" When nobody said anything, he continued, "Valerie, what happened after this proposal was approved?"

"We spent the next nine months building the system."

"How much did you have to do with that, Larry?" asked Washington.

"Not a lot, sir. Valerie occasionally popped into my office to clarify requirements. She showed me sample reports, files, and screens. Obviously, she was making progress, and I had no reason to believe that the project was off schedule."

"Were you on schedule, Valerie?" asked Washington.

"I don't believe so, Mr. Washington. My team and I were having some problems with certain business aspects of the system. Larry was unfamiliar with those aspects and he had to go to the account clerks and the accountants for answers."

Washington replied, "I don't get it! Why didn't *you* go to the clerks and accountants?"

Gene responded, "I can answer that. I designated Larry as Valerie's contact.

I didn't want her team wasting my people's time—they have jobs to do!"

"Something about that bothers me, Gene. In any case, when this project got seriously behind, Larry, why didn't you consider canceling it, or at least reassessing the feasibility?"

"We did," answered Larry. "Gene expressed concern about progress about seven months into the project. We called a meeting with Valerie. At that meeting, we learned that the new database system wasn't working properly. We also found that we needed more memory and storage on the microcomputers. And to top it off, Valerie and her staff seemed to have little understanding of the business nature of our problems and needs."

Valerie interrupted, "As I already pointed out, I wasn't permitted contact with the users during the first seven months. Besides, we were asked by Larry to start programming as quickly as possible so that we would be able to show evidence of progress."

Washington interrupted, "I'm no computer professional; however, my engineering background suggests that some design or prototyping should have been done first."

Valerie said, "Yes, but that would have required end-user participation, which Larry and Gene did not support."

Larry continued, "As I was saying, we considered canceling the project. But I pointed out that $150,000 had already been spent. It would be stupid to cancel a project at that point. I did reassess feasibility, and concluded that the project could be completed in four more months for another $50,000."

"Nobody asked me if I wanted to spend that extra money," argued Washington.

Larry added, "We realize that the system hasn't worked out as well as we had hoped. We are trying to redesign . . ."

"As well as you hoped? That's an understatement!" Washington interrupted. "Let me read you some excerpts from Gene's last monthly report. Customer accounts have mysteriously disappeared, deleted without explanation. Later, we discovered that data-entry clerks didn't know that the F2 key deletes a record. Also, customers have been legally credited for payments that were never made! Customers have been double billed in some cases! Reports generated by the system are late, inaccurate, and inadequate. Cash flow has been decreased by 35 percent! My Sales Manager claims that some customers are taking their business elsewhere. And the Legal Department says we may be sued by two customers and that it will be impossible to collect on those accounts where customers received credit for nonpayments. You tell me, what would you do if you were in my shoes?"

Discussion

1. What would you do if you were in Mr. Washington's shoes? How would you react to Larry's performance? Valerie's? Gene's?

2. What did Valerie do wrong? Was she in control of her own destiny on this project? Why or why not?

3. What did Larry or Gene do wrong? Can either be held responsible for the failure of a computer project when they have limited computer literacy or experience?

4. What was wrong with the feasibility report? Did Valerie and Larry meet often enough? Was the input to that report sufficient? Did the team commit to a solution too early? Did programming begin too soon? Why or why not?

5. Why were Valerie and her staff uncomfortable with the business problem and needs?

6. Should the project have been canceled? What about the $150,000 investment that had already been made?

7. If you were Valerie or Larry, what would you have done differently?

What Will You Learn in This Chapter?

This chapter introduces a systems development life cycle (SDLC) as the framework for systems analysis and design. You will understand the SDLC when you can

1. Describe six basic principles of systems development.

2. Describe where systems development projects come from.

3. Describe a phased approach to systems development. Also, for each phase
   a. Describe the purpose of that phase.
   b. Describe the inputs and outputs of that phase.

4. Describe activities that overlap with the entire life cycle.

Systems development is not a hit-or-miss process! The stakes are high. The moral of the chapter's minicase is that information systems don't just happen. As with any product, they must be carefully developed. Had the A/R project been better controlled, the result might have been different. Successful systems development is governed by some fundamental, underlying principles. We will study these principles in this chapter.

We also introduce a **systems development life cycle** as a disciplined approach to developing information systems. Although such an approach will not guarantee success, it will improve the chances of success. Most experts agree that there is a life cycle. Beyond that, there's little agreement. There are as many versions of the SDLC as there are authors and companies. Although their terminology differs, they are more often alike than different.

Essential Principles for Successful Systems Development

A **systems development life cycle (SDLC)** is a process by which systems analysts, software engineers, and programmers build systems. It is a project management tool, used to plan, execute, and control systems development projects.

Before we study the life cycle, let's introduce some principles (adapted from Benjamin, 1971) that should underlie all systems development. And because the characters in the minicase violated most of those principles, we'll frequently refer back to that minicase.

Principle 1: The System Is for the End-User

Information systems belong to the end-users who will benefit from them. Analysts and programmers frequently refer to "my system." This attitude has, in part, created an "us-versus-them" environment. Although programmers and analysts work hard to create technologically impressive solutions, those solutions often backfire because they don't address the real business problems or they introduce new business or technical problems. For this reason, end-user involvement is an absolute necessity for successful systems development. We must make time for end-users, insist on end-user participation, and seek agreement from end-users on all decisions that may affect them.

In the minicase, Valerie, Larry, and Gene all failed to recognize this principle. A single end-user contact (Larry) was established. Other A/R personnel didn't get to see the system until it was too late to make inexpensive changes. We can all learn from Century's mistakes. When developing information systems, the systems analyst should involve *all* affected personnel.

Misunderstandings continue to be a significant problem in systems development. End-user involvement and education will minimize such misunderstandings. They are also crucial to winning end-user acceptance of new ideas and change. Because people tend to resist change, the computer is often viewed as a threat. Through education, information systems and computers can be properly viewed by end-users as *tools* that will make their jobs less mundane and more enjoyable.

Principle 2: Establish Phases and Tasks

Most SDLCs consist of phases. In its simplest form, the SDLC consists of four phases (Figure 4.1): systems analysis, systems design, systems implementation, and systems support.

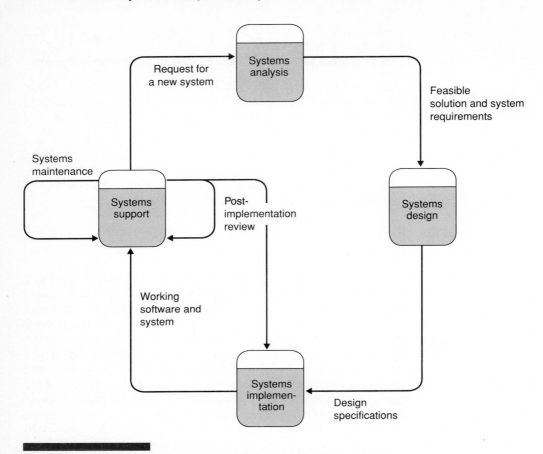

FIGURE 4.1   The Traditional Systems Development Life Cycle   Systems go through cycles of improvement characterized by these classic phases.

Because projects may be quite large and each phase usually represents considerable work and time, the phases are usually broken down into tasks that can be more easily managed and accomplished. In this chapter, we will discuss phases. In later chapters, we will expand coverage to identify typical tasks for each phase.

In the chapter minicase we see little evidence of a phased SDLC, save perhaps the default phases illustrated in the following cartoon. This haphazard approach is all too common—and the results are also common. If nothing else, a phased approach to development allows us to measure our progress and ensure that steps are not overlooked!

Wild enthusiasm   Disillusionment   Total confusion   Search for the guilty   Punishment of the innocent   Promotion of nonparticipants

The Century System Development Process?   (From William M. Taggart, Jr., *Information Systems: An Introduction to Computers in Organizations.* Copyright © 1980 by Allyn and Bacon, Inc. Reprinted with permission.)

Principle 3: Systems Development Is Not a Sequential Process

Generally speaking, the phases of a project should be *completed* in sequence. Unfortunately, most life cycles leave you with the impression that you must finish one phase or task *before* starting the next. In reality, and as Figure 4.2 demonstrates, phases of systems development can overlap. It pays to think and work ahead whenever it's practical. On the other hand, if you do work ahead, that work is tentative since phases must at least be completed in sequence. Thus, until a prerequisite phase is completed, any other work is *subject to change*. Reworking can, however, be abused. A phase or task must be completed sometime, or the system will never be implemented. Perfection is rarely achieved. Sometimes it is best to put a change on hold until after the system has been implemented, unless that change is essential to the system. This is often a judgment decision.

Did Century violate the nonsequential principle? It isn't really clear that the staff used an SDLC. Therefore, we don't really know if they approached their tasks sequentially. However, judging from the number of errors in the final system, we can guess that the Century staff adopted the sequential attitude "We can always fix it later."

Principle 4: Systems Are Capital Investments

Systems are capital investments, no different from a fleet of trucks or a new building. Even if management fails to recognize the system as an investment, you should not! When considering a capital investment, two issues must be addressed.

First, for any problem, there are likely to be several possible solutions. The analyst should not accept the first solution that comes to mind. The analyst who fails to look at alternatives is an amateur! Second, after identifying alternative solutions, the systems analyst should evaluate each possible solution

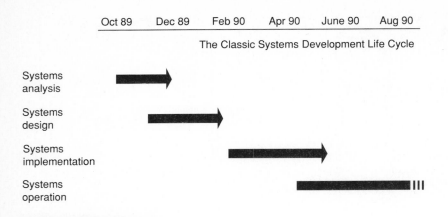

| Oct 89 | Dec 89 | Feb 90 | Apr 90 | June 90 | Aug 90 |

The Classic Systems Development Life Cycle

Systems analysis

Systems design

Systems implementation

Systems operation

FIGURE 4.2   Nonsequential Nature of the Classic Phases   Each bar represents a period of time spent on a phase. Although the phases are clearly completed in sequence, the phases can overlap one another. See, for instance, systems analysis and systems design. However, note that systems implementation is usually dependent on completion of systems design. Systems support is an ongoing activity.

for feasibility, especially for cost-effectiveness. **Cost-effectiveness** is defined as striking a balance between the cost of developing and operating a system and the benefits derived from that system. Cost-benefit analysis is an important skill to be mastered.

If you look back at the A/R project at Century, you will see that alternatives were not evaluated. In fact, it seems that the technology was selected early and that the choice was not motivated by the problem. With respect to cost-effectiveness, the costs were grossly underestimated, which leads us to our next principle.

Principle 5: Don't Be Afraid to Cancel

A significant advantage of the phased approach to systems development is that it provides several "go" or "no go" decision opportunities. There is a temptation *not* to cancel a project because of the investment already made. Century made this mistake and threw more money into a system that still failed! In the long run, canceled projects are less costly than implemented disasters.

We advocate a *creeping commitment* (Gildersleeve, 1976) to systems development. At any "go" or "no go" checkpoint, all costs are considered *sunk* (meaning *irrecoverable*). They are irrelevant to the decision. The project should be reevaluated at the checkpoint to determine if it is *still* feasible. The concept of sunk costs is familiar to some financial analysts and managers, but it is frequently forgotten or not practiced by the majority of practicing analysts and end-users.

Principle 6: Documentation
Is a Product of All Phases

Failure to develop *working* documentation is one of the most frequent and critical errors made by analysts. Most students and practitioners talk about the importance of documentation, but talk is cheap! When do you really place *comments* in your computer programs? After you finish, of course! We all tend to *post*document. Unfortunately, we often carry this bad habit over to systems development.

Documentation should be a working by-product of the entire systems development effort. Documentation reveals strengths and weaknesses of the system to others—*before* the system is built. It stimulates end-user involvement and reassures management about progress. Be wary of any SDLC that has a documentation phase, a planned postdocumentation approach that causes communication breakdowns.

Century's staff failed to document their efforts. They probably didn't even document their programs. This book teaches technique, but it also teaches documentation. Learn to use the tools and techniques to communicate with end-users *during* the life cycle, not after!

There you have it—six principles that should underlie any SDLC. These principles, summarized in Figure 4.3, can be used to evaluate any life cycle, including ours!

Where Do Systems Projects Originate?

Before we study the phases, it is important to understand how projects get started. End-users initiate most projects since they are closer to the business activities that need improvement. Analysts, on the other hand, are frequently expected to survey the business for possible improvements. Finally, and increasingly more common, many projects are identified by an intensive information systems planning activity called **information resource management** (IRM). Using IRM methods, experienced analysts and noncomputing managers chart an information systems direction that mirrors corporate plans.

The impetus for most projects is a problem, opportunity, or directive. Problems may either be current, suspected, or anticipated. **Problems** are undesirable situations that prevent the business from fully achieving its purpose, goals, and objectives. For example, an increase in time required to fill an order can trigger a project to reduce that delay.

Principles to Guide Systems Development

1   The system is for the end-user. Get the end-user involved throughout systems development.

2   Establish phases and tasks to better manage systems development projects.

3   Systems development tasks are not strictly sequential. They can overlap. Also, you may need to backtrack to correct mistakes.

4   Systems are capital investments. They should be economically justified as such.

5   Establish checkpoints to reevaluate feasibility, and don't be afraid to cancel infeasible projects despite sunk costs.

6   Documentation should be the natural, working by-product of systems development. Don't postdocument phases or tasks.

FIGURE 4.3   Principles for System Development   These six principles can be used to evaluate any systems development process or strategy, including our own systems development life cycle.

An **opportunity** is a chance to improve the business even in the absence of specific problems. For instance, management is always receptive to cost-cutting ideas, even when high costs are not currently considered a problem.

A **directive** is a new requirement that's imposed by management, government, or some external influence. For example, the Equal Employment Opportunity Commission, a government agency, may mandate that a new set of reports be produced each quarter. Or management may dictate support for a new product line or policy. Some directives are technical. For instance, systems may be strategically directed to convert from batch to on-line processing or from conventional files to database. This is appropriate if the current technology is obsolete, difficult to maintain, slow to perform, or cumbersome to use.

There are far too many potential problems, opportunities, and directives to list them in this book. However, James Wetherbe (1984, p. 114) has developed a useful framework for classifying problems, opportunities, and directives. He calls it **PIECES** because the first letter of each of the six categories, when put together, spells the word *pieces*. These categories are

- The need to improve **performance**
- The need to improve or control **information** (or data)
- The need to improve **economics** or control costs
- The need to improve **control** and security
- The need to improve **efficiency** of people and machines
- The need to improve **service** to customers, partners, employees, and the like

Figure 4.4 expands on each of the categories.

The categories of the PIECES framework are related. Any given project can be characterized by one or more categories. Furthermore, any given problem, opportunity, or directive may have implications in more than one PIECES category. PIECES is a practical framework, not just an academic exercise!

The PIECES framework is significant because it teaches you to always examine project triggers in terms of their bottom-line impact on the business. When you begin a systems project, consider using Figure 4.4 to either list the problems, or better still, identify problems that the end-user has yet to see.

The Phases of a Systems Development Life Cycle

A nine-phase SDLC is illustrated in Figure 4.5. The rounded rectangles represent project phases. The arrows represent inputs and outputs (working documentation) for a phase. The person figures are people and organizations with whom the analyst may interact. Finally, the shaded diamonds indicate checkpoints at which the end-user should reevaluate feasibility. We'll walk through each phase of the SDLC for this project, describing what you have to do in those phases.

Walking Through the Life Cycle Phases

Given the project request, we can walk through the phases used to develop an information system. Throughout this discussion, we suggest you mark Figure 4.5 for quick reference.

Survey Project Scope and Feasibility   We begin a typical project with a preliminary analysis of project scope and feasibility. The **survey phase**, also called a *preliminary investigation* or *feasibility study*, determines whether or not significant resources should be committed to the other phases of the life cycle. During the survey phase, you will define the scope of the project, including all end-users (at all levels of responsibility), perceived problems and opportunities, business and technical constraints, perceived project goals, and possible solutions. Then you assess the *initial*—we stress the word *initial*!—feasibility of the project.

FIGURE 4.4   PIECES Problem-Solving Framework and Checklist

The following checklist for problem, opportunity, and directive identification uses Wetherbe's PIECES framework. Note that the categories of PIECES are not mutually exclusive; some possible problems show up in multiple lists. Also, the list of possible problems is not exhaustive. The PIECES framework is equally applicable to both computerized and manual aspects of a system.

I. The need to improve performance
  A. Improve throughput, the amount of work performed over some period of time
  B. Improve response time: the average delay between a transaction and a response to that transaction
  C. Throughput and response time should be evaluated separately and collectively

II. The need to improve information and data
  A. Improve information (the outputs of the system—used for planning, control, and decision making)
    1. Lack of any information
    2. Lack of needed information
    3. Lack of relevant information
    4. Too much information
    5. Information that is not in a useful form
    6. Information that is not accurate
    7. Information that is difficult to produce
    8. Information that is not timely
    9. Illegal information
  B. Data (the inputs to the system)
    1. Data is not captured
    2. Data is captured but not in a timely fashion
    3. Data is not accurately captured
    4. Data is difficult to capture
    5. Data is captured redundantly
    6. Too much data is captured
    7. Illegal data is captured
  C. Stored data
    1. Data is stored redundantly
    2. Data is not accurate
    3. Data is not consistent in multiple stores
    4. Data is not secure against accident
    5. Data is not secure against sabotage
    6. Data is not well organized
    7. Data organization is too inflexible to meet information needs
    8. Data cannot be easily accessed to produce information

III. Economics: The need to reduce costs
  A. Costs are unknown or untraceable to source
  B. Costs are excessive

IV. The need to improve control and security
  A. There is too little control
    1. Input data is not adequately edited
    2. Crimes are committed against data
      a. Fraud
      b. Embezzlement

3. Ethics are breached based on data or information
4. Redundantly stored data is inconsistent in different files
5. Privacy of data is being violated
6. Processing errors are occurring
7. Decision-making errors are occurring
8. System is deviating from planned performance
  B. Too little security
    1. People get unauthorized access to space or facilities
    2. People get unauthorized access to computers
    3. People get unauthorized access to data or information (manual or computer)
    4. People execute unauthorized updates of data
  C. Too much control or security
    1. Bureaucratic red tape slows the system
    2. Controls inconvenience end-users or customers
    3. Controls cause excessive processing delays
    4. Controls result in lost transactions

V. The need to improve efficiency
  A. People or machines waste time
    1. Data is redundantly input or copied
    2. Data is redundantly processed
    3. Information is redundantly generated
  B. Machines or processes waste materials and supplies
  C. Effort required for tasks is excessive
  D. Materials required for tasks are excessive

VI. The need to improve service
  A. The system produces inaccurate results
  B. The system produces inconsistent and, hence, unreliable results
  C. The system is not easy to learn
  D. The system is not easy to use
  E. The system is too complex
  F. The system is awkward
  G. The system is inflexible to situations and exceptions
  H. The system is inflexible to change (new needs or requirements)
  I. The system does not interface well to other systems
  J. The system does not work with other systems
  K. The system is not coordinated ("left hand does not know what right hand is doing")

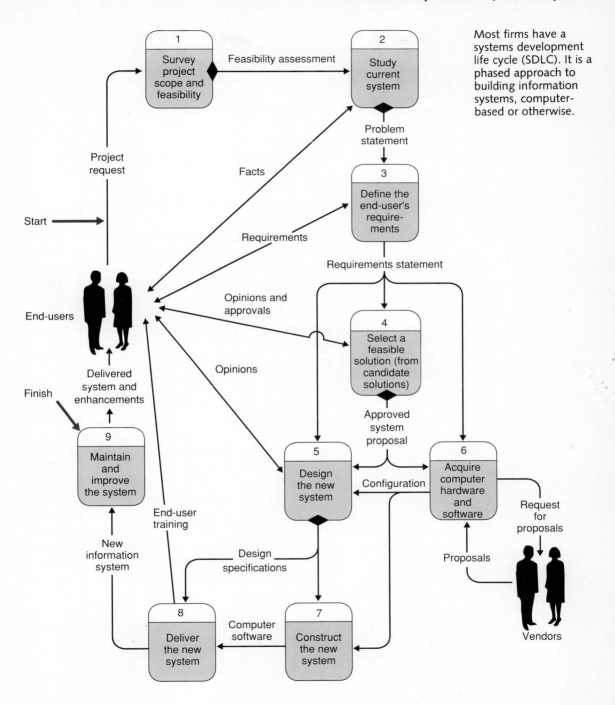

Most firms have a systems development life cycle (SDLC). It is a phased approach to building information systems, computer-based or otherwise.

FIGURE 4.5   A Systems Development Life Cycle

The output of the survey phase is a **feasibility assessment** that presents findings and recommendations. It usually includes a preliminary cost-benefit analysis that determines if the project (*not* any specific systems solution) is feasible. At this stage of the project, feasibility is rarely more than a statement that the problems, opportunities, or directives are worthy of being addressed. Detailed or accurate cost and benefit estimates are rarely possible, a concept that cannot be overstressed to the end-user. Premature commitments to budgets, expectations, and solutions should be avoided since you have yet to fully study the current system, define complete requirements, or identify and analyze alternatives.

Frequently, the findings must be reviewed by management or a *steering committee* to prioritize the project relative to other potential projects.

Study and Analyze the Current System   There's an old saying that suggests, "Don't fix it until you understand it." You need to understand the existing system, manual or computerized, before you design and build a new system. Therefore, we now study and analyze the current system. Do the problems and opportunities really exist? If so, how serious are they? Many times, the initial problems were mere symptoms, frequently of more serious or subtle problems. During the **study phase**, you want to address the causes and effects of problems, opportunities, and directives.

To analyze problems, opportunities, and directives, you need to understand *how* the existing system operates. Thus, the study phase educates you about the business application. Only after studying how a system works should you identify or analyze the problems and opportunities.

The findings of the study phase are documented as some form of **problem statement** for the next phase of the life cycle. This problem statement may be either a formal report or an updated feasibility assessment. Based on the findings, the end-user can either cancel the project or approve continuing to the next phase.

A useful analogy to an information systems development project is the construction of a house. Consider the problem of building your dream home. The architect is the systems analyst of the housing industry. Does the architect do a study phase? You bet! The architect surveys the land and soil, studies other houses in the area, and learns the building codes that are pertinent to the property. Essentially, the architect, like the systems analyst, is studying the existing environment.

Define the End-User's Requirements   Now you can design a new system, right? No, not yet! What should the new system provide for the end-user? What performance level is expected? Careful! We didn't say *how*. We said *what*. The next phase of our SDLC is to define end-user requirements. You go to the end-users and find out what they need or want out of the new system. Essentially, the purpose of the **definition phase** is to define the inputs, files, processing, and outputs for the new system, but to do so without expressing computer alternatives and technical details. The **definition phase** is also known as the *general design* or *logical design* (as opposed to *physical design*) phase in other versions of the SDLC.

The most popular approach to requirements definition is **modeling**. The analyst will draw diagrams of the system, from different points of view and at different levels of detail. You are probably familiar with at least *program models* such as structure charts, flowcharts, and pseudocode. They model the modular structure and logic of a program. You will learn several *systems* modeling tools and techniques in this book.

An alternative approach to requirements definition is **prototyping**. When end-users have difficulty defining requirements, the analyst uses powerful new computer tools to build a prototype, or small-scale working model of the final system. The end-user can then react to the prototype to help the analyst establish the requirements more easily. In other words, the end-users may recognize their requirements when they see the prototypes, good or bad. Chapter 6 expands somewhat on this approach.

The entire requirements specification is organized into a document called a **requirements statement**.

Let's return to our house project. The architect will also define the home-owner's requirements before designing the house. How many people are in the family? What are their ages? How important is privacy? Where do the family members spend most of their time? How many full bathrooms are wanted? This is only a sample of questions the architect uses to learn the homeowner's requirements.

Select a Feasible Solution (from Candidate Solutions)   Now can we design the new system? Let's think about it for a moment. Design is a detailed, technical, and time-consuming activity. But there are numerous alternative ways to design the A/R system. How much of the system should be comput-erized? Should we purchase software or build it ourselves (called the *make-versus-buy* decision)? If we decide to make the system, should we design an on-line or batch system? Should we design the system for a mainframe com-puter or microcomputer? The next phase of our SDLC is to select a feasible solution. Implicit in this phase is the need to identify candidate solutions.

The **selection phase** determines *how* the new system is to be designed but only at a very high level—no details! After defining candidates, each candidate is evaluated for

- *Technical feasibility.* Is the solution technically practical? Does our staff have the technical expertise to design and build this solution?

- *Operational feasibility.* Will the solution fulfill the end-user's require-ments? To what degree? How will the solution change the end-user's work environment? How do end-users feel about such a solution?

THE NEXT GENERATION

Systems Integration: The Next Challenge for the Life Cycle

"The challenge: to be able to incorporate the new without losing investment in the old" [Gantz, 1987].

Systems integration has become the new challenge for analysts of the next decade. It is not enough to purchase or build systems that fulfill requirements. It is not enough to purchase or build systems that are easy to use and maintain. We must integrate or interface those systems to the myriad other systems that are essential to the business.

Consider the portfolio of information systems in the typical company. There may be dozens of large systems and hundreds of smaller systems. Many systems are built around purchased software, from numerous different vendors, and using varying information technology of different computing eras. Many systems were built in-house, some many years ago with outdated methods and technologies. Systems may be implemented on different computers, with different operating systems and using different file and database technology. Some systems are on mainframe computers. Others are on minicomputers or microcomputers.

Now, consider the implications. You are building or buying a new system. The system needs to use data already stored in other systems to reduce duplicate data entry and inconsistent data between the systems. Today, many analysts find themselves spending more time analyzing and designing such interfaces than analyzing and designing the system itself.

As Gantz (1987) notes, "Today, systems integration is something done out of necessity, if not desperation. Either competitors are doing it, certain applications demand it, or the economics are too good to ignore."

Much of the need for systems integration is born out of our increased dependence on purchased software. To reduce development time, we purchase software instead of building it. Unfortunately, we often spend as much time modifying or augmenting the software to interface with other required systems as we would to design and build the system in-house.

In-house systems are not immune to problems. They too must be interfaced with other systems that may use dramatically different technology, techniques, and file structures.

What is the answer? How will the life cycle change to address this problem? Only the next generation of systems analysis and design will tell. However, in our view, the life cycle doesn't really change. Only the scope of activities changes.

For instance, during the study phase, we need to spend more time studying constraints imposed on us by *other* systems to which our system interfaces. During the definition phase, we need to define interface requirements to a greater extent than we have in the past. During the selection phase, we need to evaluate a new kind of technical feasibility, the technical implications of connecting systems. During the acquisition phase, we must actively consider the impact of software package technology on our interfaces. And during the design phase, we will continue to spend more time designing interfaces. Hopefully, the time spent in these phases will make design easier.

What can you do to get ready for this change in perspective? For one thing, you should learn about the technology of systems integration—data communications, networking, distributed processing, and the like. Second, you need to expand your perspective on all projects. Don't restrict your view of the project to the system itself. Consider how the system fits into the federation of systems of which it is a part.

One thing is certain. The problems of systems interfacing will increase as businesses increase their use of computers.

- *Economic feasibility.* Is the solution cost-effective (as defined earlier in the chapter)?

Infeasible candidates are eliminated from further consideration; however, several alternatives usually prove to be feasible. You are looking for the *most* feasible solution—the solution that offers the best *combination* of technical, operational, and economic feasibility. The selection phase output is a formal **systems proposal** to the end-users, who will usually make the final decision. The **approved systems proposal** is input to the next two phases.

Does our architect evaluate alternatives? Yes, the architect will examine the ranch, bilevel, and trilevel alternatives. Is the exterior to be finished in brick, stone, wood, or aluminum? Will there be gas or electric appliances? These decisions will affect the ultimate design of the house.

Acquire Computer Hardware and Software (if called for in the approved solution)   Here is a phase that is missing from many SDLCs. College graduates are often shocked to discover the high percentage of computer software that is purchased or leased rather than built. Recall that the *make-versus-buy* decision was made in the last phase. Also, any new system may present the need to acquire additional hardware, such as computer terminals.

During the **acquisition phase,** you need to determine which specifications are important for the equipment and software to be purchased. Needs are communicated in a **request for proposals** that is sent to vendors who may be able to fulfill those needs. After receiving **proposals** from the vendors, your job is to acquire the proposed hardware or software configuration that best meets your needs at a reasonable cost.

Returning to our house-building analogy, does it go through an acquisition phase? Yes, the architect and contractor must acquire building materials at the lowest possible cost without sacrificing quality. Thus, building materials must be carefully chosen, usually through a bidding process similar to a request for proposals to select computer technology.

Design the New System   Given the approved, feasible solution from the selection phase, you can now design the new system—finally! You understand *what* the requirements are from the definition phase and *how* you want to fulfill those requirements from the selection phase. Thus, you can justify the time and cost to design the new system.

Traditionally, computer outputs are designed first because output design can affect the design of inputs, files, and methods. Then the analyst designs files (or databases) and inputs. Finally, methods and procedures that maintain files (or databases) and transform inputs into outputs must be designed. Throughout design, the analyst must specify internal controls to ensure that the system is reliable and secure. Analysts use numerous tools to design outputs, inputs, files, methods, and procedures. You'll learn how to use such tools in Part Three of this book.

The **design phase** should be somewhat familiar to most of you who, in programming assignments, responded to the output of the design phase, **design specifications**. You may have also found yourself redoing those specifications when they were incomplete or unclear.

An alternative to written design specifications is **design by prototyping**. This technique has emerged as the preferred method for many analysts. Introduced earlier as a requirements definition technique, it is even more powerful as a design technique. Using higher-level programming languages and application generators, such as Ashton-Tate's dBASE IV or Information Builder's FOCUS, an analyst can more quickly build and modify working prototypes of the new system than he or she could using languages like COBOL, FORTRAN, and Pascal. End-users can react to those prototypes. And prototypes can literally replace the traditional paper specification as a blueprint of the final system. Chapter 12 will greatly expand on prototyping as a design technique.

We should note that our architect can similarly design or blueprint a house to fulfill the homeowner's requirements.

Construct the New System   Now we come to the SDLC phase with which you are probably most familiar—constructing the new system. Actually, you are most familiar with the principal activity of this phase, programming. The **construction phase** is frequently the most time-consuming and tedious phase of the life cycle. However, the time required for construction is often longer than it should be because the preceding phases were completed hastily or not at all. Programmers should work from specifications that have been developed and refined through the earlier phases. If the specifications are unclear, incomplete, inaccurate, or otherwise faulty, the construction phase will be much more complicated and time consuming! The output of this phase is **computer software** that has been thoroughly debugged and tested.

Alternatively or additionally, the construction phase may involve the installation and modification of purchased software packages. These packages were chosen during the purchasing phase and often need to be modified to run on your computer system.

Note that, in the construction phase, the principal figure is the programmer, not the analyst. The analyst is still involved to the extent that specifications may need to be clarified; the perfect specification probably doesn't exist. And just as the programmer replaces the analyst as the principal figure during information systems construction, the contractor replaces the architect as the principal figure during house construction.

Deliver the New System   What's left to do? Now we must deliver the new system into operation. The new systems solution represents a departure from the way things are currently done; therefore, the analyst must provide for a smooth transition from the old system to the new system and help the end-users cope with normal start-up problems. The **delivery phase** also involves training and the writing of manuals.

In the house-building example, you would now close on your house (sign the mortgage papers) and move into the house.

Maintain and Improve the System   Once the system is placed into operation, the analyst's role changes from development to support. This may include periodic evaluations and reviews. In addition, the analyst will respond to maintenance and improvement requests submitted by the end-users. **Maintenance** is the correction of errors and omissions that is essential to the system's usefulness. Most such errors are software bugs that were not caught during testing in the construction and delivery phases. **Improvements** are the addition of new capabilities such as reports, new interfaces with other systems, and new features such as better screen or report layouts. Enhancements usually constitute an abbreviated SDLC and build on documentation created when the system was originally developed.

It cannot be overemphasized that, as changes are made to the system, documentation needs to be maintained. Well-maintained documentation can significantly accelerate the study phase when the system is due for an overhaul.

And our house-building analogy offers parallels. You would ask the builder to correct building errors (maintenance). And over a period of years, either you, the builder, or other builders would improve the house—for example, build a room(s) addition, panel a room, build in shelves, install new windows, and the like.

That's just about it for the basic SDLC. The similarity between our SDLC and the design and construction of a house is striking, but not coincidental. You see, we cannot take full credit for our SDLC. Then again, neither can the housing industry. Both the SDLC and the house-building process were problems. We wanted something that we didn't have. Each process is an example of a classic problem-solving approach. Compare, for instance, the SDLC in Figure 4.5 with the building construction life cycle in Figure 4.6. They are more alike than different.

The Phases Are Not Really Sequential

Earlier, you learned that the life cycle is not strictly sequential. A nonsequential view of our life cycle is illustrated in Figure 4.7. It demonstrates how phases *might* be overlapped. Note that the completion of phases, annotated with numbered bullets in Figure 4.7, is still sequential. This diagram only depicts opportunities for overlapping phases. The actual overlap in any project will depend on the size and complexity of the project and the number of analysts and programmers assigned to it.

Although the SDLC is not strictly sequential, most phases do have some prerequisites. These prerequisites are shown with vertical, dashed lines. For instance, because the survey phase determines project feasibility, it is shown as a strict prerequisite for the study. And because end-user requirements

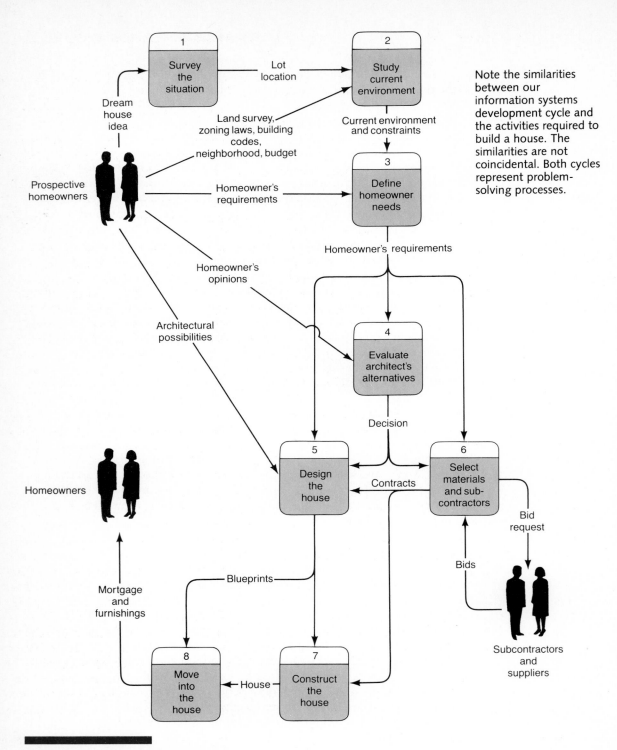

Note the similarities between our information systems development cycle and the activities required to build a house. The similarities are not coincidental. Both cycles represent problem-solving processes.

FIGURE 4.6 House-Building Example

Activity

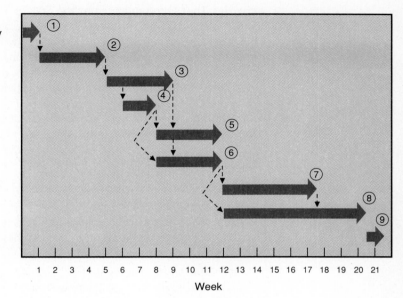

Survey project scope and feasibility

Study current system

Define end-users' requirements

Select a feasible solution

Acquire new computer
  hardware and software

Design the new system

Construct the new system

Deliver the new system

Maintain and improve the system

Week

FIGURE 4.7   Opportunities to Overlap SDLC Phases   This chart depicts opportunities to overlap systems development phases. The horizontal axis is time. The phases are represented by bars. Where the bars overlap, the phases overlap. The actual overlap in any given project depends on the project's size and the resources committed to the project.

should address verified problems and opportunities, the study phase is suggested as a prerequisite to the definition phase.

On the other hand, as soon as general requirements become known, analysts can begin the selection phase, identifying and evaluating candidate solutions. Meanwhile, detailed requirements are still being specified. Clearly, as Figure 4.7 shows, design should not begin before commitment to a specific technical solution.

Especially when using the prototyping technique, you can be simultaneously using prototypes to define requirements and present design layouts. Note that design and acquisition overlap even though the system cannot be implemented until purchased technology is received and installed.

Theoretically, construction should not begin until design is completed. This is true because most programs are costly to rewrite to changing specifications. On the other hand, prototyping techniques have allowed design and construction to frequently blend into a single activity. Delivery usually begins shortly after construction.

Activities That Overlap Much of the SDLC

There are several activities that overlap many or all phases of the life cycle. For this reason, we choose not to call them phases. They include *fact finding*, *documentation and presentation*, *feasibility analysis*, and *project management*. Let's briefly examine each of these activities.

Fact Finding   Fact finding—or *data collection*—is the formal process of using research, interviews, questionnaires, sampling, and other techniques to learn about the system. Fact finding occurs during all of the following phases:

- *Survey phase.* The analyst collects general facts about the problems, opportunities, directives, environment, end-users, and so forth.
- *Study phase.* The analyst collects facts about how the current system functions, and about problems and opportunities. This phase normally requires the most fact finding.
- *Definition phase.* The analyst collects facts about the end-user's requirements and expectations. This phase involves the second most extensive fact-finding process.
- *Selection phase.* The analyst collects facts about candidate solutions, end-user opinions about candidates, and costs and benefits.

Documentation and Presentation   Recall that one of our principles stated that documentation should be a working by-product of the SDLC. If you study Figure 4.5 carefully, you will see that each named arrow represents documentation. In reality, all documentation to a project is kept in a workbook or project dictionary for easy reference and modification. Today, the analyst can use computer-assisted tools to create and maintain systems documentation. These tools will be described and presented throughout this book.

Presentations of documentation are also numerous in the life cycle. Each named arrow in Figure 4.5 also represents documentation that should be presented and reviewed with end-users. Presentations may be oral or written. Word processors, spreadsheets, and other microcomputer tools have greatly simplified the analyst's presentation tasks.

Feasibility Analysis   Too many projects call for premature solutions and estimates. We feel that this approach results in an overcommitment to the project. If we are so accurate in our feasibility estimates, why are so many information systems projects late and over budget? Systems analysts tend to be overoptimistic in the early stages of a project. They underestimate the size and scope of a project because they haven't yet completed a detailed study.

For these reasons, we only assess "worthiness" during our survey phase. Then we use the *creeping commitment* approach to reevaluate feasibility at appropriate checkpoints (indicated by small diamonds in Figure 4.5) during the life cycle.

Project Management   Systems development projects usually involve a team of analysts and programmers who work together. **Project management** is the ongoing process by which the team leader plans, delegates, directs, and controls progress to develop an acceptable system within the allotted time and budget. The SDLC provides the basic framework for project management.

These SDLC common activities are formally presented in the *Modules* that make up Part Four of the book. The modules include

Module A—Project Management

Module B—Fact-Finding Techniques

Module C—Communications Skills for the Systems Analyst

Module D—Feasibility and Cost-Benefit Analysis Techniques

As soon as you've completed this chapter, you will have sufficient background to study any of the modules, in any order. You may study entire modules or parts of modules.

Systems Analysis, Design, Implementation, and Operation

We conclude Chapter 4 with Figure 4.8, which graphically depicts the relationship between our SDLC phases and the classic phases: *analysis*, *design*, *implementation*, and *support*.

Summary

Systems development projects are usually triggered by problems, opportunities, and directives. Problems exist when the current system is not fulfilling the business's purpose, goals, and objectives. Opportunities are chances to improve a system despite the absence of problems. Directives are decisions imposed on a system by management or government. All problems, opportunities, and directives can be evaluated in terms of their bottom-line impact relative to performance, information and data, economy, control and security, efficiency, and service.

Six basic principles apply to all systems development projects. First, we should actively involve the end-users in systems development. Second, we should use a phased approach to systems development. Third, we should recognize that phases can overlap and that we will also need to backtrack from time to time. Fourth, we need to appreciate that systems are capital investments and should be justified as such. Fifth, we should establish check-

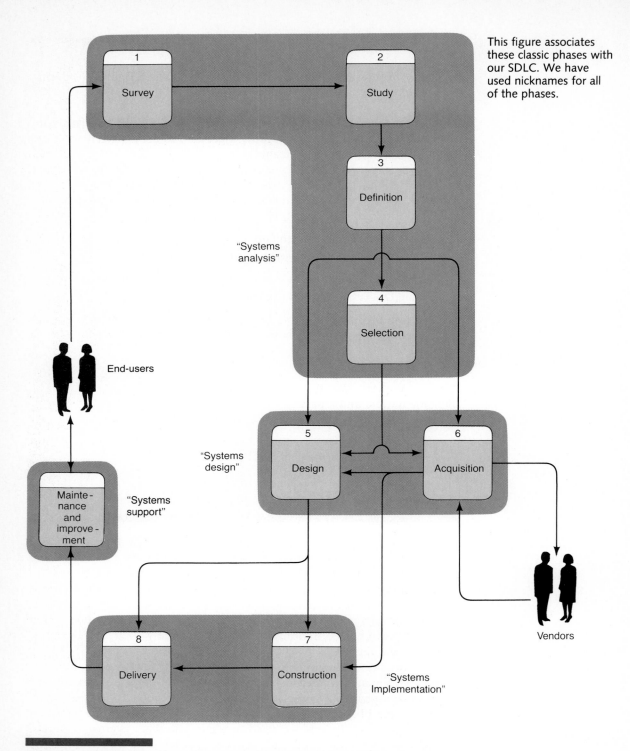

This figure associates these classic phases with our SDLC. We have used nicknames for all of the phases.

FIGURE 4.8   Systems Analysis, Design, Implementation, and Support

points to allow us the option of canceling the project if it has become infeasible. And sixth, we should document a system while we develop that system.

A systems development life cycle (SDLC) is a disciplined approach to developing information systems. Given a project request, the analyst should initially survey project scope and feasibility. The survey phase determines if systems development resources should be committed to the project. Next, the analysts will study and analyze the current system. This detailed study of the current business environment identifies and evaluates problems, opportunities, and constraints. Once the current system is fully understood, the definition phase specifies the end-user's requirements. End-user requirements should be specified independently of the ways that the computer might be used. Given the end-user requirements, analysts can identify and analyze alternative solutions. These phases are known collectively as systems analysis.

If new computer hardware or software is required, the next phase requires analysts to acquire new computer technology, interacting with computer vendors to select the best technology for the best price. The next phase is traditional design, including outputs, inputs, files, databases, terminal screens and dialogue, methods and procedures, and internal controls. These phases are known collectively as systems design.

Next, during the construction phase, software is written, installed, modified, and tested. Finally, the analysts and programmers must deliver the new system into operation, training end-users and providing for a smooth transition from the old system to the new system. These phases are known collectively as systems implementation.

From that point forward, the system is said to be in operation. During systems support, the analyst will support the system, providing maintenance and improvements as needed.

Fact finding, documentation and presentations, feasibility analysis, and project management are all ongoing activities of the life cycle.

Problems and Exercises

1. What are the six fundamental principles of systems development? Explain what you would do to incorporate those principles into a systems development process.

2. Make an appointment to visit a systems analyst at a local information systems installation. Discuss the analyst's current project. What triggered the project—a problem, opportunity, or directive? How does it relate to the PIECES framework?

3. Using the PIECES framework, evaluate your local course registration system. Do you see problems or opportunities? (*Alternative*: Substitute any system with which you are familiar.)

4. Assume you are given a programming assignment that requires you to make some modifications to a computer program. Explain the problem-solving approach you would go through. How is this approach similar to the phased approach of the systems development life cycle presented in this chapter?

5. Which phase of the SDLC presented in this chapter do the following tasks characterize?

    a. The analyst demonstrates a prototype of a new sequence of work order terminal screens.

    b. The analyst observes the order-entry clerks to determine how a work order is currently processed.

    c. The analyst develops the internal structure for a database to support work order processing.

    d. An analyst is teaching the plant supervisor how to inquire about work orders using the new microcomputer.

    e. A plant supervisor is describing the content of a new work order progress report that would simplify tracking.

    f. The analyst is reading an inquiry concerning whether or not a computer system might solve the current work order tracking problems.

    g. The analyst is installing the microcomputer and database management system needed to write work order processing programs.

    h. The analyst is reviewing the company's organizational chart to identify who becomes involved in work order processing and fulfillment.

    i. The analyst is comparing the pros and cons of a software package versus writing the programs for a new work order system.

    j. An analyst is testing a computer program for entering work orders to the system.

    k. The analyst is correcting a program to more accurately summarize weekly progress.

    l. The analyst is comparing a variety of microcomputers and available software that can be purchased.

6. Visit a local information systems installation. Compare that SDLC with the one in Figure 4.5. Evaluate the company's SDLC with respect to the six systems development principles. (*Alternative*: Substitute the SDLC used in another systems analysis and design book, possibly assigned by your instructor.)

7. Management has approached you to develop a new system. The project will last seven months. It wants a budget next week. You will not be allowed to deviate from that budget. Explain why you shouldn't overcommit to early estimates. Defend the *creeping commitment* approach as it applies to cost estimating. What would you likely be forced to do if management insisted on the up-front estimate with no adjustments?

8. You are an independent consultant. Write a proposal letter that offers your services as a consultant to solve the problems in the Century minicase.

The end-user, who is skeptical of all computer people, is turned off by computer buzzwords. Be sure your proposal explains in step-by-step fashion how you will build a system that meets this end-user's needs.

9. You have an end-user who has a history of impatience, encouraging short-cuts through the systems development life cycle and then blaming the analyst for systems that fail to fulfill expectations. By now, you should understand the phased approach to systems development. For each phase, compile a list of possible consequences to use when the end-user suggests a shortcut through or around that phase.

Projects and Minicases

1. Jeannine Strothers, Investments Manager, has submitted numerous requests for a new investment tracking system. She needs to make quick decisions regarding possible investments and divestments. One hour can cost her thousands of dollars of profit for her company.

   She has finally given up on Information Systems for not giving her requests high enough priority to get service. Therefore, she went to a computer store and bought a microcomputer along with spreadsheet, database, and word processing software. The computer store salesperson suggested that she build a database of her investments and options, subscribe to a computer investments databank (accessed via a modem in the microcomputer), feed data from her database and the bulletin board into the spreadsheet, play "what if" investment games on the spreadsheet, and then update the database to reflect her final decisions. The word processor will draw data from the database for form letters and mailing lists.

   After discussing her plans with Jeff, a systems analyst at another company, he suggested she take a systems analysis and design course before beginning to use the spreadsheet and database. The local computer store, on the other hand, said she doesn't need any systems analysis and design training to be able to develop systems using the spreadsheet and database programs. Their reasoning is that spreadsheets and database tools are not programming languages; therefore, you don't need analysis and design to build systems with them. Is the computer store correct? Why or why not? Can you convince her to take the systems analysis and design course? What would your arguments be?

2. Jeannine Strothers, the impatient manager in minicase 1, did not take Jeff's advice. She built the new system, but she can't get top management to allow her to use it. And she's run into a number of other problems.

   First, the Financial Comptroller has been reevaluating company investment strategies and policies. She didn't know that. The new system does not account for many of the policies that are being considered.

Her own staff has rejected the investment and divestment orders generated by the system. She used Information Systems' existing file structure to design those orders, only to find out that her clerks had abandoned those files two years ago because they didn't include the data necessary to execute order transactions. Her staff is also critical of the design, saying that minor mistakes send them off into the "twilight zone" with no easy way to recover.

The computer link to the investment databank has been useless. The data received and its format were not compatible with systems requirements. Although other databanks are available, the current databank has been prepaid for two years. Additionally, Jeannine is now skeptical of such services.

Some of her subordinate managers are insisting on graphic reports. Unfortunately, neither her database management nor spreadsheet package supports graphics. She's not sure how to convert the data of either package to a graphic format (assuming that it is even possible).

To top off her problems, she isn't sure that her existing database structure can be modified to meet new requirements without having to reprogram all programs, even those that appear to be working. And her boss is not sure that he wants to invest the money in a consultant to fix the problems.

Her analyst friend, Jeff, was not very sympathetic to her problems: "Jeannine, I don't have any quick answers for you. You've taken so many shortcuts through the project life cycle. When we do a system, we go through a carefully thought-out procedure. We thoroughly study the problem, define needs, evaluate options, design the system and its interfaces, and only then do we begin programming."

Jeannine replied, "Wait a minute. I only bought a microcomputer. It's not the same as your mainframe computer. I didn't see the need for going through the ritual you guys use for mainframe applications. Besides, I didn't have the time to do all those steps."

Jeff's parting words were philosophical. "You didn't have the time to do it right? Where will you find the time to do it over?"

What principles did Jeannine violate? Why do so many people today fall prey to the belief the life cycle for a business application is somehow different when using microcomputers? What conclusions can you draw from Chapters 2 and 3 (the information systems chapters) that might help Jeannine learn from her mistakes? What would you recommend to Jeannine if she were to decide to approach her manager with a plan to salvage the system?

3. Evaluate the following scenarios using the PIECES framework. Do not be concerned that you are unfamiliar with the application. That isn't unusual for any systems analyst. Use the PIECES framework to brainstorm potential problems you would ask the end-user about.

a. The staff benefits and payroll counselor is having some problems. Her job is to counsel employees on their benefit options. The company has just negotiated a new medical insurance package that requires employees to choose from among several health maintenance options (HMOs). The HMOs vary according to employee classifications, contributions, deductibles, beneficiaries, services covered, and service providers permitted. The intent was to provide the most flexible benefits possible for employees, minimize costs to the company, and control costs to the insurance agency (which would affect subsequent premiums charged back to the company).

 The counselor will be called on to help employees select the best plan for themselves. She currently responds manually to such requests. But the current options are more straightforward than those under the new plan. She can explain the options, what they do and do not cover, what they cost and may cost, and the pros and cons. However, current employee distrust of the new plan suggests that she will need to provide more specific suggestions and responses to employees.

 She may have to work up scenarios—possibly worst-case scenarios—for many employees. The scenarios will have to be personalized for each employee's income, marital and family status, current health risks, and so on. In working up a few sample scenarios, she discovered first that it takes one full day to get salary and personnel data from the Information Systems Department. Second, employee data is stored in many files that are not always properly updated. When conflicting data become apparent, she can't continue her projections until that conflict has been resolved. Third, the computations are complex. It often takes one full day or more to create investment and/or retirement scenarios for a single employee. Fourth, there are some concerns that projections are being provided to unauthorized individuals, such as former spouses or nonimmediate relatives. Finally, the complexity of the variations in the calculations—there are a lot of "If this, do that" calculations—results in frequent errors, many of which probably go undetected.

b. The manager of the tool and die shop needs help with job processing and control. Jobs are currently processed by hand. Processing occurs as follows. First, a job number is established. Next, the job supervisor estimates time and materials for the job. This is a time-consuming process, and delays are common. Then, the job is scheduled for a specific day and estimated time.

 On the day the job is to be worked on, materials orders have to be issued to Stores. If materials aren't available, the order has to be rescheduled.

 Time cards are completed in the shop when workers fulfill the work order. These time cards are used to charge back time to the customer. Time cards are processed by hand, and the final calculations are entered on the work order. The work orders are checked for accuracy and sent to CIS, where accounting records are updated and the customer is billed.

The problem is that the customer frequently calls to inquire on costs already incurred on a work order, but it's not possible to respond because CIS sends a report of all work orders only once a month. Also, management has no idea of how good initial estimates are or how much work is being done at any tool or machine, or by any worker.

c. State University's Development Office raises funds for improving instructional facilities and laboratories at the university. It has uncovered a sensitive problem: the data is out of control.

The Development Office keeps considerable redundant data on past gifts and givers as well as prospective benefactors. This results in multiple contacts for the same donor—and people don't like to be asked to give to a single university over and over!

To further complicate matters, the faculty and administrators in most departments conduct their own fund-raising and development campaigns, again resulting in duplication of contact lists.

Contacts with possible benefactors are not well coordinated. While some prospective givers are contacted too often, others are overlooked entirely. It is currently impossible to generate lists of prospective givers based on specific criteria (for example, prior history, socioeconomic level, and so on), despite the fact that data on hundreds of criteria have been collected and stored. Gift histories are nonexistent, which makes it impossible to establish contribution patterns that would help various fund-raising campaigns.

Annotated References and Suggested Readings

Benjamin, R. I. *Control of the Information System Development Cycle*. New York: Wiley-Interscience, 1971. Benjamin's 16 axioms for managing the systems development process inspired our adapted principles to guide successful systems development.

DeMarco, Tom. *Structured Analysis and System Specification*. Englewood Cliffs, N.J.: Prentice-Hall, 1978. Chapter 1 describes many problems that plague projects.

Gantz, John. "Systems Integration: Living in a House of Our Own Making." *Telecommunication Products + Technology*, vol. 5, no. 5, May 1987, pp. 32–54. This "management white paper" (or tutorial) is an excellent introduction to the problems of systems integration and its data communications implications.

Gildersleeve, Thomas. *Successful Data Processing Systems Analysis.* Englewood Cliffs, N.J.: Prentice-Hall, 1976. We are indebted to Gildersleeve for the creeping commitment approach.

London, Keith. *The People Side of Systems.* New York: McGraw-Hill, 1976.

Matthies, Leslie H. *The Management System: Systems Are for People.* New York: Wiley, 1976.

Taggart, William. *Information Systems: An Introduction to Computers in Organizations.* Newton, Mass.: Allyn and Bacon, 1980.

Wetherbe, James. *Systems Analysis and Design: Traditional, Structured, and Advanced Concepts and Techniques.* 2d ed. St. Paul, Minn.: West, 1984. We are indebted to Wetherbe for the PIECES framework.

CHAPTER FIVE

The

Structured

Methodologies

Minicase

For a change of pace, let's start the chapter minicase with the discussion questions. Then consider the following fable. This fable came from a book called *Rethinking Systems Analysis and Design*, Jerry Weinberg's philosophical look at the subject.

Discussion

1. What is the moral of the fable?
2. Methodologies are specific strategies and techniques for developing systems. Each uses specific tools and steps and each claims to address the needs of systems development better than the other methodologies. Today's analyst is faced with an almost overwhelming choice of methodologies. Before we study and evaluate those choices, how might the fable and its moral relate to the study of methodologies?

The Three Ostriches: A Fable

Three ostriches had a running argument over the best way for an ostrich to defend himself. Although they were brothers, their mother always said that she couldn't understand how three eggs from the same nest could be so different. The youngest brother practiced biting and kicking incessantly, and held the black belt. He asserted that "the best defense is a good offense." The middle brother, however, lived by the maxim that "he who fights and runs away, lives to fight another day." Through arduous practice, he had become the fastest ostrich in the desert—which you must admit is rather fast. The eldest brother, being wiser and more worldly, adopted the typical attitude of mature ostriches: "What you don't know can't hurt you." He was far and away the best head-burier that any ostrich could recall.

One day a feather hunter came to the desert and started robbing ostriches of their precious tail feathers. Now, an ostrich without his tail feather is an ostrich without pride, so most ostriches came to the three brothers for advice on how best to defend their family honor. "You three have practiced self-defense for years," said their spokesman. "You have the know-how to save us, if you will teach it to us." And so each of the three brothers took on a group of followers for instruction in the proper method of self-defense—according to each one's separate gospel.

Eventually, the feather hunter turned up outside the camp of the youngest brother, where he heard the grunts and snorts of all the disciples who were busily practicing kicking and biting. The hunter was on foot, but armed with an enormous club, which he brandished menacingly as the youngest brother went out undaunted to engage him in combat. Yet fearless as he was, the ostrich was no match for the hunter, because the club was much longer than an ostrich's leg or neck. After taking many lumps and bumps, and not getting in a single kick or bite, the ostrich fell exhausted to the ground. The hunter casually plucked his precious tail feather, after which all his disciples gave up without a fight.

When the youngest ostrich told his brothers how his feather had been lost, they both scoffed at him. "Why didn't you run?" demanded the middle one. "A man cannot catch an ostrich."

"If you had put your head in the sand and ruffled your feathers properly," chimed the eldest, "he would have thought you were a yucca and passed you by."

The next day the hunter left his club at home and went out hunting on a motorcycle. When he discovered the middle brother's training camp, all the ostriches began to run—the brother in the lead. But the motorcycle was much faster, and the hunter simply sped up alongside each ostrich and plucked his tail feather on the run.

That night the other two brothers had the last word. "Why didn't you turn on him and give him a good kick?" asked the youngest. "One solid kick and he would have fallen off that bike and broken his neck."

"No need to be so violent," added the eldest. "With your head buried and your body held low, he would have gone past you so fast he would have thought you were a sand dune."

A few days later, the hunter was out walking without his club when he came upon the eldest brother's camp. "Eyes under!" the leader ordered and was instantly obeyed. The hunter was unable to believe his luck, for all he had to do was walk slowly among the ostriches and pluck an enormous supply of tail feathers.

When the younger brothers heard this story, they felt impelled to remind their supposedly more mature sibling of their advice. "He was unarmed," said the youngest. "One good bite on the neck and you'd never have seen him again."

"And he didn't even have that infernal motorcycle," added the middle brother. "Why, you could have outdistanced him at half a trot."

But the brothers' arguments had no more effect on the eldest than his had had on them, so they all kept practicing their own methods while they patiently grew new tail feathers.

From Gerald M. Weinberg, *Rethinking Systems Analysis and Design*, pp. 23–24. Copyright © 1982 by Scott, Foresman and Company. Reprinted by permission.

What Will You Learn in This Chapter?

This chapter introduces you to **systems development methodologies,** a subject separate from but related to the SDLC (Chapter 4). Most of the chapter focuses on the so-called "structured techniques"—the most commercially popular of the methodologies. You will understand the structured techniques and methodologies when you can

1. Compare and contrast the systems development life cycle and systems development methodologies.
2. Compare and contrast the following methodologies:
   a. Process modeling methodologies
      —Structured Programming
      —Structured Systems Analysis and Design
   b. Data modeling methodologies
      —Information Engineering
   c. Working model methodologies
      —Prototyping

   For each methodology, you should be able to describe its principal supporters, strategy, tools, and techniques in great detail.
3. Describe how computer tools can be made to support systems development methodologies.
4. Justify the integrated use of several methodologies as opposed to adopting one methodology.
5. Make a lifelong commitment to avoid ostrich camps (refer to the minicase that opened this chapter).

Of the many methodologies available, industry has most embraced the so-called structured techniques. We'll briefly examine those techniques in the following pages. It is not our intent to teach you the techniques in this chapter. But you will learn to directly use many of them in Parts Two and Three of this book. We also want to expand on *prototyping,* one of the techniques introduced in Chapter 4. Prototyping is being hailed in some circles as the replacement for the structured techniques. We'll try to evaluate its strengths and weaknesses.

What Are Systems Development Methodologies?

Systems development methodologies are frequently confused with the life cycle. Is there a difference? Some experts claim that systems development methodologies are a substitute for the life cycle. In this section, we hope to dispel that myth and demonstrate that systems development methodologies are intended to complement the life cycle, not replace it!

Long Live the Life Cycle!

Some experts argue that the life cycle has become obsolete. They are wrong! The life cycle is not dead! Recall the intent of the life cycle—to plan, execute, and control activities for a project. The life cycle is a project management tool. It defines the phases and tasks that are *essential* to systems development, no matter what type or size of system you may try to build. For instance, we should always assess feasibility of a project (the *survey phase*) before committing resources to the project. Similarly, we should always identify problems (the *study phase*), define requirements (the *definition phase*), evaluate alternative solutions (the *selection phase*), *design* the system, and so forth.

Where then, did all the myths about the life cycle's death get started? The answer is *methodologies.* Let's formally define that term:

> **Methodologies** are specific, step-by-step strategies for completing one or more of the phases of the systems development life cycle. Methodologies impose their own tools and standards on the SDLC.

Note the words *one or more!* No methodology, including the structured techniques, currently addresses all phases of the life cycle. For instance, most methodologies have not specifically addressed the make-versus-buy issue relative to software (our *acquisition phase*). Why? Because most methodologies were born out of the software engineering discipline, which assumes that you are going to build, not buy, your software. Most methodologies also ignore classic issues such as feasibility analysis, fact finding, creativity, and problem solving.

Why, then, do people think the life cycle is extinct? The answer is that different methodologies have placed such varying emphasis on specific phases of the life cycle that many people don't recognize it anymore. For instance, before the structured methods we placed our greatest emphasis on the programming phases, paying little attention to the front-end analysis and design phases. The structured techniques shifted that attention, first to the design phases, and then to the analysis phases.

In any case, until a methodology comes along that supports the entire life cycle, the life cycle will remain a viable and essential foundation for the systems analyst and software engineer. Furthermore, if a methodology ever achieves this rather lofty goal, it seems to us that it will have to "reinvent the wheel," that wheel being the life cycle.

What Are the Structured Techniques?

In this section we introduce the popular structured techniques. We'll survey the following approaches:

- Structured Programming
- Structured Design (various approaches)
- Structured Analysis
- Data Modeling

> Note: *We'll only introduce the concepts that underlie these approaches*. Entire books have been written on each approach (see the annotated bibliography at the end of the chapter). You won't *fully* understand the techniques (or examples). That's all right; you'll learn them soon enough.

A Framework for the Structured Methodologies

In Chapters 2 and 3, we introduced a pyramid model to conceptualize information systems. A pyramid model (Figure 5.1) has also been used to characterize the structured methodologies. The model suggests that systems can be described from two points of view, processing and data.

The process dimension is based on the input-process-output (IPO) concept you learned in Chapter 3. Process-oriented methodologies build models of systems based on studying the flow of inputs to processes to outputs. *Structured Design* and *Structured Analysis* are examples of such methodologies.

The data dimension is based on studying the data that describes the business application. Data-oriented methodologies build models of systems based on the ideal organization and access of data, independent of how that data will (or might) be used to fulfill information (output) requirements. *Information Engineering* and *Object-Oriented Design* are examples of such a methodology.

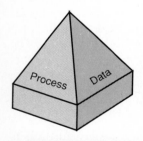

FIGURE 5.1 Modeling Dimensions for a System One basis for methodologies is the systems dimensions that they model. Most popular methodologies seek to build either models of systems processing or systems data or both.

Structured Programming: The Roots of the Structured Revolution

Structured Programming was the first structured technique, but it wasn't a *systems* technique. Instead, it focused only on computer programs. Structured Programming has evolved into a de facto (meaning unwritten) standard in much of the computing industry. Still, after all these years, the technique is often misunderstood.

Structured Programming is not *top-down* programming; it has nothing to do with the so-called top-down strategy. Structured programming deals only with program logic and code. It suggests that a well-structured program is written exclusively with various combinations of three control structures, called *restricted control structures.* These structures, as you are probably aware, are

- a *sequence* of instructions or group of instructions
- a *selection* of instructions or group of instructions based on some decision criteria (this construct is often referred to as the *if-then-else* or *case* construct)
- an *iteration* of instructions or group of instructions based on some criteria (this construct comes in two basic forms: *repeat-until* and *do-while*)

These constructs can be repeated within one another. An important characteristic of the preceding structures is that they *must* exhibit a *single-entry*, *single-exit* property. This means that there can only be one entry point into the structure and one exit point from the structure. Structured code reads like this page, top to bottom, with no backward references. This makes the code easier to read, test, debug, and maintain. The structured programming constructs and property are illustrated in the flowchart in Figure 5.2.

The structured programmer will design logic, using modeling tools such as flowcharts, box charts, language-independent pseudocode, and action diagrams—carefully restricting the structures to those described above. Then, the programmer will code that structure, carefully preserving the restricted control structures and single-entry, single-exit property.

Depending on the language used, the programmer may still use GOTO statements or their equivalents. Contrary to some popular belief, Structured Programming is *not* GOTO-less programming. It seeks only to control the frequently undisciplined use of the GOTO statement, especially the GOTO that sends control backward (bottom to top) through the program.

As a structured technique, Structured Programming supports only the *construction*, *delivery*, and *maintenance/improvement* phases of our life cycle. Structured programs are easier to write, much easier to test and debug, and much easier to maintain (due to their start-to-finish, booklike readability). Structured Programming cannot compensate for inadequate designs or analysis, however.

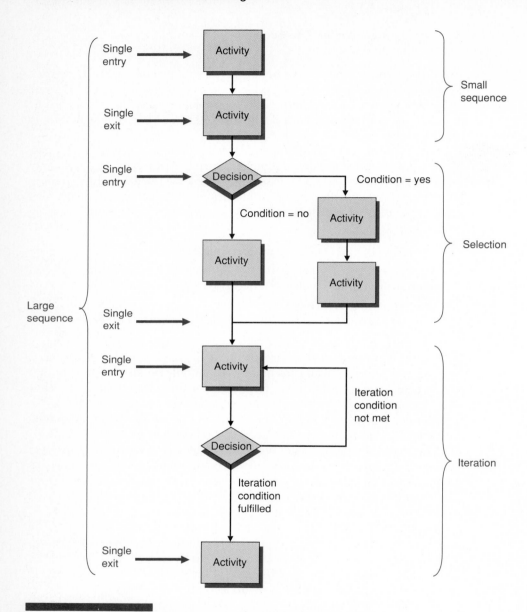

FIGURE 5.2 Structured Programming Concepts This flowchart demonstrates the basic concepts behind structured programming, *restricted control structures* and *single-entry*, *single-exit* flow.

Structured Design

Structured Design, also called *Structured Systems Design* and *Functional Decomposition*, was the second of the structured techniques to be developed. Its principal advocates include Ed Yourdon, Meiler Page-Jones, Jean Dominique Warnier, Ken Orr, and Michael Jackson. It is a technique for factoring computer programs into independent modules. The concept is simple. Design a program as a top-down hierarchy of modules. A module is a group of instructions—a paragraph, block, subprogram, subroutine, or the like. These modules are designed so that they serve one and only one function. Separate schools of thought have developed on the proper technique for accomplishing a well-structured design. They include

- *Yourdon-Constantine.* This technique derives the ideal structure from the flow of data through necessary program functions.
- *Warnier-Orr.* This technique derives the ideal structure from the contents of the outputs and inputs.
- *Jackson.* This technique also derives the ideal structure from the contents of the outputs and inputs.

These techniques use different tools to represent the top-down structure of modules. Because the technique is more widely known and practiced, the remainder of this discussion will focus on the Yourdon approach.

Yourdon Structured Design seeks to factor a program into the top-down hierarchy of modules with the following properties:

- Modules should be highly *cohesive*; that is, each module should accomplish one and only one function.
- Modules should be loosely *coupled*; in other words, modules should be minimally dependent on one another. This is achieved by minimizing data flow and parameter passing between modules.

The specification or model derived from Yourdon Structured Design is called a *structure chart* (Figure 5.3). The structure chart is derived by studying the flow of data through the system. Structured Design supports a very limited portion of the *design* phase of our life cycle, namely, *program design*. It does not support other design activities such as file design, database design, or internal controls design. Furthermore, it does not support any of the other classic life cycle phases, although it simplifies both *construction* and *delivery* phases through its top-down structure.

The benefits of a structured design are numerous. First, programs that are factored according to Structured Design can be more easily written and tested by multiple programmer teams. Why? Because the interfaces between modules are both well defined and limited by rules, the modules that test correctly by themselves should test correctly when brought together as a system. Top-down program structures also simplify programming effort because

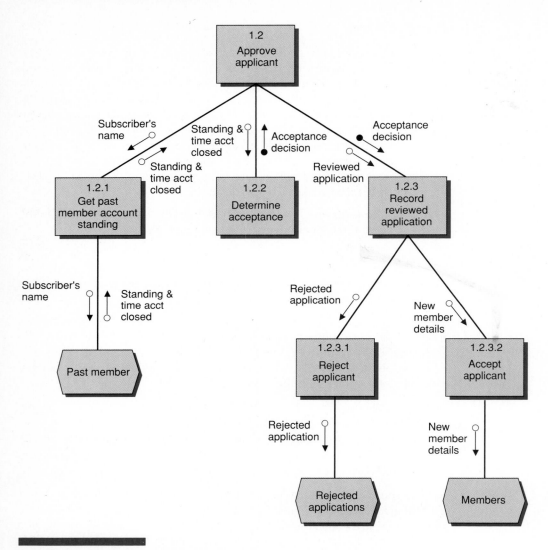

FIGURE 5.3 Process Model This structure chart is typical of the process models that characterize methodologies like *Structured Design*.

they lend themselves to top-down coding and stub testing (a familiar technique to experienced programmers).

Second, program modules that are developed according to Structured Design tend to be reusable. This is because they are built to be cohesive. Code reusability has become a major issue in industry because nobody can afford to repeatedly "reinvent the wheel." Modules will not be reusable unless they are intentionally designed for reusability. Structured Design is a major step in the right direction.

Third, systems and programs developed with Structured Design are more easily maintained. The major difficulty in program maintenance is called the *ripple effect*. It occurs when a change in one module requires changes in numerous other modules. Structured Design intentionally seeks to reduce the ripple effect by minimizing intermodule connections and dependence.

This book will teach you Structured Design techniques in Chapter 19.

Structured Analysis

Structured Analysis, also called *Structured Systems Analysis*, is the most popular and most widely practiced of the structured *systems* techniques. Its principal advocates are Tom DeMarco, Chris Gane, Trish Sarson, and Ed Yourdon. The Structured Analysis technique is simple in concept. The new systems specification evolves from a series of flow models called data flow diagrams or DFDs (Figure 5.4). These are not flowcharts! They do not explicitly show control of flow through a system. They only show flow of data, storage of data, and the processes that respond to and change data. Because of its dependence on data flows and processes, Structured Analysis is typically referred to as a *data flow approach*; however, it conforms more closely to what experts call the *process-oriented* process.

The analyst produces several sets of DFDs in pure Structured Analysis. These DFDs differ with respect to

- whether they model the current system or the system to be built
- whether they model the implementation details of the system (the so-called *physical* system) or the essence of the system (the so-called *logical* system)

The concept of a *logical system,* sometimes called an *essential system*, was created by and is crucial to Structured Analysis. It addresses the following problem. We tend to damage our own creativity by prematurely thinking about a new system in terms of *how* it should work (called the *physical system*). Structured analysis requires the analyst to define *what* the system should do (the logical system) before deciding *how* the system should achieve these goals. Advocates insist that by reducing the system to its logical essence, the following benefits are realized:

- The analyst more accurately defines end-user requirements by not prematurely worrying about technology.
- The analyst is more inclined to conceive more creative alternative solutions instead of solutions that are based on the existing system (the "we've always done it that way" syndrome).

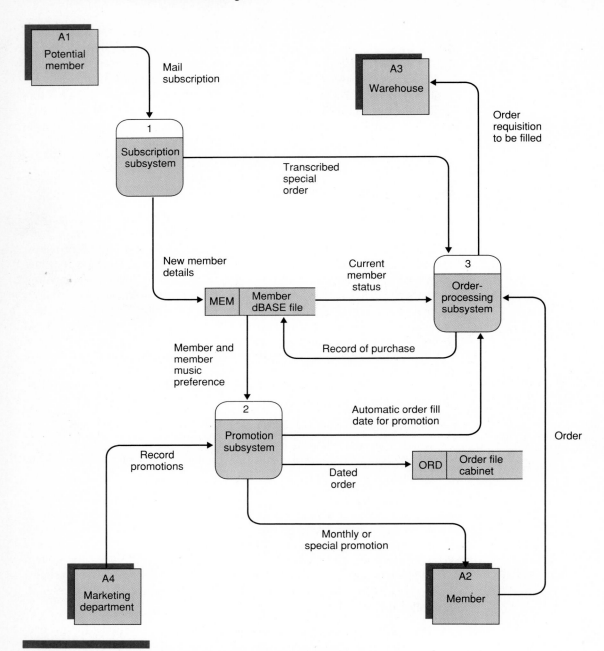

FIGURE 5.4   Process Model   This data flow diagram is typical of the process models that characterize methodologies like *Structured Analysis*.

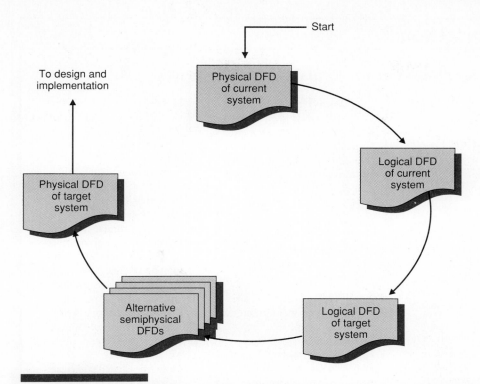

FIGURE 5.5   Logical and Physical Models in Structured Analysis   Structured Analysis uses data flow diagrams (DFDs) as its modeling tool. Various models of the existing and new system are drawn. Physical models show how the system is, could be, or will be implemented. Logical models show what the system does, independent of how it is implemented. The intermediate logical models are said to encourage the analyst to seek creative solutions in the absence of bias toward existing or easy solutions.

We don't want to get too far ahead of ourselves since we plan to teach you DFDs in Chapters 5, 7, and 9. Still, at the risk of oversimplifying the methodology, the analyst first draws physical DFDs for the existing system as part of the study phase. The physical DFDs contain references to people, machines, forms, and other physical properties of the existing system. These physical properties make the DFDs easy to verify. The analyst then transforms the DFDs into their logical equivalents. Next, the analyst carefully changes and supplements the logical DFDs to model the logical requirements of the new system. Then the analyst considers alternative physical implementations of those requirements. The "best" alternative is documented as physical DFDs to clearly show *how* the system will be implemented. This general approach is graphically summarized in Figure 5.5.

THE NEXT GENERATION

Structured Development for Real-Time Systems

The most recent of the structured techniques addresses a new and rapidly growing class of applications, *real-time systems*. Real-time systems exhibit the following important properties (adapted from Ward and Mellor, 1986):

- The systems frequently contain sensors that monitor the environment—for example, temperature, altitude, motion, presence of substances, electromechanical stimuli like keystrokes at a terminal, and the like.

- The systems frequently contain devices that, under automatic control of the system, can modify the environment—for example, furnaces, air conditioners, traffic lights, robots, printers, and the like.

- The systems frequently require programs that can simultaneously execute different instructions in response to different inputs and stimuli. This is rare, if even existent, in *mere* on-line information systems.

Languages such as Ada support this type of system.

- The systems require even better response time than on-line systems. Response times may be fractions of a second.

- The systems typically require greater precision than other computer-based systems. Errors and omissions tend to be more damaging.

Examples of such systems are numerous. They include operating systems, teleprocessing monitors, network software, computer-integrated manufacturing systems, environmental control systems, air traffic control systems, missile tracking and guidance systems, spacecraft control systems, and robots.

Where does the computer information systems analyst fit in? Our recent experience suggests that many new graduates and alumni are working on computer-integrated manufacturing systems since those systems are being interfaced to and

integrated with computer information systems. Computer specialists are finding themselves teamed with engineers and technologists!

Structured development for real-time systems is a new methodology, similar in many respects to structured analysis and design. For instance, it uses a new type of data flow diagram that can show the system's interface to devices and machines that the system seeks to employ and control. These DFDs also show the processes that use and control the devices.

However, structured development for real-time systems specifies an important new dimension of systems—*events*. Specifically, the technique models states of a system and the events that change the system from one state to another. Figure 5.6, called a *state-transition diagram*, illustrates the model.

At the risk of oversimplifying the technique, it is similar to Structured Analysis in that it

Structured Analysis was the first structured technique to specifically address the *survey*, *study* (through the physical and logical models of the existing system), *definition* (through the logical model of the new system), and *selection* (through the physical models of the new system) phases of our life cycle.

THE NEXT GENERATION

Structured Development for Real-Time Systems, Continued

suggests that you first build a logical model and that this model will eventually evolve into the implementation or physical model. As you progress into design, various specifications that can only be produced by engineers and technologists would be added to the model (e.g., circuit diagrams). Structured Design techniques are used for software components of the system.

It is difficult to assess strengths and weaknesses of this recently developed technique. Suffice to say, it does present important new features for a growing class of applications.

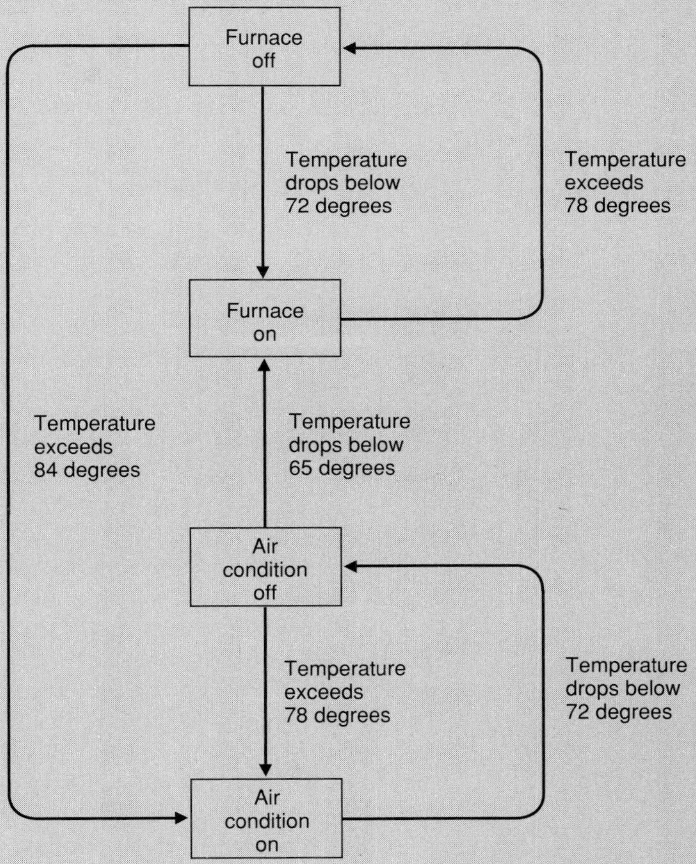

FIGURE 5.6   Event Model
This state-transition diagram is typical of the event models that characterize methodologies like *structured development of real-time systems.*

Furnace off

Temperature drops below 72 degrees

Temperature exceeds 78 degrees

Furnace on

Temperature exceeds 84 degrees

Temperature drops below 65 degrees

Air condition off

Temperature exceeds 78 degrees

Temperature drops below 72 degrees

Air condition on

But they only address those phases from the specification or modeling viewpoints. The analyst must rely on classic techniques to collect facts, analyze the system for problems and opportunities, identify candidate solutions, and evaluate those candidates for feasibility. This reality is not well understood by many who have tried and failed with the technique.

Structured Analysis may be the most popular of the structured techniques, but on closer examination, we note that few, if any, analysts apply the technique with the rigor suggested in the literature. For instance, few analysts draw the four or more sets of DFDs described in the literature. In fact, Ed Yourdon recently updated classical Structured Analysis to eliminate the first step, a physical model of the current system (see references). For one thing, it would take too long. Also, the transitions between the various physical and logical DFDs are frequently difficult to specify. But many analysts do base their designs on DFDs. DFDs are covered extensively in Chapters 7 and 9.

Data Modeling Approaches

Although data modeling methods have been with us since the advent of database methods, systems development techniques based on data modeling are relatively new. Examples include *Information Engineering*, as created by James Martin and Clive Finkelstein, *Information Modeling*, as created by Matt Flavin, and *Object-Oriented Design,* as created by Stephen Mellor and Sally Shlaer. These methodologies fall into the data-centered approach classification described earlier.

Through experience, many analysts concluded that systems based on data flows and processes—in other words, Structured Analysis—are not always adequate. Such systems are not always complete, meaning that DFDs may not ensure complete and accurate data and information. Also, while systems derived from DFDs and Structured Analysis meet current requirements, those same systems are not always flexible enough to adapt to tomorrow's requirements.

In looking for an answer to these problems, many analysts began turning to techniques based on database modeling. Database modelers build systems by studying the system's data, *independently of how the data is currently used.* They model the data for completeness, stability, and adaptability. These are ideal goals for all systems, not just those that use database technology. Hence, a new structured technique was born.

The data modeling techniques are briefly described as follows. Identify business entities about which the business or application collects data—for instance, CUSTOMERs, ORDERs, PARTs, EMPLOYEEs, BINs, and the like. These entities are sometimes called *subjects* or *objects* of data. Then, identify the data elements that describe these entities or subjects (for example, CUSTOMER NUMBER, CUSTOMER NAME, BALANCE DUE, and CREDIT RATING all describe the entity CUSTOMER). Then, identify the natural relationships that exist between entities—for instance, customers PLACE orders FOR parts. Draw a picture or *data model* of the entities and relationships (Figure 5.7). Finally, use formal techniques to ensure that the data model will be flexible enough to adapt to current and future requirements based on the same data.

The benefit of this technique is important. If the files and databases for the new system are built according to the data model, they will

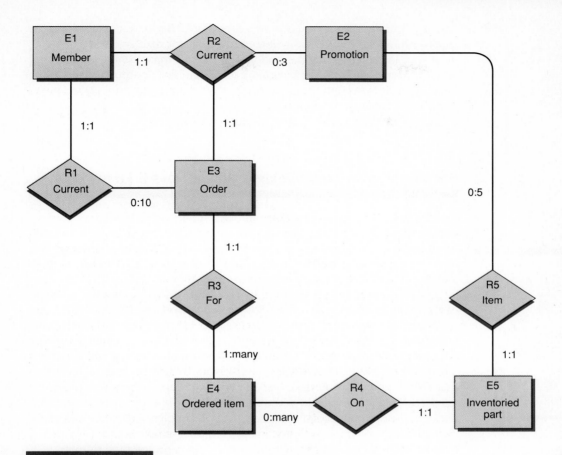

FIGURE 5.7 Data Model This model, called an *entity-relationship diagram*, is typical of the data models that characterize methodologies like *Information Engineering*.

- contain accurate and up-to-date data
- meet all of today's output requirements
- meet future output requirements without drastic changes to the system, because the data is already there or can be easily added to the appropriate entities

This technique is becoming increasingly popular. Unfortunately, there are problems to overcome. Most books represent data modeling as a *database* technique. Actually, it works equally well with conventional files and distributed data. Most books advocate building very large, corporate databases and systems. Actually, the technique works equally well for incrementally building small systems that can be easily integrated at a later date.

It should be noted that methodologies such as *Information Engineering* and *Structured Analysis* are beginning to merge data and process modeling techniques. It is becoming impossible to classify any methodology as "pure" data modeling or process modeling.

You'll study data modeling techniques in Chapter 8.

The Prototyping Methodology: Where Does It Fit In?

**DATABASE
✓LANGUAGES
FOR PROTOTYPING**

FOCUS
RAMIS
ADS/O
IDEAL
NATURAL
SAS
MANTIS
R:BASE SYSTEM V
dBASE IV

Prototyping is an engineering discipline that has found its way into computer systems development. The idea, at least in the engineering sense, is to build small-scale, working models of a product or its components. When it applies to computer systems development, it means that the analyst builds a small-scale working model of the system or a subsystem.

Computer systems prototyping is the result of the increased availability of *database languages* (also called *fourth-generation languages* or *4GLs*, and *applications generators*). These languages provide powerful tools for quickly generating small-scale, working models of files, databases, screens, reports, and the like. The 4GLs tend to be less procedural than traditional languages like BASIC, COBOL, FORTRAN, C, and Pascal. By less procedural we mean that the code is more English-like and in many cases allows you to specify *what* the system should do without specifying *how* to do it. This makes it possible to develop prototypes more quickly. Examples of database languages suitable for prototyping are listed in the margin. A sample screen (Figure 5.8) from FOCUS demonstrates a menu-driven tool for prototyping outputs.

How do you prototype systems? Prototype systems or subsystems tend to focus on files, inputs, processes, and outputs—the basic systems components. Usually, factors such as throughput, response time, scratch files, and internal controls and security are ignored. Prototypes, once developed, are tested by the end-users. The end-users suggest changes, and the prototype is revised to accommodate those changes. The conciseness and improved readability of 4GL code makes this possible. The cycle is repeated until the end-users accept the system.

The benefits of prototyping are numerous. End-users become more active participants in systems development. They tend to be more excited by working prototypes than paper design specifications. Requirements definition is simplified through realization that many end-users will not understand or be able to state their requirements until they see a prototype. The likelihood that end-users will approve a design on paper only to reject the implementation of that design is greatly reduced. Finally, design by prototyping is said to reduce development time; however, some experts dispute these savings.

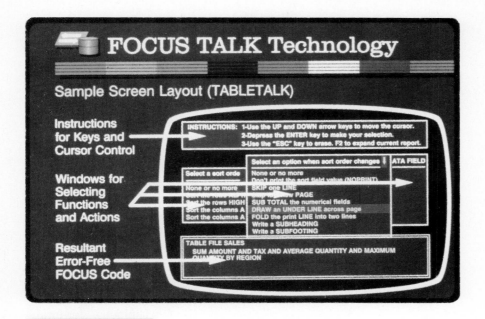

FIGURE 5.8 Prototyping This screen, from the prototyping language FOCUS, allows analysts (or end-users) to quickly generate a working prototype of a new output using a prototype database with sample data (TableTalk window-driven report generator from Information Builders, Inc., New York).

But prototyping can be dangerous. What are the disadvantages? The first danger is a trend by computer systems prototypers to skip through analysis and design too quickly. This is a return to the disastrous "code, implement, and repair" methodology of the prestructured era. Overreliance on prototyping also tends to lead to development of technological approaches that don't always solve problems and fulfill requirements. Systems implemented from prototypes frequently suffer from lack of flexibility to adapt to changing requirements; they were developed "quick and dirty." And they're not always easy to change. Several experts have noted the growing libraries of poorly designed, unstructured, unreadable, and inadequately documented 4GL code. Finally, prototyping often stifles creativity. We tend to implement the first solution that comes to mind.

A fundamental question is "Can you prototype without a specification?" Based on how 4GLs are being used by many analysts, you might get the impression that paper specifications—for example, the structured techniques—are obsolete. The answer, however, is an emphatic no. No competent engineer would prototype without a specification. Consequently, we will teach you prototyping in the context of complementing the structured techniques.

CASE: Computer-Assisted Software (or Systems) Engineering

Computer-Assisted (or -**Aided**) **Software** (also called **Systems**) **Engineering** or **CASE** is the latest technique. But it is not so much a technique as it is an *enabling technology* for the structured techniques. In other words, CASE provides computer-automated support for the structured techniques. Synonyms for CASE include *computer-assisted design/computer-assisted programming (CAD/CAP)* and *the analyst/programmer workbench.*

CASE helps analysts overcome one of the key disadvantages of the structured technique: lower productivity. The very rigor required by the structured techniques usually slows project progress. CASE increases that productivity through computer tools. At the same time, it improves quality in both systems and documentation.

CASE products are built around the concept of a project dictionary (or encyclopedia) that stores all facets of the systems specification. The dictionary is complemented by a host of facilities (Figure 5.9) that support various methodologies and techniques. These facilities typically include

- Graphics tools for various structured and classic techniques—for example, data flow diagrams, flowcharts, data models, structure charts, state-transition diagrams, and the like.

- Dictionary tools to record, maintain, and report on systems details—for example, contents of files, inputs, and outputs; properties of data elements; and logic rules for processes.

- Prototyping tools for external designs of inputs, screens, forms, and outputs. Some CASE tools can be interfaced to fourth-generation languages and applications generators.

- Automatic quality checking for graphics and dictionary specifications. The computer can proof models and specifications for consistency and completeness errors.

- Code generators to eliminate or reduce programming effort required to transform the specifications into working systems.

- Cost-benefit analysis tools and interfaces to spreadsheets.

- Project management tools for planning and controlling the activities of the systems development life cycle.

- Documentation assemblers to combine various specifications into reports for various audiences, technical and nontechnical. This facility frequently interfaces to popular word processors.

No single CASE product includes all of these facilities. Examples of popular CASE products include Excelerator and Excelerator/RTS (RTS stands for Real-Time Systems) from Index Technology, Design Aid from Nastec, Information

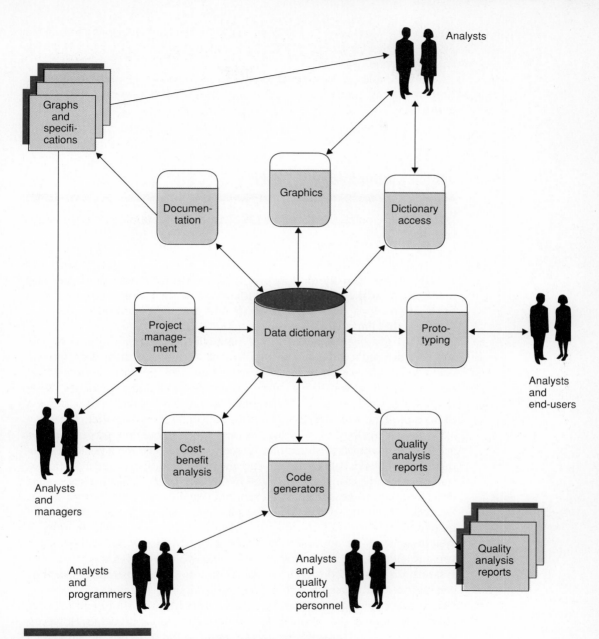

FIGURE 5.9   Computer-Assisted Software Engineering (CASE)   CASE tools provide the analyst with a computerized dictionary for their structured specification. Facilities provide access to and quality checks on project data stored in the dictionary.

Engineering Workbench from KnowledgeWare, Structured Architect from Meta Systems, Analyst/Designer Toolkit from Yourdon, and System Developer's Prokit from McDonnell Douglas.

Many samples of documentation in Parts Two and Three of this book were prepared using Excelerator. CASE is one of the fastest growing technologies at this time.

Which Methodology Is Best?

Did you figure out the moral of the fable in the chapter minicase? As Weinberg stated it:

It's not know how that counts. It's know when.

In other words, you should look for the strengths in all methodologies and determine how they can be applied to each individual project at appropriate times. If you look hard enough, you'll find that most methodologies have strengths that improve almost any project.

We already suggested that prototyping without a specification is wrong. Which structured technique best complements prototyping, then? Both process-oriented and data-oriented methods have their strong points. We believe the question is moot. You should always use both modeling approaches. Why? Because, as we have seen, all systems have at least a data and a process dimension. One is not necessarily more important than the other.

A useful analogy to this suggestion is the accounting profession. No accountant would suggest that management choose between a position statement (also called a balance sheet), income statement (also called a profit-loss statement), and a cash flow statement. Why? Each of these distinct statements describes the business from a different but equally important viewpoint.

Additionally, the structured techniques and prototyping not only complement one another. They also complement the classic tools and techniques of systems analysis, design, and programming. For this reason, we will not teach you any specific, single methodology. Instead, we will teach a wide variety of systems analysis and design tools and techniques, demonstrating how they can and should be used together to develop superior systems.

We are not trying to start a new camp of ostriches. You should make every effort to keep up to date about new methodologies and approaches. However, don't blindly align yourself with any one camp, as the ostriches did when they steadfastly followed the favorite brother. Look for ways to integrate and complement methodologies. And always remember, the basic life cycle is always there to guide your use of all methodologies!

Summary

Systems development methodologies are used to implement or complement the systems development life cycle. The most popular methodologies are the structured techniques: Structured Programming, Structured Design, Structured Analysis, and Information Engineering. Most structured techniques are based on either process-oriented or data-oriented models.

Structured Programming was the first structured technique. Through restricted control structures and single-entry, single-exit structures, Structured Programming gave rise to more readable and maintainable programs. It also ushered in a host of complementary systems techniques.

Structured Analysis and Structured Design are companion, process-centered methodologies. Structured Analysis builds process models, called data flow diagrams, for a system. Structured Design transforms the process model for a system into a top-down, process model for the programs that comprise the system.

Information Engineering is a data-centered methodology. It is especially well suited to database applications, but it can benefit all types of applications. Systems that are built from stable data models are said to be more easily adapted to new requirements over their lifetimes.

Although it is not a structured technique, prototyping has become a very popular complementary methodology. Based on an engineering approach, prototyping builds small-scale working models of a system. These models, called prototypes, help end-users more quickly define requirements, feasibility, and designs. Prototyping can be detrimental when it is used to take shortcuts around systems analysis, though.

Methodologies are gaining powerful new support from CASE, Computer-Assisted Software Engineering. CASE provides automated tools to improve both productivity and quality when the structured techniques and prototyping are used.

In the final analysis, the modern analyst should be familiar with all popular methodologies, especially structured methodologies and prototyping. But instead of choosing a methodology, the analyst should integrate new and old approaches, taking advantage of the strengths of each.

Problems and Exercises

1. Define the term *methodology*. Why don't most methodologies address the make-versus-buy issue relative to software?

2. Explain why some experts argue that the systems development life cycle has become obsolete.

3. Differentiate between a process-oriented versus a data-oriented systems development methodology. Give an example of each.

4. Assume you are being interviewed by a college recruiter. The college recruiter is annoyed by the liberal use of buzzwords in the computing field. The recruiter is particularly annoyed at hearing people use the word *structured* all too loosely. How would you define "structured programming" to the recruiter?

5. What are the three restricted control structures? What is one important characteristic common to the three structures?

6. Briefly describe the concept of structured design. Identify three separate schools of thought on structured design. Which is the more popular of the three? What benefits does structured design offer to systems development?

7. The Yourdon Structured Design technique seeks to factor a program into the top-down hierarchy of modules that is highly cohesive and loosely coupled. What is meant by the term *highly cohesive*? What is meant by the term *loosely coupled*?

8. Briefly describe the concept of structured analysis. What benefits does structured design offer to systems development? What systems development life cycle phases are addressed by structured analysis? What are some of the disadvantages of the structured analysis technique?

9. Briefly describe the concept of information engineering. What benefits does information engineering offer to systems development? What systems development life cycle phases are addressed by information engineering? What are some of the disadvantages of information engineering?

10. Briefly describe the concept of prototyping. What benefits does prototyping offer to systems development? What are some of the dangers or disadvantages of prototyping?

11. Define the acronym CASE. What are some of the facilities typically included in a CASE product?

Projects and Minicases

1. In Chapter Four you learned that systems projects originate because of the need to address one or more problems, opportunities, and directives. In this chapter you learned about a number of systems development methodologies and their advantages and disadvantages. For each systems development methodology, characterize a systems project that that would be ideally suited for applying the methodology. Characterize a systems development project that might benefit from an integrated application of one

or more of the methodologies. Finally, assume your boss has assigned you the responsibility of studying existing systems development methodologies and making a recommendation about which systems development methodology the organization should adopt. What recommendation would you make?

2. You are teaching a course on the systems development methodologies presented in this chapter. The students are somewhat confused over the objectives of each methodology. Each uses different tools and techniques and seemingly focuses on achieving different results. Prepare a study guide that briefly describes each methodology in terms of its concepts, advantages, and disadvantages. Using the systems development life cycle (SDLC) as a framework, describe the scope of emphasis that each methodology places on the SDLC phases.

Annotated References and Suggested Readings

Connor, Denis. *Information System Specification and Design Roadmap.* Englewood Cliffs, N.J.: Prentice-Hall, 1985. This book provides an excellent survey of all of the structured techniques we introduced, plus a few that we didn't discuss.

DeMarco, Tom. *Structured Analysis and System Specification.* Englewood Cliffs, N.J.: Prentice-Hall, 1978. This is the classic book on the Structured Analysis technique. As we go to press, we hear that this book is being updated. DeMarco is a gifted writer, and we encourage readers to seek out the second edition when it becomes available.

Falvin, Matt. *Fundamental Concepts of Information Modeling.* New York: Yourdon Press, 1981. A data modeling methodology is described by the author.

Gane, Chris, and Trish Sarson. *Structured Systems Analysis: Tools and Techniques.* Englewood Cliffs, N.J.: Prentice-Hall, 1979. Another popular book on the original Structured Analysis technique.

Martin, James, and Clive Finkelstein. *Information Engineering.* Vols. 1 and 2. Savant Institute, 1981. These are the classic books on the Information Engineering technique.

Mellor, Stephen J., and Paul Ward. *Structured Development of Real-Time Systems* (3 volumes). New York: Yourdon Press, 1986.

McMenamin, Stephen M., and John F. Palmer. *Essential Systems Analysis.* New York: Yourdon Press, 1984. This book does an excellent job of describing the difference between physical and logical systems and how to use the concepts in Structured Analysis.

Page-Jones, Meiler. *A Practical Guide to Structured Systems Design.* New York: Yourdon Press, 1980 (possibly available through Prentice-Hall). This book is still our favorite book for learning the Structured Design technique.

Ward, Paul. *Systems Development Without Pain.* New York: Yourdon Press, 1984. This was the first practical book that clearly suggested the integrated use of process-centered and data-centered methods.

Weaver, Audrey. *Using the Structured Techniques.* Englewood Cliffs, N.J.: Prentice-Hall, 1987. This excellent paperback is a case study of the structured techniques using both process-centered and data-centered approaches.

Weinberg, Gerald. *Rethinking Systems Analysis and Design.* Haughton-Mitlon, 1979. Source of the ostrich story, this book is full of similar and interesting stories and anecdotes that cause the reader to rethink the subject.

Whitten, Jeffrey L., and Lonnie D. Bentley. *Using EXCELERATOR for Systems Analysis and Design.* St. Louis: Times Mirror/Mosby College Publishing, 1987. Chapter 1 is a detailed introduction to the concept of computer-aided software engineering.

Yourdon, Edward. *Modern Structured Analysis.* Englewood Cliffs, N.J.: Prentice-Hall, 1989. This is the long-awaited update to DeMarco's classic book. It substantially extends Structured Analysis to include data modeling, real-time tools, and reduced emphasis on current system models.

Yourdon, Edward, and Larry Constantine. *Structured Design.* Englewood Cliffs, N.J.: Prentice-Hall, 1976. This is the classic book on Structured Design.

PART TWO

Systems Analysis Tools and Techniques

How to Analyze an Information System: A Problem-Solving Approach

Process Modeling with Physical Data Flow Diagrams

Data Modeling with Logical Entity Relationships Diagrams

Systems analysis—a process performed by many but practiced by few. Does that sound like a contradiction? It really isn't. Systems analysis is considered an essential activity by most data processing professionals. However, few of those professionals practice systems analysis with rigor or according to well-defined standards. Why? Simply stated, because they don't know how. We tend to spend the least time on those activities with which we are the least familiar, no matter how important they are! How important is systems analysis?

Systems analysis is the most critical process of information systems development. It is during systems analysis that we learn about the existing business system, come to understand problems, define objectives and priorities for improvement, define business requirements, and evaluate alternative solutions. Clearly, the quality of the subsequent systems design and implementation is dependent on a good system analysis. In fact, the best technical design and implementation are useless if they don't solve the correct problems, fulfill objectives, and meet requirements in a cost-effective fashion! So why is systems analysis the most short-changed of the systems development processes? Because most analysts are not well schooled in use of systems analysis tools and techniques, that's why!

The purpose of this unit is to introduce you to the systems analysis process and to some useful tools and techniques for performing that process. Six chapters make up this unit. The first chapter introduces the systems analysis process. That process consists of the four phases: survey, study, definition, and evaluation.

The next five chapters develop systems analysis skills in use of tools and techniques that are especially useful during systems analysis. These five chapters develop expertise with the following tools: hierarchy charts, data flow diagrams, entity-relationship diagrams, data dictionaries, Structured English, and decision tables. Hierarchy charts are useful for breaking a system into

Process Modeling with Logical Data Flow Diagrams

Defining Logical Data and Information Requirements in a Project Dictionary

Defining Logical Policies and Procedures in a Project Dictionary

subsystems and tasks for easier understanding. Data flow diagrams depict information system problems and solutions in terms of the flow of data through the system and the work performed on that data. Entity-relationship diagrams model the data and structures that describe a system. Data dictionaries document the content and structure of data and information in a system. Finally, Structured English and decision tables are useful for documenting business policies and procedures. Although none of these tools are exclusive to the systems analysis process, they are most useful to it. We will also examine possible uses for systems design and implementation.

In addition to hierarchy charts, data flow diagrams, entity-relationship diagrams, data dictionaries, Structured English, and decision tables, you may elect to skim or read (or you may be assigned one or more) of the modules in Part Four. These modules survey skills that are not restrictive to the systems analysis process but that are nevertheless extremely important to effective systems analysis.

CHAPTER SIX

How to Analyze an

Information System:

A Problem-Solving Approach

Collins Financial Services

Senior systems analyst Fred McNamara was meeting with Ken Borelli, the MIS manager at Collins Financial Services. Fred had just completed an evaluation of a new software package, a fourth-generation programming language (4GL). In addition to evaluating the 4GL, Fred had been asked to learn about a new, popular systems development approach—called prototyping—that makes use of 4GLs to build working models of systems. Fred is meeting with Ken to give him his assessment of the 4GL product and prototyping as an alternative approach to systems development.

"So what do you think about that new software product—is it worthy of being called a 4GL?" Ken asked as he gestured for Fred to take a seat.

Fred replied enthusiastically, "Without a doubt. So-called third-generation programming languages such as COBOL can't be compared to it! I think this product can do wonders for the systems staff. It is very user friendly. It provides a number of facilities that assist in developing a complete system. To give you some idea, I used a facility to develop a database containing actual data, another facility to produce a relatively complex printed report, and other facilities to develop menus and other input and output screens—all in a fraction of the time that would have been required with COBOL. The facilities led me through a series of questions or prompts. All I had to do was answer the questions. The 4GL generated program code, which I subsequently executed. The productivity implications are tremendous."

"Sounds like the product is a good investment. And what about prototyping? Do you think prototyping is something we should consider doing as an alternative to our current approach to developing systems?"

"Well, prototyping certainly takes advantage of tools such as 4GLs. The strategy is very simple. It emphasizes the development of a working model of the target system, instead of traditional paper specifications. You begin building the model by first defining the database requirements for the new or desired system. Identifying the database requirements is relatively simple, since the data for most systems already exists, either in computer files or manual forms. Afterward, you would then build and load a database using the 4GL. Once you have the database built, the rest is easy. You can use various facilities to quickly generate the menus, reports, and input and output screens. The analyst needn't worry about whether the working model is totally complete or accurate. The end-user would be encouraged to review the model and provide the analyst with feedback. If the model, say, a report, is not acceptable, the analyst simply makes requested changes and reviews it again with the end-user at some later time. This repetitive process and active end-user participation are considered essential and to be encouraged."

"Now that's a new one!" Ken commented sarcastically. "I certainly agree with the idea of encouraging end-user participation. But this attitude of encouraging or expecting to keep redoing work would be difficult to adjust to." Ken reflected for a moment. "Well, it sounds like this new 4GL and prototyping should be pursued further. Both 4GL and prototyping seem to offer some productivity gains. And I've been looking for a way to get the end-users more involved in the systems development process, and I think this prototyping approach is the answer. There is one other thing: I suspect my staff's morale would be improved. I believe my staff would be motivated by this new 4GL product and prototyping's emphasis on building a model as a basis for performing systems development."

Fred jumped in. "I agree! I can't wait to start my next project. I've got just the project picked out. I received a request for a new Employee Benefits system from the Personnel Department. I thought I'd use the 4GL in conjunction with the prototyping approach to complete this project. I won't be needing any programmers since I'll be developing the system myself while I'm working with the end-users. I've already drafted a memo asking Personnel to provide me with some sample records, forms, reports, and other materials that will help me identify their data storage requirements. From those samples, I'll be able to build a database for the new system. Then I'll start meeting with the end-users to define and implement screens and reports."

"Hold your horses, Fred," instructed Ken. "I do have some concerns about this approach to systems development. Neither the tool nor the approach justifies a departure from the systems development life cycle concept we follow here at Collins. And that's exactly what you're proposing. You're proposing to select a project, define some basic requirements, and jump right into the design and construction of a new system. That I won't have. I want you to reconsider things."

Ken reached for a systems development standards manual on his bookshelf. He turned to a figure that depicted the systems development life cycle phases. "Notice that our systems development life cycle includes several systems analysis phases, including a survey or preliminary investigation phase. Do you fully understand why we require this first phase?"

"Sure, that's where we perform a very quick study of the proposed project request. We try to gain a quick understanding of the size, scope, and complexity of the project," Fred replied, wondering what Ken was getting at.

"You're half right, but why must we complete the phase?" Ken didn't even give Fred a chance to respond. "We receive numerous project requests from our end-users! I have a limited number of resources. I can't take on all the requests. It is the purpose of this phase to address the seriousness of the problems and to prioritize the project request against other requests."

Fred conceded, "I see what you're saying. I sort of jumped the gun by picking this Personnel project without considering other project requests that should be given higher priority."

"Good," said Ken. "Now, let's consider the second phase, the detailed study phase. If the project is to be pursued further, we then conduct a detailed study or investigation of the current system. We want to gain an understanding of the causes and effects of all problems, and to appreciate the benefits that might be derived from any existing opportunities. We don't want to bypass this phase. This phase ensures that any new system we propose solves the problems encountered in the current system."

"That brings us to the definition phase, right?" Fred interrupted.

"Right," confirmed Ken. "For all practical purposes this is where you were proposing to begin your project. You were going to define data storage requirements for a new Employee Benefits system. I assume you would also attempt to identify other requirements. What particularly bothered me was that I didn't get an impression that you intended to study the database requirements to ensure data reliability and flexibility of form. And what about completeness checks? What were you going to do to ensure that processes were sufficient to ensure data will be properly maintained?"

Fred looked dejected. "I'm beginning to lose my confidence in this prototyping approach. I was about to make some big mistakes by trying to bypass several important problem-solving phases and tasks."

"There's no reason to write off prototyping," Ken replied. "So long as we base our prototypes on some sound design principles, I think we can still achieve all the benefits that you described earlier. There's one more analysis phase, selection. Here, we look at alternative solutions such as manual versus computer-based systems, on-line versus batch systems, and the like. These options are evaluated for technical, operational, and economical feasibility. It seems to me that this activity should precede extensive prototyping. We don't want to prematurely commit to less feasible or infeasible solutions."

"I understand," said Fred. "It looks like we still have potential here, especially in the definition phase of analysis. And I suspect that prototyping will greatly accelerate our design and implementation phases."

Ken nodded agreement. He could tell that Fred did indeed understand the need to analyze a system as part of the systems development life cycle. "I think we both understand that we need to give this prototyping approach some more thought. I want to establish a task force to study prototyping and its implications for our internal systems development standards. And I want you to be in charge of that committee. In the meantime, I see no reason not to try the new 4GL in conjunction with our usual development standards."

Discussion

1. Do you think Fred did a thorough evaluation of the fourth-generation software product? What benefits do you believe can be derived from using such a tool?

2. Did Fred view prototyping as an alternative to the traditional systems development life cycle? If so, how should he have viewed it?

3. What systems analysis phases would have been skipped by the prototyping approach Fred proposed to follow? What do you think would have been the results of the Employee Benefits project if Fred had approached the project in the manner he originally prescribed?

What Will You Learn in This Chapter?

In this chapter, you will learn more about the first four phases in the systems development life cycle: *survey*, *study*, *definition*, and *selection*. These four phases are collectively referred to as *systems analysis*. You will know that you understand the process of systems analysis when you can

1. Define *systems analysis* and relate the term to the survey, study, definition, and selection phases of the life cycle.

2. Describe the survey, study, definition, and selection phases of the life cycle in terms of
   a. Purpose and objectives
   b. Tasks or activities that must or may be performed
   c. Sequence or overlap between tasks or activities
   d. Techniques used
   e. Skills you need to master

3. Describe systems analysis in terms of problem solving.

4. Describe the roles of fact-finding, interpersonal communications, and cost-benefit analysis in systems analysis.

What is systems analysis? We already know what a system is. But what is analysis? Analysis is the process of studying a problem to find the best solution to that problem. The result? Systems analysis gives us the target for design and implementation. Thus, we note that systems analysis is problem solving!

In this chapter, we expand on four systems analysis phases: (1) surveying the feasibility of the project, (2) studying and analyzing the current system, (3) defining end-user requirements for an improved system, and (4) selecting, from alternatives, a feasible systems solution (see Figure 6.1). For each phase, we study the *purpose* and *objectives* of the phase, the *tasks* that should be performed, and important *skills* to be mastered. We also address the issue of

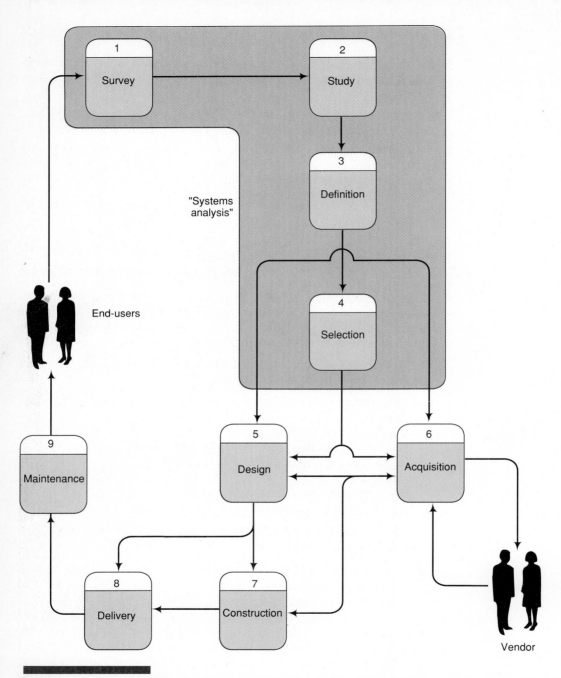

FIGURE 6.1 The Systems Analysis Process Systems analysis is defined as the survey, study, definition, and selection phases of the systems development life cycle.

cost-benefit and feasibility analysis, an important concern to management during the systems analysis phases. Your study of skills in the subsequent chapters and modules will be easier if you understand this systems analysis process.

Systems Analysis Is Problem Solving

In Chapter 1, you learned about skills that are essential in becoming a successful systems analyst. One of those skills was problem solving. This skill is especially important during the systems analysis phases. In fact, the entire systems analysis process is a problem-solving process.

The natural problem-solving process may be viewed as consisting of three steps. Step 1 is the identification of the problem or situation needing to be solved. Step 2 involves defining the desired outcome or situation. Step 3 involves identifying and selecting one from among a number of alternative solutions that address the problems and would result in the desired outcome or situation. This process is mirrored in systems analysis.

Recall that systems projects are suggested by either end-users or systems analysts and are triggered by the identification of problems, opportunities, or directives. Let's draw a distinction between problems, opportunities, and directives. **Problems** are undesirable conditions or situations that prevent or can be expected to prevent the business from fully achieving its business purpose, goals, objectives, and policies. For example, an increase in time required to fill an order can trigger a project to reduce that delay. An **opportunity** is a chance to improve the business system even though the existing system is performing acceptably. For instance, management is always receptive to cost-cutting ideas, even when costs are not currently considered a problem. A **directive** is a new requirement that is imposed by management, government, or some other external influence. For example, the Equal Employment Opportunity Commission, a government agency, may mandate that a new set of reports be produced each quarter. Or management may dictate support for a new product line or corporate policy.

Recall that Chapter 4 introduced the PIECES framework for identifying and classifying problems, opportunities, and directives (see margin). We will refer to the PIECES framework as we learn about the problem-solving phases of the systems analysis process. But first, let's examine each of the PIECES categories in greater detail.

PIECES FRAMEWORK

Performance
Information (Data)
Economic
Control (Security)
Efficiency
Service

Performance Analysis

Performance problems occur when business tasks are performed too slowly to achieve objectives. Performance opportunities occur when someone recognizes a way to speed up a business task that is otherwise achieving objectives. A performance directive may occur if management decides that all transactions are to be done *on-line* to the computer to improve performance.

Performance is measured by throughput and response time. **Throughput** is the amount of work performed over some period of time. Most throughput projects are concerned with transaction processing throughput. Consider the following scenario:

> A local credit union has been studying data about consumer loan applications. Over the past year, loan applications have increased 124 percent. The manager realizes that, if this growth rate continues, the current loan officers will not be able to keep up with the demand. The throughput of the current system must be increased.

Response time is the average delay between a transaction and a response to that transaction. The following example illustrates the point:

> A construction company has been contracted to perform repairs and improvements for a large corporate site consisting of many buildings. The corporate site submits work orders to the construction company. The work orders go through a processing cycle that may include Information Systems, Purchasing, Accounting, Personnel, and Operations. Currently, an average delay of 62 days occurs between the submission of the order and the arrival of the work crew to fulfill the order. Management wants to reduce this response time as much as possible.

Although throughput and response time are considered separately in the preceding discussion, they should also be considered jointly. For instance, one way to improve throughput in our credit union example would be to improve the loan officers' average response time for each loan application by giving those officers timely credit information.

Information and Data Analysis

Information is a crucial commodity for end-users. The information system's ability to produce useful information can be evaluated for problems and opportunities. Improving information is not a matter of generating large volumes of information. In fact, information overload is a major problem in many businesses—one that is easily recognized by piles of computer outputs!

Situations that call for information improvements include

- *Lack of any information concerning the decision or current situations.* For example, the Accounting Department suspects that air travel reimbursements do not reflect minimum costs and bargains that could be obtained. However, it has no information to support its suspicions; therefore, it cannot justify possible changes to its procedures.

- *Lack of relevant information concerning the decisions or current situations.* For example, a personnel manager must allocate scarce overtime dollars to the supervisors of three manufacturing departments. The report that predicts the amount of work to be done does not break the information down to the department level.

- *Information that is not in a form useful to management.* For example, an inventory control clerk for a large printing business must reorder paper and supplies each Monday. The clerk is given an inventory report. However, the report includes all 3,000 inventory items. The clerk has to compare quantity in stock and projected usage for each item on the report—just to identify items that need to be reordered. An exception report that has already made the comparison between supply and demand and reports only those items that need to be reordered would be more convenient.

- *Lack of timely information.* Consider this example: A hotel chain allows customers to make reservations for any hotel in the chain from any other hotel in the chain. However, when a reservation is made or canceled, it takes three days to get that information to the hotel that is affected. Meanwhile, that hotel may overbook or underbook rooms because the information is not timely.

- *Too much information.* This was previously referred to as information overload. End-users frequently complain about receiving numerous, large reports that contain unnecessary data.

- *Inaccurate information.* Information contains errors that lead to bad decisions or other problems.

Information may also be the focus of a directive; the classic example of this is a new reporting requirement imposed by a local, state, or federal government agency.

Whereas information analysis examines the outputs of a system, data analysis examines the data stored within a system. Problems frequently encountered include:

- *Data redundancy.* The same data is captured and/or stored in multiple places. Data redundancy consumes valuable storage and creates problems with data integrity. Consider, for instance, a manager who receives notification that he has eight registered copies of a word processor. These copies are eligible for a free upgrade to the new version. He sends in the order and is promptly notified that the software vendor cannot upgrade the copies because he is not legally registered. Obviously, the notification and order were processed across data files with inconsistent data.

- *Data inflexibility.* The data is captured and stored; however, it is organized such that certain reports and inquiries are impossible or difficult. Consider a large collection of data describing researchers who have had contact with radioactive isotopes. The files—in numerous file cabinets—are organized by researcher. Now the government requests data on every researcher who has used a particular isotope. The request can only be fulfilled by going through each of the researchers' files one by one. (Did you also notice the performance problem in this scenario?)

Economic Analysis

Economics is perhaps the most common motivation for projects. The bottom line for most managers is dollars and cents. Economic problems and opportunities pertain to costs. Consider the following:

> The Marketing Department needs to establish the new prices for products in its catalog. In order to establish a price that will recover manufacturing costs and overhead and provide an acceptable profit margin, it needs a cost breakdown by product, including materials, direct labor, and overhead (for instance, utilities and plant maintenance). Although the department has access to budgeted cost standards, it needs historical data on actual costs because those costs may be exceeding the budgets.

Or consider this example:

> A purchasing manager has been ordered to reduce the costs of raw materials. There are two ways to reduce costs: first, by carefully comparing the alternative pricing structures offered by different suppliers; second, by taking advantage of bulk quantity discounts. However, it is necessary to strike a balance between reducing purchasing costs and increasing inventory costs (yes, it costs money to store and handle inventory!). An information system can greatly assist the decision-making process.

Control and Security Analysis

Business tasks need to be monitored and corrected when substandard performance is detected. Controls are installed to improve systems performance, prevent or detect systems abuses or crime, and guarantee the security of data, information, and equipment. Two types of control situations trigger projects: too few controls and too many controls.

A system that has too few controls may result in discrepancies between the information system and the business system. The following example is typical:

> A distribution warehouse for farm machinery parts is experiencing a stock problem. The computer information system is releasing orders after checking the inventory file to ensure that the products ordered are in stock. However, when the warehouse clerk tries to fill the order, the parts are not always in stock. An analysis reveals that, when the stock clerks place new inventory on the shelves, they do not count that inventory.

They simply accept the supplier's word that the quantity shipped is accurate. This system has too few controls to ensure the accuracy of inventory counts.

But we can go overboard! A system with too many checks and balances slows the throughput of the system. The red tape associated with decision making in some firms causes long delays and other problems.

Society's greatest concern about the Information Revolution is the privacy and security of data and information. This is a classic example of security-motivated projects. Data and information have become important resources. Access to data and information may be controlled to some extent by government legislation and corporate policy. Because policies and laws are in a constant state of change, many systems projects are triggered by such changes.

Efficiency Analysis

Efficiency can be confused with **economy**. However, whereas economy is concerned with the amount of resources used, efficiency is concerned with how those resources are used with minimum waste. Efficiency is defined as output divided by input; therefore, efficiency problems and opportunities seek to increase output, decrease input, or both. The commodities to be increased or decreased can be people, money, materials, or any other resource. The idea is to get more from less or at least to get more out of what you have. Consider this example:

> A manufacturing facility consists of 125 workstations of various types. Different products go through different types of workstations during production. Management is concerned with the need to expand production, but there is no money to expand facilities (did you recognize the throughput nature of this problem?). Management has observed two major limitations in current operations. First, separate orders for the same product are not consolidated. This causes workstations to be set up and broken down for the same product several times each day. Second, management has noticed that some workstations seem to be idle during some parts of the day and overworked during other parts of the day. Obviously, the production scheduling and control system is not making efficient use of production workstations.

Service Analysis

Service improvements represent a diverse category. Projects triggered by service improvements seek to provide better service to the business, to its customers, or to both. Improved services are intended to increase the satisfaction of customers, employees, or management. Like our other categories, service improvements may be intended to solve specific service problems, exploit opportunities to improve service, or fulfill a management directive. Service improvements are intended to improve one or more of the following:

- *Accuracy.* Accuracy is concerned with correctness of processing results. For instance, a company may want to reduce the number of billing errors on customer invoices. We've all heard stories about customers who get a $100,000 phone bill by mistake. Another example is order processing. Instead of sending the customer 10,000 pencils, the system sent 10,000 boxes of pencils, each box containing 100 pencils. Or instead of sending part A-4666-L-G (man's tweed sportcoat, size 46 long, color gray), a customer receives D-4666-L-G (woman's dress, size 6, long sleeve, color gray).

- *Reliability.* No, reliability is not the same as accuracy. Whereas accuracy is concerned with *correctness*, reliability is concerned with *consistency* of processing and results. For example, an order-processing system may be denying credit to some customers but allowing credit to others with equivalent credit ratings and payment histories.

- *Ease of use.* Today, much concern is expressed about the *user friendliness* (or, as is more often the case, lack of user friendliness) of computer-based systems. A system, whether manual or computerized, should be as easy to use as possible. Many projects are initiated to improve the ease of use of computer systems. Similarly, many projects are initiated to overcome manual systems that have become too complex and awkward.

- *Flexibility.* Flexibility is concerned with a system's ability to handle exceptions to normal processing conditions. For instance, management may encounter a situation in which installment payments are promptly posted to customer accounts. However, prepayments or overpayments against accounts are sent through a lengthy process that delays customer orders even though those customers are not delinquent in their payments. The system isn't flexible enough to respond to special payment alternatives.

- *Coordination.* A business system consists of many functions that must coordinate their activities to achieve goals and objectives. The desired result is *synergy*, meaning that the whole organization receives a benefit greater than the sum of the parts. A classic example of coordination occurs between the production and inventory functions of a business. When production is scheduled, it is important that raw materials be delivered to the workstation at the scheduled time. However, the workstations may not have enough floor space to hold an entire day's requisition for raw materials. Clearly, an information system is needed to coordinate the flow of raw materials to workstations.

It should be obvious from the preceding discussion that the categories of the PIECES framework are related. Any given project can be characterized by one or more categories. Furthermore, the cause of any specific problem may be another problem itself. PIECES is a practical framework. As we study the analysis phases, you'll learn how to apply it.

How to Survey Feasibility of the Project and Study the Current Information System

An engineer would not design a bridge without thoroughly studying the environment that the new bridge will occupy. An architect would not design a building without studying the land and the laws that govern the use of that land. A good business manager will not impose numerous policy changes without understanding existing policies and their implications. Unfortunately, many systems analysts and programmers try to build new information systems without thoroughly studying existing systems. As a result, many information systems fail to solve existing problems, meet the needs of end-users, or provide cost-effective solutions to problems.

Purpose and Objectives of the Survey and Study Phases

The survey and study phases serve the same basic purpose—to understand the current system and its problems and to determine if solving the problems would be beneficial. The only difference between the two phases is the degree of understanding desired and the amount of time spent in the phase; the survey phase is a **preliminary investigation** of the system, whereas the study phase is a much more **detailed investigation**. To draw a clearer distinction between the survey and study phases, let's examine a few trends.

The computing industry is experiencing several trends. The cost of microcomputers is decreasing, and thus microcomputer sales are rising. Also, the average end-user's level of computer literacy is increasing. Because of these trends, the computing industry is currently experiencing an increase in the number of applications project requests. This will only add to the burden of most organizations, which already have a three- to four-year backlog of project requests. One might ask how decisions are made as to which project requests will be undertaken and which will be placed in backlog. Surely the steering committee doesn't make that decision solely from the project request itself. A typical project request (see Figure 6.2) is usually inadequate for making decisions concerning whether to dedicate scarce computing resources. To help the steering committee better evaluate a project request, systems analysts are frequently required to conduct a preliminary investigation or survey on the project request. This investigation is intended to gain a very general feel for the size and scope of the requested project and degree of urgency for solving problems, exploiting opportunities, or fulfilling directives. The analysts typically spend a limited amount of time—for example, 7 to 10 days—conducting the preliminary investigation. Afterward, the analysts' findings and recommendations are presented to the steering committee, which then makes a more informed decision as to whether the project should be rejected, accepted but placed in backlog, or immediately followed up with a detailed study.

```
REQUEST FOR SYSTEM SERVICES                    FORM 100

SUBMITTED BY  Barbara Rushin              DATE    7-25-90

DEPARTMENT    Transportation Fleet Services

TYPE OF REQUEST  [x ] New system
                 [  ] System redesign
                 [  ] System modification

PROBLEM STATEMENT (attach additional documents if necessary)

We would like a cost accounting system that makes every

vehicle in the fleet a cost center. Currently we have no

way of attributing all direct and indirect departmental costs

to any vehicle. Hence, we are unable to determine when a

vehicle's costs are exceeding the benefits for that vehicle.


REQUEST FOR SERVICE TO   Develop a system that allows all direct

and indirect costs incurred by this department to be assigned to

specific vehicles. System should generate monthly vehicle

costing reports and summaries.


ACTION (to be completed by Steering Committee)

   [xx] Request Approved   Assigned to  Wayne Tatlock

                           Start Date   as soon as possible

   [  ] Request Delayed (backlogged) Until

   [  ] Request Rejected for Reason:
```

◄ **FIGURE 6.2 Typical Project Request** A typical project request is often inadequate for making decisions concerning whether to dedicate scarce computing resources. To help the steering committee better evaluate a project request, systems analysts are frequently required to conduct a preliminary investigation into the project request.

Because of their similar purposes, to gain an understanding of the current system, these two phases are examined together in this section.

Identify End-Users in the Current System During the survey and study phases, it is important to identify all end-users in the current system. The formal organization structure is frequently documented in the form of an organization chart. The analyst must be careful to update organization charts instead of accepting them as true indications of the reporting relationships in the business. It's important to note that most organization charts do not show the clerical and service staff workers, important participants in any project.

Each end-user's role and interest in the system should be understood. The analyst must identify two specific groups of end-users: those who *use* the system and those who are or might be *affected by* the system. The latter group includes all those whose jobs are affected by the inputs and outputs of the system being studied. For instance, the accounting system in most businesses will be affected by virtually every other system. It is important to take these people into account because they may be affected by any changes you make to the current system. All of them may offer valuable information.

Analyze the Business Aspects of the Current System It is important for the analyst to identify and analyze the purpose, goals, objectives, and policies of business systems. Recall that these elements form the business dimension of the pyramid model. Even more important, the analyst should determine how well the current information system supports that business. Are the objectives directed toward achieving the goals? If not, why not? Are the activities of the end-user consistent with the business? The objective of the study phase is to isolate points at which the information system is inconsistent with the business. For example, most production systems establish cost standards for labor, materials, and overhead. If these costs are being exceeded, the analyst should try to determine if the standards or goals are reasonable or if production management's policies cause excessive costs through lack of control.

It is equally important to understand whether the information system being studied is in harmony with the business as a whole. A system might be *self-serving* and not in the best interest of the business as a whole. Consider, for instance, a purchasing system. To reduce inventory costs, a purchasing manager may be ordering excessive quantities of materials, thus obtaining lower unit prices via quantity discounts. This policy may adversely affect the inventory system, which will experience a higher inventory carrying cost because materials will remain unused for a longer period of time.

Identify and Analyze the Information Systems Functions Provided by the Current System During the survey and study phases, the analyst should identify and analyze the information systems functions provided by the current system. The analyst must identify all of the transactions currently processed and any problems, opportunities, or constraints that exist relative to these transactions. The analyst should not restrict the study to transactions processed on the computer. Manually processed transactions are equally important!

Analysts should also study all the information and reports currently being generated and used by and for end-users in the current system. Emphasis should be placed on how the information is generated and used—or not used—as well as on specific problems with the information. For instance, information may be termed incomplete, inaccurate, untimely, or inadequate.

Finally, the analyst should study the decisions made by the end-users when performing their jobs. How do these workers get the information needed to make those decisions? How do they use this information? Is the information sufficient for all decisions? Be on the lookout for data that is being collected and stored for no specific reason. This data may be necessary to support the ad hoc decision-making needs of the end-users.

Identify and Analyze the Components of the Current Information System

Recall that *analysis* is formally defined as the separation of a whole into its parts. The decomposition makes it easier to critique the whole. During the study phase, we'll perform analysis by identifying and evaluating the system components. What are the responsibilities of each person in the system? How do all the end-users interact? What methods—computerized and manual— are used to process data and information? What step-by-step procedures are used? What files and databases—manual and computerized—exist in the system? What hardware supports the existing system? Is the hardware adequately supporting the system? What software supports the current system? Are the end-users satisfied with these programs? If not, why not? It is extremely important to understand how all these components interact to make the current system work or not work.

How to Complete the Survey and Study Phases

So far, we've discussed *what* you need to do in the survey and study phases. In this section, we want to discuss *how* you perform the two phases. Let's identify and discuss the specific tasks, documentation, and skills for these phases.

Figure 6.3 depicts the survey and study phase tasks and their documentation. Each horizontal bar is a task. A vertical line indicates one specific point in time, usually a date. Immediately, you should notice that the tasks are not sequential; they overlap. On any given date, you will likely be doing more than one task. Let's study each task in greater detail.

Task: Survey the Feasibility of the Project   The survey phase was previewed in Chapter 4. Recall that the survey phase is sometimes called a *preliminary study* or *feasibility study.* The survey phase ranks project requests so that time

Survey and study phases

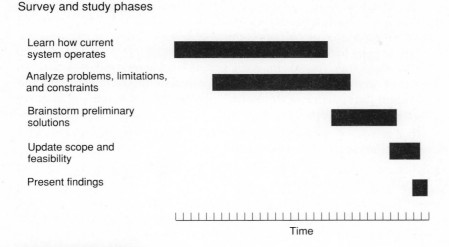

FIGURE 6.3 **Feasibility Survey and Study Phase Tasks** The feasibility survey and study phases can be broken down into the tasks suggested on this diagram.

is spent on those projects most valuable to the business. In many organizations, a **steering committee** of business managers decides which will be developed and which projects will be backlogged for later dates. The purpose of the survey phase is to provide the steering committee with information with which to make those decisions.

The survey phase is triggered by a project request. The analyst is asked to present a feasibility survey to the steering committee. How is the survey accomplished?

The analyst normally conducts a first-contact interview or meeting with the key managers or end-users of the current system. At this meeting, the analyst attempts to "get a feel" for the current system and the problems. Specifically, the analyst seeks answers to the following types of questions:

- What is the nature of the problems, opportunities, and directives? What led the end-user to make the request? Note that the PIECES framework may prove useful for categorizing the problems, opportunities, and directives.
- What are the end-user's goals for and expectations of the project?

- What is the scope of the system? (This may change during the detailed study, but you should try to define what you think the scope to be at this time.) Scope can be thought of as follows:

  —To what key events, inputs, or data must the system respond? Try to get a sense of the timing and volume.
  —What responses or outputs is the system expected to produce? Again, try to get a feel for timing and volume.
  —To what other systems must the system—computerized or not—connect? All systems are part of a bigger system, the business itself. The bigger the system you are studying, the greater the number of its likely connections to other systems.

- What personnel are directly or indirectly involved with the system? Who will be the principal end-user contact(s)?

- What are the possible negative effects of the problem or the anticipated benefits of the opportunity?

- Are there any constraints on the project—such as deadlines, policies, contracts, budget, or the like—that can't be broken, or computer systems that can't be changed?

- Does anyone have any potential ideas for solving the problems or exploiting the opportunities? (It is important to stress that any ideas will be treated as *preliminary* ideas!)

Given the answers to these questions, the analyst performs the first of many cost-benefit analyses on the project. The purpose of a cost-benefit analysis is to determine if the business benefits to be realized by continuing the project are greater than the costs that are likely to be incurred if the project is continued.

At this early stage of the project, costs are difficult to pinpoint. Why? Because we haven't (and shouldn't have) addressed solutions. Still, we know how much the problems and limitations are costing the business. We can assume that, by solving the problems and eliminating the limitations, we will realize a benefit *at least* equal to those costs. And we can estimate whether those benefits, measured over some reasonable lifetime for the new system, will exceed the costs of developing the new system. If so, the project should be continued. Thus, the cost-benefit analysis, at least for the survey phase, is a measure of the urgency or worthiness of the project, both as a stand-alone project and relative to other proposed projects.

The feasibility survey phase also serves other uses. It helps measure or establish the level of management commitment to the proposed project. It also provides the analysts with a sense of the computer literacy of the users.

The findings of the survey should be presented to management or the steering committee in the form of a report called a *feasibility survey.*

Task: Learn How the Current System Operates   The next phase is a *detailed* study of the current system. The first task of the detailed study is to learn how the current system operates. Note in Figure 6.3 that this task doesn't start until the previous task is completed and approved. We need to understand what the current system is doing and how the system is doing it before we try to analyze problems and opportunities or define solutions.

A significant amount of factual data about the current system, problems and opportunities, and needs should be collected during this task. Because the end-users can't anticipate which facts you need and when you need them, recent evidence suggests that **fact-finding** is a very important skill for this task! In particular, you should use fact-finding techniques. These techniques are not specific to analysis; therefore, they are covered in Part Four, Module B of the book.

Fact-finding is virtually useless without **verification**, and so it is frequently useful to document the current system in some fashion. Documentation techniques for systems are abundant; however, the evidence now suggests that you must be careful not to overdocument existing systems. Still, some verification documentation of data and processing is considered mandatory for analysis. You'll learn several useful documentation tools in the next few chapters.

Task: Analyze Problems, Limitations, and Constraints in the Current System   Referring again to Figure 6.3, you will see the next task: analyze problems, limitations, and constraints in the current system. Note that this task, like fact-finding, also occupies your time through most of the study phase. You are constantly analyzing your facts.

You might be asking, "Weren't problems identified in the survey phase?" Yes, they were. But as is usually the case, these problems were only symptoms, frequently symptoms of other problems, perhaps not well known or understood by the end-users.

Problem analysis is a difficult skill for beginning analysts. Experience indicates that most new analysts as well as many experienced analysts try to solve problems without analyzing them. If we were to ask you to state a problem, your response would probably include the words *we need to* or *we want to*. Do you see what's wrong? You are stating the problem in terms of a solution. You must learn to state the problem, not the solution. Furthermore, you need to analyze the problem in terms of its causes and effects.

The PIECES framework becomes most useful at this point during the project. As you collect facts, note problems and limitations according to PIECES. Remember, a single problem may have implications in more than one category of PIECES. Don't restrict yourself to problems and limitations noted by end-users. You may also see potential problems! Next, for each problem, limitation, or opportunity, ask yourself the following questions and record answers to them.

1. What is causing the problem? What situation has led to this problem? Understanding *why* is not as important. Existing systems usually evolved randomly, and it is pointless to dwell on history (although you must respect it since your end-users may have played an important role in the development of these systems).

2. What are the negative effects of the problem or failure to exploit the opportunity? Learn to be specific. Don't just say, "excessive costs." How excessive? You don't want to create a $20,000 system to solve a $1,000 per year problem.

3. If the effect is another problem, repeat steps 1 and 2.

This technique is called *cause-effect analysis*. Why do it? Because through cause-effect analysis you identify true problems, justify their seriousness, and improve your own credibility. If you learn to do this kind of analysis well, and learn to present it effectively, you will find end-users much more receptive to your solutions. You will probably also be labeled as management material, since this type of true analytical ability is much sought after.

Task: Brainstorm Preliminary Solutions   Even during this early phase of systems development, it can be useful to generate some ideas about possible solutions to problems. Be careful not to commit yourself to solutions this early, though. By definition, *brainstorming* defines possible solutions without evaluating them. No solution is considered too outlandish or infeasible. Brainstorming can help you educate the end-users about the potential for computing. Although commitment is discouraged, end-users' ideas should not be discouraged or analyzed. Note them for use in later phases.

Task: Update Scope and Feasibility   Sometimes the study phase reveals that the system is more complex than originally believed. Consequently, scope should be reevaluated at this point. If the project is larger than we first thought, we have two basic options. First, the scope could be reduced so that the project might still be completed on time and within budget. Scope reductions may take the form of dealing exclusively with highest-priority subsystems or most costly problems, presumably with an eye toward eventually expanding the system to include those subsystems or problems that were dropped. Second, we could adjust the schedule and budget to allow for greater time and costs.

This task also includes the second of our cost-benefit or feasibility analyses. The detailed study phase has taught us more about the current system, and we can now update feasibility estimates. Is the project still feasible?

We haven't committed to any solution, so we still don't know what the new system will cost. But we now have a thorough understanding of the problems, limitations, and opportunities for improvement. And we should have a better estimate of how much the problems and limitations in the

Study Phase Report

I. Introduction
   A. Purpose of the report
   B. Background of the project
   C. Scope of the project and the report
   D. Structure of the report

II. Tools and techniques used to complete the study

III. Findings
   A. Walkthrough and evaluation of the current system
   B. Summary of problems, limitations, and constraints

IV. Recommendations
   A. Preliminary solutions and ideas
   B. Updated feasibility assessment

V. Conclusions

VI. Appendixes (may include detailed systems documentation and sample forms)

FIGURE 6.4   The Study Phase Report   This is an outline of a typical report written by the analyst at the end of the study phase.

current system are costing the business, including intangible costs. The sum of those costs, projected over the lifetime of the new system, equals the updated benefits to be derived from the new system.

Let's say, for example, that the new system will provide $100,000 per year of benefits. If the projected lifetime of the new system is five years, we can ask, "Is the solution likely to cost us more than $500,000 [$100,000 $\times$ 5]?" Once again, we see that the ability to perform a cost-benefit analysis is an important skill for this task. As a result of this analysis, we can cancel the project, reduce its scope to make it more feasible, or continue it.

Task: Present Findings   The end product of the study phase tasks is the **problem statement**. These findings may be presented orally or in writing, perhaps both. A suggested outline of a study phase report is presented in Figure 6.4. Be careful not to offend end-users in the presentation. Tact is necessary when discussing problems. A good strategy is to always precede critical analysis with positive analysis—systems strengths before systems weaknesses! Once again, we see that presentation and communications skills are very important. The approved problem statement is passed on to the definition phase.

How to Define End-User Requirements for an Improved System

Many analysts make a critical mistake after completing the study of the current information system. The temptation at that point is to begin looking at alternative solutions, particularly computer solutions. The most frequently cited error in new information systems is "The system doesn't do what we wanted (or needed) it to do." Did you catch the key word? It's *what*! Analysts are frequently so preoccupied with the computer solution that they forget to define the business requirements for the solution. The phase *defining end-user requirements* is critical to the success of any new information system!

Purpose and Objectives of the Definition Phase

The purpose of the definition phase is to identify *what* the improved information system must be able to do without specifying *how* the system could or will do it. We can achieve the goal of the definition phase by accomplishing the following objectives.

Involve All End-Users in the Information Systems Requirements   During the definition phase, the analyst should actively involve all of the end-users who were identified during the study phase. It is especially important to give end-users at every level of the organization the opportunity to define goals, objectives, and information systems needs.

You will probably discover that, like you, end-users will have difficulty specifying what they want or need without considering how to meet those needs. Don't discourage the discussion of computer alternatives, but filter those ideas into a mental suggestion box and focus on the underlying business requirements. Also, learn to always ask *why* something is needed or wanted. For instance, if a manager asks for a report that compares budgeted sales for each product against actual sales, ask why that information is needed and how the information will be used. If the reason is to identify those products not selling as well as projected, perhaps an exception report identifying only those products would be better.

Review and Refine the Business Requirements for the New Information System   How will you know if the new system, when implemented, is successful? Think about it. End-user requirements can be fulfilled and still lead to failure. Requirements are for the information system. The information system is for the business! A better evaluation criterion for the information system is to ask if that system helps fulfill the business's purpose, goals, objectives, and policies.

We need some criteria by which to evaluate end-user requirements. The data and information that the end-users request should be evaluated against those criteria. If an end-user says, "I need certain information on overdue accounts," you should ask why. The end-user may respond, "So I can identify and reduce credit losses!" The reduction of credit losses is a *business* objective, albeit a general one. By defining objectives, we can then measure the value of specific systems requirements. Therefore, during the definition phase, the analyst should reevaluate and refine the business purpose, goals, objectives, and policies. Then we can define information systems objectives that are consistent with the business objectives and policies.

Define the Information Systems Functions to Be Provided by the New Information System   Can end-users define their data and information requirements? For many of you, one of the most aggravating experiences will be the "I don't really know" response to your "What do you need?" question. Be honest! You probably assumed that end-users *can* define their requirements. Right? Not always. End-users *can* define most *transactions* because they already exist. And many *reports* and *inquiries* can be defined because the end-users already use or know how they'll use the information. These are **information requirements** for the new system.

But there are reports and inquiries the end-users cannot define. Every manager encounters situations that could not be predicted and that call for decisions. Because the decisions that will be called for can't be predicted, the information needed to help the manager make those decisions can't be predicted. Fortunately, you can probably determine what types of data describe the business environment. That data, in some format, will probably be the solution to the manager's needs. These are the **data requirements** for the new system.

Define the Noncomputer Components of the New Information System   Careful! Many of the components of the information system should *not* be considered during the definition phase. For instance, we should not consider the computer equipment or program components at all—they specify how a system works, not what it does. Furthermore, information systems methods should similarly be ignored. And here's a real surprise—it is also premature to consider the role of people in the new system. Do you see why? We are only interested in what the system must do, not who or what does it. The only relevant components are data, data storage, procedures, and information.

What data (transactions) must be processed? What information—reports and inquiries—must be produced? What data must be stored? We do not

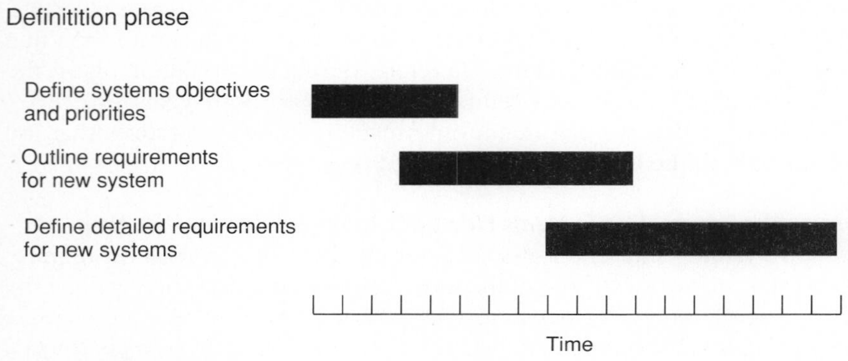

Definitition phase

Define systems objectives
and priorities

Outline requirements
for new system

Define detailed requirements
for new systems

Time

FIGURE 6.5 Definition Phase Tasks The definition phase can be broken down into the tasks suggested here.

consider how or where that data is to be stored. Finally, what procedures and policies transform the data into information? For instance, a credit policy may exist for approving orders. These are the only information systems components that concern us during the definition phase.

How to Complete the Definition Phase In this section, we identify and discuss the specific tasks, documentation, and skills for the definition phase. Figure 6.5 illustrates the definition phase tasks. Once again, we call your attention to the fact that the tasks are not necessarily sequential, with opportunities for overlap. Also notice that we have placed the tasks into time perspective with the study phase tasks.

Task: Define Systems Objectives and Priorities Your attention should now shift from the current system to an improved system. Define systems objectives and priorities first. The input to this task is the problem statement from the study phase. The *problem statement* identifies constraints that might limit the requirements and solutions. *Constraints* are situations that cannot be controlled or changed. Given the constraints, identify specific objectives to support the business application and its end-users.

Objectives should not be stated in terms of inputs, outputs, or processing. Instead, objectives should be precise, measurable statements of business performance. The objectives define the expectations for the new system. For example:

Reduce the number of uncollectible customer accounts by 50 percent within the next year.

Increase by 25 percent the number of loan applications that can be processed during an eight-hour shift.

Decrease by 150 percent the time required to reschedule a production lot when a workstation malfunctions.

Did you notice how the objectives stated the specific outcome? This gives us a standard against which we can measure the data and information requirements that will be defined in the next task. Your objectives should address the problems and limitations identified during the study phase.

You should also have noticed that the objectives do not specify computers or a computer solution. We know the end-users want a computer. But how will that computer improve the business? That is the more fundamental question.

Try using the PIECES framework for objectives. It worked well for identifying problems and opportunities. Presumably, the requirements should address those problems and opportunities, with appropriate allowances for changes in project scope since the study phase. What are the specific performance objectives? Information and data objectives? And so forth.

Finally, you also need to consider the possibility that not all objectives may be met by the new system. Why? The new system may be larger than expected, and you may have to reduce the scope to meet a deadline. Rank the objectives in order of importance. Then, if the scope must be reduced, the higher-priority objectives will tell you what's most important.

Task: Outline the Requirements for the New System   How do you write an English composition or research paper? Hopefully, you outline the paper first. Why? So you can maintain a perspective on the *whole* paper. Otherwise, the paper may have no theme or direction. An information system presents the same dilemma. Therefore, the second task in the definition phase is to outline the requirements for the new system. Avoiding details, we define what inputs (data), outputs (information), and processes are needed in the new system.

Although the outline can be in narrative form, analysts have found pictures or **models** more useful. A picture is worth a thousand words. **Systems modeling** is one of the most important skills to be learned by the analyst. Models communicate our understanding of requirements back to the end-users. They also serve as a starting point for identifying possible solutions. There are two widely used systems analysis modeling techniques: data modeling and process modeling. All systems store data. It pays to understand and organize that data. This skill, called *data modeling*, is taught in Chapter 8. Similarly, all systems

THE NEXT GENERATION

Group Analysis: Joint Application Design, a Technique for Accelerating the Analysis Process

Systems analysis, as defined in this book, can be the most frustrating process in the systems development life cycle. On the one hand, we know that shortcuts in the survey, study, definition, and selection phases can have serious detrimental impact on later phases. The cost of correcting errors during analysis increases exponentially as you progress to design and implementation.

On the other hand, analysis is the most abstract phase—that is to say, the product of analysis is, for the most part, a paper specification. Users can become frustrated with the process since it does not produce a product that they can identify as "progress" toward a new system.

Question: Is there any technique or method that can accel-erate the analysis process without taking shortcuts through the analysis phases? Answer: There are several techniques that change the user-analyst interaction model to accelerate the analysis (and design) processes. Arguably, the most well-known of these techniques is IBM's JAD, Joint Application Design.

JAD-like techniques attempt to overcome the most serious bottleneck in analysis, *fact finding*. Fact-finding, covered extensively in Part Four, Module B, is the process of collecting facts, opinions, and priorities from the user community. During analysis, these facts, opinions, and priorities address the problems and opportunities in the current system (survey and study phase), user requirements (the definition phase), and technical solutions (the selection phase).

What bottlenecks? Fact-finding generally consists of some combination of research, document sampling, work sampling, observation, questionnaires, and interviews (again, for more information, see Part 4, Module B). The bottleneck occurs on two fronts:

- The time required to carry out these activities. There are usually many more users and managers than analysts.

- The conflicting information gathered as a result of these activities. It is natural for people to have conflicting views, opinions, and priorities. Analysts generally compensate with follow-up interviews, which further delay the process.

These are hardly new problems. And neither are the class

process incoming data to produce outgoing data. This skill, called *process modeling* or data flow modeling, is taught in Chapters 7 and 9.

What other skills are necessary to perform this task? Because you need to solicit objectives and requirements from end-users, fact-finding techniques are once again essential. Finally, because we need to review models with the end-users, written and oral presentation techniques are also critical.

Task: Define Detailed Requirements for the New System   Given the general requirements identified in the previous task, we should also define detailed requirements for the new system. We still want to place our emphasis on what the end-user wants rather than on how to do it! For example, if the end-

THE NEXT GENERATION

Group Analysis: Joint Application Design, a Technique for Accelerating the Analysis Process, Continued

of solutions proposed in this box feature. What is new is that *group-oriented techniques* seem to be enjoying renewed interest for the next generation of systems analysis and design.

JAD-like techniques replace much (but not all) of conventional fact finding with group-oriented approaches. The concept is simple. Get all (or most) of the users, managers, analysts, and other interested parties in a single room for extended group sessions that more quickly isolate facts, opinions, and priorities. Analysts serve as the organizers, recorders, leaders, and mediators of the sessions. Ideally, major phases can be reduced to a few work sessions.

To succeed, JAD-like techniques require several conditions. First, management must be willing to release workers from their day-to-day jobs (or

pay overtime) to participate in sessions (usually ½ day per session). Second, management must be willing to participate in the sessions themselves. Management must also foster an environment of cooperation and listening when working with subordinates during the sessions. Finally, recorders and leaders must be well trained to focus and redirect discussion, and mediate conflicts and disputes.

The renewed interest in JAD-like techniques can be attributed to the impact of improved modeling methods (to be covered in Chapters 7 through 9), and CASE (computer-assisted systems engineering) technology.

JAD-like techniques are making use of CASE technology to develop and display models during the actual work sessions

(using large-screen projectors connected to CASE workstations). JAD-like techniques are also being revised to incorporate the latest modeling techniques and strategies.

If this renewed interest continues, we may soon witness a new de facto standard in systems analysis—rapid problem analysis, requirements definition, and technology choice through extensive user participation in group-oriented systems development methods. Some people believe that analysts could very well become employees of the user community as opposed to the information systems community. Their job would be defined as bringing together the two groups to perform analysis and design. Only time will tell.

user identified a need for information on overdue customer accounts, we should now specify the details of this requirement. The following elements might be included in overdue account information:

CUSTOMER ACCOUNT NUMBER

CUSTOMER NAME

CUSTOMER ADDRESS

BALANCE DUE

DUE DATE

DATE OF LAST PAYMENT

PURCHASES SINCE LAST PAYMENT

Notice that we have not specified the media or format of the information. Similarly, we need to define the procedure for identifying the overdue accounts. For example:

> An account is considered overdue if there has been no payment in the past 30 days and there is a positive account balance due. Do not report accounts with a balance due that are being properly disputed by the customer.

Thus we see that the systems models from the last task must be expanded to communicate details about data and processing. These skills are taught in Chapters 10 and 11.

It is in detailed requirements definition that you frequently encounter end-user difficulties. End-users may understand that they need information but may not be able to explain the content of the information. Through **prototyping**, a technique introduced in prior chapters, analysts can frequently get end-users to *recognize* their requirements in lieu of defining them. However, as the minicase to this chapter suggests, the technique is not a magic answer. Prototyping will, however, be a continuing theme in this book.

Requirements are formatted as a **requirements statement** or **project dictionary** that is passed on to the design and selection phases of our systems development life cycle. At a minimum, a requirements statement should specify enough detail to make it possible to proceed to the design and selection phases. Requirements statements are often believed to be large documents that are difficult to prepare and tedious to read. This is unfortunate, but you will learn how to use special tools for documenting and presenting requirements. In addition, you need to learn how to present requirements in pieces, rather than in large, intimidating documents. Finally, it is important to document requirements in terms the end-user can understand.

How to Select a Feasible Information Systems Solution

Given the business requirements for an improved information system, we can finally address how the new system—including computer-based alternatives—might operate. The final solution must also be feasible. Let's examine the selection phase as we did the survey, study, and definition phases. The overriding theme of this section is that you should never automatically go with your first hunch. Good analysts always identify options, analyze options, and then sell feasible solutions based on the analysis.

Purpose and Objectives of the Selection Phase

The purpose of the selection phase is to identify the best solution (*possibly* computer-based) to our end-user requirements. We can define the following objectives for the selection phase.

Define Alternative Solutions in Terms of Information Systems Components During the selection phase, we define those information systems components ignored during the definition phase. These include

- Methods by which inputs and outputs will be implemented—for instance, an information requirement could be implemented as a printed report or a terminal screen display
- Roles and functions of the end-users in the system (how will job descriptions change?)
- Information systems methods and procedures—for instance, on-line versus batch processing
- Computer equipment required to implement the alternative solutions
- Computer programs that must be programmed or purchased to implement the alternative solutions
- Files and databases that will need to be designed, built, and loaded for the alternative solutions
- Internal controls that need to be installed to ensure security and reliability of the solutions

Evaluate the Impact on the End-User As was the case in the definition phase, all end-users identified in the study phase should participate in the selection phase. In one sense, this can be more difficult. Why? Because this is the first phase in which we talk about how the computer might be used. Also, many end-users are not familiar with the capabilities and limitations of the computer. Therefore, the analyst must often educate the end-user during the selection phase. On the other hand, some end-users possess some knowledge of computers—perhaps too much knowledge. Such end-users often want to leap before they look. This is increasingly true of end-users knowledgeable about microcomputers. The importance of involving end-users in the feasibility analysis of alternatives cannot be overstressed! They must be willing to live with the final solution. And they usually must justify the expenditure.

Evaluate the Impact on the Business The analyst's principal concern with the business during the selection phase is to ensure that the proposed information system fulfills goals, objectives, and policies of both the business and the project. This is made easier by ensuring that the requirements defined in the previous phase were also consistent with the business.

Evaluate How Well Alternative Solutions Provide Required Information Systems Functions The transaction-processing, management reporting, and decision support requirements were specified during the definition phase. In the selection phase, you are simply defining different ways to implement those requirements. For example, an information requirement might be fulfilled by a printed report or a terminal-displayed report. A specific transaction

Selection phase

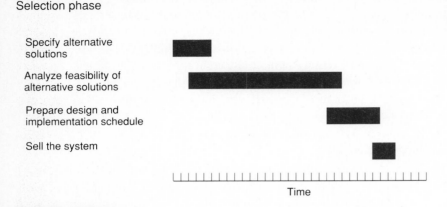

Specify alternative solutions	
Analyze feasibility of alternative solutions	
Prepare design and implementation schedule	
Sell the system	

Time

FIGURE 6.6   Selection Phase Tasks   The selection phase can be broken down into the tasks suggested in the figure.

can be processed as a batch or directly processed on-line. Given such alternatives, you evaluate how well the solutions fulfill the functional requirements. This evaluation can be complicated by end-user preferences and conflicting interests.

How to Complete the Selection Phase

Let's identify and discuss the specific tasks, documentation, and skills necessary for the selection phase. Figure 6.6 depicts the tasks of the selection phase.

Task: Specify Alternative Solutions   There's *always* more than one way to implement the requirements of an information system. Our first task is to specify alternative solutions. This normally involves defining alternative person-machine boundaries. A person-machine boundary separates those tasks that will be performed by people from those that will be performed by computer. At least four levels of decisions must be made:

1. Where should we draw the person-machine boundary? Different alternatives should define different degrees of automation. Which tasks will be manual? Which will be automated?

2. How should the person-machine interface of each individual alternative be handled? Will data be input as a batch or input on-line? Will the information be printed or displayed? There are numerous batch and on-line combinations possible in most information systems.

3. Will the solution be centralized or distributed? In other words, what portion of the alternative will be implemented on your mainframe or minicomputer? What portion will be done on microcomputers? Will data have to be passed between computers? If so, how will this be done?

4. What technology—hardware and software—must be purchased to support each of the alternatives? What technology—hardware and software—might be purchased to support the alternatives? Note that the *make-versus-buy* issue is first addressed at this time. By *make* we mean to design and write the programs in-house. Alternatively, we could *buy* software packages that closely approximate out needs.

The media for all inputs and outputs should be specified. The hardware and software required should be specified. The manner in which data will be stored should be defined. And the methods and procedures for operating the system must be identified. Many of the skills developed for documenting the current system and defining the new systems requirements are also valuable for defining alternative solutions.

Task: Analyze Feasibility of Alternative Solutions The next task is to analyze the feasibility of each alternative solution. This is the third assessment of cost-benefit feasibility during systems analysis. Prior cost-benefit analyses were limited by our not having defined specific technical solutions. Consistent with our creeping commitment approach to feasibility, we can now perform a more thorough feasibility analysis.

But this feasibility analysis should not be limited to costs and benefits. How should you analyze the feasibility of a potential solution? Most analysts evaluate solutions against three sets of criteria:

1. *Operational feasibility* is defined as the suitability of the solution from the standpoint of the people who will have to use that solution.

2. *Technical feasibility* is defined as the availability of suitable technology to support the solution and adequate expertise to develop the solution. Also, it must be possible to implement the solution within a reasonable time frame.

3. *Economic feasibility* is the classic cost-benefit analysis. Because we are now looking at specific technical solutions, the costs to develop those solutions can be estimated with greater precision. Additionally, the costs to operate the new information system (supplies and expenses) can now be estimated. Finally, the benefits to the business can be estimated with greater

Systems Proposal

I. Introduction
   A. Purpose of the report
   B. Background of the project leading to this report
   C. Scope of the project
   D. Structure of the report

II. Tools and techniques used
   A. Solution generation
   B. Feasibility analysis (cost-benefit)

III. Information systems requirements

IV. Alternative solutions and feasibility analysis

V. Recommendations

VI. Appendixes

FIGURE 6.7 Systems Proposal Report This is an outline for a systems proposal, the result of the selection phase. The system proposal presents the feasibility of alternative information systems solutions and recommends a course of action that will guide systems design and implementation activities.

precision. Whichever alternative solution offers the greatest lifetime benefits minus lifetime costs is the best *economic* solution.

The ability to perform a feasibility analysis, specifically a cost-benefit analysis, is a major skill you need to learn! Because feasibility analysis is a skill that overlaps many other phases and tasks, it is taught in Part Four, Module D.

Task: Prepare a Design and Implementation Schedule for the Recommended Solution Given a recommended solution, you should plan a schedule for the phases that are still to be completed in the project: design, selection, construction, and delivery. Experience will help you estimate the time required to complete the tasks in these phases (described fully in Chapters 12 and 20).

Task: Sell the System Proposed solutions must be accepted prior to final commitment to design and implementation. Both oral and written presentations are frequently required. If you've been using PIECES, then the system can be sold not only on economic grounds (the last task), but also on sound business grounds: *This is how the system will improve performance, information and data, costs (economics), control and security, efficiency, and service*! It can be an excellent strategy.

These recommendations are normally presented in the form of a **systems proposal** or **feasibility report**. This proposal or report is presented to management for approval. A suggested format for the system proposal or feasibility report is presented in Figure 6.7.

Once again, oral and written communications skills, taught in Part Four, Module C, play a significant role in your success.

Let's conclude the selection phase tasks by taking another look at Figure 6.6. Note that the selection phase actually begins before the definition phase tasks are completed. Why? Only the general requirements from the definition phase are needed to trigger the selection phase. Furthermore, the completion of detailed requirements is usually dependent on final scope adjustments made during the selection phase. Thus, the selection phase should be completed before the detailed requirements definition task.

Figure 6.8 on page 168 depicts the document and data flows between all of the tasks that comprise the analysis phases. The tasks are represented by rounded rectangles. The documentation passed between tasks is represented by arrows.

Summary

Systems analysis is the most critical process of information systems development. Systems analysis can be defined as problem solving; it consists of four phases that can be successfully completed by applying appropriate skills and carefully addressing each dimension of the information system.

The problem-solving aspects of systems analysis can be augmented through a special framework called PIECES. PIECES categorizes problems and opportunities according to the following groups: performance, information and data, economy and costs, controls and security, efficiency, and service. The chapter demonstrated the use of PIECES for identifying, analyzing, and solving problems in various phases and tasks of systems analysis.

The purpose of the survey—the project feasibility phase, also referred to as a *preliminary study* or *feasibility study*—is to determine the initial feasibility of a project request. The product of this phase is the feasibility survey that is presented to a steering committee for a decision on whether the project should be developed. If the project is approved, the next phase is the study of the current information system. The purpose of that phase is to learn how the current system operates. The analyst documents the current system in some fashion and presents the findings to end-users for verification. The

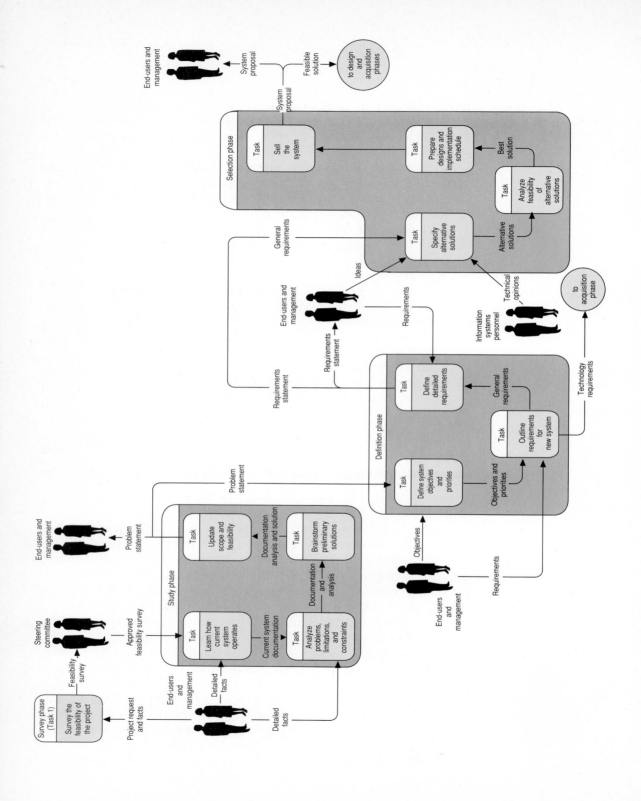

◄ FIGURE 6.8 Documentation Flow Between Analysis Tasks This picture shows each task as a rounded rectangle and data and documentation flows as named arrows.

analyst then studies the current system to identify problems, limitations, and constraints and brainstorms preliminary solutions. Finally, the analyst updates the feasibility estimates and presents the findings as a problem statement or formal study phase report.

The third phase of systems analysis is to define end-user requirements for a new information system. The purpose of this phase is to identify what the new, improved information system must be able to do. The problem statement from the study phase is used to define systems objectives and priorities. General requirements for the new system are outlined in terms of inputs, processing, and outputs. These general requirements are verified with the end-users before specific details—such as content, timing, and volume—are added. The product of this phase is the requirements statement.

The fourth phase of systems analysis is to select a feasible solution from alternative information systems candidates. Several alternative solutions are identified and evaluated in terms of operational, technical, and economic feasibility. The analyst will recommend the best solution to management for approval. The recommendation may be presented in the form of a systems proposal or feasibility report.

A cost-benefit analysis is an integral part of the survey, study, and selection phases. The cost-benefit analysis determines if the expected systems development and lifetime costs for a new system will be offset by the benefits of the new system. The repetition of a cost-benefit analysis allows management to gradually commit to the new system, thus reducing the possibility of implementing costly failures.

In the following chapters, you will be introduced to tools and techniques for accomplishing the systems analysis tasks presented in this chapter.

Problems and Exercises

1. Identify and briefly describe the purpose of the four systems analysis phases.
2. Explain why systems analysis is the most critical process in the systems development life cycle. List some potential consequences of a poor systems analysis.
3. Differentiate between problems, opportunities, and directives. Give examples of each.
4. Characterize each of the following situations as problems, opportunities, or directives. Remember, problems and opportunities are sometimes hard to distinguish. In those cases, describe how the situation might be either a problem or an opportunity.
   a. Management has decided to offer credit to regular customers. Previously, all sales required prepayment.

b. A bank's manager wonders if bank machine transactions executed at other banks can be posted to customer accounts in fewer than the current average of two days. If so, the bank could save several thousand dollars' worth of interest payments each day.

c. A baseball manager's competitive edge might be improved by access to information about a hitter's history against specific pitchers.

d. When a manufacturing workstation breaks down, management finds it difficult to modify the production schedule to reassign work from the broken workstation to workstations that might have capacity. Thus, products don't get produced.

e. Management has decided that all computer files should be integrated into databases to improve access to and flexibility of data.

f. Total cash from customer payments does not equal the sum of payments posted against customer accounts.

5. Using the PIECES framework, evaluate your local course registration system. Do you see problems or opportunities? (*Alternative*: Substitute any system with which you are familiar.)

6. Evaluate the following situations according to the PIECES framework. Do not be concerned that you are unfamiliar with the application. That isn't unusual for any systems analyst. Use the PIECES framework to brainstorm potential problems you would ask the end-user about.

a. The investments officer for ABC Co. has a problem. Currently, its bank contacts ABC's accounting department each day to relay information regarding deposits, check clearing, current and legal bank account balances, and float on recently issued checks. Accounting notifies the investments officer, who does a cash flow projection. Using pencil and multicolumn ledger sheets, this projection takes a couple of hours to complete. These projections are used to estimate capital available for investment in stocks, bonds, futures, and real estate, among other things. Investments are made on a daily basis. The trouble is that the company is too dependent on the bank clerk phoning in the information. If the cash projections go too slowly, valuable time is lost (the stock market opens early in the morning, but ABC is rarely ready to make investments at that time). Considerable capital can be made or lost in less than one hour.

b. National Fund-Raising, Inc., an independent fund-raiser for various nonprofit clients, has uncovered a major problem with its data. It keeps considerable redundant data on past donors and prospective donors. This results in multiple contacts for the same donor—and people don't like to be asked to give to a single annual fund multiple times! Contacts with donors and follow-up are not well coordinated. While some prospective donors are contacted too often, others are overlooked entirely. It is currently impossible to generate lists of prospective donors based on specific criteria, despite the fact that data on hundreds of criteria has been collected and stored. Donor histories are nonexistent, which makes it impossible to establish contribution patterns that would help various fund-raising campaigns.

7. Apply the PIECES framework to each of the following situations. Remember, the categories of PIECES overlap; therefore, any given situation may have implications in more than one category.

a. During the processing of room assignment applications, much of the same data is typed onto different forms at different times.

b. A sales manager knows total sales for each region but can't tell how well each product line is doing in each region.

c. A new record-keeping system allows students to see their test and project scores for any course; however, the system also allows a student to see all of the scores for any student in the same course.

d. The cost of manufacturing a specific product has dramatically reduced that product's profit margin.

e. Warranty claims have increased because defective parts are being used in products.

f. Manufacturing produces too much of some products and not enough of other products.

g. The chief accountant's office must consolidate the accounting statements for several years, calculating a percentage change (plus or minus) for each item on the statements.

h. The cost-accounting office needs to determine what percentage of a product's increasing costs are attributed to higher pay scales, which would be out of management's control, and what percentage to lower labor productivity, which is controllable by management.

i. The marketing department needs to improve the sales staff's ability to make changes to orders that haven't already been shipped.

8. Differentiate between the survey and study phases of the systems analysis process. Why not just begin with the study phase?

9. What components of an information system should not be considered during the definition phase? Why?

10. Differentiate between the definition and selection phases of the systems analysis process.

11. Describe three criteria that should be used to evaluate solutions during the selection phase.

12. What skills are important for a systems analyst to be able to successfully perform systems analysis? How are these skills used in each phase of the systems analysis process?

13. How does the information systems pyramid model aid in systems analysis? Examine each face of the pyramid separately.

14. What is the end product of each systems analysis phase? Explain the purpose and content of each of these products.

15. A systems project has been approved by a steering committee. The time allocated for completing the project is one year. How much time would you allocate to the analysis, design, and implementation processes? Explain your rationale for distributing the time.

16. Explain how the time spent completing the systems analysis of a project can best be managed.

Projects and Minicases

1. A company is considering awarding your consulting firm a contract to develop a new and improved system. But at the beginning, it only wants to commit to systems analysis. It is concerned about your ability to understand its problems and needs. And most of all, it wants to see what kind of computer-based solutions you propose before it contracts with you to do the design and implementation of a new system. Write a letter of proposal that will address their concerns.

2. You have been developing an improved sales tracking system for Kimberly Auto Sales. You have already studied the existing sales tracking system to identify problems and opportunities. You have also defined new requirements for an improved system. Finally, you have defined two alternative solutions that will fulfill the requirements. The first alternative merely fulfills the minimum requirements, processing sales transactions as a batch and producing the required reports. The second alternative processes transactions on-line, with the salesperson entering additional data about the negotiations that preceded the sale. The on-line option also provides immediate access to customer reports and inquiries about types of customers who buy cars, features that are selling well, negotiation strategies that do and don't seem to work, and so forth. After hearing the two strategies described, Mr. Kimberly immediately requested that the second option be designed and implemented. On what feasibility criterion was Mr. Kimberly's decision being made? Why is that decision based on incomplete analysis? What other criteria and questions should Mr. Kimberly address before making the final decision?

Annotated References and Suggested Readings

Wetherbe, James. *Systems Analysis and Design: Traditional, Structured, and Advanced Concepts and Techniques.* 2d ed. St. Paul, Minn.: West, 1984. We are indebted to Dr. Wetherbe for the PIECES framework.

SOUNDSTAGE

The Existing System Is Analyzed for Problems and Opportunities

When we left Sandra and Bob, they had completed a preliminary study of the member services problems at SoundStage Entertainment Club. Now they must do a detailed study of the current member services system.

EPISODE 3   This episode begins shortly after the Steering Committee has approved Sandra and Bob's preliminary study report and authorized a more detailed study of the current system.

Fact-Finding Strategy

Sandra and Bob are planning strategy with David Hensley, Member Services Manager, in Sandra's office.

"Good morning, David!" Sandra was genuinely excited about working on the new project. "I asked you to join us so that Bob and I could explain our strategy for the first phase of our project."

"Great! How are we going to do this?" David asked as he took a seat.

"This phase will be completed in four steps," explained Sandra. "First, we will review the document and work flow in your current member services system. To do this, we will have to gather factual data from your people. After

verifying the facts, we can begin the second step, which is an analysis of the problems, opportunities, and constraints in the current system. The third step will be to generate some preliminary solutions. But we must emphasize that these solutions are only *tentative* at this point in the project and may be revised or discarded later. Finally, we will reevaluate feasibility in order to determine if the project is still feasible. This entire phase should take about two weeks."

David shifted in his chair. "That sounds good, but I do have a concern. My people have jobs to do and those jobs are important. It sounds like you will be requiring a significant amount of their time. We'll cooperate as much as we can, but we have to process the promotional materials, member orders, and membership transactions."

Bob responded, "I'm glad you brought that up! Sandra and I wanted to discuss that issue with you at this meeting. There can be no substitute for involving your people. However, we have a fact-

finding strategy that *will* minimize the amount of time that your people have to spend away from their jobs. We won't take any more of their time than is absolutely necessary. First, we want to collect as much information as we can without interviews. We'd like you to collect copies of all the forms and documents used in member services. Also, if you have any procedure manuals or written instruction guidelines, we'd like copies of them, too. It would be especially helpful if the forms you give us are in various stages of completion. We can learn a lot about the current system by studying these forms."

"And the best part is that my people don't have to give up any of their time!" David added. "I'll put that package together and get it to you today."

Sandra nodded. "That would be great. After we've studied the forms, we'll ask for your permission to observe a couple of your clerks as they do their jobs. It would be a good idea if they know

Episode 3, continued ▶

that we aren't there to evaluate their performance; we just want to see how they handle orders and membership requests."

"I don't think that will be a problem," said David. "I've already explained that the new system will not eliminate anybody's job, but just make them easier to do. They seem excited about having a better way to do things."

Sandra continued, "We'll try to avoid asking questions during the observations. After we've studied the documents and spent some time observing the current system in operation, we'll collect additional facts by using questionnaires and holding interviews. The questionnaires will be done first and will help us to prepare for the interviews. The nice thing about questionnaires is that they can be completed when your clerks aren't busy. We'll give you the questionnaires and let you distribute them. We'd appreciate it if you could return the completed questionnaires within three days. After collecting as many facts as possible by observing and studying the questionnaires, we'll conduct interviews. You can see how this strategy minimizes the amount of time your people have to spend away from their jobs while still providing us with the data we need."

"That sounds great!" David exclaimed. "But don't you have to verify those facts you collect?"

"Certainly!" Sandra replied. "We will verify the facts we collected by drawing pictures of your system. Then, we'll schedule meetings with you and some of your staff to verify our understanding and correct any mistakes. We

promise not to waste your people's time, but you must always remember that the new system is ultimately *their* system."

(Our study continues one and a half weeks later.)

The fact-finding strategy has been working very well. Except for Bob and Sandra's presence in the Member Services Department, there hasn't been any difference in the normal, day-to-day operations.

By studying the forms, documents, and procedure manuals that David provided, Sandra and Bob were able to understand the general *shape*, as they called it, of the system. They were able to document the normal procedures for processing a membership application, a club promotion, and a member order, provided there were no unusual exceptions or abnormalities. Observation of several clerks over a period of time cleared up some misunderstandings and identified a number of unusual conditions that could occur. In order to collect additional facts, questionnaires were designed and distributed. These questionnaires were intended only to eliminate some of the need for direct interviews. Also, the responses helped Sandra and Bob to profile the attitudes of some of the personnel involved.

A couple of interviews have already been conducted. Throughout the entire process, Sandra and Bob have noted problems, opportunities, complaints, bottlenecks, and constraints. But they have not offered their opinions or discussed

solutions. The big picture has unfolded very nicely. With their documentation in hand, Sandra and Bob join some of the Member Services personnel for a walk-through of the current system. We'll sneak in on the meeting . . .

Review of the Current System

Sandra called the meeting to order. "If everyone will take a seat, we can begin. Before we get started, I'll introduce everyone since you work in different areas and may not know each other. Ann Martinelli is the Membership Director. Joe Bosley is the Director of Promotions, and Sally Hoover is the Order-Processing Director. The purpose of this meeting is to review the current data and work flow through the member services system. Bob and I have drawn some diagrams of the system. What we want to do today is to discuss these diagrams with you. It is very important that you point out any errors, omissions, or contradictions that you see." [Note that Sandra did not use intimidating terms like *data flow diagram* that the systems users would not have understood.]

Sandra placed a diagram on the overhead projector (see Figure E3.1). "This is a diagram of the document and work flow through the first-time membership subscription processing portion of the current system. Remember that just because some of the activities are connected, that doesn't mean they have to follow each other; they could be happening at the

Episode 3, continued on page 176 ▶

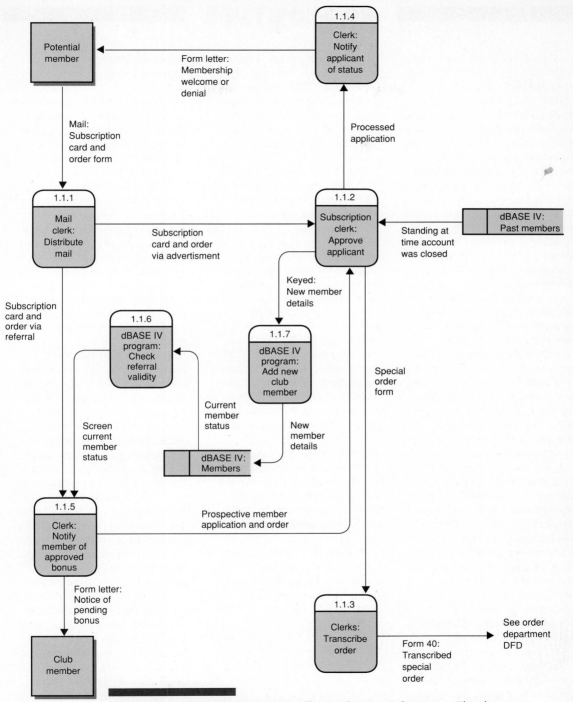

FIGURE E3.1 SoundStage New Member Application Subsystem This diagram illustrates the current new member applications processing functions of SoundStage Entertainment Club. Notice the physical implementation details.

same time. Let's walk through this diagram. A SUBSCRIPTION CARD AND ORDER FORM from a POTENTIAL MEMBER arrive in the mail. A mail clerk sorts the mail into two groups, SUBSCRIPTION CARDS and ORDERS VIA ADVERTISEMENTS and SUBSCRIPTION CARDS and ORDERS VIA REFERRALS. The SUBSCRIPTION CARDS and ORDERS VIA ADVERTISEMENTS are forwarded to the subscription clerk who uses a dBASE IV PAST MEMBERS' FILE to determine their standing at the time the account was closed, if they were previously a member."

"Excuse me," interrupted Ann. "We keep information about past members in file folders in a filing cabinet, not a dBASE file."

"Thanks for calling that to my attention, Ann," Sandra replied. "That's my mistake. I remember your telling me that. It was the current member data that is on a dBASE file. That is exactly the kind of thing we want to verify in this meeting," Sandra said as she noted the correction on the diagram so the change could be made to the next set of formal diagrams. Sandra continued, "If the applicant is approved, the NEW MEMBER DETAILS are keyed into the MEMBERS' FILE using a dBASE IV program, is that right?" Noting a nod of approval, she continued, "The PROCESSED APPLICATION is forwarded to a clerk who notifies the new member with a MEMBERSHIP WELCOME FORM LETTER."

"I think I forgot to tell you that we store the processed applications in an applications notebook for future reference," Ann interrupted again.

"That's easy to fix," Bob said while making the change. "The subscription clerk then sends a SPECIAL ORDER FORM to another clerk, who transcribes the order onto a FORM 40: TRANSCRIBED SPECIAL ORDER, which is then forwarded to the ORDERS DEPARTMENT for processing."

When everyone had agreed that this portion of the diagram was correct, Sandra continued, "The processing of a club member referral subscription is a little bit different. The SUBSCRIPTION CARD AND ORDER VIA REFERRAL are sent to a clerk who verifies that the referring member is a current member in good standing by using a dBASE IV program to check the CURRENT MEMBER STATUS. A PENDING BONUS FORM LETTER is sent to the club member and the PROSPECTIVE MEMBER APPLICATION AND ORDER are sent to the subscription clerk who processes them as we discussed earlier."

"What about the current member's bonus order?" asked Sally. "Normally, the membership staff tear off the bottom half of the referral application form and send it to us for transcription onto a Form 40."

"Oops," said Sandra. "I completely forgot about that. I'll fix it so that the existing member will get his or her referral bonus order."

(You should have the idea by now. We can quietly sneak out of the meeting.)

Analyzing Problems and Opportunities

Three more days have passed. Sandra and Bob have completely revised and verified the Member Services System. Let's join them in Sandra's office and see what they're doing.

"Well, Bob, I think we're just about done documenting the current system. Now we need to review and discuss some of the current system's problems and limitations before we submit our problem statement report."

"Yes," answered Bob. "Let's look at one subsystem at a time, beginning with the membership system. I'll take some notes for the report."

"One of the problems I noticed was the amount of time necessary to look up a potential member in the members' filing cabinet. When I was watching, it took an average of five minutes. Ninety percent of the time the applying member had not been a previous member and wasn't in the filing cabinet. That results in a lot of wasted time," Sandra concluded.

"Noted!" Bob responded. "I thought that the mail clerk spends too much time sorting subscriptions into advertisements and referrals. This results in two different problems. First, some subscriptions are incorrectly sorted and delivered to the wrong person for processing. They have to be sent back, and that increases the amount of time necessary to process the application. Second, when the volume of subscriptions

SOUNDSTAGE

is high, it can take the mail clerk several hours to sort and deliver the subscriptions to the appropriate person. There should be some way we could reduce the number of hours wasted there."

(We will leave the room now.)

Sandra and Bob will do a similar analysis on each subsystem and prepare a formal report that presents their findings to management. This report will include an updated feasibility analysis and an updated list of alternative solutions. The updated list of alternative solutions will focus on the ability of each alternative to solve the problems identified during the study phase. Costs will not become a significant factor until the evaluation phase.

Where Do We Go from Here?

This episode of the running case presented some of the important skills used during the study phase. Before Sandra and Bob tried to figure out what's wrong with the system, they tried to learn as much as possible about the way the system works. First, they collected facts about the system. Fact-finding is a crucial skill for a systems analyst. Because fact-finding is done during several other phases of the systems development life cycle, we have chosen not to include training on this skill in this unit of systems analysis. Instead, you will find the necessary material in Part Four, Module B. You already have all the prerequisites you need to study that module right now. Second, in order to learn about the

system, Sandra and Bob drew diagrams. In Chapter 7 you will learn that the diagram they drew is called a *data flow diagram*—a very important tool that all systems analysts should master. But data flow diagrams are only tools. That's easy to forget. Keep in mind that systems analysis is a people-oriented activity. Also notice that Sandra and Bob had to present their findings. If you count the written report, they made two presentations of their understanding and ideas. Presentations, like fact-finding, require skills that are not limited to systems analysis. Therefore, you will find the module on presentation in Part Four of this book. Again, you have the necessary preparation for understanding this material, which is presented in Part Four, Module C.

CHAPTER SEVEN

Process Modeling

with Physical

Data Flow Diagrams

The Internal Revenue Service

The following case, adapted from "The IRS: How Your Return Is Processed," *USA Today*, January 8, 1986, p. 7A, describes how the typical IRS regional center processes your tax return. The adaptations are intended to simplify the challenge that follows the minicase.

Initially, postal trucks bring mail to the loading docks of the regional center. The envelopes are sent to a mailroom, where they are sorted by a special machine that can handle more than 30,000 pieces per hour. The machine reads bar codes to sort envelopes by type of return—for example, long form versus short form, and whether the envelope contains checks (easily identified by their magnetic ink codes). The sorted envelopes are sent to Receipt and Control, where they are further separated into 27 types falling into three general categories: short forms needing refunds, long forms needing refunds, and returns containing tax payments. From this point, processing requirements are described separately.

Short Forms (1040EZ) Needing Refunds

1. Operators submit forms to an optical-mark reader that scans the returns and stores the data to magnetic tapes for later processing.

2. Tapes are read by the main computer. It determines the correct tax; decides whether a refund should be sent; updates taxpayers' files; and prints letters, notices, liens, and the like. Refund information is sent to the National Computing Center in Martinsburg, West Virginia, which subsequently triggers the Treasury Department to issue the actual refund checks. Letters, notices, and other communications are sent to local IRS sites around the country, from which appropriate information is sent to taxpayers.

Long Forms (1040) Needing Refunds

1. Returns needing refunds are sorted by weight into blocks (20) of batches (100 per block) to be processed as single units, sorted on a movable cart called a *batching cart*.

2. Batches are computer numbered to ensure that no returns are lost or excessively delayed. Batches are forwarded to examiners.

3. Examiners check for and correct errors, code the returns for computer processing, and send back to the taxpayers any returns with incomplete or uncorrectable data.

4. Clerks stamp a document locator number of each return for additional tracking capability as the return moves through the system. Returns, physically on the batching carts, are placed in a holding area.

5. Returns are input to the computer system via CRT terminals. Data is stored on magnetic tape for subsequent processing.

6. Tapes are read by the main computer. It determines the correct tax; decides whether a refund should be sent; updates taxpayers' files; selects returns for possible tax audits; and prints letters, notices, liens, and the like. Refund information is sent to the National Computing Center in Martinsburg, West Virginia, which subsequently triggers the Treasury Department to issue the actual refund checks. Notices and information regarding audits are sent to local IRS sites around the country, from which appropriate information is sent to taxpayers.

Returns (1040EZ and 1040) Containing Payments

1. Examiners check for and correct errors, code the returns for computer processing, and send back to taxpayers any returns with incomplete or uncorrectable data.

2. Returns are entered into the computer via terminals. The computer immediately checks taxpayer calculations and amounts, assigns document locator numbers, and stores the data.

3. Step 2 is repeated using different terminal operators. Data is checked against the first set for accuracy. Error reports are sent to examiners. Accurate data is stored for subsequent processing.

4. Checks are collected for daily deposit into the Federal Reserve Bank.

5. Examiners check for errors [*authors' note*: presumably from step 3], correct the errors they can—correcting the data in the files—and writing the taxpayers for missing information.

6. At this point, the returns follow identical processing as described in steps 5 and 6 under "Long Forms (1040) Needing Refunds."

Challenge

One of the challenges for systems analysts is to communicate their understanding of systems requirements and solutions. To this end, analysts have learned that pictures often communicate better than words. On a single $8\frac{1}{2}$ by 11 sheet of paper, draw a picture to describe the preceding system. If you have difficulty, don't be too concerned. This chapter introduces you to a useful tool.

Discussion

This case study is based on information presented in a newspaper article. You will undoubtedly identify missing facts. What kinds of "what if" and "what happens if" questions can you think of? In the final analysis, the article points to the difficulties of describing systems verbally (even though a floorplan was depicted in the original article).

What Will You Learn in This Chapter?

This is the first of three graphic systems modeling chapters. Modeling was introduced in Chapter 6 as a systems analysis tool. In this chapter you will learn how to use one of the most popular systems modeling tools, *data flow diagrams* (DFDs)—specifically *physical* DFDs. Physical DFDs can be used to document processes and the flow of data and information through an existing system. You will understand physical data flow diagrams as a system analysis tool when you can

1. Document the flow of data through a system using physical data flow diagrams that show not only what the system does, but how it is currently implemented.
2. Describe how data flow diagrams document an information system and how they can be used during the systems analysis phases.
3. Use leveling (also called *explosion*) to break large or complex systems into subsystems that can be more easily verified and understood.

In the last chapter, you learned the importance of studying a system and its problems and opportunities before defining requirements or designing solutions. You can't analyze a system if you don't understand it. Consequently, you study a system to learn it. We learn about a system by studying existing documentation and talking with end-users. (For a detailed discussion of fact-finding techniques, see Part Four, Module B.) But this alone will not suffice. We can't be sure that we understand the current system unless we verify our findings. How do we tell end-users what we *think* we've learned about their system?

We cannot always expect English narratives to communicate our understanding, because language is subject to considerable interpretation. Also, narratives might explain what you know, but not necessarily what you *don't* know. If the reader fails to realize what you've left out, the omission may go unnoticed until too late. We need a different tool!

In this chapter, we study a technique known as process modeling. Specifically, you will learn how to use a process modeling tool called physical data flow diagrams. Physical DFDs are pictures of systems processing and data flow without excessive concern for details. Today, DFDs are the tool of choice for modeling systems. Let's study this tool, its conventions, and its uses.

Modeling a System

As a systems analyst, you constantly deal with unstructured problems. Analysis—at least good analysis—imposes structure on an unstructured problem. One way to structure a problem is to use models. A model is a graphic representation of a system. Models can be built for existing systems—to better understand those systems—or for target systems, to define requirements and designs.

In this chapter we will specifically address physical models. Most analysts draw physical models without even thinking about it. **Physical models** (also called *implementation models*) show not only *what* a system is or does, but also *how* the system is physically implemented. This is especially useful when verifying models with end-users. To them, the system is the sum of its physical properties: the forms, computers, files, file cabinets, people, timing, and so on. A verifiable model must exhibit the same traits.

Systems have a tendency to overwhelm the best of us. Do you recall the information systems components you learned in Chapter 3? The large number of procedures, forms, reports, files, people, and machines can hinder our understanding, unless we learn tools to deal with that complexity. Because we, as computer systems analysts, may ultimately implement processes as programs, we need process models. A **process model** is a picture of the flow of data through a system and the processing performed on that data. These pictures are often easier to read than prose. Process modeling helps us grasp inputs, outputs, processing, and the relationships between processes.

The term *process modeling* comes from the depiction of processes and how they interact or interface with one another. These interactions take the form of data flows between processes. For this reason, process models are also frequently called *data flow models*.

Process modeling is a very popular technique. The most popular process modeling tool is the data flow diagram. This tool's popularity is an outgrowth of a systems analysis methodology called *Structured Systems Analysis and Design*.

Data Flow Diagram Conventions and Guidelines

A **data flow diagram** illustrates the flow of data through a system and the work—or processing—performed by that system. Physical DFDs show how the system is currently implemented, at least so far as this chapter is concerned. Physical DFDs use a visual language. Let's study that language.

Physical Data Flow Diagram Language and Symbols

Physical DFDs are quite easy to read and understand. But before we begin, we should acknowledge an ongoing debate regarding choice of symbols. There are two alternative, but equivalent, symbol sets. Figure 7.1 demonstrates the Gane-Sarson symbol set. Figure 7.2 demonstrates the DeMarco-Yourdon symbol set. The debate over which is best is actually irrelevant. In this chapter, we will use Gane-Sarson symbols; you should use whichever set you or your end-users (or instructor) find most comfortable. We don't suggest that you mix and match symbol sets, though, since this would greatly confuse analysts who are already familiar with uses of one or the other. Study either diagram or both as we describe the symbols.

Both figures demonstrate the visual language of physical DFDs. We have diagrammed a familiar system, the management of your personal finances. The Gane-Sarson and DeMarco-Yourdon symbols are also reproduced below, in the left and right margins, respectively.

Despite its namesake, the emphasis in any DFD is placed on processing—sometimes called *activities*. A **process** is represented by a rounded rectangle (Gane-Sarson) or a circle, also called a *bubble* or a *transform* (DeMarco-Yourdon). Processes transform inputs into outputs. The details of the processing—the logic or procedure—are not shown.

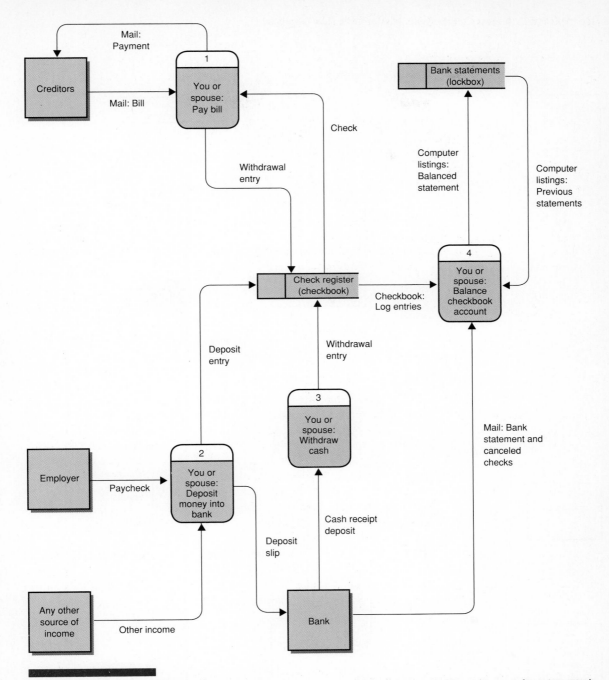

FIGURE 7.1 Simple Physical Data Flow Diagram This is a simple physical data flow diagram for a personal finance system. It uses the Gane-Sarson symbol set, equivalent to the alternate symbols used in Figure 7.2. Note that the processing in the system may occur at different dates and times, perhaps simultaneously. This distinguishes the data flow diagram from the flowchart.

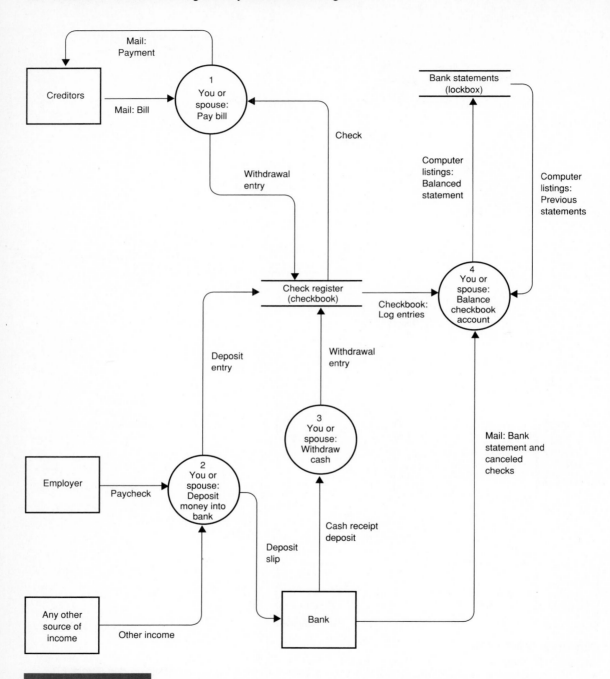

FIGURE 7.2 Simple Physical Data Flow Diagram This is a simple data flow diagram for a personal finance system. It uses the DeMarco-Yourdon symbol set, which is equivalent to the symbols used in Figure 7.1.

Processes on physical DFDs can represent activities performed by people, computers, or machines.

The square (Gane-Sarson) or rectangle (DeMarco-Yourdon) depicts **internal** or **external entities**. These entities define the boundary of the system. They provide the net inputs to, and receive the net outputs from, the system.

As you learned in Chapters 2 and 3, most systems also store data for later use. **Data stores** are depicted as open-ended boxes. Physical examples of data stores include file cabinets, desk drawers, binders and books of data or reports, index cards, in/out boxes, mailboxes, logs, and of course computer files and databases.

Finally, the named arrows depict **data flows**. Data flows represent inputs or outputs such as reports, forms, documents, terminal displays, computer-to-computer data transmissions, letters and memos, data retrievals, file updates, phone calls, and other conversations.

Don't confuse DFDs with flowcharts! Most of you are reading this book after some exposure to computer programming. Program design frequently involves the use of flowcharts. *But data flow diagrams are very different from flowcharts!* Let's summarize the differences:

- Processes on a DFD can operate in parallel. Several processes may be working simultaneously. For instance, you may be balancing your checkbook at the same time as your spouse is paying the bills. This is a key advantage over flowcharts, which tend to show only sequences of processes. In a realistic business system, activities and processing overlap with one another.

- DFDs show the flow of data through a system. Flowcharts show the sequence of steps in an algorithm. DFDs, unlike flowcharts, do *not* explicitly show looping and decision (if-then-else) constructs.

- DFDs can show processes that have dramatically different timing. For example, you make many deposits and withdrawals to your bank account before you balance your account.

Noting these differences, let's study physical DFD symbols and conventions in greater detail.

The Internal or External Entity Every system has a boundary. The first decision you must make in any project is to define the *scope*, sometimes called the *context*, of the system. We'll discuss the technique later; however, internal and external entities define the system's boundaries. These entities are sometimes called *sources* (of data, the net inputs to the system) or *destinations* (of information, the net outputs from the system).

Internal and external entities are people, organizations, and other systems with which your system interacts. They are almost always one of the following:

- An office, department, division, or similar organization unit within a company that, while not directly using the system, either provides inputs to the system or receives outputs from the system. These are *internal* entities. For instance, if you are building a system for the purchasing department, it will likely interface with the people in the accounts payable and inventory control departments.

- Organizations, agencies, or individuals that are outside your company but that provide inputs to or receive outputs from your system. These are *external* entities. Examples include customers, suppliers, government agencies, subcontractors, banks, consultants or consulting agencies, and the like.

- Another system—possibly, though not necessarily, computer-based—that is separate from your system but with which your system must interact. We cannot overemphasize the importance of this type of entity. When building a system, you must realize that it will become part of a federation of systems to which it must connect. Thus, if your system must access files or outputs from another system, especially files and outputs whose contents and structure cannot be changed—a very common constraint—then that other system and its files should appear in your physical DFDs as an external entity. For instance, a purchasing system may draw on the files and outputs from an inventory control and an accounts payable information system. Other systems may be either internal or external to your company.

- One of your own system's end-users or managers. In this case the end-user or manager is a source of an input(s) to or an output(s) from the system. This type of *internal* entity is common to most information systems projects. Let's clarify this point. Most systems produce reports. Some of those reports, called external outputs, leave the system to go to people, departments, and organizations other than your end-users. But other outputs, called internal outputs, are produced specifically for your own end-users and managers. These reports would be used for internal control. A similar analogy exists for net inputs that originate from *within* your system.

A common question is "Should data processing services be depicted as external entities?" If your system is currently supported by information systems, then that computer system and its programs and procedures are within the system's scope, not external to it, and should be depicted by process symbols.

How should external and internal entities be labeled? Labels should be descriptive. If the entity represents a person, use the person's title. Names can be used; however, titles are better since people frequently change positions. If the entity represents a system, label it "<name> system."

To avoid crossing data flow lines on a DFD, it is permissible to duplicate internal and external entities on DFDs. If internal or external entities are duplicated on or between pages of DFDs, they should all be marked as indicated in the margin. This mark alerts the reader to expect additional occurrences of this same entity on the same page or other pages.

As a general rule, internal and external entities should be located on the perimeters of the page, consistent with their definition as a systems boundary. This is not a strict rule, however.

Internal/
external
entity
name

The Data Flow   What is a data flow? Think of the **data flow** as a highway down which packets of data of known composition are allowed to travel. A more subtle type of data flow—those involving flows to and from data stores—requires special explanation. These data flows are used to demonstrate additions to, deletions from, modifications to, and data retrievals from the data stores.

Data flows must be inputs to or outputs from processes. In any case, the key to defining a data flow is *known composition*. Occurrences of the data flow must contain data. This principle is one of the most consistently violated rules of DFDs! Why? The temptation is to use a data flow to illustrate step 1, step 2, and so on. This is *flowchart* thinking! Such thinking results in unnamed data flows with no data composition. How can you avoid this mistake? The following rules will help:

1. Label or name all data flows as *noun clauses*. If you can't name the data flow, it probably doesn't exist. For example, look at Figure 7.3. What should we name the data flow in question? How about GET NEXT REQUISITION? That won't work. Why? Because GET NEXT REQUISITION is a verb clause— it describes a step, not data.

2. Try to describe the data elements that make up the data flow. If you can't name or describe those elements (for example, REQUISITION NUMBER, DATE, MATERIAL REQUESTED, QUANTITY), then the data flow doesn't exist.

Physical data flow names should be descriptive and meaningful to your end-users, who will have to verify your understanding of the current system. Names should include any acronyms and form numbers that will help the end-users immediately recognize that data flow in the model. Suppose, for example, that end-users explain their system as follows: "We fill out a '23' in triplicate and send it to. . . ." A descriptive data flow name might read FORM 23: COURSE REQUEST FORM. Suppose, for another example, we take cus-

FIGURE 7.3 Data Flows Should Have Data Content All data flows should contain data. If they don't, they are probably not data flows and shouldn't be shown. The questioned data flow on this diagram is probably not data; it looks like it might be an instruction like GET NEXT REQUISITION. If so, it should be deleted.

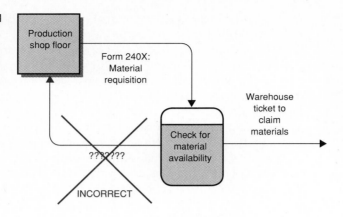

tomer orders by phone. A good data flow name might be PHONE CALL: CUSTOMER ORDER. Again, when drawing *physical* models, we want to communicate the implementation so that it can be recognized by the end-users.

In addition to focusing on nouns, data flow names should contain appropriate adjectives and adverbs to describe the status of flows as they move through systems processes. For instance, consider the data flows to and from a credit check process. The incoming flow might be called FORM 2: ORDER AWAITING CREDIT CHECK. The outgoing flow might be called FORM 2: APPROVED CREDIT ORDER.

In addition to naming conventions, there are other common rules for data flows. Recall that we described a data flow as a *packet* of data. The packet concept is critical. Data that should travel together should be shown as a single data flow, no matter how many documents are physically involved. For example, suppose the phone company sends you an itemized INVOICE of your monthly charges *along with* a PAYMENT CARD that must be returned with your PAYMENT. The INVOICE and PAYMENT CARD should be depicted as one data flow, not two (Figure 7.4). The PAYMENT CARD and PAYMENT are similarly shown as one data flow.

In implemented systems, businesses frequently make use of multipart carbon or carbonless forms. The classic example is the sales draft you sign when you use a credit card. One copy is yours. Another is the merchant's. The third is sent to the bank that gave you the card. At various points during processing, multipart forms *diverge* to follow separate processing paths, often exiting from the system entirely. Show the original and copies as a flow that diverges from a common point on a process. Each copy proceeds to a different destination—process, data store, internal entity, or external entity (see Figure 7.5).

FIGURE 7.4   The Data Flow Packet Concept
If two or more documents travel together, they should be shown as a single data flow.

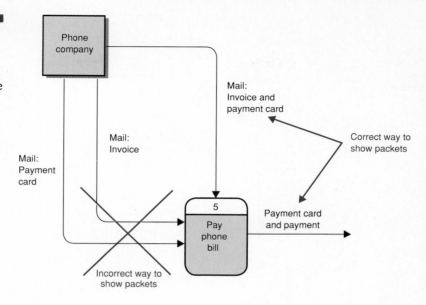

Finally, *all data flows must begin and/or end at a process, because data flows either initiate a process or result from a process!* Consequently, all of the data flows depicted on the left side of Figure 7.6 are illegal. The corrected flows are shown on the right side.

The Process   Processes are work or actions that are performed by people, machines, or computers on incoming data flows to produce outgoing data flows. Processes can be performed by people, departments, robots, machines, or computers (specifically, computer programs). The following naming conventions address the essence of the process.

Process names depend on the level of detail you are trying to depict on the DFD. Some processes represent the entire system, one subsystem, or one function. Other processes represent detailed tasks. It depends on whether the physical DFD is depicting the big picture or a small portion of the total system. Process names should be chosen accordingly. Process names on the general DFDs will, of necessity, be equally general. Examples of general process names are listed in the margin.

On the other hand, the processes on the detailed DFDs should be given more specific names consisting of a *strong action verb* followed by an *object clause* that describes what the work is performed on or for. Examples of detailed names are listed in the margin. Additionally, process names on physical DFDs should identify the processor: person(s), location(s), machine(s), or computer program(s).

✓ **GENERAL PROCESS NAMES**

Accounts Receivable System

Billing Subsystem

Credit Function

✓ **DETAILED PROCESS NAMES**

Credit Customer Account

Record Amount in Dispute

Update Customer Credit Rating

Identify Poor Credit Risks

Summarize Receipts

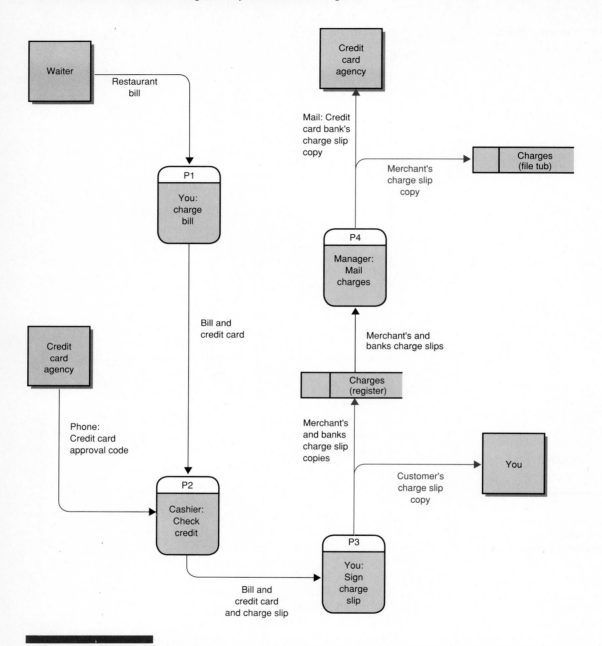

FIGURE 7.5   Diverging Data Flows   Diverging data flows represent copies or subsets of the same data flow that go to different destinations.

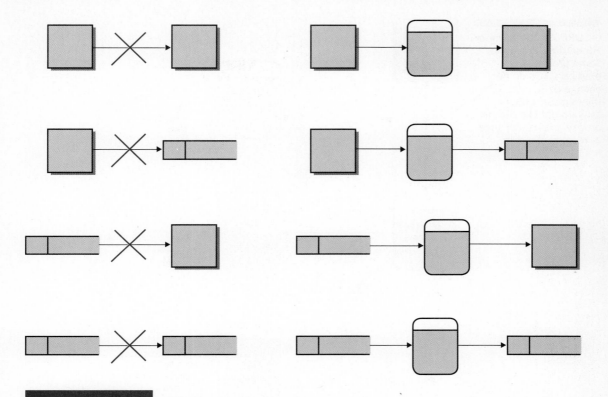

FIGURE 7.6 Common Mechanical Errors All data flows must begin and/or end at a process. All of the examples on the left side violate this rule. The likely corrections are shown on the right side of the figure.

For physical DFDs, you also include processes that do nothing more than route data flows to their next destination. For example, Figure 7.7 indicates that a secretary routes transactions to appropriate locations for subsequent processing.

You may have noticed that a process can have many incoming and outgoing data flows. Some may occur every time the process is executed. Others may occur optionally, under certain conditions. The DFD does not show which data flows are mandatory and desirable or in what combinations they occur. These details are deferred until the policies and procedures for the processes have been specified (see Chapter 10). On the other hand, all processes must avoid the following errors, demonstrated in Figure 7.8:

- What is wrong with process 1? The process has inputs but no outputs. We call this a *black hole* because data enters the process and disappears. If you identify a black hole, you probably just forgot the output.

FIGURE 7.7 Routing or Forwarding Processes
Some processes on physical DFDs do not change data. Instead, they simply route or forward data to another process that will change or react to the data.

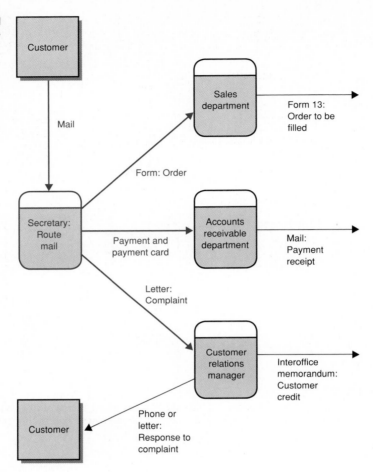

- What is wrong with process 2? The process has output but no input. Unless you are Merlin the Magician, this is a *miracle*! In this case, the input data flows were likely forgotten.
- What is wrong with process 3? The process is what we call a *gray hole*. The inputs to the process are insufficient to produce the outputs.

We conclude our discussion by noting that the inputs for any given process must contain all the data necessary to produce the outputs. This is the final test of any DFD. It can be embarrassing for a programmer to notify you that one or more fields you've requested on an output cannot be produced from the inputs and files that you specified for the program (a program is a process).

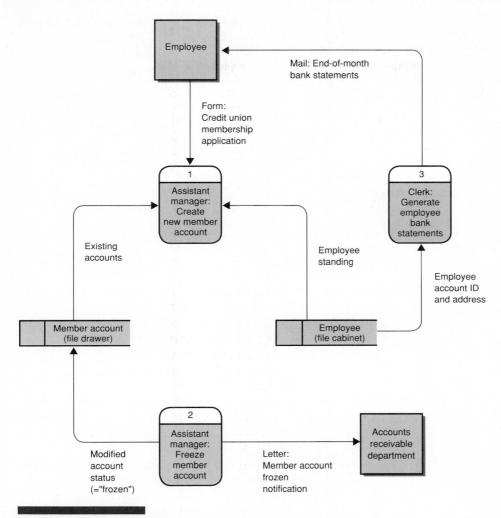

FIGURE 7.8 Common DFD Errors   Process 1 has inputs but produces no outputs. This is called a *black hole*. Process 2 produces outputs but receives no inputs. This is called the *miracle*. Process 3 has inputs and outputs; however, the inputs are not sufficient to produce the outputs.

The Data Store   Data stores are inventories of data. As data is input, it is frequently stored for later use. Think of data stores as the *memory* of the system. Any place that data accumulates is a data store.

The term *data* should not be part of the name because it is implicit that a data store stores data. Additionally, names should imply both content and implementation media. For example, ORDERS (IN/OUT BOX) refers to stored order forms that are awaiting processing. EMPLOYEE MASTER FILE (VSAM) describes a computerized file using the VSAM file organization method. ESTI-MATES (dBASE IV) describes a collection of records that make up an esti-mating database implemented with dBASE IV.

Names may also include appropriate adjectives or clauses to better describe the data store. Adjectives can be used to distinguish between similar but distinct stores of subsets of data (for instance, PENDING ORDERS versus CLOSED ORDERS, or MEMBERS versus PAST MEMBERS).

To avoid crossing data flow lines on a DFD, it is permissible to duplicate data stores. If data stores are duplicated on or between pages of DFDs, they should all be marked as indicated in the margin. This mark tells readers that they can expect additional occurrences of this same data store on the same page or other pages.

Data flows to and from data stores are significant and sometimes difficult to grasp. The trick is not to think in computer terms. Our guidelines, illus-trated in Figure 7.9, are as follows:

1. Only processes may connect to data stores. Only processes can read and update data stores. But be careful. On physical DFDs, only computerized processes may interact with computerized data stores. Similarly, only man-ual processes may interact with noncomputerized data stores.

2. Double-ended arrows for data flows are discouraged on detailed DFDs because the direction of arrows is significant. Data flow direction is inter-preted as follows:

   a. A data flow from a data store to a process means that the process uses that data. Notice that we said *uses*, not *reads*. Of course we have to read a file to use it! The *read* is assumed. Name the flow to reflect the portion of the data required by the process.

      If a form is physically removed from the data store—as is the case when forms are placed somewhere until a specific event occurs—use the standard data flow naming conventions described earlier.

   b. A data flow to a data store means that the process updates the data store. Updates may include any or all of the following:

      • *Adding or storing new* records or forms—for instance, adding a new customer to the CUSTOMERS data store—or adding a new entry to a log.

      • *Deleting or removing old* records or forms—for instance, deleting inactive customers from a store.

      • *Changing existing* records or forms—for instance, changing the credit rating, address, or balance of an existing customer.

3. As programmers, you know that in computer programs, reads and writes to files occur in pairs. You can't modify a record without first reading it. You can't delete a record without first locating it. Don't depict this level of detail! It clutters the diagrams by requiring every interaction with every

Data store name

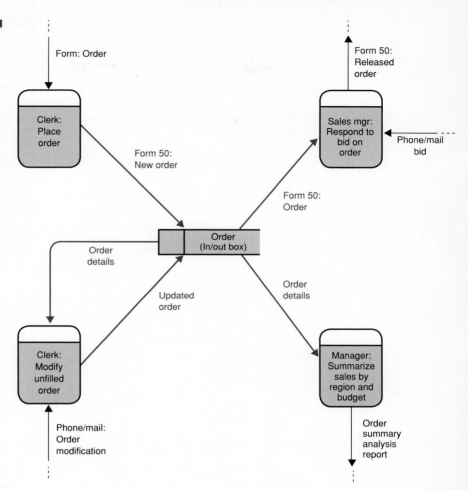

FIGURE 7.9 The Significance of Data Flows to and from Data Stores Be careful with data flows to and from data stores. The direction of the arrow may signify use of data, removal of data, creation of data, deletion of data, or modification of data. Read the guidelines carefully to make sure you know the difference. DFDs should never show the reading of data for their own sake!

data store to have two flows. Show only the *net* or final data flow. In most cases this means the add, delete, or change. Still . . .

4. Some processes both use and update data stores. The key word is *use*. For example, calculations or decisions might be necessary before you update the data stores. If so, as a general rule, use separate data flows for the use and the update.

The Value of Physical Data Flow Diagrams

Why are data flow diagrams so popular? They offer several advantages. First, they force analysts to communicate their understanding of systems to end-users. Second, they require analysts to examine the interfaces between systems, subsystems, and the rest of the world. Finally, they model a system in physical terms that are familiar to end-users. The documentation value of physical DFDs can be measured against our information systems pyramid model. After all, it is the information system that we want to document. Physical DFDs clearly depict all of the IPO components of the information system. The components of the information system correspond to the DFD symbol as follows:

IPO Component	DFD Component
Data (input)	Data flow
Information (output)	Data flow
End-users	Internal and external entity and process performed by or for end-user
Methods and procedures	Process
Data storage	Data store
Computer equipment	Process and data store
Computer programs	Process
Internal controls	Data flow and process

Additionally, DFDs can document all information systems functions, including transaction processing, management reporting, and decision support (see margin). In fact, in general, most systems follow the patterns illustrated in Figure 7.10.

Computer-Assisted Systems Engineering for DFDs

Computer-Assisted Systems Engineering (CASE) was introduced in Chapter 5 as an emerging, enabling technology for systems analysis and design methodologies. Virtually all CASE products support computer-assisted data flow diagramming. Furthermore, CASE takes the drudgery out of drawing and maintaining the diagrams.

Using a CASE product, you can easily create and link professional, readable DFDs without the use of paper, pencil, erasers, and templates. The DFDs can easily be modified to reflect corrections and changes suggested by end-users; you don't have to start over! Also, most CASE products provide powerful analytical tools that can check your DFDs for mechanical errors, consistency errors (between diagrams), and completeness errors. The potential time and quality savings are substantial.

One popular CASE tool for drawing DFDs (and other models) is Excelerator. Figure 7.11 illustrates typical Excelerator screens for drawing data flow diagrams.

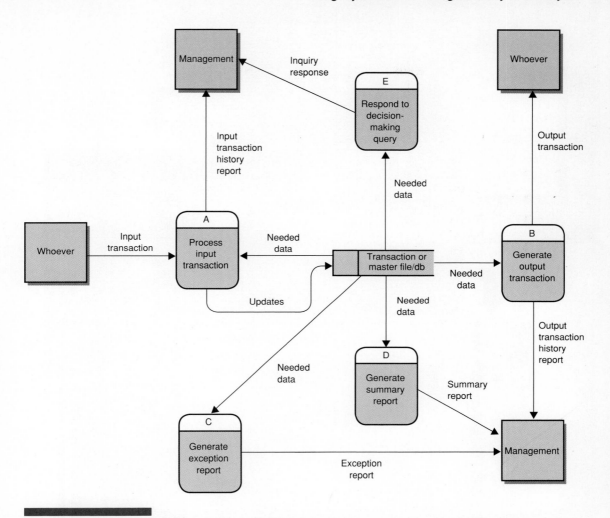

FIGURE 7.10 A General Model for a Typical Information System This DFD demonstrates that the tool effectively models virtually any modern capability of an information system. Most processing in information systems corresponds to one or more of the processes depicted here.

Using Physical Data Flow Diagrams for Systems Analysis

The use of physical data flow diagrams as a systems analysis tool is continually evolving, primarily in association with the Structured Systems Analysis methodology, discussed in Chapter 5. Let's walk through each of our analysis phases

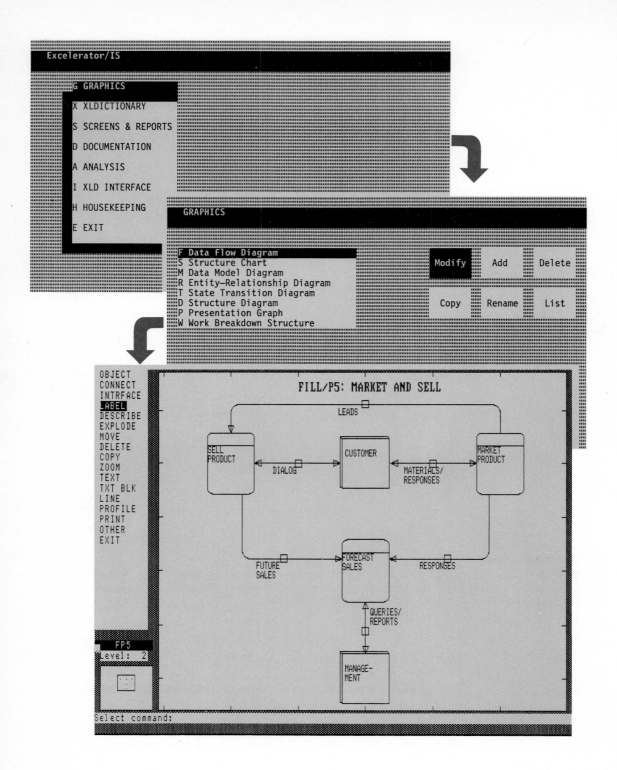

and describe the current state of the art, as well as past practices that are no longer recommended.

Structured Systems Analysis: The Data Flow Methodology

During the **study phase** of systems analysis, Structured Systems Analysis and DFDs are useful for modeling the scope of a systems project in terms of its interfaces (depicted as data flows) with other systems, the business as a whole, and the outside world—all depicted as internal and external entities.

Historically, Structured Analysis requires extensive modeling of the existing system as a study phase activity. Although this is still recommended in most books, study phase DFDs are being deemphasized by many experts (see the "Next Generation" feature for this chapter). They have noted that most analysts spend far too much time modeling existing systems. This leads to what they call *analysis paralysis*, a term that describes a project stuck in the study phase.

This doesn't mean physical DFDs are obsolete or useless. As already noted, they are very useful for modeling the general flow through the current system, as a learning device. Just don't fall into the trap of spending inordinate time to draw perfect physical DFDs for existing systems that will be redesigned anyway!

Looking ahead, we eventually want to strip data flow diagrams of their physical, implementation-dependent properties. This is particularly true during the **definition phase**, when our focus should be on *what* the system will do instead of *how* the system will accomplish the requirements. In Chapter 9, you will learn how to draw *logical* DFDs that serve this purpose. And looking further ahead to the **selection phase**, we may once again draw physical DFDs to reflect the final implementation decisions made regarding the new system.

How to Perform Process Modeling with Data Flow Diagrams

You already know enough about physical DFDs to read them. But as a systems analyst or knowledgeable end-user, you must learn how to draw them. We will use the Entertainment Club systems project to teach you how to draw physical DFDs.

The scenario is as follows. We have completed our survey phase of the life cycle. With management approval to continue, we began collecting facts about the current system. We sampled business forms, documents, and procedures manuals. We interviewed all end-users, some several times. We observed various aspects of the operational system. Now, to confirm that we understand the system, we draw physical DFDs to model the facts (actually, we probably

◄ **FIGURE
7.11 Computer-Assisted Data Flow Diagramming** These screens (sequenced by the arrows), from a CASE product called *Excelerator*, are typical of the trend toward computer-assisted drawing tools being used for data flow diagrams.

THE NEXT GENERATION

Structured Systems Analysis: Version 2.0

Version and release numbers have traditionally been used to identify a software package's evolution and improvement. In this box feature, we adopt that approach to analyze a popular systems analysis methodology and predict its future.

Structured Systems Analysis, as we know it today, was introduced to the information systems community in 1978 by Tom DeMarco, Chris Gane, and Trish Sarson. It enjoyed a rapid and widespread growth in popularity through the mid-1980s. In recent years, however, we have heard less about Structured Analysis. This is because newer methodologies such as those based on data or information modeling are getting increased attention, and Structured Analysis is maturing.

Structured Analysis has always been considered a *data flow* approach to systems analysis. The popularity of data flow diagrams is attributed to the emergence of the methodology. In classic Structured Analysis,

the analyst draws five complete sets of data flow diagrams:

1. physical data flow diagrams of the current system

2. logical (implementation-independent) equivalents of the physical diagrams drawn in step 1

3. logical data flow diagrams for the new system

4. alternative semiphysical data flow diagrams for different implementation options

5. physical data flow diagrams of the target system, reflecting the final implementation chosen

To complement the DFDs, analysts compiled data dictionaries and procedural specifications, which are discussed later.

How well have Structured Analysis and data flow diagrams worked? For most analysts, the results have been mixed. On the one hand, data flow diagrams have proved easy to draw and even easier to read. The methodology, although rarely performed with the originally intended rigor, has improved the results of systems analysis. On the other hand, we've learned that data flow diagrams and Structured Analysis cannot by themselves ensure completeness, consistency, and accuracy in systems development. You can still miss problems and requirements! Furthermore, the very act of going through all the various stages of data flow diagramming slows systems analysis—in an era when productivity has become an issue.

What's wrong with classical Structured Analysis, and can it be corrected? First, the methodology may have introduced too much rigor, though this is understandable since before Structured Analysis appeared, systems analysis methods were essentially haphazard.

Second, we have learned that Structured Analysis was an incomplete methodology. Data flow diagrams, its principal tool, model only one dimension of

continued ▶

sketched the diagrams throughout the fact-finding process). All the while, we have recorded various problems, opportunities, constraints, and general concerns that end-users have communicated.

We intend to carefully walk you through examples and provide a future reference for the day you use this popular modeling tool.

THE NEXT GENERATION

Structured Systems Analysis: Version 2.0, Continued

systems: processing. James Martin and Carma McClure (1985)—two of the best-known authorities and lecturers on methods and tools—offer the following analysis of DFDs (our own comments are bracketed):

Data flow diagrams are easy for both analysts and end users to understand. They have an important role to play . . . [but] a concern with data flow diagrams is that they are deceptively simple. They can look correct and give the analyst [and end-users] a comfortable feeling, when a closer examination would reveal a different picture [errors, inconsistencies, and so on]. . . . The appropriate role of the data flow diagram needs to be recognized. It is a useful form of overview sketch. [But] it needs to be used in conjunction with thorough data analysis and data modeling [also called information modeling].

Though we take exception to Martin and McClure's description of DFDs as mere "sketches," their analysis of DFDs is otherwise accurate. Furthermore, Structured Analysis is evolving to correct its overdependence on DFDs.

We now recognize that systems have at least two separate but related dimensions: processing and data. The process dimension can be at least partially modeled by DFDs; indeed, this chapter refers to DFDs as *process models*. The data dimension is usually documented with *data models*, which you will encounter in Chapter 8. Data modeling has gained momentum because of the trend toward database applications. Structured Analysis must adapt to the trend.

The next generation of Structured Analysis, according to one of its principal advocates, Ed Yourdon, will also address the simultaneous use of process and data models. The methodology will also address analysis

paralysis, the problem caused by spending too much time diagramming the current system with DFDs. The idea of separating physical and logical diagramming will survive—arguably, this was the most significant contribution of the original Structured Analysis. DFDs for existing systems will be de-emphasized; however, we hope they are not eliminated. As this chapter demonstrates, they are especially useful as a tool for learning about the existing system.

There will be other improvements, to be sure. New types of DFDs are expected (see the "Next Generation" feature for Chapter 9). New secondary tools are also expected. When the history books are finally written, we may very well liken the next generation of Structured Analysis to a child entering puberty. Only time will tell what the adulthood of Structured Analysis may bring. (Also see Suggested Readings by Yourdon at the end of the chapter.)

Physical Data Flow Diagrams as a Study Phase Tool

The primary usefulness of physical data flow diagrams in the study phase is to verify facts about how the current system works. You should not let the act of drawing physical DFDs obscure your goal in studying the current system, namely, the identification and analysis of problems, opportunities, and constraints.

Physical DFDs can be drawn in various levels of detail, ranging from depicting an entire system's "big picture" to showing the detailed processing of a single transaction. These various levels of detail can be drawn top down to unveil the system a little at a time. The entire collection of DFDs is called a *leveled set*. Let's walk through some examples of physical DFDs at various levels.

Step 1: Draw a Context DFD to Describe the System's Relationship to the Rest of the World

Many analysts find it useful to begin by drawing a simple DFD that depicts the system being studied as it relates to other systems, the business, and the outside world. This picture is usually called a **context diagram**.

How much detail to depict is a matter of preference; however, we have found it best to restrict the context diagram to depicting the main inputs and outputs of a system. Less common inputs, outputs, and their sources and destinations can be deferred until lower-level DFDs are drawn.

Figure 7.12 illustrates the context diagram for the SoundStage Entertainment Club. The system being studied has been named MEMBER SERVICES. The DFD suggests that the primary purpose of the system is to process MEMBER ORDER COUPONS that are received in response to CLUB PROMOTIONS AND CATALOGS. Recall that because PROMOTIONS and CATALOGS travel together, they are shown as a single data flow.

We can immediately see the main boundaries of our project. We are not concerned with how the WAREHOUSE fills orders or with how the MARKETING DEPARTMENT selects monthly promotions.

Notice that the data flow names are physical; they suggest their own implementation as it currently exists.

A context diagram's primary purpose is to confirm project scope with top-level management. It doesn't show enough detail to be used for verification of systems processing. In fact, all systems processing has been collapsed into a single process to hide that detail. Some people call this the *black box* concept: hiding details where they are not crucial to the model's purpose.

Step 2: Draw Physical DFDs to Depict Interfaces Between Organizational Units

The context diagram is too simple for verification. But we can draw a more detailed picture of the context DFD process to show greater detail. The technique, called *exploding the process*, is illustrated in Figure 7.13.

Figure 7.13 illustrates another valuable use of physical DFDs—the way they depict the interfaces between organizational units or individuals. We see that the MEMBER SERVICES are actually performed by three organizational units: the SUBSCRIPTION, ORDERS, and PROMOTION DEPARTMENTs. These three departments interact to provide the full range of member services

FIGURE 7.12 The Context Diagram The context diagram is the most general picture of a system. The system is depicted as a single process. Its inputs and outputs convey the main purpose of the system.

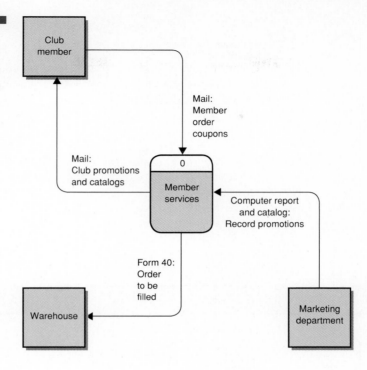

required. These interactions are shown as data flows between the three processes on the physical DFD.

Notice that all of the data flows on the parent process (the one we exploded) have been carried down to this diagram. This is called **balancing** the diagrams. Balancing ensures consistency between levels. Also note that some new data flows were added. Every process must have at least one input and one output; otherwise it would be a black hole. The depicted inputs and outputs are considered the main ones for these processes. We may defer less common data flows until we draw physical DFDs for each department.

Note that data flows don't have to balance up. For instance, MAIL: SUBSCRIPTION CARD AND ORDER FORM appears on this physical DFD but not on the context DFD. Why? Because SUBSCRIPTIONS represent fewer than 5 percent of the total transaction throughput of the system.

Notice that we have also introduced our first data store. Other data stores undoubtedly exist. When should data stores first be introduced? They should be introduced at whichever level of DFD where they are shared by two or more processes. Other stores will be deferred until we draw more detailed physical DFDs.

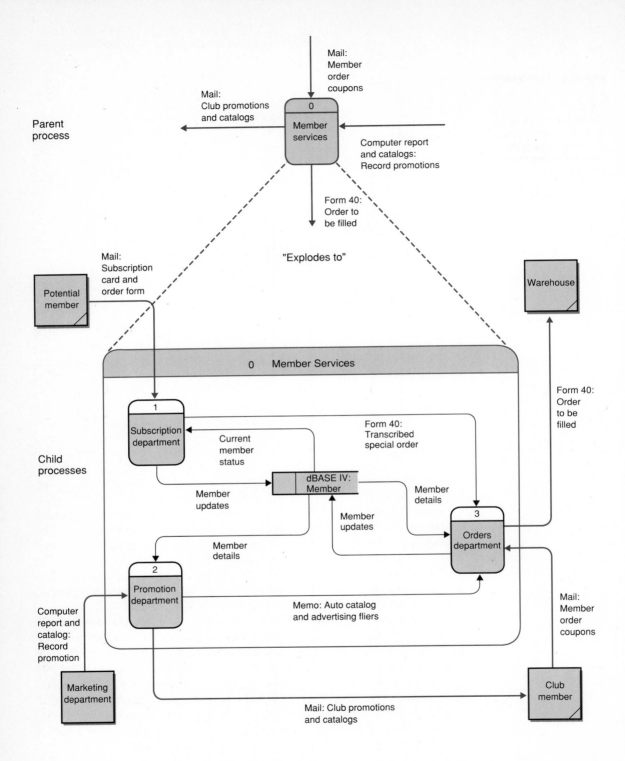

Parent process

Mail: Member order coupons

Mail: Club promotions and catalogs

0
Member services

Computer report and catalogs: Record promotions

Form 40: Order to be filled

"Explodes to"

Mail: Subscription card and order form

Potential member

Warehouse

Form 40: Order to be filled

0   Member Services

Child processes

1
Subscription department

Current member status

Form 40: Transcribed special order

dBASE IV: Member

Member updates

Member details

Member updates

Member details

3
Orders department

2
Promotion department

Memo: Auto catalog and advertising fliers

Computer report and catalog: Record promotion

Mail: Member order coupons

Marketing department

Club member

Mail: Club promotions and catalogs

A diagram of this type would be most useful in a meeting with the managers of the three units. You would confirm that the interactions depicted actually occur and that no important interactions have been overlooked.

It should be noted that some small projects deal exclusively with a single organizational unit. If so, this second physical DFD is inappropriate. In such cases you would explode the context DFD process directly into the DFD described below in step 3.

Step 3: Draw Physical DFDs to Show the Work Performed in a Single Organizational Unit Next, we can explode each of our department processes to reveal still further detail. One useful approach is to depict individual transactions performed by the department in question. A single process is designated for each transaction. The input data flow would be the transaction. The output data flows would be the response(s) to the transaction. This approach is illustrated in Figure 7.14.

Figure 7.14 is exploded from the SUBSCRIPTION DEPARTMENT process in Figure 7.13. Notice that we no longer border our page with the parent process symbol; it was for learning purposes only. How can you tell what process this new physical DFD was exploded from? Look at the number above each process. It always begins with "1." This means that all of these processes are part of process 1 in Figure 7.13.

Once again, new data flows have been introduced to communicate additional detail. Still, trivial or less common data flows may still be hidden from view at this level. Also, new data stores are introduced according to the sharing rule described in the last section. Note that process names conform to the *processor name* plus *process description* conventions described earlier in the chapter. Finally, note that all data flow names continue to reflect their actual implementation.

This diagram serves to identify individual transactions for which detailed processing must be studied. The physical DFD would be verified by a manager or supervisors, who can then direct the analyst to those individuals who might best provide details about each transaction's processing.

A physical DFD similar to Figure 7.14 must be drawn for each organizational unit.

Step 4: Draw Physical DFDs to Depict the Detailed Processing for Each Transaction An example of the processing required for a single transaction, SUBSCRIPTION CARD AND ORDER FORM—received via MAIL—is shown in Figure 7.15. This is the lowest-level physical DFD we will draw. It would be used to verify this transaction's processing methods and procedures. As we collect the facts needed to draw this physical DFD, we also solicit opinions and facts concerning problems, opportunities, and constraints, reinforcing

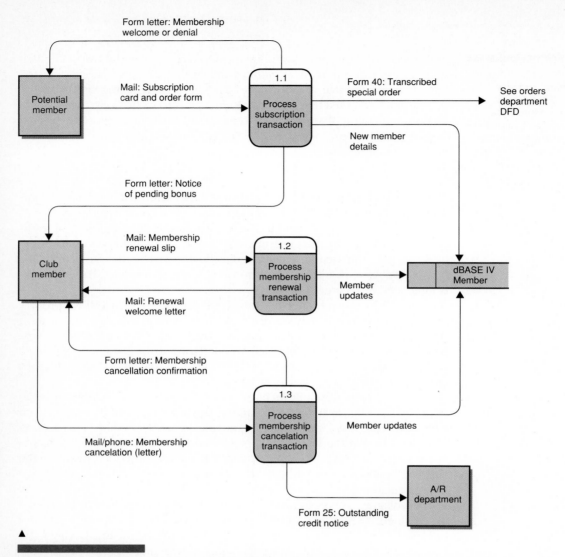

FIGURE 7.14 Middle-Level DFD to Depict Work Performed by a Single Organizational Unit
This diagram shows the work depicted by one department or subsystem.

FIGURE 7.15 Physical DFD for a Single Transaction (See facing page.) This DFD depicts the work ▶
performed on a single transaction. Similarly, a DFD could show the work required to generate a single report or
inquiry response.

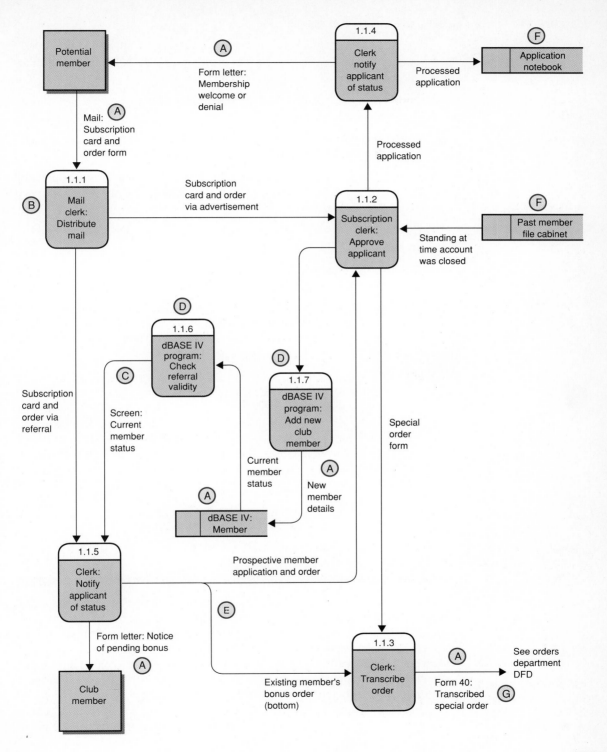

the primary purpose of drawing study phase DFDs. We call your attention to the following letters on the diagram:

Ⓐ Once again, this diagram's data flows balance with those of its parent process in Figure 7.14.

Ⓑ Note that subscriptions come to a mail clerk from where they are routed. Routing processes are shown on physical DFDs.

Ⓒ An inquiry is actually a two-way flow. There are a question (input) and an answer (output). But to keep the DFD simple, we show only the *net* data flow, the output.

Ⓓ The MEMBER data store is a computer-based file; thus all connecting processes must be computer programs.

Ⓔ The diverging data flow in this case represents the routing of a two-part perforated ORDER FORM.

Ⓕ Neither of these data stores is shared by more than one process. This seems to violate our rule about when to introduce data stores, except that this is the lowest-level DFD we will draw. Thus, in the interest of completeness, they must be introduced at this time.

Ⓖ FORM 40: TRANSCRIBED SPECIAL ORDER appears to be going nowhere. This is called an *interface data flow*. The interface data flow exits from this diagram and enters another diagram. Which diagram? We need to refer to the parent diagram (Figure 7.13). There we see that this data flow travels from the SUBSCRIPTION DEPARTMENT to the ORDERS DEPARTMENT. Thus, the detailed physical DFD for the ORDERS DEPARTMENT will include an input interface data flow of the same name and the processing of that data flow.

Looking Ahead: Data Flow Diagrams for Systems Analysis and Design

We have drawn physical DFDs for the existing system. These diagrams can be verified through individual and group discussions with appropriate end-users. If necessary, the physical DFDs can be revised; however, analysts should avoid spending too much time documenting a system that will soon be redesigned (remember the *analysis paralysis* problem). Also, too much time spent on the existing system's details can bias the analyst toward current techniques and methods, reducing creativity and open-mindedness.

Once the analyst sufficiently understands the existing system, time is best spent on problem and opportunity analysis, not DFD maintenance. The physical DFDs for the current system should be set aside as you make the transition from the current system to the new system. The next two chapters will address modeling tools and techniques for defining new systems requirements, one of which is *logical* DFDs.

Summary

A physical data flow diagram (DFD) is a tool for drawing a picture of an information system. As a systems analysis tool, it is primarily useful for documenting the existing system to ensure that the analyst truly understands that system. Physical DFDs illustrate the flow of data and work through a system. There are only four symbols that can appear on a DFD: the process, the internal or external entity, the data store, and the data flow. With these symbols, you can model processing for virtually any information system. Physical DFDs should not be confused with systems flowcharts. The key difference is that physical DFDs, unlike flowcharts, can show parallel processes—processes that can operate simultaneously. Data flow diagrams can effectively model all transaction-processing, management reporting, and decision support functions in a system.

Analysts can draw both very general physical DFDs (to depict subsystems, functions, and their relationships) and detailed physical DFDs (to depict flow for specific transactions, documents, and reports).

Problems and Exercises

1. Why is a *data flow* diagram called a *process* model?
2. Explain to an information systems recruiter what makes physical DFDs physical. Explain why you should include implementation details when drawing a physical DFD. Doesn't it clutter the diagram? Can you think of any circumstances in which physical details might be distracting?
3. A manager who has noted your use of physical data flow diagrams to document an existing system has expressed some concern because of his prior, less-than-satisfying experience with such information systems tools as flowcharts. Defend your use of physical DFDs. Explain why they are useful and how they are superior to flowcharts. Concisely explain the symbolism and how to read a physical DFD. (*Note*: The answer to this exercise should be a standard component in any report that will include DFDs. You cannot be certain that the person who reads this report will be familiar with the tool.)

4. Explain why a systems analyst might want to use physical data flow diagrams to document the automated portion of an existing information system rather than simply accepting the existing technical information systems documentation, such as systems flowcharts and program flowcharts.

5. Draw a physical data flow diagram to document the flow of data in your school's course registration and scheduling system.

6. Draw a physical data flow diagram for some day-to-day "system" that you use or observe in use—for instance, your morning routine; making your favorite meal, including appetizer, entree, side dishes, and dessert; constructing something from scratch.

7. Draw a physical data flow diagram for the inventory control system that was described in the minicase that opened Chapter 2.

Minicases

1. Given the following narrative description, draw a physical context data flow diagram for the portion of the activities described.

   The purpose of the TEXTBOOK INVENTORY SYSTEM at a campus bookstore is to supply textbooks to students for classes at a local university. The university's academic departments submit initial data about courses, instructors, textbooks, and projected enrollments to the bookstore on a TEXTBOOK MASTER LIST. The bookstore generates a FORM 17: PURCHASE ORDER, which is sent to publishing companies supplying textbooks. Book orders arrive at the bookstore accompanied by a PACKING SLIP, which is checked and verified by the receiving department. Students fill out a BOOK REQUEST FORM that includes course information. When they pay for their books, the students are given a paper tape CASH REGISTER SALES RECEIPT.

2. Given the following narrative description, draw a physical context data flow diagram for the portion of the activities described.

   The purpose of the PLANT SCIENCE INFORMATION SYSTEM is to document the study results from a wide variety of experiments performed on selected plants. A study is initiated by a researcher who submits a STUDY PROPOSAL on a STUDY MASTER SHEET. After a panel review by a group of scientists, the researcher is required to submit a detailed study description on a FORM 11. An FDA PERMIT REQUEST FORM is sent to the Food and Drug Administration, which sends back a preprinted, numbered PERMIT CARD that is placed in a Rolodex file. As the experiment progresses, the researcher fills out and submits DETAILED EXPERIMENT NOTES that are stored in folders in a filing cabinet. At the conclusion of the project, the researcher's results are reported on a PLOT ANALYSIS DIAGRAM.

3. Given the following narrative description of a physical system, draw a simple physical data flow diagram to illustrate the system's "big picture."

The purpose of the production scheduling system is to respond to PRODUCTION REQUEST by generating a daily PRODUCTION SCHEDULE (one copy each for the plant manager and the shop line supervisor), generating MATERIAL REQUESTS (sent to the STORES DEPARTMENT) for all production orders scheduled for the next day, and generating JOB TICKETS for the work to be completed at each workstation during the next day (sent to the SHOP LINE SUPERVISOR). The work is described in the following paragraphs.

The production scheduling problem can be conveniently broken down into three functions: routing, loading, and releasing. For each product on a PRODUCTION REQUEST, we must determine which workstations are needed, in what sequence the work must be done, and how much time should be necessary at each workstation to complete the work. This data is available from the PRODUCT ROUTE SHEETS, which are kept in loose-leaf binders. This process is referred to as ROUTING THE ORDER and results in a ROUTE TICKET.

Given a ROUTE TICKET (for a single product on the original PRODUCTION REQUEST), we then LOAD THE REQUEST. Loading is nothing more than "receiving" dates and times at specific workstations. The reservations that have already been made are recorded in another loose-leaf binder, which is labeled WORKSTATION LOAD SHEETS. Loading requires us to look for the earliest available time slot for each task, being careful to preserve the required sequence of tasks (determined from the ROUTE TICKET).

At the end of each day, the WORKSTATION LOAD SHEETS for each workstation are pulled from the binder. A PRODUCTION SCHEDULE is created from these worksheets. JOB TICKETS are prepared for each task at each workstation. The materials needed are determined from the BILL OF MATERIALS FILE—a notebook—and MATERIAL REQUESTS are generated for appropriate quantities.

4. Health Care Plus is a supplemental health insurance company that pays claims after its policyholders' primary insurance benefits through their employer or another policy have been exhausted. The following narrative partially describes its claims-processing system. Draw a physical data flow diagram for the portion of the systems activities described.

Policyholders must submit an Explanation of Health Case Benefits (EOHCB) form along with proof that their primary health policy claim has been paid. All claims are mailed to the claims-processing department.

All claims are initially sorted by the claims screening clerk. With a form letter, this clerk returns all requests that do not include the EOHCB or EOHCB reference number. For those requests returned, a PENDING CLAIM TICKET is created, dated, and stored in a file cabinet by date. Once each week, the clerk deletes all tickets that are more than 45 days old and sends a form letter to the policyholders notifying them that their case has been closed. Requests that include the EOHCB are then sorted according to type of claim, and the PENDING CLAIM TICKET is destroyed at this time. Requests that include an EOHCB reference number are matched up with an EOHCB form, which is pulled from the OPEN CLAIMS file cabinet. At the end of each day, all these claims are forwarded to the preprocessing department.

In the preprocessing department, clerks screen the EOHCB for missing data. They complete the form if possible. Otherwise, a copy of the claim is returned to the policyholder with a letter requesting the missing data. The original EOHCB is placed in the OPEN CLAIMS file cabinet, and a PENDING CLAIM TICKET is sent to the claims screening clerk. Completed claims are assigned a claim number, and the claim is microfilmed and filed for archival purposes.

A different clerk checks to see if the proof of primary health care policy payment was included or is on file in the PRIMARY PAYMENT file cabinet. If it is not available, the policyholder is sent a letter requesting the proof. The EOHCB is placed in a PENDING PROOF file cabinet. Claims are automatically purged if they remain in this file for more than 14 days (a letter is sent to policyholders whose claims have been purged).

If proof is available, another clerk pulls the policyholder's policy record from the POLICY file cabinet, records policy and action codes on the EOHCB, and refiles the policy record. At the end of the day, all preprocessed claims are forwarded to Information Systems.

5. Given the following narrative description, draw a physical context data flow diagram and the next-level data flow diagram for the portion of the activities described.

The purpose of the GREEN ACRES REAL ESTATE INFORMATION SYSTEM is to assist agents as they sell houses. Sellers contact the agency and an agent is assigned to help the seller complete a SALES REQUEST FORM. Information about the house and lot taken from that form is stored on a computer disk file. Personal information about the sellers is copied by the agent onto a SELLERS PERSONAL INFO SHEET and stored in a filing cabinet. When a buyer contacts the agency, he or she fills out a BUYER REQUEST. Every two weeks, the agency send prospective buyers a homes-for-sale magazine and a key for the magazine containing street addresses. Periodically, the agent will find a particular house that satisfies most or all of a specific buyer's requirements, as indicated in the BUYERS REQUIREMENTS REPORT distributed weekly to all agents. The agent will occasionally photocopy a picture of the house along with vital data and send the MULTIPLE LISTING SERVICE (MLS) SHEET to the potential buyer. When the buyer selects a house, he or she fills out a PURCHASE AGREEMENT that is forwarded through the real estate agency to the seller, who responds with an

OFFER ACCEPTANCE or a COUNTEROFFER. After a PURCHASE AGREE-MENT has been accepted, the agency sends an APPRAISAL REQUEST to an appraiser, who appraises the value of the house and lot. The agency also notifies its finance company with a FORM 17: FINANCE REQUEST.

6. Given the following narrative description, draw a physical context data flow diagram and the next-level data flow diagram for the portion of the activities described.

The purpose of the OPEN ROAD INSURANCE SYSTEM is to provide automotive insurance to car owners. Initially, customers are required to fill out an INSURANCE APPLICATION REQUEST FORM, which is kept in a notebook until a policy is issued. A DRIVER'S RECORD REQUEST is sent to the local police department, which sends back a DRIVER'S RECORD REPORT that is stored in a filing cabinet. Also, a VEHICLE REGISTRATION REQUEST CARD is sent to the Department of Motor Vehicles, which supplies a VEHICLE REGISTRATION SHEET, which is also kept in a file cabinet. POLICY CONTRACTS are sent in by various insurance companies and entered on a computer file using dBASE IV. The agent determines the best policy for the type and level of coverage desired, and gives the customer a copy of the INSURANCE POLICY along with an INSURANCE COVERAGE CARD. The customer information is now entered into a dBASE IV file. Periodically, a computer program generates a FEE STATEMENT, which—along with ADDENDUMS TO POLICY—is sent to the customer, who responds by sending in a PAYMENT with the FEE STUB.

Annotated References and Suggested Readings

DeMarco, Tom. *Structured Analysis and System Specification.* Englewood Cliffs, N.J.: Prentice-Hall, 1978. This is the classic book on the Structured Systems Analysis methodology, which is built heavily around the use of data flow diagrams. DeMarco is a gifted writer who expresses complex ideas with ease.

Gane, Chris, and Trish Sarson. *Structured Systems Analysis: Tools and Techniques.* Englewood Cliffs, N.J.: Prentice-Hall, 1979. An early Structured Analysis methodology book. Not as thorough as DeMarco but relatively easy to read and grasp. Their DFD symbology is identical to ours. A methodology called STRADIS has been developed around this approach. STRADIS is marketed through a St. Louis–based company called Improved Systems Technologies.

Keller, Robert. *The Practice of Structured Analysis: Exploding Myths.* New York: Yourdon Press, 1983. A nice, concise overview of the first generation Structured Analysis methodology.

Martin, James, and Carma McClure. *Diagraming Techniques for Analysts and Programmers.* Englewood Cliffs, N.J.: Prentice-Hall, 1985.

Yourdon, Edward. "Whatever Happened to Structured Analysis?" *Datamation*, June 1, 1986. This article provides a first glimpse at the evolution of Structured Analysis as it will appear in forthcoming works from Yourdon Press, the largest library of titles on Structured Analysis. From this article, we learn that Structured Analysis will deemphasize current systems modeling.

Yourdon, Edward. *Modern Structured Analysis.* Englewood Cliffs, N.J.: Yourdon Press, 1989. This is the long-awaited update to DeMarco's classic book. It includes all of the suggested updates to the methodology as described in this chapter's Next Generation feature. It paints a dismal future for this chapter's use of physical DFDs for the current system.

SOUNDSTAGE

Requirements Analysis for the New System (Part I)

When we left Sandra and Bob, they had completed their study of the current member services system. Now they must define the business requirements for a new and improved system.

EPISODE 4 We begin this episode shortly after SoundStage Entertainment Club's Steering Committee has approved Sandra and Bob's study phase report. The committee has agreed that the projected benefits of solving Sound-Stage's member services problems exceed the projected costs for continuing the project. We join David Hensley, Manager of Customer Services, in a meeting with Sandra and Bob.

Objectives for the Target System

"Hi, Bob. I really enjoyed our basketball game yesterday. I hope your ankle is getting better. What are we supposed to accomplish today?" David asked.

Bob took a seat. "Today, I'd like for us to agree on objectives for the new system. The objectives we come up with shouldn't have anything to do with computer solutions or specific reports the system should produce. Instead,

we want to establish a set of business objectives that will serve as criteria for evaluating alternative solutions. Keep in mind that we want to be sure to establish objectives that will address the problems and opportunities we uncovered with the current system. An example of such an objective might be that we want to respond to a membership application request within five days. Notice that this objective addresses the problem we identified with the current system's unacceptably slow turnaround time for processing membership requests. In fact, this objective also addresses an opportunity. Though we may consider the current level of customer satisfaction to be acceptable, the fulfillment of this objective will likely have a positive influence on the customer. Do you understand?"

"Sure, I understand. But I really wanted to talk about the computer. I had some good ideas about what we might be able to accomplish using the computer,"

David replied with disappointment.

"We will, but first we should concentrate on what we want the new system to accomplish, rather than solutions or ways it might be accomplished. Sandra and I believe that the business problem should define the solution, not the other way around. Thus, we'll be sure to end up with a system that best supports your needs, regardless of the way it is actually implemented. Now, how about some business objectives?"

"Well, Bob, I think decreasing the amount of time to process a member's order would be the most important objective."

Bob replied, "I'd like to get a little more specific than that. Let's try to quantify the objective. Currently the system can handle 300 orders a day. I know you want to increase that capacity, but by how much?"

"Okay, I see what you mean. I'd like to be able to process at

Episode 4, continued ▶

least 700 orders per day. In fact, management projects that their new promotional campaign will stimulate growth that could soon double the current load."

"That's the kind of information I need," encouraged Bob. "Now we have a precise, measurable target. Any new system we consider or implement can be evaluated on the basis of it being able to process 700 or more orders per day."

"I get it. It's sort of like management by objectives. Our management does this with employees. They establish measurable objectives against which they can evaluate the employees' performance. And of course in establishing the objectives, they defer discussion of how those objectives are to be achieved," explained David as Bob and Sandra smiled and nodded their heads in agreement.

"Okay, how about this one. Some managerial directives have been established that will result in an increase in our membership by 20 percent next year. It currently takes up to three hours for the mail clerk to sort new membership applications into advertisement applications and referral applications. We simply can't afford that kind of time delay. We must be able to process at least 100 membership applications per day."

"That's good!" replied Bob. "I was aware of those management directives being passed down, but they had totally slipped my mind. We'll make sure the new system takes them into account."

"Great! I've got some more. How about . . ."

(We can leave this meeting now. The discussion will continue until all objectives for the new system have been identified.)

General Requirements for the New System

Two weeks have passed. Sandra and Bob have been working on the business requirements statement for the new member services system. They have determined the general systems requirements and documented them with diagrams. We join Bob as he is leading a walkthrough to verify his understanding of the new system's data storage requirements with the Member Services staff. Sandra, Bob, David Hensley, the Member Services Manager, Ann Martinelli, the Membership Director, Sally Hoover, the Director of Order Processing, and Joe Bosley, the Promotions Director, are also at the meeting.

Bob dims the lights and places a slide on the overhead projector. "I caution you not to be overwhelmed by this diagram [Figure E4.1]. This diagram is simply a pictorial representation of the data storage requirements for the new system. The purpose of this picture is to illustrate the types of things—we'll call them entities—about which the new system must store data and relationships that may exist between those entities."

"In other words, this is a picture of the computer files that the new system will maintain," interrupted Joe.

"Not necessarily," Bob answered. "Let's stop thinking of

files, both computer files and manual files. Rather, we simply want to concentrate on identifying all entities about which the new system is to store data, independent of how that data is physically stored now or how it might be stored for the new system. In the next phase of the project, we will evaluate different ways to implement the data."

Noting a general consensus of understanding and agreement, Bob points to a rectangular symbol on the slide and begins. "The rectangles on this diagram represent those things or entities about which the new system is to store data. For example, the new system will store data that describes MEMBERS, ORDERS, PROMOTIONS, and so on. Recall that Sandra or I met previously with each of you to help you identify those entities you desired the new system to store data about. Our goal is to confirm your data storage requirements one last time. Now, are there any additional entities you feel we may have omitted or overlooked?"

"What about SUPPLIERS?" asked Sally. "We'd like to be able to determine the supplier for a particular TITLE."

"Whoops!" interrupts Sandra. "Bob, I forgot to add it to the diagram."

"No problem, would you add that to the figure? Don't forget to draw any relationships that may exist between it and any other entities." Bob turned his attention back to the end-users. "That's good. Are there any others you feel are missing?"

Ann spoke up: "I can't think of any. You've included all those

Episode 4, continued ▶

SOUNDSTAGE

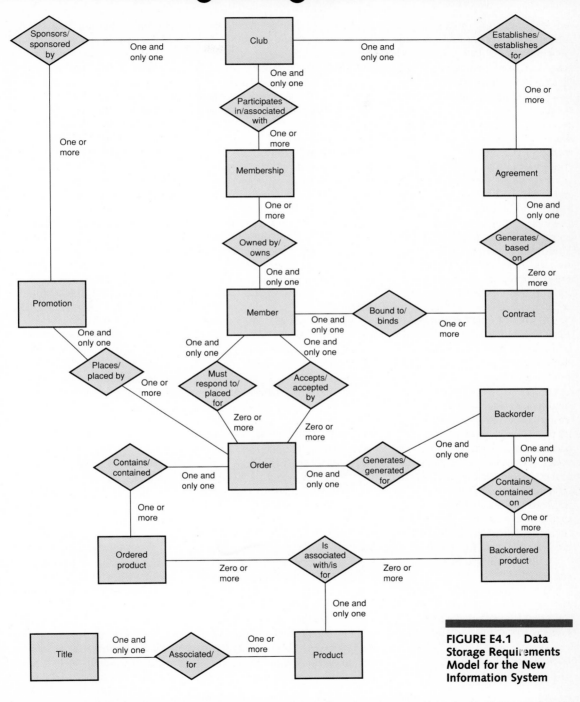

FIGURE E4.1 Data Storage Requirements Model for the New Information System

entities Sandra and I identified last week when we were discussing my people's specific data storage requirements. I assume some of these other entities are of interest to the other individuals in this room?"

"That's correct, Ann," replied Bob. "This diagram identifies data storage requirements for the entire member services system."

Pausing for a few moments and noting no additional feedback from the end-users, Bob decided to move on. "Ok, let's shift our attention toward identifying and confirming the relationships between these entities. The diamond shapes represent the relationship between groups of entities. For example, a MEMBER is BOUND TO a CONTRACT and a CONTRACT BINDS a MEMBER."

"Why are there two relationships between ORDER and MEMBER?" Sally asked.

"Let me answer that question, Bob," Sandra interrupted. "Sally, recall that earlier you explained to me that SoundStage generates automatic orders that members must respond to. The relationship on the left identifies a given MEMBER and the automatic ORDER they MUST RESPOND TO. Or vice versa, it identifies for a particular automatic ORDER which MEMBER the order was PLACED FOR."

"Ok. I understand that much," offered Sally.

"Good. Well, essentially the other diamond symbol represents a different type of relationship between an ORDER and MEMBER. For example, it represents ORDERs that the MEMBER ACCEPTS. Of course, that order

may have been the original automatic order or some variation. Likewise we can interpret the relationship in the other direction. That is why we labeled the diamond symbol with two verbs separated by a slash—so you can read the relationship more easily. It's also important to emphasize that these two relationships are mutually exclusive. That is, one or the other of them will always exist, but never both."

"I get it," replied Sally. "At any point in time, the relationship between a MEMBER and ORDER is such that the ORDER is either awaiting a MEMBER's response, or it represents an ORDER that has been responded to by the MEMBER."

"Thanks, Sandra, I don't think that I could have explained it nearly that well," Bob admitted. "Are there any other questions or comments?"

"Isn't there a relationship between a PROMOTION and a TITLE?" asked Joe. "When a particular CLUB SPONSORS a monthly PROMOTION, that PROMOTION is always for a particular TITLE."

"That's very good, Joe. I missed that relationship," Bob admitted as he noted the change.

"What's this ONE OR MORE, ZERO OR MORE, and ONE AND ONLY ONE stuff appearing on the line that connects the rectangles and diamond symbols?" inquired Ann.

"Good question, and if there are no additional questions concerning relationships I will answer that right now," Bob replied. "That portion of the diagram indicates the complexity of the relationships

between entities and provides us with a more thorough understanding of their meanings and interpretations. Follow me closely. Let's look at the relationship between PRODUCT and TITLE. A given PRODUCT is FOR ONE AND ONLY ONE TITLE. Conversely, a given TITLE is ASSOCIATED WITH ONE OR MORE PRODUCTS."

"I see. But, if that is how you interpret it, I think I see a mistake with the SUPPLIER entity that Sandra added to the diagram," Ann stated with concern. "Though it's true that a given SUPPLIER SUPPLIES ONE OR MORE TITLEs, it's not true that a TITLE is SUPPLIED BY ONE OR MORE SUPPLIERs. In fact, a particular TITLE is SUPPLIED BY ONE AND ONLY ONE SUPPLIER."

"That's a very good observation, Ann," Sandra offered. Sandra then turned to Bob. "That's great! I was worried about the implications of the complexity of that relationship."

Bob continued the walkthrough of the figure while Sandra took notes of changes to reflect the additional comments of the end-users. Once the walkthrough verification of the diagram was completed, the walkthrough shifted toward the verification of several additional slides.

"We're not through verifying the data storage requirements for the new system just yet." Bob placed the first of several slides on the overhead projector. "Now we want to verify what data we need to store about each of the entities appearing on that diagram we just finished discussing. We've sampled many of your existing forms and

Episode 4, continued ▶

documents to identify specific data that you are currently storing about those entities. We've also talked with you individually to allow you the opportunity to sug-gest additional data that you would like to start keeping about those entities. Let's confirm our list for each entity. First, let's look at the list of data we would like the new system to store about a MEMBER (see Figure E4.2). Are there any additional items that you'd like to add to this list?"

Episode 4, continued ▶

Entity: AGREEMENT

Contents: (AGREEMENT NUMBER, MAXIMUM PERIOD OF OBLIGATION, NUMBER OF PURCHASES REQUIRED WITHIN PERIOD, BONUS CREDITS/PURCHASE AFTER OBLIGATION, NUMBER OF MEMBERS ENROLLED, NUMBER OF MEMBERS WHO HAVE FULFILLED)

Entity: BACKORDER

Contents: (ORDER NUMBER + BACKORDER DATE)

Entity: BACKORDERED PRODUCT

Contents: (ORDER NUMBER + BACKORDER DATE + PRODUCT NUMBER + MEDIUM CODE, QUANTITY BACKORDERED)

Entity: CLUB

Contents: (CLUB NAME or CLUB ABBREVIATION, NUMBER ENROLLED YEAR-TO-DATE, NUMBER CANCELED YEAR-TO-DATE)

Entity: CONTRACT

Contents: (MEMBER NUMBER + AGREEMENT NUMBER, CONTRACT ENROLLMENT DATE, CONTRACT EXPIRATION DATE, NUMBER OF PURCHASES REQUIRED, NUMBER OF PURCHASES TO DATE)

Entity: MEMBER

Contents: (MEMBER NUMBER, MEMBER NAME, MEMBER ADDRESS GROUP: MEMBER STREET ADDRESS; MEMBER POST OFFICE BOX NUMBER; MEMBER CITY; MEMBER STATE; MEMBER ZIP CODE; MEMBER PHONE NUMBER, DATE OF FIRST MEMBERSHIP, BALANCE PAST DUE, BONUS CREDITS NOT USED)

Entity: MEMBERSHIP

Contents: (MEMBER NUMBER + CLUB NAME, MUSICAL/MOVIE PREFERENCE CODE)

Entity: ORDER

Contents: (ORDER NUMBER, ORDER DATE, ORDER STATUS, TOTAL ORDER PRICE, PREPAID AMOUNT)

Entity: ORDERED PRODUCT

Contents: (ORDER NUMBER + PRODUCT NUMBER + MEDIUM CODE, QUANTITY ORDERED, QUANTITY SHIPPED, PRODUCT PRICE)

Entity: PRODUCT

Contents: (PRODUCT NUMBER + MEDIUM CODE, QUANTITY ON HAND, LIST PRICE, CURRENT RETAIL PRICE)

Entity: PROMOTION

Contents: (CLUB NAME + PROMOTION DATE, RELEASE DATE, AUTOMATIC FILL DATE)

Entity: SUPPLIER

Contents: (SUPPLIER NUMBER, SUPPLIER NAME, SUPPLIER ADDRESS)

Entity: TITLE

Contents: (PRODUCT NUMBER, TITLE OF WORK, COPYRIGHT DATE, ARTIST/PRODUCER NAME, LABEL/DISTRIBUTOR NAME)

**FIGURE E4.2
Typical Data Store
Contents**

"Looks complete to me, but why are some of these names underlined?" asked David.

"The underlined items represent the data item or items that uniquely identify a particular occurrence of the entity. In other words, the data item MEMBER NUMBER uniquely identifies a single member. In some cases we use a plus sign to indicate where two or more data items are required to uniquely identify an entity."

"Some of the member reports we use include the member's AGREEMENT ENROLLMENT and EXPIRATION DATES. Shouldn't that information be kept for each agreement?" Ann asked.

"That's right, Ann," Bob answered. "But an agreement is something like 'buy 10 records in two years.' This information really describes a specific occurrence of an agreement that we called a contract. Maybe it would be better if we called this data CONTRACT ENROLLMENT DATE and CONTRACT EXPIRATION DATE."

(We can leave this meeting now. The discussion would continue until all of the objects in the system, the associated data, and their relationships had been discussed and verified.)

Several days have passed since the meeting that verified the new system's data requirements. We join Sandra as she is leading a walkthrough of the order-processing requirements with the Member Services staff in order to verify her understanding of the new system's processing requirements.

Sandra placed a picture of the proposed system on the overhead projector (see Figure E4.3). "You have already met with Bob and me to confirm the data storage requirements of the new system. Now we want to confirm what processing will be performed on that data. You'll remember seeing pictures similar to this when we were trying to learn about the current member services system. They differ slightly in that these pictures don't show details of how the processing will be performed."

Sandra dimmed the lights and drew their attention to a data flow appearing on the diagram. "Once a MEMBER ORDER is received from a CLUB MEMBER, the member's credit status is checked using CREDIT DETAILS from the MEMBERS file and, if necessary, a CREDIT DECISION from the ACCOUNTS RECEIVABLE DEPARTMENT. Moving on, if the member's credit status is poor, a REQUEST FOR PREPAYMENT is sent to the CLUB MEMBER using MEMBER DETAILS from the MEMBERS file, the PENDING ORDER STATUS is recorded in the ORDERS file, and the ORDERED PRODUCTS are recorded in the ORDERED PRODUCTS file. When the MEMBER's PREPAYMENT arrives, the ORDER PAYMENT is sent to the ACCOUNTS RECEIVABLE DEPARTMENT, the pending order status is deleted, and the RELEASED PENDING ORDER is checked for product inventory just like an approved order."

"Excuse me, Sandy," Sally interrupted. "Sometimes the MEMBER ORDER has a notation on it that indicates the member has changed his or her address. When that happens, I have to notify the membership people so that the new address can be recorded."

"That's important, Sally," Sandra responded as she made the correction to the diagram. "An APPROVED ORDER is then checked to determine whether or not the ordered products are available by checking the PRODUCT AVAILABILITY STATUS from the PRODUCTS file. If the order is unfillable, the BACKORDER DETAILS are recorded in the BACKORDERS file and the BACKORDERED PRODUCTS in the BACKORDERED PRODUCTS file, and the CLUB MEMBER is sent a BACKORDER NOTICE. If the order is fillable, the RELEASED ORDER is stored in the ORDERS file, the ORDERED PRODUCTS are recorded in the ORDERED PRODUCTS file, and an ORDER TO BE FILLED is sent to the WAREHOUSE. After the warehouse has filled the order, it returns a SHIPPING NOTICE to us so we can close the order and send a FILLED ORDER FOR BILLING notice to the ACCOUNTS RECEIVABLE DEPARTMENT."

"What about canceling the AUTOMATIC ORDER?" Joe asked. "When we receive a MEMBER ORDER from a CLUB MEMBER we have to cancel the AUTOMATIC ORDER created by the promotion subsystem. Otherwise, the member would receive two orders."

"Isn't that normally done as soon as you determine the member's credit status?" Sandy inquired. "And the AUTOMATIC ORDER is stored in the ORDERS file awaiting the promotion automatic release date, right?"

Episode 4, continued ▶

FIGURE E4.3 Process Requirements Model for the New Information System

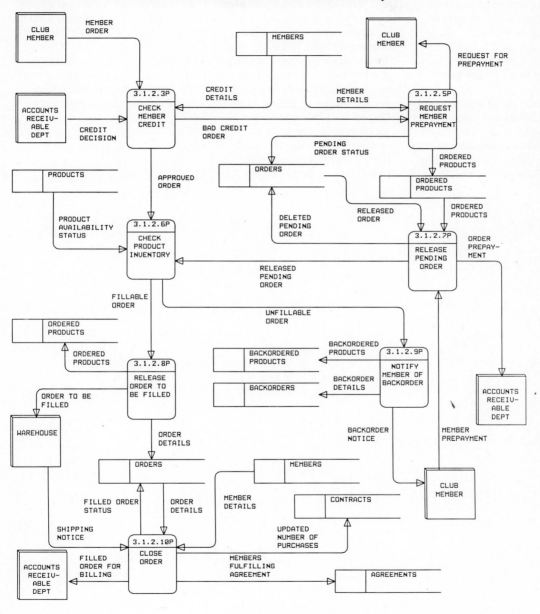

SOUNDSTAGE

(We can leave now. This meeting would continue until all of the diagrams illustrating the processing requirements of the new system had been verified.)

Where Do We Go from Here?

This episode introduced the establishment of business objectives for a new system. It also introduced the *entity relationship diagram* for modeling the stored data requirements for the new system. In addi-

tion, this episode introduced the use of *logical data flow diagrams* for modeling the process requirements for the new system. Chapter 8 will discuss entity relationship diagrams (ERDs). ERDs are used to document the system's data independent of how that data is or will be used. They also describe relationships between groups of data. Chapter 9 will review the use of data flow diagrams, with an emphasis on the use of logical DFDs rather than the physical DFDs you studied in Chapter 7.

This is done because, as we define requirements for the new system, we want to concentrate on what it should do, not how it should do it. Therefore, we eliminate the physical implementation details from the data flow diagrams and look only at the logical processing steps for defining the general requirements for the new system.

CHAPTER EIGHT

Data Modeling with

Logical Entity

Relationship Diagrams

Precious Jewels Diamond Center

Precious Jewels Diamond Center is a franchised jewelry store that specializes in diamonds and other gems, custom selected by and for customers. Gems are custom set into rings, pendants, and the like. Precious Jewels also serves as a diamond broker, providing gems to other franchises and jewelry stores. These gems are sent out on approval. The stores have the option of purchasing the gems or returning them.

Frank Burnside, a systems consultant to Precious Jewels, is meeting with Jeff Kassels, Vice President. They are discussing the possibility of hiring Frank to develop improved information systems for their IBM and Compaq microcomputers.

Jeff initiated the discussion. "Where do I start? About two years ago, we decided to purchase two microcomputers. On the recommendation of the computer store, we also bought some packages."

"What packages?" Frank asked.

"Lotus 1-2-3, dBASE IV, and WordPerfect. Unfortunately, we just didn't have the training to exploit these packages, especially the dBASE database package. Therefore, I hired some young students who were into micros; you'd probably call them hackers. They didn't have much experience."

"And what happened?" Frank asked.

"They wrote some dBASE and Lotus programs for inventory control and sales. The programs seemed to work. We entered lots of data into the system and it generated several reports. On the other hand, we later realized the need for new reports and inquiries. We tried to generate them ourselves, but I guess we just didn't understand the report writer in dBASE. The original students were unavailable, so we hired a woman who does dBASE programming on the side. She couldn't seem to generate the reports from the data. I know the data is in there because we put it there and it does come out on the original reports. I don't see how you can put data in and not be able to get it out."

Frank explained, "It's not an unusual phenomenon. Actually, it's quite common. Inexperienced analysts—no, even many experienced analysts—tend to design systems to meet today's needs. They fail to recognize that the structure of stored data can make it difficult or impossible to adapt to changing needs and situations. Your data is probably not well organized. Poorly designed computer files are no different from poorly organized manual files."

"That's not all," replied Jeff. "There are growing problems with data already in the system. As I scan the original reports, I note that records exist that should have been deleted a long time ago. For instance, I find records of purchase orders for gems that have been sold and paid for a year ago. To make matters worse, I find records of gems for which I can't find a purchase order. I need that purchase order to compare my valuations and prices to the valuation and prices I was charged by my suppliers."

"Again," Frank replied, "this situation is typical of poorly designed systems. Before a system is built, I put considerable effort into understanding its data and the complex business relationships between different sets of data. I try to understand when data needs to be created, when it should be changed, and when to delete or not to delete it. For instance, I suspect that you keep customer data. You probably wouldn't want me to delete a customer for which you have outstanding balances, would you? By studying your data, I come to understand your values, needs, requirements, and policy constraints."

"Can I add new fields to existing files?" Jeff asked. "The second consultant told me that she'd have to rewrite many of the existing programs."

"Unfortunately, that's probably true. You see, the data files are too closely associated with the programs that use them. The programs expect the data in specific files. If you now realize that some of that data belongs in different files, the original programs must be modified to reflect the new location."

Jeff frowned. "Are you telling me that this is an unavoidable by-product of using computers? If so, I may want to scrap this whole idea."

"Absolutely not," Frank said. "Data can be structured independently of the programs that will use it. You can easily minimize the likelihood of extensive modifications to programs."

"Okay. But can you solve this problem? I'm getting reports from the system that show conflicting data. I don't see how that can happen."

"It's another design flaw," Frank answered. "I suspect that many data elements are stored redundantly in different files. When you store data in more than one location, you increase the chances that you will modify data in one location and not in the other locations. Consequently, you find data conflicts in the reports generated from the different files. Data redundancy should be minimized at all costs."

"This is interesting," Jeff said. "When our system was designed, the students just sat down and drew some sort of flow diagram based on the reports I needed. I never considered the strategy of focusing on the raw data."

Frank smiled. "If you think about it, it makes more sense. If I understand your data, help you get control of and capture that data, organize it in a way that is flexible and adaptable, you'll realize two important advantages. First, you'll still be able to generate the reports you need now. But second, you'll be better able to create new reports as the need arises. After all, the data has been captured and stored in a flexible format. We can fix this system. Give me the chance to show you how!"

Discussion

1. How does Frank's approach to identifying requirements differ from the traditional "Tell me what outputs you want" approach?
2. What benefits do you think can be derived from studying data before you study output needs and processing requirements?
3. Why do you think that consultants—and experienced analysts—so frequently ignore or do not adequately consider the future implications of the systems they design?

What Will You Learn in This Chapter?

This is the second of three graphic systems modeling chapters. In this chapter you will learn how to use a popular data modeling tool, *entity relationship diagrams*, to document a system's data, independently of how that data is or will be used—that is, independently of inputs, outputs, and processing. You will know data modeling as a systems analysis tool when you can

1. Define data modeling and explain why it is important.
2. Identify the data entities in a system that should be described by data.
3. Draw an entity relationship data model that depicts the data entities and the natural relationships between those data entities.
4. Identify the data elements that describe each data entity.
5. Explain the need to analyze an entity relationship data model for simplicity, redundancy, and flexibility.
6. Using data analysis, modify the entity relationship data model so that every data element is associated with one and only one data entity, called *third normal form*.

Beginning with this chapter, we change our emphasis from the study of the current system (study phase of systems analysis) to the definition of requirements for an improved system (definition phase of systems analysis). The improved system is sometimes called the *target system*.

How do you accurately describe and verify end-user requirements? That question has haunted analysts for many years. Requirements analysis has often been described as elusive. Why? Because we tend to focus on systems outputs and the processing required to produce those outputs. The output requirements for any system are constantly changing to adapt to new situations, needs, and problems. But requirements can be defined and analyzed. The trick is to learn how to *model* systems requirements in a form that can fulfill both today's and tomorrow's needs.

Model the system! You know how to do that! In the last chapter, you modeled the existing system with data flow diagrams. Why not use data flow diagrams to model the processes for the target system? Actually, we will do that, though not just yet.

Modern systems analysts have learned that process modeling is simplified after the data that describes the system is fully understood. Think about it. What do we do as data processors? We input *data*. We store *data* for later use. We update *data*. We process *data*, both for storage and to produce new *data* in the form of useful information. When end-user requirements change, those changes usually manifest themselves as the need for new information to be generated from *data* that already exists. *Data* is the information system's center of the universe. As Figure 8.1 demonstrates, processing develops around the data we capture and store, no matter how we choose to store it.

In this chapter, you will learn to identify, analyze, and model data to be stored in the target system, independently of the many ways that you might choose to store all or portions of that data—for example, as forms in file cabinets, traditional computer files, or modern computer databases. Our data models will also be derived independently of how the data will eventually be processed. In fact, if you properly model a system's data, the resulting system should be flexible enough to adapt to any future process and output requirements that an end-user could imagine: a lofty but attainable goal for systems development!

Logical Modeling: The Essence of a System

In the last chapter, you were introduced to systems models. Recall that a model is a graphic representation of a system. In this chapter we want to focus on models for target systems, especially those that model systems requirements—in this chapter, stored data requirements.

An important concept, in this chapter and the next, is the distinction between *logical* and *physical* models. You learned to draw physical models in Chapter 7 (physical data flow diagrams). Actually, most analysts draw physical models without even thinking about it. **Physical models** show not only what a system is or does, but also how the system is physically implemented.

FIGURE 8.1 Data Modeling Concept
Data is the center of any system's universe. Processing evolves around data. Some processing serves to capture, store, and maintain accurate data. Other processing produces useful information from stored data. In any case, understanding and modeling data allow processing to evolve as needed.

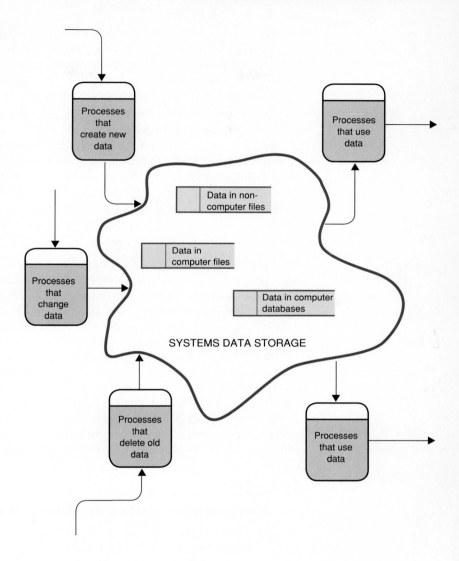

Thus a physical model of data might describe the format of the data (for example, $$$$9.99, a COBOL PICTURE clause for a data element), the media on which the data is recorded (for example, preprinted form or computer-generated report), or how the data is organized (for example, as an indexed file or a database record).

Physical models are acceptable. However, analysts have learned that requirements should ideally be specified in an implementation-independent fashion. There are several reasons for this:

- Implementation-independent models remove biases that are the result of the way the existing system is implemented or the way that any one person thinks the system might be implemented. Thus, we overcome the "we've always done it that way" syndrome. Consequently, implementation-independent models encourage creativity.

- Implementation-independent models reduce the risk of missing functional requirements because we are too preoccupied with technical details. Such errors can be costly to correct after the system is implemented. By separating *what* the system must do from *how* the system will do it, we can better analyze the requirements for completeness, accuracy, and consistency.

- Implementation-independent models allow us to communicate with end-users in nontechnical language. We frequently lose requirements in the technical jargon of the computing discipline.

Implementation-independent models are usually called **logical models**. But sometimes they are called **essential models**, because they model the **essence** of a system. During the requirements definition phase, they model the **essential requirements** of the system, those requirements that must be fulfilled no matter how we might implement the system. This chapter will present data modeling as a logical modeling technique for data storage requirements. We want to avoid the physical implications and concerns about files and databases, at least until the file and database design chapters.

Data modeling is a very popular logical modeling technique. Its origins are in the area of database design; however, data modeling has evolved into a logical systems analysis technique that can complement process models (data flow diagrams). There are numerous data modeling tools; however, we will concentrate on one, the *entity relationship diagram*.

Entity Relationship Diagrams

An **entity relationship diagram** illustrates "data at rest." This contrasts somewhat with data flow diagrams, which for the most part illustrate data in motion, as data flows. Actually, DFD data stores show data at rest relative to the processes that use and update those stores. You should think of an ERD as a more detailed picture of those data stores, independent of the processing performed with those data stores. Let's introduce the tool using a simple example.

All systems contain data—usually lots of data! Data describes persons, objects, events, and other things of interest to the business and its workers. In Figure 8.2, we see three objects of interest in a typical order-processing system: customers (who are *persons*), orders (which are *events*), and parts (which are *objects*). This figure also demonstrates that the "things" we describe

FIGURE 8.2   A Simple, Artistic Data Model   A data model depicts things about which we store data and the relationships between those things. Since most of us are not artists, we need a less artistic version of this data model.

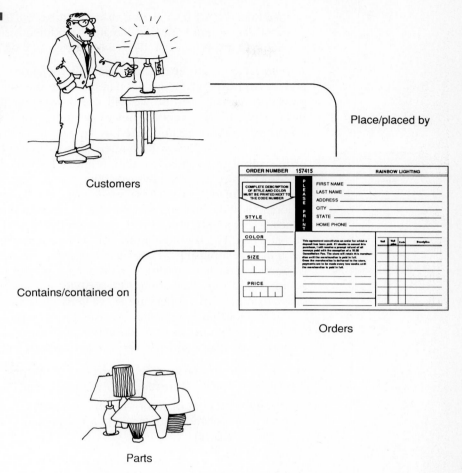

Customers

Place/placed by

Orders

Contains/contained on

Parts

with data are also naturally associated with one another. For example, a customer "places" an order (or the inverse association: an order is "placed by" a customer). Figure 8.2 can be thought of as an artistic version of an entity relationship diagram. Let's study a more formal language.

Entity Relationship Diagram Conventions and Guidelines

Like DFDs, entity relationship diagrams are quite easy to read and understand. There is, however, one basic concept that you must commit to memory:

Entity relationship diagrams do *not* depict flow or processing. They should not be read like data flow diagrams or flowcharts. Entity relationship diagrams depict data at rest, data being stored. They also do not imply how data is implemented, created, modified, used, or deleted. If you remember this concept, ERDs will be easy to read and interpret.

As with DFDs, there are various symbolic notations suggested by different authors and experts. Figure 8.3 demonstrates the popular Chen entity relationship diagram, the one we will use throughout this chapter. First, let's briefly examine the symbols. Then we will teach you how to read the diagram. The Chen symbols are duplicated in the margin for your reference.

A **data entity**, which will be referred to as an **entity** from now on, is the main symbol on an ERD. An entity is anything, real or abstract, about which we want to store data. For any given entity, the real world will contain multiple occurrences of that entity. Most entities correspond to persons, objects, events, or locations in the business environment. For the time being, we have recorded, beneath each entity, typical data elements that would describe occurrences of that entity. The list of elements is by no means exhaustive.

A **data relationship**—called a **relationship** from now on—is a natural association between entities. For instance, Figure 8.3 demonstrates that, in addition to storing data about CUSTOMERs and ORDERs, customers PLACE orders (or, inversely, orders are PLACED BY customers).

All relationships are further described by words or symbols that indicate the number of occurrences of one entity that can exist for a single occurrence of the related entity, and vice versa. There are three general possibilities:

- one-to-one (1:1)—for one occurrence of the first entity there can exist only one related occurrence of the second entity and vice versa

- one-to-many (1:M or M:1)—for one occurrence of one entity there can exist many related occurrences of a second entity; it doesn't matter which is first or second

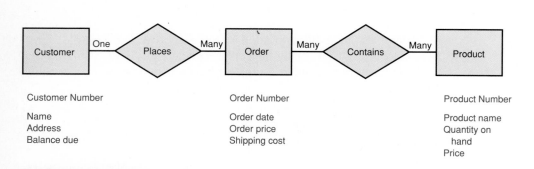

FIGURE 8.3   Entity Relationship Diagram as a Data Model   One of the most popular data modeling tools is Peter Chen's Entity Relationship Diagram. Entities (rectangles) are described by data elements, which in this example are written below their corresponding entities.

- many-to-many (M:M)—for one occurrence of the first entity, there can exist many related occurrences of the second entity, and for one occurrence of the second entity there can exist many occurrences of the first entity

Now, how do you read ERDs? *If properly labeled*, an ERD should be read as simple sentences. Figure 8.4 duplicates our ERD to demonstrate this easy-to-learn approach. The sentences describe the natural associations between our entities, which, as you should recall, are described by data. Notice that for every relationship there are two sentences, one written above and one written below the relationship. Thus, all relationships have *two interpretations*—one in each direction.

This bidirectionality is important. For example, we need to know all parts on an order (ORDER CONTAINS PART); that's what orders are all about. But similarly, we may want to find all orders for a specific part (PART CONTAINED ON ORDER). The ERD forces us to look at these natural associations in both directions.

FIGURE 8.4 How to Read an Entity Relationship Diagram Reading an ERD is easy. If properly labeled, an ERD reads as a series of sentences. Sample sentences have been written beside entities and relationships to demonstrate this concept.

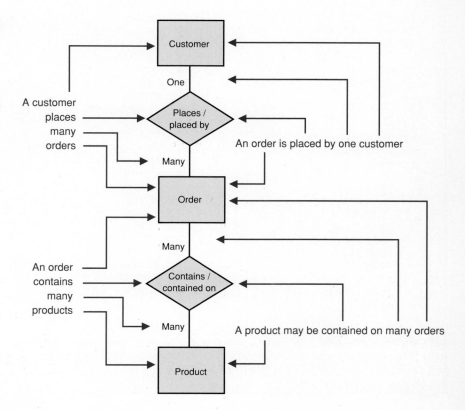

Classes and Examples of Entity Types

Entity types that describe a person, group of persons, or other living entity:

ACCOUNT
AGENCY
ANIMAL
APPLICANT
BORROWER
CHAPTER
CHILD
CLASS
CONTRACTOR
CORRESPONDENT
CLIENT
CREDITOR
CUSTOMER
DEPARTMENT
EMPLOYEE
EMPLOYER
GROUP
JOB OPENING
JOB POSITION
OFFICER
SALESPERSON
STUDENT
SUPPLIER
TEAM
VENDOR

Entity types that describe an object or group of objects:

BOOK
CHEMICAL
COURSE
EQUIPMENT
MACHINE
MATERIAL
PART
PRODUCT
SUBSTANCE
VEHICLE

Entity types that describe events (notice that most events have corresponding business forms used by companies to capture data about the event. Events are also characterized by the fact that they *happen* and/or have *duration):*

AGREEMENT
APPLICATION
APPOINTMENT
ASSIGNMENT
BACKORDER
BUDGET
CLAIM
CONTRACT
DEFECT RETURN
DEPOSIT
DISBURSEMENT
FLIGHT
FORECAST
INVOICE
JOB
LICENSE
MEETING
PAYMENT
PROJECT
PURCHASE ORDER
QUOTE
REQUISITION
RESERVATION
RESUME
SALES ORDER
SEMESTER
SHIPMENT
STAY
STEP
TASK
TEST
TIME BUCKET
WORK ORDER

Entity types that describe locations:

BRANCH
BUILDING
CITY
COUNTRY
COUNTY
ROOM
ROUTE
SALES REGION
SCHOOL ZONE
STATE
STORAGE BIN
ZONE

Note: Many of the above entity types may actually be data elements in some systems. For example, STATE may be an element that describes the entity type SUPPLIER. When does state become an entity type? When the system needs to store two or more data elements that specifically describe a STATE.

Now that you understand the basic symbols and how to read the diagram, let's study entity relationship diagramming conventions in greater detail.

Entities   Once again, an entity is anything that can be described by data. All business systems seek to capture and store data about various entities, regardless of whether computers are used. Entity types tend to fall into four classes: persons, objects, locations, and events. Examples are listed in Figure 8.5.

Notice that we name entities with *nouns* that describe the person, object, event, or location they represent. Names should be singular so as to distinguish the logical concept of the entity from occurrences of the entity. Names may include appropriate adjectives or clauses to better describe the entity—for instance, *PENDING* ORDER, to distinguish it from *CLOSED* ORDER.

We've said that entities are anything described by data; however, we've said little about the data itself. **Data elements** are characteristics that are common to a particular entity. In Figure 8.6 we see a collection of data elements that describe a typical entity, PART. Underneath this logical picture of the entity, we demonstrate multiple occurrences of the PART entity and the values its data elements take on. Usually, at least one data element takes on a unique value for the entity. In our example, PART NUMBER is the key. We will call this (these) data element(s) the **key**. A key uniquely identifies one, and only one, occurrence of the entity. We'll discuss keys in more detail later in this chapter. They will prove very important when we analyze our data model.

Relationships   Relationships are natural associations between one or more entities. These associations can be relatively quickly identified after the entities are identified. Relationships are important. Whenever we store data in forms, files, and databases, we must understand the potential implications of updating that data. For instance, we shouldn't delete a CUSTOMER for which there are outstanding INVOICEs, unless of course we don't care that we might not collect payment from that customer for the delivered goods. The study of such updating requirements is enhanced through inspection of the data model.

Recall that entities were named with nouns. Relationships should be named with *verbs*. This enhances the *simple sentence* interpretation described earlier in the chapter. Although relationships could be named with two verbs to illustrate both directions, this practice clutters the diagram. We will name only one direction—the other should be considered implicit.

There are several types of relationships. Most relationships exist between two different entities. Such a relationship exists between the entities COURSE and STUDENT (Figure 8.7a). Notice, also, that many relationships can be established between the same two entities. In this case, one relationship describes the courses a student is currently taking (present tense). The other describes courses that the student has taken (past tense).

Relationships may also exist between different occurrences of the same entity. The classic example is a bill of materials (Figure 8.7b). This relationship

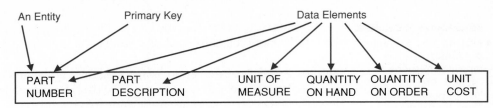

PART NUMBER	PART DESCRIPTION	UNIT OF MEASURE	QUANTITY ON HAND	QUANTITY ON ORDER	UNIT COST
567M	3/4 inch lug nut	Dozen	43	20	8.98
567P	1 inch lug nut	Dozen	0	100	9.98
568A	1/4 inch fitted nut	Each	134	0	5.98
•					
•					
•					
689	1/2 inch hose	Foot	90	0	3.98
690	3/4 inch hose	Foot	7	100	4.39
691	1 inch hose	Foot	66	0	5.98
•					
•					
•					
745	1/2 inch hose clamp	Each	107	0	.98
746	3/4 inch hose clamp	Each	21	20	.98
747	1 inch hose clamp	Each	0	30	1.98
•					
•					
•					

FIGURE 8.6 Data Elements That Describe an Entity   Every entity is described by data elements that pertain to that entity. For occurrences of these entities, the data elements take on values. One element, the key, takes on a unique value for each occurrence of the entity.

shows that an occurrence of PART may be built from occurrences of different PARTs. Similarly, a PART may be used to build one or more other PARTs.

Relationships can also be dependent on one another. The word *OR* in Figure 8.8(a) demonstrates that both relationships cannot exist for a single occurrence of ORDER. An ORDER may be placed for PARTs or SERVICE, but not both. Thus, there may exist an occurrence of the relationship between an ORDER and PARTs **or** an ORDER and a SERVICE.

FIGURE 8.7 Types of Relationships

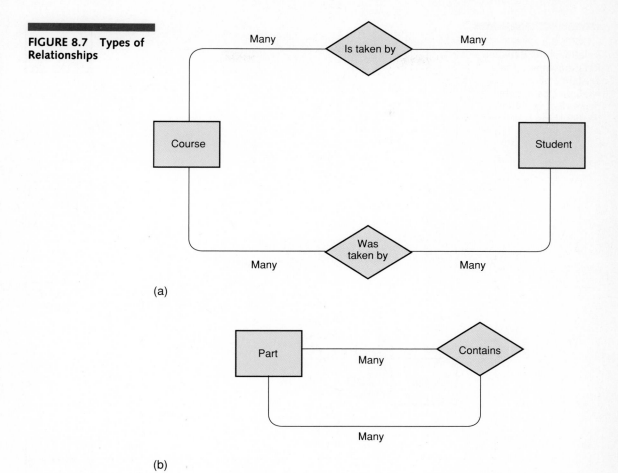

(a)

(b)

Similarly, the word *AND* in Figure 8.8(b) illustrates a relationship where for a specific ORDER, if it has been shipped (SHIPMENT), it must also be billed (INVOICE). As you can see, relationships demonstrate important associations that must be maintained between different collections of data to be stored!

Earlier we suggested that "for one customer there could be many orders." This was intentionally oversimplified. You could correctly argue that, at any given point in time, "for one customer, there might be no (zero) current orders." Obviously, we wouldn't delete customer data just because that customer has no current orders. We need a more complete notation. Figure 8.9 demonstrates a richer vocabulary for describing the complexity of relationships. A shorthand notation is also offered (in color). Study each relationship to see if you understand its complexity (or simplicity, as the case may be).

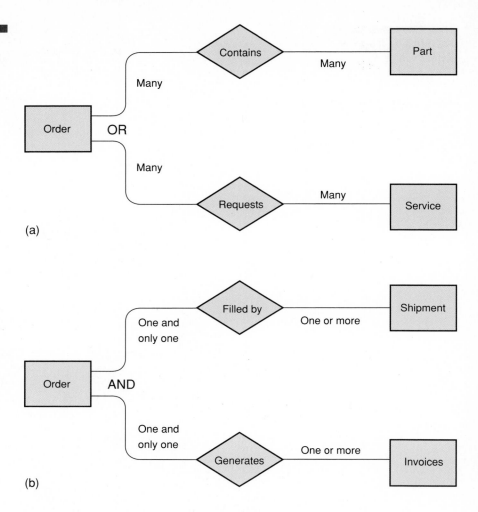

(a)

(b)

The chart has been annotated to help you properly interpret the complexity of the relationships. Now, check your interpretation against ours:

- For a CUSTOMER, there can be **zero or more** ORDERs. On the other hand, for an ORDER, there must be **one and only one** CUSTOMER.

- For an ORDER, there must exist **one or more** PARTs on that order. Similarly, for a PART, there can be **zero or more** outstanding ORDERs for that part.

How did you do?

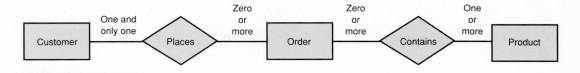

FIGURE 8.9   Richer Vocabulary for Relationships

Computer-Assisted Systems Engineering (CASE) for ERDs

Computer-Assisted Systems Engineering (CASE) was introduced in Chapter 5 as an emerging, enabling technology for systems analysis and design methods. Virtually all CASE products support computer-assisted data modeling. Most CASE products specifically support entity relationship diagrams and/or equivalent data modeling tools. CASE takes the drudgery out of drawing and maintaining the diagrams.

Using a CASE product, you can easily create and link professional, readable ERDs without the use of paper, pencil, erasers, and templates. The ERDs can be easily modified to reflect corrections and changes suggested by end-users; you don't have to start over! Also, most CASE products provide powerful analytical tools that can check your ERDs for mechanical, completeness, and consistency errors. Some CASE products perform or verify the data analysis of a data model. (Data analysis, an important but time-consuming technique, will be described later in this chapter.) The potential time and quality savings are substantial.

All of the ERDs in this book were created with a popular CASE product called Excelerator. Only the color screens and annotations were added by an artist. All of the entity types on our ERDs were automatically cataloged into a dictionary so that we can add detailed information, such as data elements, as the project moves through analysis, design, and implementation. Figure 8.10 demonstrates a sequence of Excelerator screens for computer-assisted data modeling.

Using Entity Relationship Diagrams for Systems Analysis

Entity relationship diagrams have many uses. ERDs or equivalent models have become an integral part of several methodologies, including Information Engineering (Martin), Information Modeling, Object-Oriented Design (Mellor), (Yourdon), and most recently, Modern Structured Systems Analysis.

Data Modeling as a Structured Systems Analysis Technique

We've already described data modeling as a definition phase tool for systems analysis. Let's place the role of data modeling into better perspective. Recall that systems analysis consists of four phases: survey, study, definition, and selection. During the **survey** and **study phases**, you are primarily defining problems and opportunities. But during these phases, you also begin collecting samples and examples of documents, forms, files, and reports that will eventually be used for data modeling. These samples teach you the following:

- the entities about which the system is currently storing data
- what data is currently used and not used (on or in the documents, forms, files, and reports)
- how the data is currently used
- the vocabulary or language of the system and its end-users—perhaps the most important knowledge gained through studying forms and files
- how the data is currently captured and stored
- rules concerning the data and its use and storage

The **definition phase** looks at requirements independently of the ways that we will implement those requirements (we've been calling this *logical modeling*). This is the phase when data modeling is actually performed. You identify the entities, data elements, and relationships. You draw the data model. Furthermore, you improve the data model through a technique called *data analysis*, which is discussed later in this chapter.

As will be demonstrated in the next chapter, data models complement process models that are normally developed as part of the Structured Systems Analysis Methodology. In fact, modern Structured Analysis considers the data model to be an essential complement to data flow diagrams.

The **selection phase** of systems analysis identifies and analyzes alternative solutions. The following selection phase issues are addressed:

- How much of the data model and data elements is already stored in computer files and databases? To what degree can such files or databases be used or modified? The answers define the new files or database that must be designed.
- How much of the data should be automated? How much will be stored in manual files? Will any data require both approaches? Sometimes we choose to computerize the most frequently used data to speed access. The full set of data is then stored in paper files or the equivalent.
- For that portion of the data model that represents new *computer data storage*, should the data be centrally stored in a database or distributed

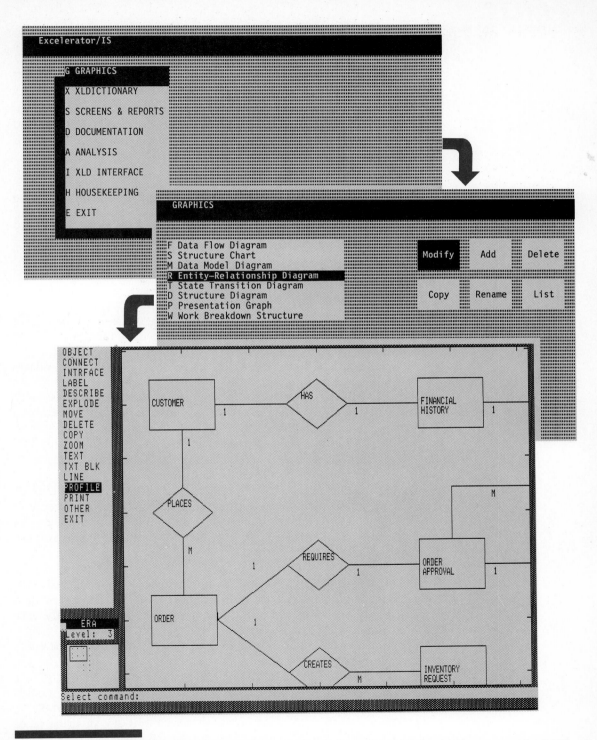

FIGURE 8.10   Computer-Assisted Tools for ERDs   This sequence of screens demonstrates a popular CASE tool's usefulness in drawing data models.

between separate files or databases? Should it be stored on one computer or several computers?

- How should the data be physically stored? For instance, files can be organized as sequential, relative, or indexed files. Databases can be organized according to a variety of methods and products, such as IMS, IDMS, dBASE IV, and many other database and file managers.
- How will relationships between data entities be stored?

Notice that the above questions demonstrate our progression from the logical, implementation-independent model into a more physical, implementation-dependent model. As the project moves into systems design and implementation phases, the model will become more and more physical.

How to Perform Data Modeling with Entity Relationship Diagrams

You already know enough about ERDs to read them. But as a systems analyst or knowledgeable end-user, you must learn how to draw them. We will use the Entertainment Club systems project to teach you how to draw ERDs.

The scenario is as follows. We have completed the survey and study phases for our Entertainment Club system. We fully understand the current system's strengths, weaknesses, limitations, problems, opportunities, and constraints. We discovered that the member services system processes orders for both audio/video products and general merchandise (for example, posters, T-shirts, and the like). However, because general merchandise accounts for only 3 percent of total sales revenue and 5 percent of total transactions, it has been eliminated from the scope of the continuing project.

We begin the definition phase by defining data storage requirements for the new system. Consistent with our logical modeling theme, these requirements will be implementation-independent. We intend to carefully walk you through examples and provide a future reference for the day when you use this popular modeling tool.

Step 1: Identify Entities   The first task in data modeling is relatively easy. You need to identify in the system those entities that are or can be described by data. You should not restrict your thinking to entities about which the end-users know they want to store data.

A good quality check for each entity is to ask yourself whether the entity has multiple occurrences. If not, it isn't an entity.

Another technique for identifying entities is to study the forms and files sampled during the study phase. Be careful, though! Some forms identify event entities. Examples include ORDERs, REQUISITIONs, PAYMENTs, DEPOSITs, and so forth. But most of these same forms also describe other entities. Consider a course request form for your school's course registration system. A COURSE REQUEST is itself an event entity. But the average course request form also contains items that describe other entities such as STUDENT (a person), COURSE (an event), INSTRUCTOR (a person), ADVISER (a combination), SECTION (a somewhat abstract location), and the like. These same entities would also be derived from other course registration system forms and files.

After brainstorming and studying the forms for the Entertainment Club's member services system, the entities in Figure 8.11 were identified. The list will change as we walk through the remaining steps.

PRODUCT (an object). Audio and video titles sold to members through the club: records, tapes, compact discs, videodiscs, and other audio/video media.

CLUB (a group of persons). One of the clubs to which members can belong - for instance, Records & Tapes, Compact Disc, Video, and Games.

MEMBER (a person). A member of one or more clubs.

AGREEMENT (an event). A contract whereby a member agrees to purchase a certain number of products within a certain time period. After fulfilling that agreement, the member will receive bonus credits, as specified in the agreement, for each additional purchase.

PROMOTION (an event). Promotions are monthly or quarterly events whereby automatic, dated orders are created for all members in a club. The promotion specifies a "Selection of the Month" that will automatically be filled unless it is canceled or revised by the member within a specified time period.

ORDER (an event). An automatic, dated order generated in response to a promotion. The order may be approved, revised, or canceled via the member's response.

INVOICE (an event). An invoice generated in response to a shipped or partially shipped order.

FIGURE 8.11 Entities for the Entertainment Club

Step 2: Identify or Define Keys for Each Entity   Next, identify the data element or elements that uniquely identify one and only one *occurrence* of each entity. You know these as **primary keys**. In many cases, these keys already exist in either computer files or printed forms. Some special terms expand on the concept of keys.

Some entities have more than one key. These are called **candidate keys**, since each element, by itself, is a candidate for uniquely identifying one and only one occurrence of the entity. For instance, both MEMBER NUMBER and MEMBER NAME are candidate keys for the entity MEMBER. (MEMBER NAME is allowed as a key since it is unique in most, but not all, cases.)

Some entities may not have explicit keys. If so, you have two options. First, you should establish a key using the combination of two or more existing elements that, together, always identify one and only one occurrence of the entity. We'll call this a **combination key**, though database experts may use the term *concatenated key*. For instance, each occurrence of PRODUCT is uniquely identified by the combined values of PRODUCT NUMBER and MEDIA CODE (thus, "27654 + cassette" is a different occurrence of PRODUCT than "27654 + compact disc").

Alternatively, you should consider inventing a simpler candidate key as an alternative to the above combination key. Simpler keys—discussed fully in Chapter 10 under the topic of *codes*—can be extremely useful, not only for computer files but also for easier inquiries and forms.

What do you do if you have two or more entities that have the same key? You may find situations where different entities legitimately share the same key. Consider the event entity ORDER. Other entities, such as BACKORDER (an entity we forgot to define but will need), are triggered by ORDERs. It is not uncommon for a business to assign the same key, in this case ORDER NUMBER, to all of the related entities. This makes the ORDER easier to track. Each entity is a distinct event. You could, however, create a combination key for each entity to more clearly distinguish between the entities. For example, we can give BACKORDER the combination key ORDER NUMBER + BACK-ORDER DATE.

The keys for our entities are depicted in Figure 8.12. We have adopted the common practice of underlining primary keys.

If you cannot define keys for an entity, it may be that the entity doesn't really exist; that is, *multiple* occurrences of the so-called entity do not exist.

Step 3: Draw a Rough Draft of the Entity Relationship Data Model   The next task in data modeling is to build a first draft of the entity relationship diagram. We complete the draft model by brainstorming relationships between our entities. We have completed this step in Figure 8.13. Since this is your first substantial ERD, let's walk through it:

Ⓐ   A CLUB—for example, record, cassette, compact disc, and so on—ESTAB-LISHES one or more membership AGREEMENTs. Members will learn

FIGURE 8.12   Keys for Entities

PRODUCT -->	PRODUCT NUMBER + MEDIA CODE (note 1)
CLUB -->	CLUB NAME or CLUB ABBREVIATION (note 2)
MEMBER -->	MEMBER NUMBER or MEMBER NAME (note 3)
AGREEMENT -->	AGREEMENT NUMBER
PROMOTION -->	CLUB NAME + PROMOTION DATE (note 1)
ORDER -->	ORDER NUMBER
BACKORDER -->	ORDER NUMBER + BACKORDER DATE (note 4)

Note 1   A combination key.

Note 2   CLUB ABBREVIATION does not currently exist. We are proposing it as a candidate key.

Note 3   MEMBER NAME is a candidate key that is not truly unique (there could be two John Smiths). In most systems it is, however, considered unique enough to qualify as a key.

Note 4   We have added the BACKORDER entity we originally overlooked in step 1. You can add entities at any time that you discover them. The actual key is ORDER NUMBER. SHIPMENT DATE was combined with ORDER NUMBER to more clearly distinguish occurrences of BACKORDER from ORDER.

about these agreements through advertisements in newspapers and magazines.

A specific AGREEMENT is ESTABLISHED FOR one and only one CLUB.

Ⓑ A CLUB ENROLLS zero or more MEMBERs. Why zero? The club may be new and awaiting receipt of its first membership application.

A MEMBER may ENROLL IN one or more CLUBs. Note that this is our first many-to-many relationship.

Ⓒ A MEMBER is BOUND TO one or more AGREEMENTS, depending on how many clubs that member has joined.

An AGREEMENT BINDS zero or more MEMBERs. Again, the agreement may be awaiting its first member enrollment.

Ⓓ Each month, a CLUB SPONSORS one or more marketing PROMOTIONs.

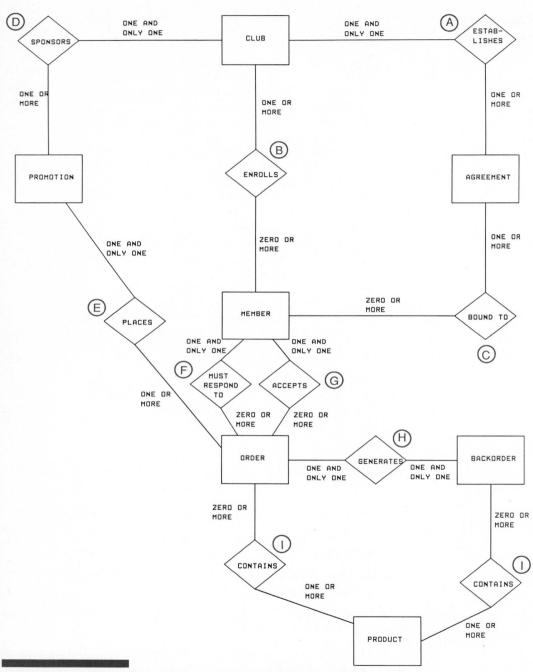

FIGURE 8.13 Initial Entity Relationship Data Model for the Entertainment Club

Each PROMOTION, however, is SPONSORED BY one and only one CLUB.

(E) A club's PROMOTION automatically PLACES one or more ORDERs for members in that club.

Members will have to respond by a specific date to cancel the order or have it filled. Most, but not all, ORDERs are PLACED BY one and only one PROMOTION. The lone exception is those ORDERs placed as a result of a member first joining the club (a bonus for joining).

(F) A promotion's ORDER is PLACED FOR one and only one MEMBER.

A MEMBER MUST RESPOND TO zero or more outstanding ORDERs. Note that at any given time, a member may have no outstanding orders to which he or she must respond. Alternatively, if a member belongs to more than one club, he or she may have several orders.

(G) Members will respond to orders by canceling, approving, or altering the orders. If the member doesn't respond to the order, the order will automatically be shipped. Thus, unless the order is canceled, a MEMBER ACCEPTS zero or more ORDERs (he or she may not currently have any orders to accept or reject).

An ORDER is ACCEPTED BY one and only one MEMBER.

Note that the relationships marked (F) and (G) are exclusive; that is, an ORDER may either be awaiting the member's response or it has been accepted, but not both.

(H) If one or more products cannot be filled due to lack of stock, an ORDER GENERATES one and only one BACKORDER.

Similarly, a BACKORDER is GENERATED FOR one and only one ORDER. This is the first one-to-one relationship we've encountered.

(I) An ORDER (or BACKORDER) CONTAINS one or more PRODUCTs, usually the product being advertised as the "Selection of the Month."

At the same time, a specific PRODUCT may be CONTAINED ON zero or more ORDERs or BACKORDERs (zero if the product is not currently selling well). Notice how this inverse relationship could help us track sales of all products. Also note that these are many-to-many relationships.

If you read each of the preceding items carefully, you probably learned a great deal about the Entertainment Club. Indeed, on many occasions end-users and beginning analysts have noted that ERDs seem to explain the total business picture better than data flow diagrams. They may be right! Some analysts use ERDs as a study phase (current system) tool. It wouldn't increase the definition phase workload. Why not? Data models of current and target systems tend to be nearly identical because the types of data stored tend *not* to change significantly over time—quite the opposite of process models for current and target systems.

Step 4: Identify Data Elements This may seem like a trivial task; however, analysts not familiar with the data modeling method frequently encounter problems. To accomplish this task, you must have a thorough understanding of the data elements for the system. By studying the forms, files, and reports, you identify those data elements that are essential to the system you are developing. We suggest the following strategy, which is best accomplished by working *with* your end-users:

1. Pick a form, document, printout of a file, report, or other sample of data. We will demonstrate the remaining steps using a sample form.
2. Circle each unique item on the form (Figure 8.14). Notice that there is not always a one-to-one correspondence between an item on the form and its associated element. For instance, check boxes frequently represent a single element since they are actually different values for that element.
3. Draw an X through circled items that won't be stored by the new system.
4. Draw an X through extraneous items such as signatures (assuming that you circled them).
5. Put an X through constant information. In other words, if every occurrence of the form has the same value for a field, that field is not an element. For instance, the "Remit Payment To" entry will always have the value "SoundStage Entertainment Club." It is therefore not an element.
6. Verify your elements with your end-users, especially if there is any question as to whether some of the elements are really needed.

Another recommended technique for identifying elements is **brainstorming**. Pick an entity and ask yourself what characteristics (data elements) describe that entity.

You must also logically name the data elements. We did so on the sample form. Don't they already have names? They do; however, forms and files usually have abbreviated element names, also called *labels*. On those forms or in those files, the names make sense. Taken out of that context, they don't always make sense. Consider the element NAME. On the order form, we know that this is a MEMBER NAME. Taken out of the context of the form, there could be other interpretations: ARTIST NAME, PRODUCT NAME, or CLUB NAME. Naming guidelines for data elements follow:

- Do not use abbreviations unless absolutely necessary. For example, a label may read COD. After studying completed forms, you find that a dollar amount is actually recorded. The correct logical name is AMOUNT TO COLLECT ON DELIVERY.

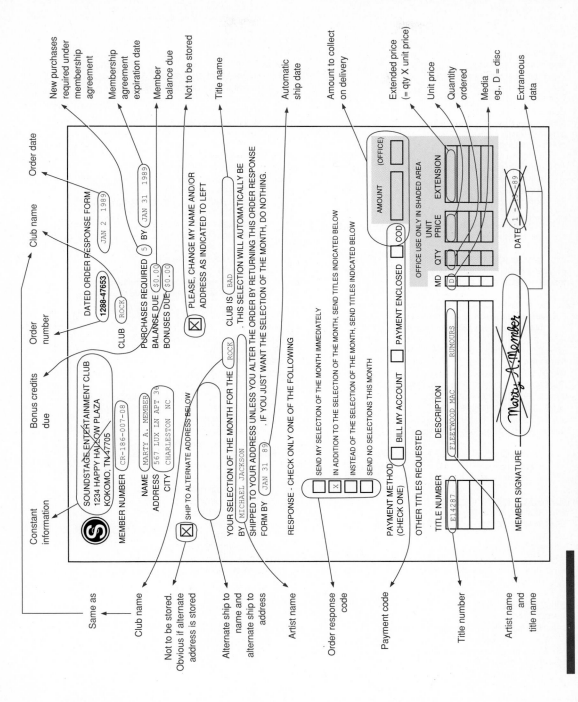

FIGURE 8.14 Forms Sampling Data elements are gleaned from sample forms and files. This form has been marked to indicate required data elements.

Entity: PRODUCT (Contents determined by studying the master product catalog)

PRODUCT NUMBER + MEDIA CODE
TITLE OF WORK
COPYRIGHT DATE
ARTIST/PRODUCER NAME
LABEL/DISTRIBUTOR NAME
QUANTITY ON HAND
LIST PRICE
CURRENT RETAIL PRICE
SUPPLIER NUMBER
SUPPLIER NAME
SUPPLIER ADDRESS GROUP
    SUPPLIER STREET ADDRESS
    SUPPLIER POST OFFICE BOX NUMBER
    SUPPLIER CITY
    SUPPLIER STATE
    SUPPLIER ZIP CODE

Entity: CLUB (Contents determined through brainstorming with Member Services Manager)

CLUB NAME or CLUB ABBREVIATION
NUMBER ENROLLED YEAR-TO-DATE
NUMBER CANCELED YEAR-TO-DATE

Entity: AGREEMENT (Contents determined by studying advertised agreements from the past five years. Additional elements added by the end-users)

AGREEMENT NUMBER
CLUB NAME
MAXIMUM PERIOD OF OBLIGATION
NUMBER OF PURCHASES REQUIRED WITHIN PERIOD
BONUS CREDITS/PURCHASE AFTER OBLIGATION
NUMBER OF MEMBERS ENROLLED
NUMBER OF MEMBERS WHO HAVE FULFILLED
NUMBER OF MEMBERS WHO HAVE NOT FULFILLED

Entity: PROMOTION (Contents determined through interviews of Marketing personnel and from promotional releases sent to members)

CLUB NAME + PROMOTION DATE
SELECTION OF MONTH NUMBER
SELECTION OF MONTH TITLE
RELEASE DATE
AUTOMATIC FILL DATE

Entity: BACKORDER (Contents determined by studying memos and notes generated in response to stock outs)

ORDER NUMBER + BACKORDER DATE
MEMBER NUMBER
BACKORDERED PART GROUP (repeats 1-n times):
    PRODUCT NUMBER
    MEDIA CODE (audio/video products only)
    PRODUCT DESCRIPTION
    QUANTITY BACKORDERED

Entity: MEMBER (Contents determined by studying a file printout from the existing dBASE IV system. A few new elements added at request of end-users)

MEMBER NUMBER or
MEMBER NAME
MEMBER ADDRESS GROUP:
    MEMBER STREET ADDRESS
    MEMBER POST OFFICE BOX NUMBER
    MEMBER CITY
    MEMBER STATE
    MEMBER ZIP CODE
    MEMBER PHONE NUMBER
DATE OF FIRST MEMBERSHIP
BALANCE PAST DUE
BONUS CREDITS NOT USED
CLUB GROUP (repeats 1-n times)
    CLUB NAME
    MUSICAL/MOVIE PREFERENCE CODE
    AGREEMENT NUMBER
    AGREEMENT ENROLLMENT DATE
    AGREEMENT EXPIRATION DATE
    NUMBER OF PURCHASES REQUIRED
    NUMBER OF PURCHASES TO DATE

Entity: ORDER (Contents determined by studying 500 order forms to be sent to members as part of promotions)

ORDER NUMBER
ORDER DATE
ORDER STATUS (added by end-users)
MEMBER NUMBER
MEMBER NAME
MEMBER ADDRESS GROUP:
    MEMBER STREET ADDRESS
    MEMBER POST OFFICE BOX NUMBER
    MEMBER CITY
    MEMBER STATE
    MEMBER ZIP CODE
    MEMBER PHONE NUMBER
ORDERED PRODUCT GROUP (repeats 1-n times):
    PRODUCT NUMBER
    MEDIA CODE
    PRODUCT DESCRIPTION
    QUANTITY ORDERED
    QUANTITY SHIPPED
    PRODUCT PRICE
    EXTENDED PRODUCT PRICE
TOTAL ORDER PRICE (sum of extended prices)
PREPAID AMOUNT
TOTAL ORDER BALANCE

- Expand on generic labels. A label may say DATE, but date of what? Try ORDER DATE. This distinguishes the element from dates for other event entity types. Other examples include CUSTOMER NAME instead of NAME, QUANTITY ORDERED instead of QUANTITY, and ORDER AMOUNT DUE instead of DUE.

- Add a question mark to data elements whose values are *yes* and *no*. For example, the element name CHANGE OF ADDRESS? is more descriptive than CHANGE OF ADDRESS. The first clearly implies that the value of the element answers the question. You can't be sure what the values of the second element name are.

- Some data elements on forms are not labeled. Recall the example of a multiple-choice element. The form describes the values of the element instead of the element itself. You must name the element (see MEDIUM on the sample form).

Step 5: Map Data Elements to Entities   Next, you need to map all data elements from step 4 to the entities. We suggest the following approach:

1. Pick an entity from your data model.
2. Find the forms, file printouts, reports, and so on whose data describes the entity.
3. Record the elements for that entity.
4. If you can't find any sample data for an entity, interview end-users to identify appropriate or desired data elements.
5. Repeat steps 1 to 4 for each entity.

Review these contents of entities with appropriate end-users and managers. Brainstorm new elements wherever possible. In general, elements need not be deleted since the data model is not a design commitment.

The initial mapping of our data elements to entities is illustrated in Figure 8.15. Pay particular attention to the comments that describe where the data elements came from. They illustrate a great variety of factual sources.

Step 6: Data Analysis—An Introduction   We've now created a data model including ERDs and a list of data elements that describe each entity. But how do we know if our data model is "good?" Admittedly, this is a subjective term. Let's define "good" for a data model:

- A good data model is *simple*. As a general rule, the data elements that comprise any entity should describe only that entity. Look at the contents of our ORDER entity (Figure 8.15). Does MEMBER NUMBER describe an ORDER? No, it really describes a MEMBER. Our model is not as simple as it could be.

    Also, an entity is considered simple if, for one occurrence of the entity, all its data elements assume one and only one value. Look again at the

◄ **FIGURE 8.15   Initial Mapping of Data Elements to Entities They Describe**

contents of the ORDER entity. For a single order, all of the data elements that make up the group ORDERED PRODUCT will assume many values, one value for each product on the order. Thus, we again see that our model can be simplified.

- A good data model is *nonredundant*. This means that no data element describes more than one entity. We definitely have some problems here. Referring again to Figure 8.15, MEMBER ADDRESS has been mapped to the MEMBER and ORDER entities. Such redundancy, if physically implemented in our eventual design, can wreak havoc with data consistency. Suppose we get a change of address for a member. We must redundantly change three records or risk having different packages, letters, and so on sent to different addresses.

  We also have some implicit redundancy in our model. PRODUCT DESCRIPTION (in the ORDER entity) may be logically equivalent to TITLE OF WORK (in the PRODUCT entity).

- A good data model should be *flexible* and *adaptable* to future needs. We tend to design files and databases to fulfill today's requirements. Such a data model is sometimes called an *applications* data model because it is conceived to support today's applications needs. Then, when new needs arise, we can't change the files or databases without rewriting many or all of the programs that used those files and databases. We can't change the fact that we tend to work on applications; however, we can make our data model as application-independent as possible to encourage the eventual design of files and databases that are not dependent on changing data and information requirements. We call this a *subject* data model.

Fortunately, there is a technique for correcting these problems and deriving a "good" subject data model for the application we are currently building. The technique is called **data analysis**. It involves the study of the data elements in the entities to refine our first-draft data model. Data analysis uses a procedure called **normalization** to simplify entities, eliminate redundancy, and build flexibility and adaptability into the data model. Normalization is a three-step procedure that places the data model into what we call **third normal form (3NF)**. To get there, we must first place our entities into **first normal form (1NF)**, and then **second normal form (2NF)**. Don't be intimidated by the terminology; it's easier than it sounds. Let's walk through the steps of data analysis and normalization.

Step 6(a): Place Entities into 1NF The first step in data analysis is to place each entity into **first normal form**. Simply stated, an entity is in 1NF if there are no elements or groups of elements that repeat for a single occurrence of the entity. Any elements that repeat actually describe an occurrence of a separate entity, possibly an entity that we haven't yet defined in our data model.

Look again at Figure 8.15. Which entities are not in 1NF? You should find three such entities: MEMBER, BACKORDER, and ORDER. Each contains a group of elements that "repeats 1–n times" for a single occurrence of the entity. For example, a MEMBER can contain data about membership in more than one CLUB. An ORDER or BACKORDER can contain data about more than one ordered PRODUCT. How do we fix these three entities to place them in 1NF?

Let's begin with an easy case, the ORDER entity. Figure 8.16(a) demonstrates the procedure. The unnormalized entity is on the left-hand side of the page. The entities in 1NF are on the right-hand side. Also, since normalization changes the graphic data model, all of these normalization figures show not only the redistribution of elements, but also the portion of the entity relationship data model that has changed.

First, we take the repeating group of elements out of the ORDER entity. That alone places ORDER in 1NF. But what do we do with the removed group? We created a new entity, ORDERED PRODUCT. Each occurrence of the elements describes one PRODUCT on a single ORDER. Thus, if an order contains five products, there will be five occurrences of the ORDERED PRODUCT entity. The key of ORDERED PRODUCT is similarly created by combining the key of the original entity, ORDER NUMBER, with the key element of the ex-repeating group, PRODUCT NUMBER.

It may help to think about what we've done from a graphic data modeling viewpoint. Look again at Figure 8.16(a). We have added the new entity ORDERED PRODUCT between the entities ORDER and PRODUCT. This structure replaces the original, direct relationship between ORDER and PRODUCT. ORDERED PRODUCT is sometimes called an **associative entity** since it describes data about the association between two other entities, in this case one PRODUCT's association with various ORDERs, or one ORDER's association with various PRODUCTs. Most associative entities are usually derived from many-to-many associations, such as the one that originally existed between ORDER and PRODUCT. That many-to-many relationship is no longer needed now that the associative entities and relationships are added to the data model. After we complete our discussion of data analysis, we'll redraw the entire data model.

Now, let's look at a somewhat more challenging example of 1NF, MEMBER (Figure 8.16b). The repeating group, CLUB GROUP, is easy to spot. But let's study the elements in the group. The element MUSICAL/MOVIE PREFERENCE CODE describes one MEMBER's association with one CLUB. The other elements, however, describe one MEMBER's association with a specific AGREEMENT. Thus, we create two new entities:

- MEMBERSHIP, whose data describes one MEMBER in one CLUB; thus, the key would be MEMBER NUMBER + CLUB NAME.

- CONTRACT, whose data elements describes one MEMBER as bound by one general AGREEMENT; thus, the key would be MEMBER NUMBER + AGREEMENT NUMBER.

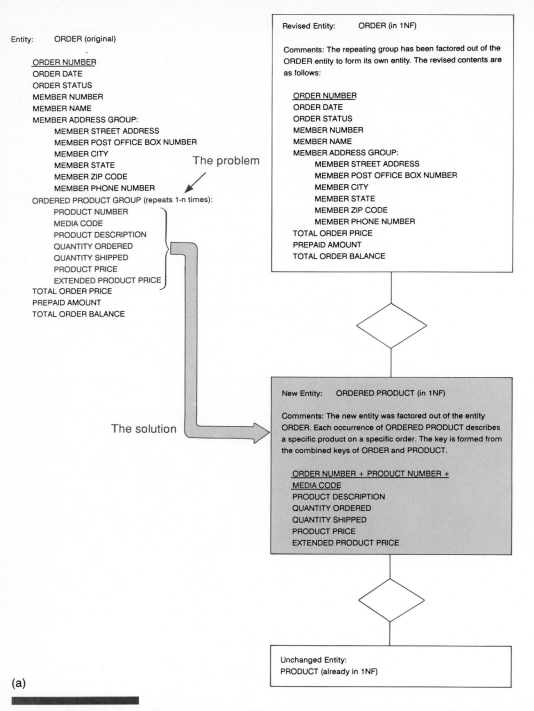

Entity: ORDER (original)

ORDER NUMBER
ORDER DATE
ORDER STATUS
MEMBER NUMBER
MEMBER NAME
MEMBER ADDRESS GROUP:
    MEMBER STREET ADDRESS
    MEMBER POST OFFICE BOX NUMBER
    MEMBER CITY
    MEMBER STATE
    MEMBER ZIP CODE
    MEMBER PHONE NUMBER
ORDERED PRODUCT GROUP (repeats 1-n times):
    PRODUCT NUMBER
    MEDIA CODE
    PRODUCT DESCRIPTION
    QUANTITY ORDERED
    QUANTITY SHIPPED
    PRODUCT PRICE
    EXTENDED PRODUCT PRICE
TOTAL ORDER PRICE
PREPAID AMOUNT
TOTAL ORDER BALANCE

The problem

Revised Entity: ORDER (in 1NF)

Comments: The repeating group has been factored out of the ORDER entity to form its own entity. The revised contents are as follows:

ORDER NUMBER
ORDER DATE
ORDER STATUS
MEMBER NUMBER
MEMBER NAME
MEMBER ADDRESS GROUP:
    MEMBER STREET ADDRESS
    MEMBER POST OFFICE BOX NUMBER
    MEMBER CITY
    MEMBER STATE
    MEMBER ZIP CODE
    MEMBER PHONE NUMBER
TOTAL ORDER PRICE
PREPAID AMOUNT
TOTAL ORDER BALANCE

The solution

New Entity: ORDERED PRODUCT (in 1NF)

Comments: The new entity was factored out of the entity ORDER. Each occurrence of ORDERED PRODUCT describes a specific product on a specific order. The key is formed from the combined keys of ORDER and PRODUCT.

ORDER NUMBER + PRODUCT NUMBER + MEDIA CODE
PRODUCT DESCRIPTION
QUANTITY ORDERED
QUANTITY SHIPPED
PRODUCT PRICE
EXTENDED PRODUCT PRICE

Unchanged Entity:
PRODUCT (already in 1NF)

(a)

FIGURE 8.16 **Transformation to First Normal Form** Repeating elements are moved to their own entity types to improve data flexibility and adaptability.

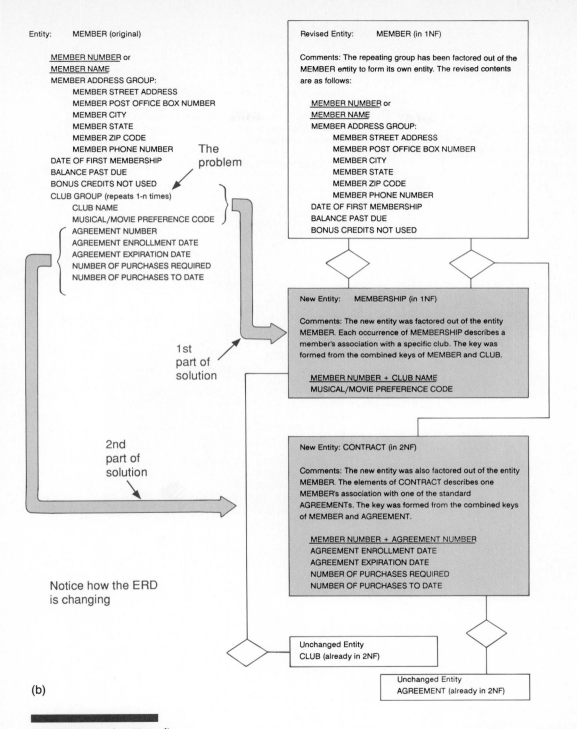

Entity: MEMBER (original)

MEMBER NUMBER or
MEMBER NAME
MEMBER ADDRESS GROUP:
    MEMBER STREET ADDRESS
    MEMBER POST OFFICE BOX NUMBER
    MEMBER CITY
    MEMBER STATE
    MEMBER ZIP CODE
    MEMBER PHONE NUMBER
DATE OF FIRST MEMBERSHIP
BALANCE PAST DUE
BONUS CREDITS NOT USED
CLUB GROUP (repeats 1-n times)
    CLUB NAME
    MUSICAL/MOVIE PREFERENCE CODE
AGREEMENT NUMBER
AGREEMENT ENROLLMENT DATE
AGREEMENT EXPIRATION DATE
NUMBER OF PURCHASES REQUIRED
NUMBER OF PURCHASES TO DATE

The problem

1st part of solution

2nd part of solution

Notice how the ERD is changing

Revised Entity: MEMBER (in 1NF)

Comments: The repeating group has been factored out of the MEMBER entity to form its own entity. The revised contents are as follows:

MEMBER NUMBER or
MEMBER NAME
MEMBER ADDRESS GROUP:
    MEMBER STREET ADDRESS
    MEMBER POST OFFICE BOX NUMBER
    MEMBER CITY
    MEMBER STATE
    MEMBER ZIP CODE
    MEMBER PHONE NUMBER
DATE OF FIRST MEMBERSHIP
BALANCE PAST DUE
BONUS CREDITS NOT USED

New Entity: MEMBERSHIP (in 1NF)

Comments: The new entity was factored out of the entity MEMBER. Each occurrence of MEMBERSHIP describes a member's association with a specific club. The key was formed from the combined keys of MEMBER and CLUB.

MEMBER NUMBER + CLUB NAME
MUSICAL/MOVIE PREFERENCE CODE

New Entity: CONTRACT (in 2NF)

Comments: The new entity was also factored out of the entity MEMBER. The elements of CONTRACT describes one MEMBER's association with one of the standard AGREEMENTs. The key was formed from the combined keys of MEMBER and AGREEMENT.

MEMBER NUMBER + AGREEMENT NUMBER
AGREEMENT ENROLLMENT DATE
AGREEMENT EXPIRATION DATE
NUMBER OF PURCHASES REQUIRED
NUMBER OF PURCHASES TO DATE

Unchanged Entity
CLUB (already in 2NF)

Unchanged Entity
AGREEMENT (already in 2NF)

(b)

FIGURE 8.16   *(continued)*

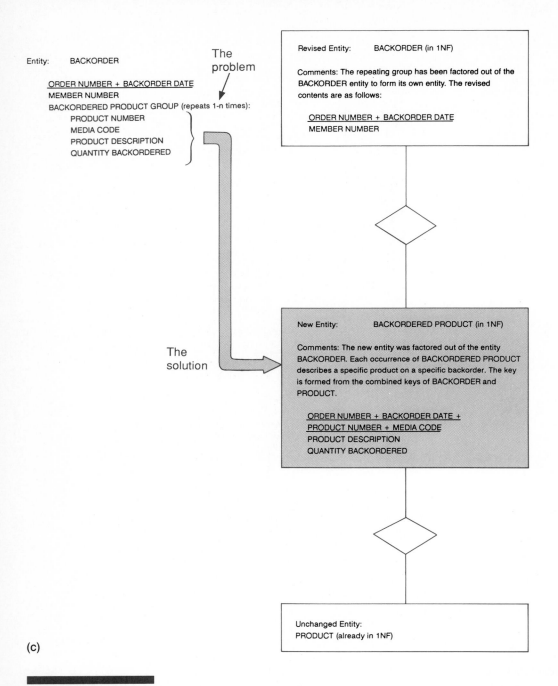

Entity: BACKORDER

   <u>ORDER NUMBER + BACKORDER DATE</u>
   MEMBER NUMBER
   BACKORDERED PRODUCT GROUP (repeats 1-n times):
      PRODUCT NUMBER
      MEDIA CODE
      PRODUCT DESCRIPTION
      QUANTITY BACKORDERED

The
problem

The
solution

Revised Entity: BACKORDER (in 1NF)

Comments: The repeating group has been factored out of the
BACKORDER entity to form its own entity. The revised
contents are as follows:

   <u>ORDER NUMBER + BACKORDER DATE</u>
   MEMBER NUMBER

New Entity: BACKORDERED PRODUCT (in 1NF)

Comments: The new entity was factored out of the entity
BACKORDER. Each occurrence of BACKORDERED PRODUCT
describes a specific product on a specific backorder. The key
is formed from the combined keys of BACKORDER and
PRODUCT.

   <u>ORDER NUMBER + BACKORDER DATE +</u>
   <u>PRODUCT NUMBER + MEDIA CODE</u>
   PRODUCT DESCRIPTION
   QUANTITY BACKORDERED

Unchanged Entity:
PRODUCT (already in 1NF)

(c)

FIGURE 8.16   *(continued)*

Convince yourself that BACKORDER has also been placed into 1NF (Figure 8.16c). The procedure is the same as was applied to ORDER and MEMBER. All other entities that did not have repeating groups of elements are, by default, already in 1NF.

Step 6(b): Place Entities into 2NF

The next step in data analysis is to place the entities in **second normal form**. It is assumed that you have already placed all entities into 1NF. An entity is in 2NF if it has a *combination* key and all non-key elements are derived by the full key, not part of it. This concept is easier to demonstrate than to explain.

We only have to check those entities that have combination keys (PRODUCT, MEMBERSHIP, CONTRACT, PROMOTION, ORDERED PRODUCT, and BACKORDERED PRODUCT). All other entities are automatically in 2NF.

First, let's check the associative entities that were created when we did our 1NF analysis. Look at the 1NF for the MEMBERSHIP entity (Figure 8.16b). The key is MEMBER NUMBER and CLUB NAME. There is only one non-key element, MUSICAL/MOVIE PREFERENCE CODE. That element's value cannot be determined if you have only a MEMBER NUMBER. Why? A member may have different preferences in different clubs to which he or she belongs. Similarly, the element's value cannot be determined if you only have CLUB NAME since, for a club, it has many members. MUSICAL/MOVIE PREFERENCE truly describes a specific member in a specific club—it requires the full key. Thus, the entity MEMBERSHIP is already in 2NF.

While you are still studying Figure 8.16(b), you can also analyze CONTRACT. The 1NF combination key is MEMBER NUMBER and AGREEMENT NUMBER. None of the non-key elements can be described if you have only MEMBER NUMBER or only AGREEMENT NUMBER. Thus, CONTRACT is also already in 2NF since all elements describe one MEMBER's promise to fulfill one specific AGREEMENT. In other words, all the values of all the non-key elements (AGREEMENT ENROLLMENT DATE, AGREEMENT EXPIRATION DATE, NUMBER OF PURCHASES REQUIRED, and NUMBER OF PURCHASES TO DATE) require the full key—MEMBER NUMBER and AGREEMENT NUMBER—as a unique identifier.

So much for cases where 1NF entities require no changes for 2NF. Now, let's examine ORDERED PRODUCT (Figure 8.17a). The key is ORDER NUMBER plus PRODUCT NUMBER. Certainly, several of the non-key elements describe one PRODUCT on one ORDER. Examples include QUANTITY ORDERED, QUANTITY SHIPPED, and EXTENDED PRODUCT PRICE (which is price times quantity). But there are non-key elements that describe PRODUCT, independent of any specific ORDER that might contain that product. For instance, you don't need an ORDER NUMBER to define values for MEDIUM CODE, MODEL/COLOR CODE, SIZE CODE, or PRODUCT DESCRIPTION. These elements truly describe a PRODUCT, not an ORDERED PRODUCT. What about PRODUCT PRICE? At first glance, it appears to describe PRODUCT, not ORDERED

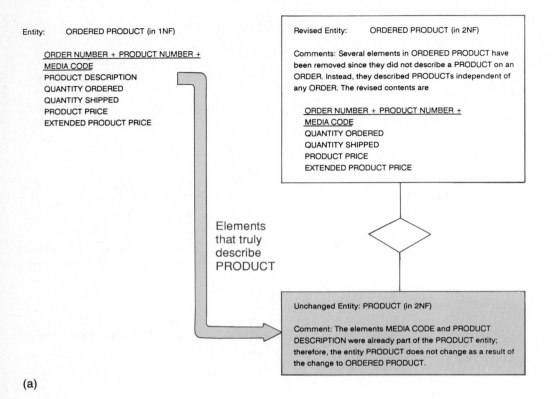

Entity: ORDERED PRODUCT (in 1NF)

ORDER NUMBER + PRODUCT NUMBER +
MEDIA CODE
PRODUCT DESCRIPTION
QUANTITY ORDERED
QUANTITY SHIPPED
PRODUCT PRICE
EXTENDED PRODUCT PRICE

Elements
that truly
describe
PRODUCT

Revised Entity: ORDERED PRODUCT (in 2NF)

Comments: Several elements in ORDERED PRODUCT have
been removed since they did not describe a PRODUCT on an
ORDER. Instead, they described PRODUCTs independent of
any ORDER. The revised contents are

ORDER NUMBER + PRODUCT NUMBER +
MEDIA CODE
QUANTITY ORDERED
QUANTITY SHIPPED
PRODUCT PRICE
EXTENDED PRODUCT PRICE

Unchanged Entity: PRODUCT (in 2NF)

Comment: The elements MEDIA CODE and PRODUCT
DESCRIPTION were already part of the PRODUCT entity;
therefore, the entity PRODUCT does not change as a result of
the change to ORDERED PRODUCT.

(a)

FIGURE 8.17   Graphic View of First Normal Form   This partial ERD demonstrates how most many-to-many relationships get modified for 1NF. The many-to-many relationship is deleted. It has been replaced by an associative entity, formed to store the repeating elements, and two one-to-many relationships back to the parent entities.

PRODUCT. But if you consider it more carefully, you discover that it describes the price at the time of order, not necessarily the current list price. Thus, PRODUCT PRICE does describe an ORDERED PRODUCT. To clarify this, we renamed the element to PRODUCT PRICE AT TIME OF ORDER.

How do we get ORDERED PRODUCT into 2NF? The left-hand column of Figure 8.17(a) demonstrates the required changes. The non-key elements that don't describe ORDERED PRODUCT were removed from ORDERED PRODUCT and merged into the PRODUCT entity. As it turned out, the elements were already contained in PRODUCT; therefore, that entity was unchanged. The entity BACKORDERED PRODUCT would undergo the same basic transformation to 2NF.

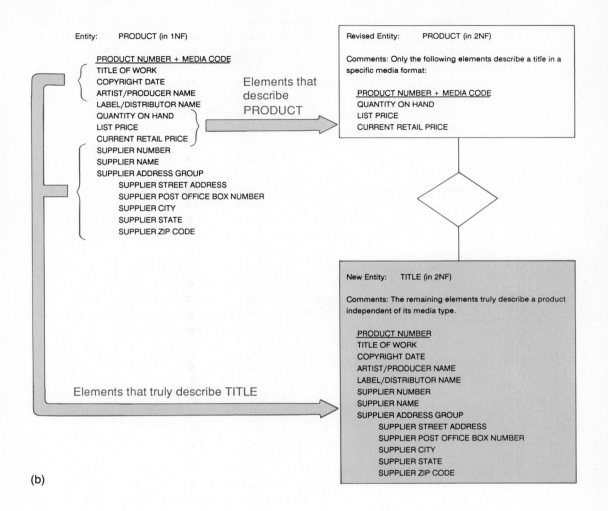

(b)

To consider one more example, let's examine the entity PRODUCT (Figure 8.17b). Although this entity was one of our original entities (not created for 1NF), it does have a combination key, PRODUCT NUMBER + MEDIA CODE. Recall that the MEDIA CODE helps differentiate between the cassette, record, and compact disc versions of the same audio PRODUCT NUMBER, as well as the VHS tape, Beta tape, 8mm tape, and videodisc versions of the same video PRODUCT NUMBER. This is where our problem lies. Some of the non-key elements' values do not depend on the value MEDIA CODE. For instance, the PRODUCT NAME for PRODUCT NUMBER V34566 is *Beverly Hills Cop II*, no matter what the MEDIA CODE value is. On the other hand, QUANTITY ON HAND requires the full key since we probably have different quantities of

Beverly Hills Cop II in the various video formats. Those elements that depend on the full combination key will remain in the PRODUCT entity. Those that depend only on PRODUCT NUMBER have been moved to a new entity—which we call TITLE—whose key is PRODUCT NUMBER. This new entity doesn't have a combination key; therefore, it is automatically in 2NF.

Before we move on to 3NF, we call your attention to what is happening. Elements are being shifted around so that they truly describe the entity in which they are placed. In some cases, this has resulted in the identification of new, simpler entities. Eventually, we will redraw our ERD to depict all these new entities and some new relationships.

Step 6(c): Place Entities into 3NF   We can further simplify our entities by placing them into 3NF. Entities must be in 2NF before beginning 3NF analysis. An entity is in 3NF if the values of its non-key elements are not dependent on any other non-key elements. Once again, the examples are simpler than the definition.

The first 3NF analysis is easy. Examine each entity. Delete any data elements whose values can be calculated or derived (through logic) from other data elements in *that entity*. For example, look at the ORDERED PRODUCT entity in Figure 8.18(a). EXTENDED PRODUCT PRICE is calculated by multiplying QUANTITY SHIPPED times PRODUCT PRICE AT TIME OF ORDER. Therefore, EXTENDED PRODUCT PRICE (a non-key element) is dependent on two other non-key elements. We simplify the entity by deleting EXTENDED PRODUCT PRICE. Figure 8.18(a) demonstrates other elements that were *deleted* through this analysis. Note that we did not depict relationships in Figure 8.18(a), since relationships have no effect on this analysis.

Be careful! Do not delete elements that can be derived from the values of elements contained in more than one entity—at least not as part of 3NF analysis.

Another 3NF analysis checks non-key elements to see if they really describe a separate entity. This analysis is only performed on those entities that do not have combination keys. For our example, this includes ORDER, TITLE, AGREEMENT, MEMBER (which has candidate keys, not combination keys), and CLUB (which also has candidate, not combination keys). Figures 8.18(b) and (c) demonstrate only those entities that change for this analysis. The other entities were already in 3NF. Let's walk through some examples.

Look at the TITLE entity. Most of the elements are dependent on the value of the key, PRODUCT NUMBER. However, the elements SUPPLIER NAME and SUPPLIER ADDRESS (a group of elements) are not dependent on the entity's key, PRODUCT NUMBER. Instead, they are dependent on the non-key SUPPLIER NUMBER. This suggests that the elements do not truly describe the entity TITLE. Actually, they describe a new entity that we'll call SUPPLIER. To correct this problem, we factor out the supplier elements and place them in a new entity, SUPPLIER. When we revise our data model (ERD), we will also add a relationship between TITLE and SUPPLIER to retain the association between the data elements that describe each.

Finally, look at Figure 8.18(c). The analysis is the same except that we don't have to create the entity MEMBER. Furthermore, all of the elements removed from ORDER were already in the MEMBER entity.

Before we leave the subject of normalization, we should acknowledge that several normal forms beyond 3NF exist. Each successive normal form makes the data model simpler, less redundant, and more flexible. However, systems analysts rarely take data models for their applications beyond 3NF. Database experts tend to step beyond 3NF when they merge data models from various applications into a single, corporate data model (or database). Consequently, we leave any further discussion of normalization to database textbooks.

The first few times you normalize a data model, the process will appear slow and tedious. However, with time and practice, it becomes quick and routine, much like the clutch on a car with a straight transmission.

Step 6(d): Further Simplify Through Inspection Normalization is a fairly mechanical process. It is dependent on naming consistencies in the original data model before normalization. When several analysts work on a common application, it is not unusual to create problems that won't be taken care of by normalization. These problems are best taken care of through **simplification by inspection.**

Examine all 3NF non-key elements that are keys in other entities. For instance, in Figure 8.18, in the AGREEMENT entity, we see CLUB NAME as a non-key element. On our ERD, we have already established a relationship between the entities AGREEMENT and CLUB; therefore, we can delete CLUB NAME from the AGREEMENT entity. (*Note:* Those who are familiar with database technology will recognize that CLUB NAME is, in fact, a *foreign key.* Foreign keys, however, are physical solutions to linking entities; they should not be included in our logical model.) The same reasoning can be used to delete MEMBER NUMBER from the BACKORDER entity.

Carefully check every entity for synonym elements—elements that have different names but are really the same thing. They can actually be created through normalization. For example, in our PROMOTION entity, we have a SELECTION OF MONTH NUMBER and SELECTION OF MONTH TITLE. These are synonyms for the TITLE entity's elements, PRODUCT NUMBER and TITLE OF WORK, respectively. We delete the synonyms from the PROMOTION entity and create a relationship on the ERD to the TITLE entity (Figure 8.19).

Also, check element names to ensure that they won't cause confusion. In the CONTRACT entity, we have two elements named AGREEMENT ENROLLMENT DATE and AGREEMENT EXPIRATION DATE, respectively. At the time, the names seemed appropriate. After normalization, they ended up in the CONTRACT entity. We will change them to CONTRACT ENROLLMENT DATE and CONTRACT EXPIRATION DATE to reduce confusion.

Entity: ORDERED PRODUCT (in 2NF)

    <u>ORDER NUMBER + PRODUCT NUMBER +</u>
    <u>MEDIA CODE</u>
    QUANTITY ORDERED
    QUANTITY SHIPPED
    PRODUCT PRICE
    EXTENDED PRODUCT PRICE

Entity: ORDERED PRODUCT (in 3NF)

Comments: The element EXTENDED PRODUCT PRICE was deleted since it can be calculated from the elements QUANTITY SHIPPED and PRODUCT PRICE.

    <u>ORDER NUMBER + PRODUCT NUMBER +</u>
    <u>MEDIA CODE</u>
    QUANTITY ORDERED
    QUANTITY SHIPPED
    PRODUCT PRICE

Entity: AGREEMENT (in 2NF)

    <u>AGREEMENT NUMBER</u>
    CLUB NAME
    MAXIMUM PERIOD OF OBLIGATION
    NUMBER OF PURCHASES REQUIRED WITHIN PERIOD
    BONUS CREDITS/PURCHASE AFTER OBLIGATION
    NUMBER OF MEMBERS ENROLLED
    NUMBER OF MEMBERS WHO HAVE FULFILLED
    NUMBER OF MEMBERS WHO HAVE NOT FULFILLED

Entity: AGREEMENT (in 3NF)

Comments: NUMBER OF MEMBERS WHO HAVE NOT FULFILLED was deleted since it could be calculated from NUMBER OF MEMBERS ENROLLED and NUMBER OF MEMBERS WHO HAVE FULFILLED.

    <u>AGREEMENT NUMBER</u>
    CLUB NAME
    MAXIMUM PERIOD OF OBLIGATION
    NUMBER OF PURCHASES REQUIRED WITHIN PERIOD
    BONUS CREDITS/PURCHASE AFTER OBLIGATION
    NUMBER OF MEMBERS ENROLLED
    NUMBER OF MEMBERS WHO HAVE FULFILLED

Entity: ORDER (in 2NF)

    <u>ORDER NUMBER</u>
    ORDER DATE
    ORDER STATUS
    MEMBER NUMBER
    MEMBER NAME
    MEMBER ADDRESS GROUP:
        MEMBER STREET ADDRESS
        MEMBER POST OFFICE BOX NUMBER
        MEMBER CITY
        MEMBER STATE
        MEMBER ZIP CODE
        MEMBER PHONE NUMBER
    TOTAL ORDER PRICE (sum of extended prices)
    PREPAID AMOUNT
    TOTAL ORDER BALANCE

Entity: ORDER (in partial 3NF)

Comments: TOTAL ORDER BALANCE was deleted since it can be calculated by subtracting PREPAID AMOUNT from TOTAL ORDER PRICE.

See next figure for completion of 3NF.

    <u>ORDER NUMBER</u>
    ORDER DATE
    ORDER STATUS
        MEMBER NUMBER
    MEMBER NAME
    MEMBER ADDRESS GROUP:
        MEMBER STREET ADDRESS
        MEMBER POST OFFICE BOX NUMBER
        MEMBER CITY
        MEMBER STATE
        MEMBER ZIP CODE
        MEMBER PHONE NUMBER
    TOTAL ORDER PRICE (sum of extended prices)
    PREPAID AMOUNT

(a)

FIGURE 8.18   Analysis and Transformation of Entities to Second Normal Form   This revision of our mapping shows how certain elements were moved from combination key entities to the entities that they really describe.

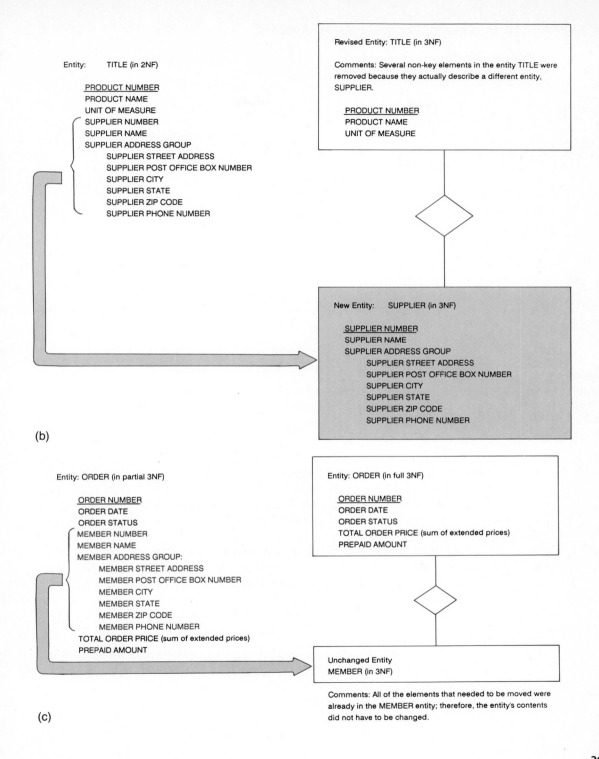

Entity: TITLE (in 2NF)

    <u>PRODUCT NUMBER</u>
    PRODUCT NAME
    UNIT OF MEASURE
    SUPPLIER NUMBER
    SUPPLIER NAME
    SUPPLIER ADDRESS GROUP
        SUPPLIER STREET ADDRESS
        SUPPLIER POST OFFICE BOX NUMBER
        SUPPLIER CITY
        SUPPLIER STATE
        SUPPLIER ZIP CODE
        SUPPLIER PHONE NUMBER

Revised Entity: TITLE (in 3NF)

Comments: Several non-key elements in the entity TITLE were removed because they actually describe a different entity, SUPPLIER.

    <u>PRODUCT NUMBER</u>
    PRODUCT NAME
    UNIT OF MEASURE

New Entity: SUPPLIER (in 3NF)

    <u>SUPPLIER NUMBER</u>
    SUPPLIER NAME
    SUPPLIER ADDRESS GROUP
        SUPPLIER STREET ADDRESS
        SUPPLIER POST OFFICE BOX NUMBER
        SUPPLIER CITY
        SUPPLIER STATE
        SUPPLIER ZIP CODE
        SUPPLIER PHONE NUMBER

(b)

Entity: ORDER (in partial 3NF)

    <u>ORDER NUMBER</u>
    ORDER DATE
    ORDER STATUS
    MEMBER NUMBER
    MEMBER NAME
    MEMBER ADDRESS GROUP:
        MEMBER STREET ADDRESS
        MEMBER POST OFFICE BOX NUMBER
        MEMBER CITY
        MEMBER STATE
        MEMBER ZIP CODE
        MEMBER PHONE NUMBER
    TOTAL ORDER PRICE (sum of extended prices)
    PREPAID AMOUNT

Entity: ORDER (in full 3NF)

    <u>ORDER NUMBER</u>
    ORDER DATE
    ORDER STATUS
    TOTAL ORDER PRICE (sum of extended prices)
    PREPAID AMOUNT

Unchanged Entity
MEMBER (in 3NF)

Comments: All of the elements that needed to be moved were already in the MEMBER entity; therefore, the entity's contents did not have to be changed.

(c)

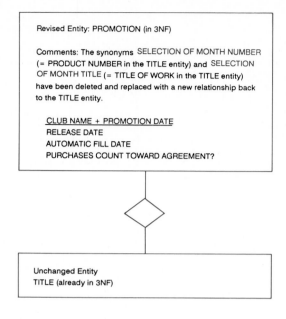

Entity: PROMOTION (last seen in 2NF but also in 3NF)

    <u>CLUB NAME + PROMOTION DATE</u>
    SELECTION OF MONTH NUMBER
    SELECTION OF MONTH TITLE
    RELEASE DATE
    AUTOMATIC FILL DATE
    PURCHASES COUNT TOWARD AGREEMENT?

Revised Entity: PROMOTION (in 3NF)

Comments: The synonyms SELECTION OF MONTH NUMBER
(= PRODUCT NUMBER in the TITLE entity) and SELECTION
OF MONTH TITLE (= TITLE OF WORK in the TITLE entity)
have been deleted and replaced with a new relationship back
to the TITLE entity.

    <u>CLUB NAME + PROMOTION DATE</u>
    RELEASE DATE
    AUTOMATIC FILL DATE
    PURCHASES COUNT TOWARD AGREEMENT?

Unchanged Entity
TITLE (already in 3NF)

FIGURE 8.19   Analysis and Transformation of Entities to Third Normal Form   Elements that can be derived from other elements in the same entity have either been deleted or removed (or merged) into a separate entity that they really describe.

If you wish, you may also check entities for elements that can be derived from elements in different entities. For instance, TOTAL ORDER PRICE (in the entity ORDER) can be calculated as the sum of QUANTITY SHIPPED times PRODUCT PRICE AT TIME OF ORDER (in the entity ORDERED PRODUCT) for each product on the order. Thus, TOTAL ORDER PRICE COULD BE DELETED.

Step 6(e): Redraw the Refined Entity Relationship Diagram   Now we can redraw our ERD to reflect the changes inspired through data analysis. Recall that the original ERD was depicted in Figure 8.13. Quickly review that ERD before you read on.

The new ERD is shown in Figure 8.20. We call your attention to the following changes:

Ⓐ We have added the associated entities that were created when we went to 1NF. In each case, we replaced the many-to-many relationship that used to exist between the associated entities.

Ⓑ We have added the TITLE entity that was created when we placed PRODUCT into 2NF. Because it was factored out of PRODUCT, we added a relationship between TITLE and PRODUCT.

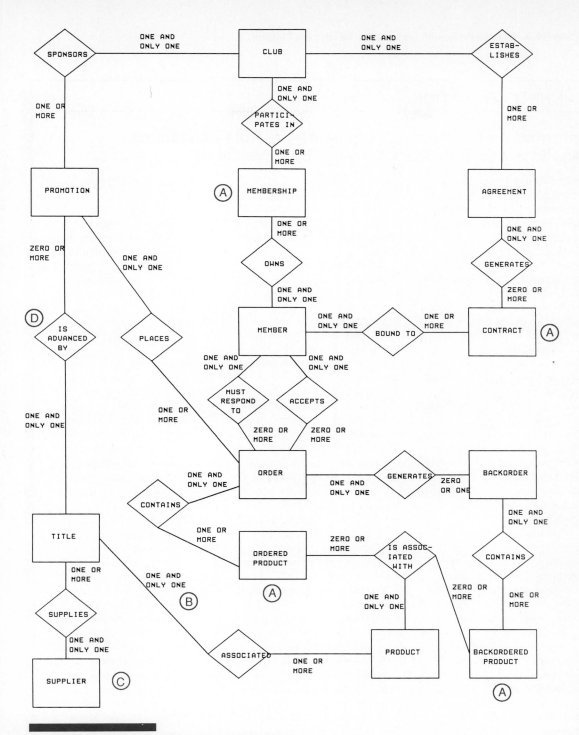

FIGURE 8.20   Final Entity Relationship Data Model

THE NEXT GENERATION

Information Engineering: Changing the Role of Systems Analysts and End-Users Through Data Modeling

Information Engineering is a complete methodology that, if fully implemented, could radically change the way systems analysts build systems. Developed by Clive Finkelstein and James Martin, Information Engineering seeks to deemphasize our traditional emphasis on processing and outputs, changing that emphasis to data storage and inputs.

Information Engineering also places much greater responsibility for systems development on end-users. The reasoning is as follows. Demand for new applications is increasing much faster than the number and productivity of computer professionals are. The only way that demand will be met is if end-users take greater responsibility for systems planning and development.

Essentially, Information Engineering attempts to redefine the roles of management, end-users, analysts, programmers, and other computer professionals. The center of the universe is stored data. Analysts, programmers, and other computer professionals will be responsible for designing all data stores and ensuring that data is captured, stored, and properly maintained. Management will be responsible for planning systems that will

evolve around data storage. Finally, end-users will design and implement all systems and programs that use stored data. Sound interesting? Here's how it works.

Management, under the direction of analysts, undertakes a comprehensive systems plan. The end result is an organizationwide data model, and various subsets of that model as they pertain to a few key systems. For the most part, management identifies the entities and relationships. Analysts and end-users identify the data elements. Analysts then normalize the model.

For each entity, analysts then define business events that create, modify, use, and delete occurrences of the entity. This guarantees that any data collected will be complete and accurate at all times. Analysts consolidate these events into procedures that will eventually be programmed.

Database and file specialists determine how the data model will be physically implemented and associate required procedures with appropriate databases and files.

Analysts and programmers implement the database and data maintenance procedures and programs. Keep in mind that all of these programs do

nothing more than capture, store, and maintain data.

To complete the systems puzzle, end-users design and implement all the programs that will use the data. How do they do this? They are able to fulfill their role by using powerful, English-like report generators, query processors, data extractors, statistical analysis software, and other easy-to-learn, easy-to-use tools.

No, we are not there yet! Actually, we're not that close to this utopia. Whether or not Information Engineering survives as a methodology, the scenario certainly makes sense. But end-user tools, despite their rapid improvement, will have to become much friendlier. Meanwhile, roles would definitely change.

Systems analysts may find their roles changing to that of *data* and *database* analysts. Other analysts will become end-user *consultants,* helping end-users find stored data and use the tools to get at that data (by definition, they will be *information center analysts*). Finally, end-users will find themselves doing more of the analysis, design, and programming . . . that is, unless they revolt. After all, "that's what we hired you computer types for, isn't it?"

Ⓒ We have added the SUPPLIER entity that was created when we placed TITLE in 3NF.

Ⓓ We have added the PROMOTION to TITLE relationship that was identified during simplification by inspection.

The final mapping of the data elements to these entities is shown in Figure 8.21. Each entity's elements describe only that entity. With the exception of keys, elements are not redundantly stored within or between entities. For one occurrence of each entity, each element occurs once at most. The model is simple. It is flexible and nonredundant.

We have made one addition to some elements. The notation *SK* means that the element can serve as a **secondary key**. A secondary key is any element whose values, though not unique to each occurrence of the entity, effectively divide occurrences of the entities into groups or subsets. The classic example is an element such as SEX that divides occurrences of an entity—for example, STUDENT or EMPLOYEE—into male and female subsets.

Step 7: Review and Refine Your Data Model   The data model (ERD) and the contents of the entities should be reviewed with appropriate end-users. Our experience with end-user acceptance of data models is excellent, especially when we explain that the data model is "a picture of the things about which the system must capture and store data and the natural business relationships between those things." End-users frequently respond by adding new and potentially helpful entities, relationships, and data elements. This becomes the basis for then defining the process requirements needed to capture, maintain, and use that data (see the next chapter).

Looking Ahead: Entity Relationship Diagrams for Systems Analysis and Design

We have completed our ERDs for an improved Entertainment Club member services system. Through drawing, analyzing, and verifying ERDs, you will thoroughly understand the vocabulary, language, and data requirements of the system to be designed.

Entity relationship diagrams will prove useful as we progress through systems analysis and design. For example, the knowledge we've gained about data will help us draw better target systems process models (data flow diagrams) in the next chapter.

During systems design, we will evaluate various computer-based techniques for implementing files and databases based on the logical data model. Some data may be gleaned—for instance, from existing files and databases. Other data will be designed or redesigned as new files and/or databases. Data

Entity: TITLE

    <u>PRODUCT NUMBER</u>
    TITLE OF WORK
    COPYRIGHT DATE
    ARTIST/PRODUCER NAME
    LABEL/DISTRIBUTOR NAME

Entity: PRODUCT

    <u>PRODUCT NUMBER + MEDIA CODE</u>
    QUANTITY ON HAND
    LIST PRICE
    CURRENT RETAIL PRICE

Entity: SUPPLIER

    <u>SUPPLIER NUMBER</u>
    SUPPLIER NAME
    SUPPLIER ADDRESS GROUP
        SUPPLIER STREET ADDRESS
        SUPPLIER POST OFFICE BOX NUMBER
        SUPPLIER CITY
        SUPPLIER STATE
        SUPPLIER ZIP CODE

Entity: CLUB

    <u>CLUB NAME</u> or <u>CLUB ABBREVIATION</u>
    NUMBER ENROLLED YEAR-TO-DATE
    NUMBER CANCELED YEAR-TO-DATE

Entity: AGREEMENT

    <u>AGREEMENT NUMBER</u>
    MAXIMUM PERIOD OF OBLIGATION
    NUMBER OF PURCHASES REQUIRED WITHIN PERIOD
    BONUS CREDITS/PURCHASE AFTER OBLIGATION
    NUMBER OF MEMBERS ENROLLED
    NUMBER OF MEMBERS WHO HAVE FULFILLED

Entity: PROMOTION

    <u>CLUB NAME + PROMOTION DATE</u>
    RELEASE DATE
    AUTOMATIC FILL DATE

Entity: BACKORDER

    <u>ORDER NUMBER + BACKORDER DATE</u>

Entity: BACKORDERED PRODUCT

    <u>ORDER NUMBER + BACKORDER DATE +</u>
    <u>PRODUCT NUMBER + MEDIA CODE</u>
    QUANTITY BACKORDERED

Entity: MEMBER

    <u>MEMBER NUMBER</u>
    MEMBER NAME
    MEMBER ADDRESS GROUP:
        MEMBER STREET ADDRESS
        MEMBER POST OFFICE BOX NUMBER
        MEMBER CITY
        MEMBER STATE
        MEMBER ZIP CODE
        MEMBER PHONE NUMBER
    DATE OF FIRST MEMBERSHIP
    BALANCE PAST DUE
    BONUS CREDITS NOT USED

Entity: ORDER
    <u>ORDER NUMBER</u>
    ORDER DATE
    ORDER STATUS
    TOTAL ORDER PRICE
    PREPAID AMOUNT

Entity: ORDERED PRODUCT

    <u>ORDER NUMBER + PRODUCT NUMBER +</u>
    <u>MEDIA CODE</u>
    QUANTITY ORDERED
    QUANTITY SHIPPED
    PRODUCT PRICE

Entity: MEMBERSHIP

    <u>MEMBER NUMBER + CLUB NAME</u>
    MUSICAL/MOVIE PREFERENCE CODE

Entity: CONTRACT

    <u>MEMBER NUMBER +   AGREEMENT NUMBER</u>
    CONTRACT ENROLLMENT DATE
    CONTRACT EXPIRATION DATE
    NUMBER OF PURCHASES REQUIRED
    NUMBER OF PURCHASES TO DATE

**FIGURE 8.21 Final
Mapping of Data Elements to Entities**

may be centralized or distributed. Many file and database organizations are usually available.

Summary

Data modeling is a technique for defining requirements for data that will be stored in a new system. This chapter presented a popular data modeling tool, entity relationship diagrams (ERDs). ERDs and data models describe the logical—as opposed to physical—requirements for a system. Logical models do not show the physical implementation of the model, even if that implementation is known or anticipated.

ERDs show entities and relationships. Entities are persons, objects, events, and locations that can be described by data elements or attributes. Examples include customers, products, orders, and regions. Relationships are natural associations between entities. For instance, customers PLACE orders that CONTAIN products.

ERDs are relatively easy to draw, since the analyst only has to define the entities, relationships, and data elements needed. Most of this can be gleaned from existing forms and files, supplemented by interviews and walkthroughs. However, data models should always be analyzed to structure the data such that it is flexible, nonredundant, and easy to understand.

Data analysis is a technique for structuring data in its simplest, most flexible form. It uses an approach called normalization to simplify the data model. Through normalization, entities are placed into first, second, and third normal form, in that sequence.

- First normal form entities contain no repeating groups of elements (for a single occurrence of the entity).

- Second normal forms contain no elements that are only partially dependent on a key that is made up from the combination of more than one element.

- Third normal form entities contain no elements that can be derived from the values of other elements in the same entity.

As you normalize the data model, the ERD may be modified to include new entities and/or relationships. Additionally, the process of normalization can delete unneeded elements or shift them to new or different entities that they truly describe.

A third normal form data model becomes the basis for an intelligent process model that will maintain stored data and meet end-user output requirements. Process models for new systems will be discussed in the next chapter.

Problems and Exercises

1. Give three reasons why many analysts believe that requirements should ideally be specified in an implementation-independent fashion.

2. Differentiate between an entity relationship diagram (ERD) and a data flow diagram (DFD).

3. Most data entities correspond to persons, objects, events, or locations in the business environment. Give three examples of each data entity class.

4. Obtain three sample business forms from a business, your school, or your instructor. What entities are described by the fields on the forms?

5. What entities are described on your class schedule form or your school's course registration form?

6. Give two examples of each of the following data relationship complexities: one-to-one (1:1), one-to-many (1:M or M:1), and many-to-many (M:M). Draw an ERD for each of your examples. Be sure to label data entities using nouns and data relationships using verbs. Annotate the graph to communicate the relationship complexity.

7. During the survey and study phases, an analyst collected numerous samples, including documents, forms, and reports. Explain how these samples will prove useful for data modeling.

8. During the definition phase, the actual data model is drawn, refined, and improved. Identify the data modeling issues that must be specified when alternative solutions are being identified and analyzed in the selection phase.

9. List and explain the seven steps for constructing a data model.

10. Given the following narrative description of entities and their relationships, prepare a draft entity relationship diagram (ERD). Be sure to state any reasonable assumptions that you are making.

    Burger World Distribution Center serves as a supplier to 45 Burger World franchises. You are involved with a project to build a database system for distribution. Each franchise submits a day-by-day projection of sales for each of Burger World's menu products—the products listed on the menu at each restaurant—for the coming month. All menu products require ingredients and/or packaging items. Based on projected sales for the store, the system must generate a day-by-day ingredients need and then collapse those needs into one-per-week purchase requisitions and shipments.

11. Write a paragraph or two explaining how you would present and verify the entity relationship diagram prepared in Exercise 10 to a group of end-users who are not familiar with computer concepts.

12. What is the difference between an entity in zero normal form (0NF) and

first normal form (1NF)? Give an example of an entity in 0NF and show its conversion to 1NF. What is an associative entity?

13. What is the difference between an entity in first normal form (1NF) and second normal form (2NF)? Give an example of an entity in 1NF and show its conversion to 2NF. What is a combination key?

14. What is the difference between an entity in second normal form (2NF) and third normal form (3NF)? Give an example of an entity in 2NF and show its conversion to 3NF. What is the difference between a candidate and a combination key?

15. It is very important that a data model be correctly specified. Describe three characteristics of "good" data models.

16. Normalize the business forms you sampled in Exercise 4. If the forms are related, consolidate your individual data models into an integrated data model.

Projects and Minicases

1. Obtain copies of all of the forms used in your school's course registration system. Your instructor may be able to supply these forms. Using the seven steps of data modeling, derive a third normal form data model to support a course registration system.

2. Given the following data element contents for several entities, indicate the normal form level for each group and explain why each group is in that particular normal form. State any assumptions you make or get an interpretation ruling from your instructor.

Microcomputer Property Accounting and Maintenance System

Entity: Equipment

Serial Number
Model Number
Manufacturer Name
Manufacturer Location
Item Description
Installed in Serial Number (if applicable)
Date Purchased
Supplier Name
Supplier Address
Purchase Price
Replacement Cost
Current Location
Current Responsible End-User

Entity: Software Package

Serial Number
Package Name
Package Type
Version Number
Registration Date
Number of End-Users Allowed
Date Purchased
Supplier Name
Supplier Address
Purchase Price
1 or more of:
    Current End-User
    Current End-User Location
(Continued)

Entity: Spare Part

    Internal Part Number
    Vendor Part Number
    Supplier Part Number
    Part Description
    Used in one or more of:
       Manufacturer Name
       Model Number

Entity: Warranty

    Serial Number
    Model Number
    Manufacturer Name
    Manufacturer Location
    Item Description
    Date Purchased
    Date Warranty Expires
       Warranty Servicer Name
       Warranty Servicer Address
       Warranty Servicer Phone

Entity: Maintenance Contract

    Serial Number
    Model Number
    Manufacturer Name
    Item Description
    Maintenance Purchase Date
    Maintenance Start Date
    Maintenance Finish Date
    Labor Covered?
    Parts Covered?
    Bill Rate for Costs Not Covered

a. Place the above entities into first normal form. For those entities that do not change, explain why.

b. Place the above entities into second normal form. For those entities that do not change, explain why.

c. Place the above entities into third normal form. For those entities that do not change, explain why.

d. Draw the 3NF data model. Describe any changes you make (simplification by inspection).

3. Given the sample form shown on the top of page 271, prepare a list of entities and their associated data elements (0NF), as determined from the document. Then, completely normalize the entities to 3NF and draw a hypothetical entity relationship diagram. Your instructor should be the final interpreter for the form.

PURCHASING REQUISITION Form 12 Rev. 1988

INSTRUCTIONS — INCLUDE IN EACH REQUISITION ONLY SUCH ARTICLES AS MAY BE PURCHASED FROM ONE FIRM. IF SPECIAL HANDLING IS DESIRED, NOTE. SEE REVERSE SIDE FOR SPECIAL COMMENTS BY REQUESTOR.

DEPARTMENT COMPLETES UNSHADED AREA | PURCHASING COMPLETES SHADED AREA | ORDER NO.

DEPT. OR FUNCTION: Computer Information Systems
COMMITMENT NO.
COMMODITY CODE | ORDER TYPE

M F C	RES. CODE	FUND	ACCOUNT NUMBER CENTER DEPT. — PROJ.	OBJECT	DEPT. REFERENCE	AMOUNT	FUND EXPIRATION DATE
1				5-6207		8,736.00	
2				5-6106		399.00	
3				5-6107		84.00	

SHIP TO STAFF MEMBER: Jonathan Doe
DEPT. 242
BUILDING & ROOM: Administration

ORDER DATE
FOLLOW UP

PRICING METHOD
☐ RQ #
☐ 1 Phone/Verbal Quote
☐ 2 Agreement/Contract
☐ 3 Price List on File
☐ 4 Repair Negotiation
☐ 5 None of the above
BUYER

REQUISITIONER'S PHONE NO.: 555-4545
MATERIAL WILL BE USED FOR
VENDOR SUGGESTED:
IBM
Main Street
Somewhere, IN 47906
VENDOR NAME
VENDOR NUMBER

FOB ☐ 1 DESTINATION ☐ 2 DESTINATION PREPAY & ADD ☐ 3 SHIPPING POINT ☐ 4 SHIPPING POINT FREIGHT ALLOWED ☐ 5 SEE BELOW VIA | TERMS

ITEM #	ITEM DESCRIPTION	MFC	QUANTITY	UNIT	UNIT PRICE	EXTENDED PRICE	DELIVER ON	EST.	COMM.
	IBM PS/2 Model 70 386 8570-121	1	1		7,995.00	4,797.00			
	IBM PS/2 2-8 MB Memory Module Expansion Option #5211	1	1		1,695.00	1,017.00			
	IBM PS/2 2MB Memory Module Kit #5213	1	3		1,395.00	2,511.00			
	IBM 8513 PS/2 Color Display	1	1		685.00	411.00			
	IBM 8770 PS/2 Mouse	2	1		95.00	57.00			
	IBM PC Network Adapter II/A #150122	2	1		570.00	342.00			
	IBM DOS 3.3	3	1		120.00	84.00			

REQUESTED — HEAD OF DEPT.: *Thomas J. Mathuen*
DATE: 6-3-89
RECOMMENDED — DEAN OR ADMINISTRATOR
DATE

APPROVED — FOR THE COMPTROLLER
DATE
APPROVED — FOR THE EXECUTIVE VICE PRESIDENT AND TREASURER
DATE

BYPASS APPROVAL REQUESTED ☐
APPROVAL SIGNATURE/DATE
OCGBA PREAUDIT
BY: DATE:

PURCHASING APPROVALS
PA | AD | DIR

Annotated References and Suggested Readings

Flavin, Matt. *Fundamental Concepts of Information Modeling*. New York: Yourdon Press, 1981. A classic book on information modeling and the entity relationship diagram approach.

Martin, James, and Clive Finkelstein. *Information Engineering*. 2 vols. New York: Savant Institute, 1981. Information Engineering is a formal, database, and fourth-generation language–oriented methodology. The method is logically equivalent; however, the authors use entity diagrams instead of entity relationship diagrams. ERDs could easily be substituted.

Mellor, Stephen. *Object Oriented Design*. Englewood Cliffs, N.J.: Prentice-Hall, 1987. This modern book presents an object(entity)–oriented approach to Structured Systems Analysis and Design.

Weaver, Audrey. *Using the Structured Techniques*. Englewood Cliffs, N.J.: Prentice-Hall, 1987. This book presents, in case study form, a structured methodology based on the complementary use of information models and process models (which we cover in the next chapter).

CHAPTER NINE

Process Modeling

with Logical Data

Flow Diagrams

Skyrocket Insurance Company

Data flow diagrams were introduced in Chapter 7; therefore, a minicase to introduce the concept would be redundant. This minicase is intended to get you thinking about why systems analysts draw *logical* data flow diagrams in addition to Chapter 7's *physical* data flow diagrams.

Skyrocket Insurance Company recently automated its claims-processing system. The old system worked; management had no qualms about how it worked. The entirely manual system used nine clerks, and these clerks did their job well. But then Skyrocket merged with two other insurance firms. The workload was simply too much for the existing system, and it seemed time to seriously examine computerization.

The analysts did their job. They thoroughly studied existing methods and learned that the system was quite simple. The claims-processing office contained nine desks, one per clerk assigned to that room. Insurance claims forms were equally distributed to the first desk in each room. Each clerk performed one step in the processing cycle. He or she placed the processed paperwork in an out box. The clerk at the subsequent desk would periodically retrieve the contents of the previous desk's out box, perform the next step, and place the resulting paperwork in his or her own out box. The processing cycle would continue until every claim passed through each desk.

Each clerk also had a work-in-process box on his or her desk. If any claim was delayed at any step—by lack of information, incomplete information, discrepancies, and so on—the paperwork would be placed in the corresponding clerk's work-in-process box after the clerk sent appropriate follow-up correspondence to the claimer, agent, or involved party. When new information was submitted, it would be passed down from desk 1 until it got to the appropriate clerk, who would remove the work in process and continue the processing cycle.

If a clerk was sick or on a very short vacation, the contents of that clerk's in box simply stacked up until the clerk returned. The clerk usually worked overtime or weekends to catch up.

It wasn't a very glamorous system, but it worked, and neither the clerks nor management had many criticisms of it. This consideration was foremost in the analyst's mind when the new system was designed.

The analyst's new system took advantage of the emergence of microcomputers and networking technology. Each clerk was given a Compaq Deskpro 386S microcomputer. The computers were networked with a Compaq 386/25 file and print server. The system worked precisely as before. Each microcomputer supported one task, placing transactions in computer files (on the server) that corresponded to the various paperwork boxes on the clerks' desks. Paperwork corresponding to the old claims forms could be printed at any time. A paper copy was always printed at the end of the processing cycle, since people tend to feel more comfortable about paper than computer files.

Interestingly, the new system has not been an overwhelming success, despite the fact that it emulated the successful system it replaced. As the workload increased due to the merger, the system could not keep up any better than its predecessor. It seemed as if the claims data frequently stalled in one of the many electronic out boxes or work-in-process boxes. Information was no easier to produce than with the old system. How could something so simple have gone so completely wrong?

Discussion

1. What do you think the analyst did wrong? (*Hint:* There is an expression that frequently haunts analysts and managers: "But we've always done it that way.")

2. Knowing what you know about them, why might *physical* data flow diagrams have been inadequate for avoiding this problem? How might they be adapted to overcome the problem?

What Will You Learn in This Chapter?

This is the third of three graphic systems modeling chapters. In this chapter you will learn how to use one of the most popular systems modeling tools, *logical data flow diagrams*, to model the processing and data flow requirements for an improved system. Logical DFDs are similar to the physical DFDs that you encountered in Chapter 7. You will understand logical data flow diagrams as a systems analysis tool when you can

1. Differentiate between physical and logical DFDs.

2. Factor a system into component subsystems, functions, and tasks and depict its structure using a decomposition diagram.

3. Document the interactions between subsystems, functions, and tasks using logical data flow diagrams.

4. Develop a set of leveled—or exploded—data flow diagrams for an information system.

5. Explain the complementary relationship between process models (logical DFDs) and information models (logical ERDs).

In the last chapter, you learned about a popular logical modeling tool, entity relationship diagrams, to model end-user requirements for *stored data*. Ultimately, however, we must define requirements for processing data into information. We call such models *process models*.

In this chapter, we discuss logical data flow diagrams, a tool for drawing pictures of systems' processing and data flow without excessive concern for details. Let's study this tool, its conventions, and its uses. We will present process models and logical DFDs not as an alternative to the data models you learned about in Chapter 8, but instead as a complement to those models.

Modeling the Essence of a System's Processes

In Chapter 7, you learned about process modeling and data flow diagrams for understanding the current system. In this chapter, we look at process modeling and data flow diagrams for understanding end-user requirements. Because the concepts of process modeling and logical modeling are so important, let's quickly review those concepts.

Most analysts draw physical process models without even thinking about it. Recall that **physical models** show not only what a system is or does, but also how it is physically implemented. Eventually, we must always specify physical models to guide the implementation of new systems. Unfortunately, we tend to become overdependent on physical models, especially those of current systems. Physical models can bias us and lead us to implement new versions of old mistakes. Consider, for example, the minicase that opened this chapter. The redesigned claims-processing system was a mirror image of the dated manual, assembly-line processing of the old system—even the bottlenecks were duplicated in the computer system. The analysts were unable to separate requirements from their current implementation!

Analysts and end-users alike are learning to build implementation-independent models. Implementation-independent models are usually called **logical models**. During the requirements definition phase, they model the essential requirements of the system: those requirements that must be fulfilled no matter how we might implement the system. This chapter will present process modeling as a logical modeling technique for end-user requirements.

All systems also process data to produce information and maintain stored data. We need to logically model those requirements. As computer professionals, we ultimately implement processes as programs; therefore, we need process models. A **process model** is a picture of the flow of data through a system and the processing that must be performed on that data. These models or pictures are normally easier to read that prose.

The term *process modeling* comes from the depiction of processes and how they interact or interface with one another. These interactions take the form of data flows between processes. For this reason, process models are also frequently called *data flow models*.

Logical process modeling defines input, processing, and output requirements independent of their implementation. Process modeling's popularity is an outgrowth of the systems analysis methodology called Structured Systems Analysis and Design.

Data Flow Diagram Conventions and Guidelines

Recall from Chapter 7 that a **data flow diagram** illustrates the flow of data through a system and the work or processing performed by that system. Let's quickly review the visual language of DFDs.

Recall also that there are two alternative, *but equivalent,* symbol sets for DFDs. We will use the Gane-Sarson symbols, as we did in Chapter 7. You can use the DeMarco-Yourdon system, which uses circles for processes, if you find it more comfortable.

Figure 9.1 is a logical process model of the personal finance system introduced in Chapter 7 as a physical process model. Notice that this DFD only shows *what* the system does, not how it does it. For comparison, you may find it useful to compare this logical DFD with the physical equivalent, reproduced as Figure 9.2.

You should already know the basics of data flow diagramming. Let's review the DFD symbols, this time focusing on logical modeling conventions.

The External or Internal Agents to the System

External or Internal Agent

External and internal entities were first introduced in Chapter 7. However, to prevent confusing the term with the entities introduced in Chapter 8, let's begin to call them by one of their popular aliases, *agents*.

External and internal agents define the system's boundaries. These agents are also sometimes called *sources* (of data, the net inputs to the system) or *destinations* (of information, the net outputs from the system). External and internal agents are people, organizations, and other systems with which your system interacts.

Agents are external if they are clearly external to your system's organization. Examples include customers, vendors, and the government. Agents are internal if they represent interfaces to other systems, organizational units, or

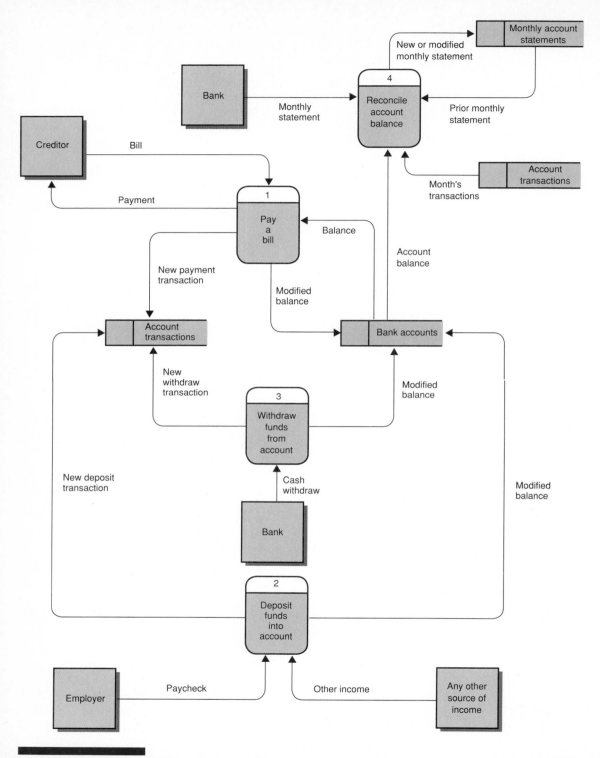

FIGURE 9.1 Simple Logical Data Flow Diagram This is a simple logical data flow diagram for the personal finance system introduced in Chapter 7. The physical equivalent is duplicated in Figure 9.2.

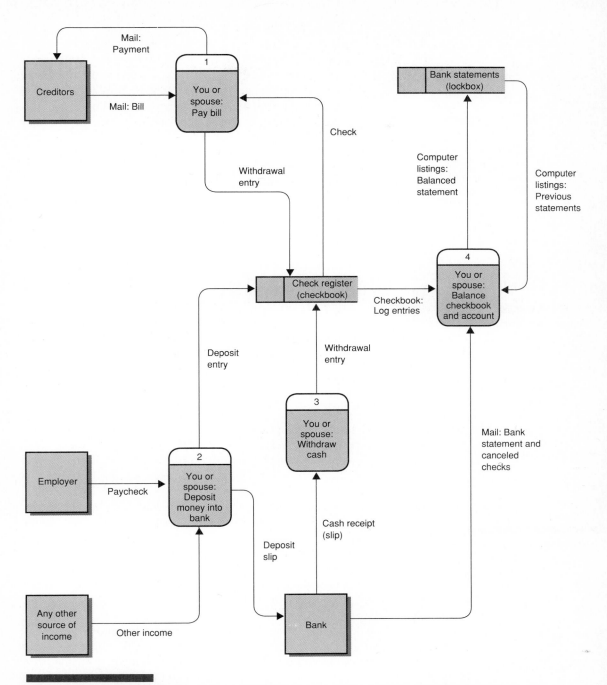

FIGURE 9.2 Simple Physical Data Flow Diagram This is the physical DFD equivalent of Figure 9.1.

people within your company. This may include your end-users, who may be the net sources of systems data or recipients of systems information.

It is important to realize that scope can change as one progresses through a project. For instance, entire subsystems that were studied and modeled as processes in physical DFDs of the current system may be removed from the scope if they are considered adequate or of lower priority. Thus, these subsystems and the people and machines that perform them can become internal agents when you model the requirements for a new system. If so, you are graphically stating that these agents' current physical implementation of interfaces to your system will *not* change.

Labeling conventions are identical to those of physical DFDs. Labels should be descriptive. Use titles but not people's names. People's names represent current physical assignments that are subject to constant change. If the agent represents a system, label it "<name> system."

Mechanical rules for external and internal agents are not changed for logical DFDs. To avoid crossing data flow lines on a DFD, it is permissible to duplicate external and internal agents on DFDs. If external or internal agents are duplicated on or between pages of DFDs, they should all be marked as indicated in the margin. This mark tells readers that they can expect additional occurrences of this same agent on the same page or other pages. As a general rule, external and internal agents should be located on the perimeters of the page, consistent with their definition as a systems boundary.

External or Internal Agent

The Logical Data Flow

What is a logical data flow? On physical DFDs, data flows correspond to things like business forms, documents, reports, terminal displays, spoken communication such as phone calls, reads and writes to files and databases, and data transmissions from one computer to another. But logical data flows are implementation-independent.

Data Flow

A logical data flow represents the minimum, essential data needed by the process that receives the data flow. By ensuring that processes only receive as much data as they really need, we reduce the **coupling** or dependence between processes. This is a very desirable property to carry forward to programs that might eventually implement the process. Why? Because it minimizes the effects of changes in one process as they might relate to other processes. Stated more simply, it leads to reduced program maintenance cost and effort!

Logical data flow names should be descriptive adjectives and nouns. However, consistent with our goal of logical modeling, data flow names should describe the data flow without describing the implementation of that data flow. Suppose, for example, that end-users explain their system as follows: "We fill out a '23' in triplicate and send it to. . . ." The real name of the "23" may actually be COURSE REQUEST FORM. Which name should you use in your DFD? If you answer COURSE REQUEST, you are correct. It eliminates two implementation terms, the identification number 23 and the reference to a paper FORM.

Remember, we are trying to understand the **essence** of the system's requirements and stimulate our creative juices. How many implementation alternatives to a paper business form can you think of? Perhaps a phone call? Or perhaps a terminal dialogue or screen. Or, if you are more futuristic, it could be accomplished through voice recognition. The name COURSE REQUEST tells us what we need to know. This concept of essence will be emphasized throughout this chapter. It truly distinguishes logical DFDs from their physical counterparts.

Finally, data flow names should be singular, as opposed to plural. We don't want to imply that occurrences are processed as a batch. Again, nonbatch implementations might be possible.

Mechanical rules for logical data flows are similar but not necessarily identical to physical data flows. Recall that a data flow is a *packet* of data. The packet concept is critical. Data that should travel together should be shown as a single data flow, no matter how many documents might be physically involved. Why? Once again, alternative implementations might split a single logical flow into two or more flows, possibly with differing implementations.

Do not use diverging data flows (see margin) on logical DFDs. Diverging flows model physical solutions like multipart forms and multiple copies of reports. But once again we emphasize the continuing concept of depicting the essence of the system. Show each potential copy as a separate data flow. Again, if you think about it, there are alternative implementations that wouldn't necessarily require multipart forms or copies.

As is the case with physical data flows, *all logical data flows must begin and/or end at a process, because data flows either initiate a process or result from a process*!

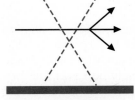

Diverging Data Flow

The Logical Process

Process

Logical processes are work or actions that are performed on incoming data flows to produce outgoing data flows. Though processes can be performed by people, departments, robots, machines, or computers, we once again want to focus on *what* work or action is performed, not who or what is doing that work or activity. The following naming conventions address the essence of the process.

Logical process names depend on the level of detail you are trying to depict on the DFD. High-level or general DFDs present a high level of the system, subsystems, functions, and interfaces (data flows) between subsystems or functions. Lower-level or detailed DFDs show the explicit processes

GENERAL PROCESS NAMES

Accounts Receivable System

Billing Subsystem

Credit Function

Process Payment

Handle Dispute

DETAILED PROCESS NAMES

Credit Customer Account

Record Amount in Dispute

Update Customer Credit Rating

Identify Poor Credit Risks

Summarize Receipts

for a small piece of the system. Logical process names should be chosen accordingly. Process names on the general DFDs will, of necessity, be equally general. Examples of general names are listed in the margin. On the other hand, the logical processes on the detailed DFDs should be given more specific names consisting of a *strong action verb* followed by an *object clause* that describes what the work is performed on or for. Examples of detailed names are listed in the margin.

For logical DFDs, we omit processes that do nothing more than move or route data, thus leaving the data unchanged. For example, omit a process that corresponds to a secretary receiving and simply forwarding a variety of documents to their next processing location. Thus, only those processes that change data should be included on a DFD. Examples of data changing processes include

- those that perform computations—for instance, calculating your grade-point average.
- those that split an incoming data flow into two or more outgoing data flows, each one a subset, possibly with similar content, of the incoming data flows—for instance, separating approved orders from rejected orders based on some decision-making criteria.
- those that combine two or more incoming data flows into a lesser number of outgoing data flows—for instance, combining requested courses with available courses to create a student schedule.
- those that reorganize incoming data flows to produce different outgoing data flows—for instance, invoice data may be filtered to report only overdue accounts, or course enrollment data might be summarized to compare demands for different courses, or data might simply be reformatted or sorted to facilitate easier use by the next process or an external or internal entity. In all these cases, the data isn't changed, but its structure is different.

You may have noticed that a logical process can have many incoming and outgoing data flows. Some may occur every time the process is executed. Others may occur optionally, under certain conditions. The DFD does not show which data flows must occur and those that occur under specific circumstances. These details are deferred until Chapter 11.

Ideally, logical processes should have as few inputs and outputs as possible. The best way to achieve this goal is to split a complex process into multiple, simpler, **cohesive** processes. A cohesive process is one that does only one thing. Cohesive processes eventually become cohesive modules in programs that are reusable and more maintainable.

As was the case with physical DFDs, you should avoid black holes (processes with no outputs), miracles (processes with no inputs), and gray holes (processes whose inputs cannot possibly be transformed into the indicated outputs).

The Logical Data Store

Logical data stores are very different from physical data stores. Both are inventories of data. Physical data stores correspond to implementations such as databases, files, file cabinets, Rolodex files, in/out boxes, log sheets, and books and binders. However, again we note that any given data store could be implemented in many different ways.

Data Store

In contrast, logical data stores represent the businesspeople, objects, and events about which we need to store data. Does that definition sound a little familiar? It should. It is the same definition as the data entity definition you learned in Chapter 8 as part of data modeling. Thus, the logical data store represents the link between our data and process models. On high-level logical DFDs, a data store may represent the entire data model (in our case, an entity relationship data model) and a subset of that model. On lower-level physical DFDs, data stores correspond to specific entities on the data model.

Generally speaking, data stores should be named after the data model or entities. The word *data* is unnecessary because it is implicit that a data store stores data. Additionally, names should not imply implementation media. For instance, avoid terms such as *file*, *database*, *file cabinet*, *file folder*, and the like. Because a data store represents all occurrences of an entity, we prefer plural names. If the data model entity is named CUSTOMER, for example, the corresponding data store on the process model would be named CUSTOMERS.

Data flows to and from data stores are interpreted the same way that physical DFDs are. Forget thinking in computer terms. Our guidelines, illustrated in Figure 9.3, are as follows:

1. Only processes may connect with data stores. Only processes can read and use data stores. Only processes can update data stores.

2. The direction of arrows is significant. Data flow direction is interpreted as follows:
   a. A data flow from a data store to a process means that the process uses that data. Notice that we said *uses,* not *reads*. The *read* is assumed. Name the flow to reflect the portion of the data required by the process. As a last resort, name the flow with the same name as the store.
   b. A data flow to a data store means that the process updates the data store. Updates may include any or all of the following:
      • *Adding or storing new records*—for instance, adding a new customer to the CUSTOMERS data store. The data flow name should not contain verbs. For example, use NEW CUSTOMER, not ADD A CUSTOMER.
      • *Deleting or removing old records*—for instance, deleting inactive customers from a store. Once again, the data flow name should not contain verbs. For example, DELETED CUSTOMER is better than DELETE A CUSTOMER.

FIGURE 9.3 The Significance of Data Flows to and from Data Stores Be careful with data flows to and from data stores. The direction of the arrow may signify use of data, removal of data, creation of data, deletion of data, or modification of data. Read the guidelines carefully to make sure you know the difference. DFDs never show the reading of data for its own sake!

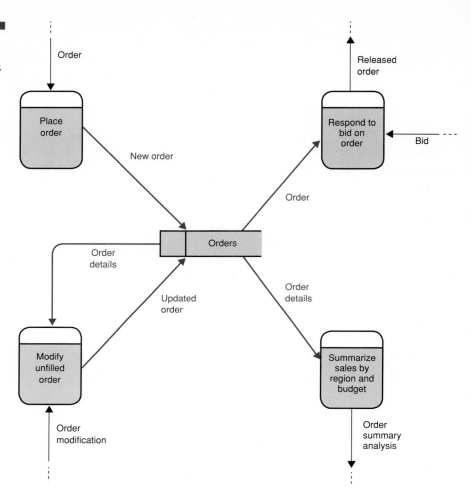

Archive data stores should not be shown on logical data flow diagrams.

- *Changing existing records*—for instance, changing the credit rating, address, or balance of an existing customer. And once again, the data flow name should not contain verbs. Use MODIFIED CREDIT RATING, not MODIFY CREDIT RATING, for example.

3. Although you know that you can't update a record without reading it, you don't depict this detail! It clutters the diagrams by requiring every interaction with every data store to have two flows. Show only the *net* or final data flow. In most cases this means the add, delete, or change. Still . . .

4. Some processes legitimately both use and update data stores. The key word is *use*. For example, calculations or decisions might be necessary before updating the data stores. If so, as a general rule, use separate data flows for the use and the update.

Computer-Assisted Systems Engineering for DFDs

As you learned in Chapter 7, most Computer-Assisted Systems Engineering (CASE) products support data flow diagramming. Using a CASE product, you can easily create and link professional, readable logical DFDs without the use of paper, pencil, erasers, and templates. These DFDs can easily be modified to reflect corrections and changes suggested by end-users; you don't have to start over! Also, most CASE products provide powerful analytical tools that can check your DFDs for mechanical errors, consistency errors (between diagrams), and completeness errors. The potential time and quality savings are substantial.

Most of the logical DFDs for the running case study in this book were created with a CASE product called Excelerator. Only the color screens and annotations were added by an artist. All of the objects on our DFDs were automatically cataloged into a dictionary so that we can add detailed information as the project moves through analysis, design, and implementation. We'll begin to look at the associated dictionary in the next chapter.

Using Logical Data Flow Diagrams for Systems Analysis

The use of logical data flow diagrams as a systems analysis tool is continually evolving, primarily in association with the Structured Systems Analysis methodology. Let's walk through each of our analysis phases and describe the current state of the art.

Modern Structured Systems Analysis

Chapter 7 introduced the Structured Analysis use of physical data flow diagrams for modeling the existing system during the **study phase**.

The **definition phase** represents the greatest potential for logical data flow diagrams as an analysis tool. This is because they intentionally model *what* the end-user wants or needs (we've been calling this the *logical* system) independent of *how* the system will be implemented to fulfill those wants and needs (we've been calling this the *physical* system). The benefits of logical modeling were discussed earlier in this chapter. We've also suggested, and will demonstrate in this chapter, how logical data flow diagrams build on and complement the data models developed in the last chapter.

During the **selection phase**, the logical DFDs from the definition phase can be studied and annotated to show alternative implementations that can subsequently be analyzed for technical, operational, and economic feasibility. Normally, different implementation alternatives are recorded on photocopies of the logical DFDs. For example, one copy of an order-approval DFD may show that all processes and data stores could be implemented on a mainframe computer. A different copy of the same DFD might suggest that some processes and stores could be implemented on a microcomputer, while others could be implemented on a mainframe. Thus, logical DFDs encourage alternative and creative solutions that can subsequently be evaluated for feasibility.

Looking ahead, DFDs can also be used in the **design** and **implementation phases**. As you might guess, design and implementation phase DFDs will return to the physical, implementation details that we discourage in this chapter.

How to Perform Process Modeling with Logical Data Flow Diagrams

As a systems analyst or knowledgeable end-user, you must learn how to draw logical DFDs to model process requirements. We will use the Entertainment Club systems project to teach you how to draw logical DFDs.

The scenario is as follows. We have completed our survey and study phases of the life cycle. We fully understand the current system's strengths, weaknesses, limitations, problems, opportunities, and constraints. We have already defined data requirements for an improved system. This resulted in the data model from Chapter 8. We will now model the corresponding process requirements.

We intend to carefully walk you through examples and provide a future reference for the day when you use this popular modeling tool.

Step 1: Draw a Decomposition Diagram to Outline DFDs   A single DFD for a typical system, even a small- to medium-sized system, would usually be very large—perhaps covering a wall in a good-sized room. Although visually impressive, such a diagram would be cumbersome to prepare, difficult to modify, and very intimidating to end-users. For this reason, we usually factor systems into subsystems, subsystems into functions, and so forth. Then we depict the system with a series of related DFDs. These diagrams are easier to prepare, read, present, and modify.

Factoring a system into subsystems and functions is an old concept—divide and conquer! Unfortunately, we tend not to do this with any rhyme or reason. The result is a collection of illogical subsystems and DFDs that, while somewhat correct, don't seem to make much sense. Consequently, we want to teach you to *plan* your factoring before you draw your diagrams. We will use an outline, a pictorial outline in this case. **Decomposition Diagrams**, also called *hierarchy charts*, show the *top-down* functional decomposition or structure of a system. They also provide us with an outline for drawing our DFDs.

There is only one symbol used on the decomposition diagram: the process symbol also utilized in DFDs. The processes are connected to form a treelike structure. Process names should conform to the naming guidelines described for DFDs. The top process, also called the *root*, represents the entire system for which you are defining requirements. The root process is **exploded** or factored into subsystems, functions, and tasks. The number of levels of subsystems, functions, and tasks is entirely dependent on the size of your project.

Figures 9.4 and 9.5 are the decomposition diagrams for the Entertainment Club project. Let's study these diagrams. First, we call your attention to the numbering scheme used in the diagram. These identification numbers will help you keep track of the subsystems and functions as you draw DFDs. It is a good idea to establish a standard numbering scheme before you draw decomposition diagrams and data flow diagrams. We follow these popular guidelines:

1. The root process is numbered 0.

2. The root process is factored into processes that are numbered consecutively, 1, 2, 3, and so on.

3. With subsequent factoring—also called leveling—of any process into subprocesses, each subprocess is numbered as a decimal of the parent process. For instance, process 1 is factored into processes 1.1, 1.2, 1.3, and so forth. Process 1.2 is factored into processes 1.2.1, 1.2.2, and so on.

This strategy is repeated throughout the decomposition diagram and the subsequent DFDs.

The following is an item-by-item discussion of the decomposition diagram in Figures 9.4 and 9.5. The letters direct your attention to specific locations on the diagram.

Ⓐ The root system process corresponds to the entire system.

Ⓑ The system is initially factored into subsystems. Before factoring a system, you might begin by studying another decomposition diagram that may already exist, the company's organization chart. Organization charts frequently show natural subsystems and functions. Sometimes, however, these charts do not depict a logical decomposition. In such cases, ignore them.

To further factor the subsystems, we have to abandon our organization chart because it goes no further. Why didn't we just factor the subsystems straight into the *detailed processing* tasks? Because there are too many of

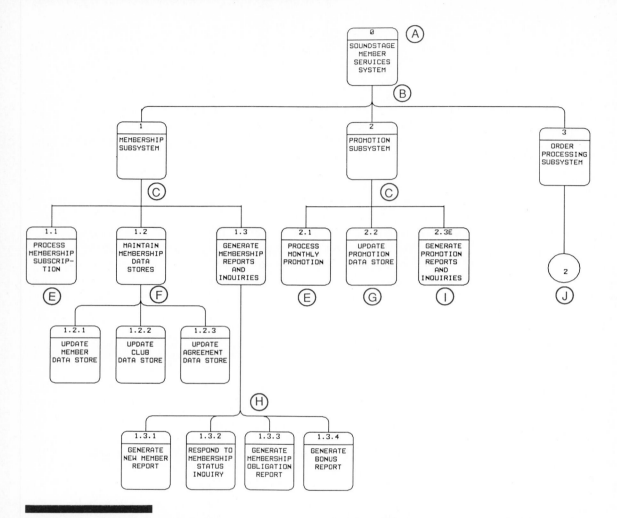

FIGURE 9.4 Decomposition Diagram   This is the decomposition diagram for the Entertainment Club system.

those tasks to place on a single data flow diagram (remember, DFDs are our goal—this is just the outline).

Is there any limit to the number of subprocesses you can factor a process into? Evidence suggests that a process should generally be factored into seven or fewer processes. Your final guideline will be the readability of the resulting DFD. Most DFDs start becoming cluttered at about five processes. Many become unacceptably cluttered at about seven processes. This guideline will make the DFDs much easier to draw by limiting to seven the number of processes that will appear on any given page.

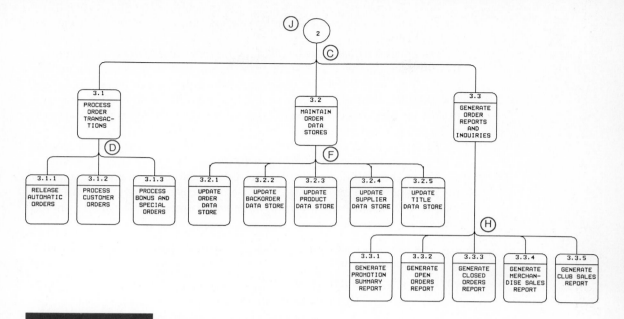

FIGURE 9.5   Page 2 of Decomposition Diagram

What about a minimum number of subprocesses for any given process? Certainly a process should never be factored into a single process, since no additional detail is revealed in such a case. Thus, if you plan to factor a process, it should be factored into two to seven child processes.

Let's return to our walkthrough of the diagram.

Ⓒ Our sample project is relatively small. There is no need to factor the subsystems into functions or subfunctions. Larger projects may benefit from such decomposition, though.

Eventually, virtually all systems get to the point where specific processing requirements can be identified. We have factored each subsystem into activities that correspond with transactions, data maintenance, and reports. *These activities are common to virtually all information systems.*

Virtually all systems respond to **transaction events**. These events just *happen*. As they happen, they bring new data into the system. Examples include orders, payments, requisitions, shipments, receipts, returns, work—recorded on time slips—and the like. Systems must respond to transaction events with appropriate outputs. Sometimes the outputs simply store the input data for later processing.

Data maintenance includes all of the custodial activities needed to ensure that accurate data is available for transaction processing. Examples include routine maintenance of data stores concerning parts, customers, employees, and the like. Most of you have probably written file maintenance programs to add, change, and delete records in master files. These programs are part of this maintenance activity.

Finally, we included a **report**ing subsystem to support management reporting and decision support activities that are equally common to most systems.

Notice the common systems pattern: transaction and maintenance activities bring new data into the system. Reporting activities generate useful information from that data. *Virtually all information systems will benefit from this factoring strategy.*

(D) Transaction-processing activities are factored into one process for each transaction event.

(E) Notice that we did not factor the membership subsystem into multiple transactions. This is because there is only one transaction for this subsystem. We simply identified that transaction as the only transaction-processing event for the subsystem.

(F) Data maintenance activities are factored into one process for each data store to be maintained. You should initially list each of the data stores to be maintained. Generally speaking, there is one data store for each fundamental data entity on your data model. We omitted maintenance for associative entities. Why? Because ORDERED PRODUCT (an associative entity) will normally be maintained by the same process that maintains ORDER (a fundamental event entity). It should be noted that more than one subsystem may share responsibility for maintaining a single entity.

(G) Note that the promotion subsystem only maintains a single data store. It is not factored.

(H) The reporting activities can be factored into outputs, or groups of outputs, in the same manner as transactions and maintenance.

(I) As demonstrated in the promotion subsystem, if the end-users are not sure about what outputs they need, just defer factoring the reporting system until a later time. It is not as crucial as the transaction and maintenance activities. Why? If we accurately identify all events and maintenance, then the data will be captured for later use in the reporting subsystem.

This concept reflects a significant trend in modern systems analysis. Traditionally, outputs were assumed to be the most important requirements to identify. Unfortunately, it has turned out that end-users are frequently unable to identify such requirements, leading to considerable frustration when they complain that implemented systems fail to meet their needs!

Consequently, we have changed the strategy to ensure that, when the end-users finally identify output needs, we will have captured and maintained the data that will be required to produce those new outputs. This strategy is made possible by the data modeling performed in Chapter 8.

Ⓙ The small circle is an off-page connector that indicates that the order subsystem is further decomposed on a page that will contain the identical off-page connector as its root. Off-page connectors allow you to draw multipage decomposition diagrams that are easier to read than a large, single-page diagram.

That's about it for our decomposition diagram. There is no need to factor beyond the transactions, maintained stores, and reports. That would be like outlining right down to the final sentences in a paper. However, our graphic outline will serve as a good outline for an integrated set of DFDs.

Step 2: Draw a Context Diagram   Logical DFDs are constructed in levels that correspond to levels in our decomposition diagram. Higher levels of DFDs depict the "big picture" in terms of subsystems and functions. Lower levels of DFDs illustrate detailed processing and data flow. Figure 9.6 illustrates this general strategy. Because each page shows a greater level of detail, we sometimes call the collective DFDs a **leveled set of DFDs**. The first DFD that is drawn—the top page in Figure 9.6—is called a **context diagram**.

A context diagram defines the scope and boundary for the system and project. The scope of any project is always subject to change; therefore, the context diagram is always subject to change. Curiously, though the context diagram is the simplest DFD of a leveled set, it tends to be difficult to draw because it carries that "definition of scope" responsibility. We suggest the following strategy for determining the systems boundary and scope (adapted from MacDonald):

1. Think of the system as a container in order to distinguish the inside from the outside.

2. Ignore the inner workings of the container. This is the classic *black box* concept of systems theory.

3. Ask your end-users what events or transactions a system must respond to. The idea of events was discussed earlier. Events happen and bring new data into the system. An example might be an order.

4. Ask your end-users what responses must be produced by the system. Examples for the order event might include backorders, picking orders, and invoices. On the other hand, some systems are intended to produce reports as their response.

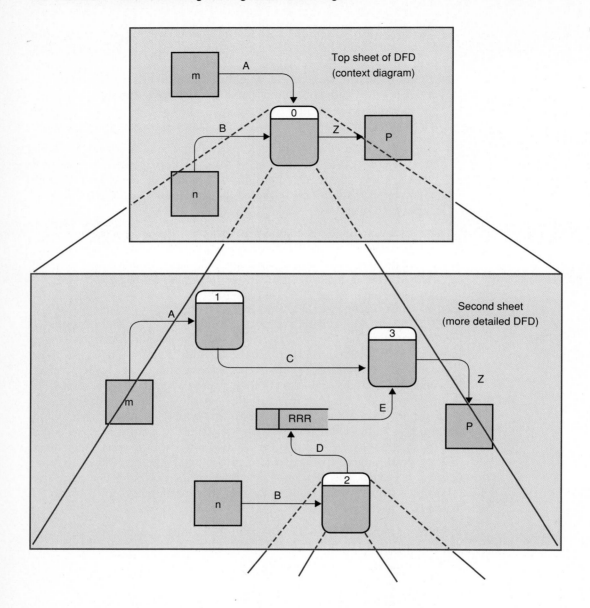

5. Identify the net sources of data about each event. These sources will be external or internal agents on the DFDs.

6. Identify the net recipients of each response or output that the system should generate. These destinations will also be external or internal agents.

7. Draw your context diagram.

◀ **FIGURE 9.6 Leveled Data Flow Diagrams** In order to deal with size and complexity, analysts should draw a series of data flow diagrams that progress from a very high-level, general view of the system to low-level, detailed views of subsystems.

The context diagram contains one and only one process, the root process from the decomposition diagram. You add the data flows and external/internal agents to the diagram to define the context. Now, if you try to include all of the inputs and outputs between a system and the rest of the business and outside world, a typical context diagram might show as many as 50 or more data flows. Such a diagram would have little, if any, communication value. Therefore, we suggest a different strategy. For the context diagram, only show those data flows that represent the *main objective* or *most common inputs and outputs* of the system. Defer less common data flows to lower-level (more detailed) DFDs to be drawn later. This strategy is applied in Figure 9.7. The main purpose of our system is to respond to membership applications, marketing promotions, and member orders. Management has also placed great emphasis on the need for sales analysis reports and inquiries. Other reports will be identified on lower-level DFDs.

One way to simplify a context diagram is to consolidate data flows. For instance, we can consolidate regular orders for records and tapes with special merchandise orders using the single data flow ORDER. Later, we can factor the consolidated flow into separate flows, on whichever lower-level DFD recognizes different processes for the flows. Similarly, we consolidate multiple sales analysis reports into a single, plural data flow. That flow will be factored into the distinct reports on a DFD that depicts separate processes to produce each report.

A few final comments are in order. A context diagram always contains one and only one process: the root process from the decomposition diagram. It is the only DFD—one per project—that will contain a single process. That process represents the entire system that will be factored or exploded into more detailed DFDs on subsequent pages. The context diagram depicts the inputs and outputs between the system, or process, and the rest of the business and outside world. The context diagram contains no data stores.

Step 3: Draw a Systems Diagram We can now explode the process on the context diagram into a more detailed picture of the system. This second DFD is usually called the **systems diagram**. It shows the major subsystems or functions and how they interact with one another. It is useful for communicating the system's *big picture*.

Note the terminology we used: *explode the process into a more detailed DFD*. Whenever you explode a process, the resulting DFD is a more detailed look at that same process. (Explosion is called *leveling* in some Structured Analysis books.) Let's demonstrate by exploding the lone process on the context diagram (Figure 9.7).

That context diagram process is exploded into the systems diagram shown in Figure 9.8. Notice that the three processes correspond precisely to the three processes on the second level of our decomposition diagram. In other words, we are following our original graphic outline (this is not to imply that

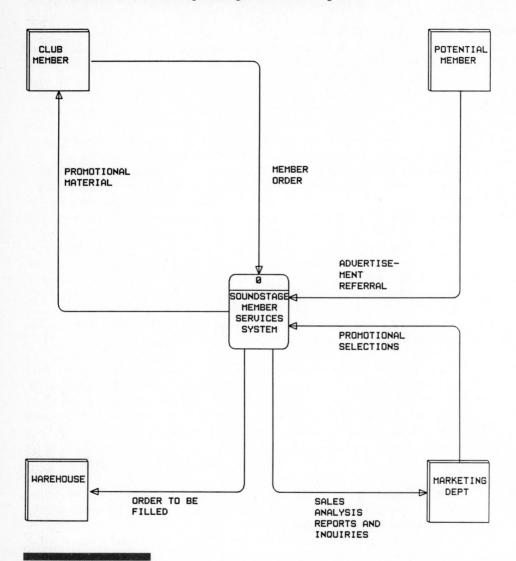

FIGURE 9.7 The Context Diagram The context diagram is the most general picture of a system. The system is depicted as a single process. Its inputs and outputs convey the main purpose of the system.

you never deviate from that outline, though). We call your attention to the following items on the systems diagram:

Ⓐ The data flows that appeared on the context diagram also appear on this diagram. This is done to maintain consistency with the previous diagram. Most books call this **balancing the diagrams**.

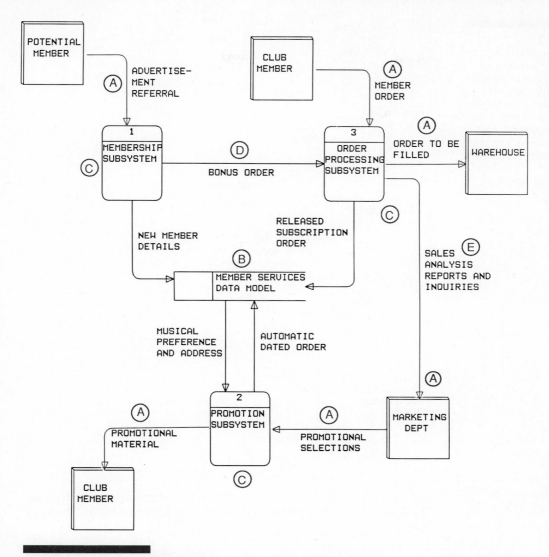

FIGURE 9.8 Systems Diagram The systems diagram depicts the major subsystems of the Entertainment Club case study.

Note that we also duplicated the external and internal agents from the context diagram. Some experts prefer not to duplicate these agents. Instead, they draw the data flow without the connecting object (as in Figure 9.9). The data flows that appear to go nowhere or come from nowhere are supposed to direct the reader's attention back to the parent DFD, in our case the context diagram. We think that this detracts from

FIGURE 9.9
Alternative Systems Diagram This diagram, although equivalent to Figure 8.12, does not communicate to end-users nearly as well. This is because the end-user is forced to return to the context diagram to see where most of the data flows come from or go to.

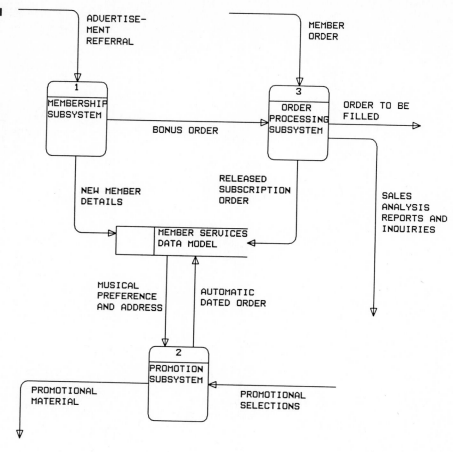

communication; therefore, we duplicate the agents. Because they are duplicated from the context diagram, we mark them with the diagonal slash.

Ⓑ The subsystems or functions on most systems diagrams share data stores, usually several data stores. To simplify the systems diagram, we generally consolidate all data stores into a single data store that represents shared data. Using a single data store keeps this diagram readable. It also avoids premature commitment to any specific implementation—for example, specific file structures.

This particular data store also conveniently explodes directly into the information model that we developed in Chapter 8. Thus, the data store represents all entity types and relationships that describe the stored data for this system.

As we draw more detailed DFDs, we will explode the common data store to reveal detailed data stores that describe specific logical collections of data.

Ⓒ The three subsystems correspond to the same three subsystems that appeared on the second level of our decomposition diagram.

Ⓓ The DFD should show all of the data flows that occur between the subsystems or functions shown on the physical DFD. This is the principal benefit of high- and middle-level DFDs: they show the interfaces between groups of processes.

Ⓔ An inquiry is actually a two-way data flow. There are a question (input) and an answer (output). But to keep the DFD simple, we show only the *net* data flow, the output.

By way of summary, this diagram illustrates the three key subsystems of the Entertainment Club's member services system: membership, promotion, and order processing. The data store represents the entire system's data model, as defined by the entity relationship diagram from Chapter 8.

The membership subsystem responds to subscriptions. The promotion subsystem generates promotions and the dated orders. The order-processing system responds to orders as received from members and new subscribers.

Step 4: Draw Middle-Level Diagrams   After drawing the systems diagram, we can explode each of the processes on that DFD to reveal still greater detail about the subsystems. Any process on any DFD can be exploded to reveal a more detailed DFD for that process. Explosion is continued until we have depicted a sufficient level of detail. All but the lowest-level DFDs are frequently called **middle-levels**. Let's walk through some of the middle-level diagrams.

Figures 9.10 through 9.12 are explosions of the subsystem processes on the systems diagram (Figure 9.8). Each is very similar, which is to be expected since we used a similar factoring strategy for all subsystems on our decomposition diagram. Once again, we factored precisely to that decomposition diagram outline. Let's walk through Figure 9.10, the explosion of the membership subsystem process (process 1) from the systems diagram.

Ⓐ The data flows that are printed in color are those that balance back to the parent process in Figure 9.8. Remember, balancing preserves consistency between levels of DFDs.

Ⓑ The BONUS ORDER data flow appears to go nowhere. It is called an *interface data flow*. It had to be added to preserve balancing between this DFD and the parent process (process 1) in Figure 9.8. An interface data flow indicates that the flow goes to a process in another subsystem. You can always find out where it goes by looking at the parent diagram

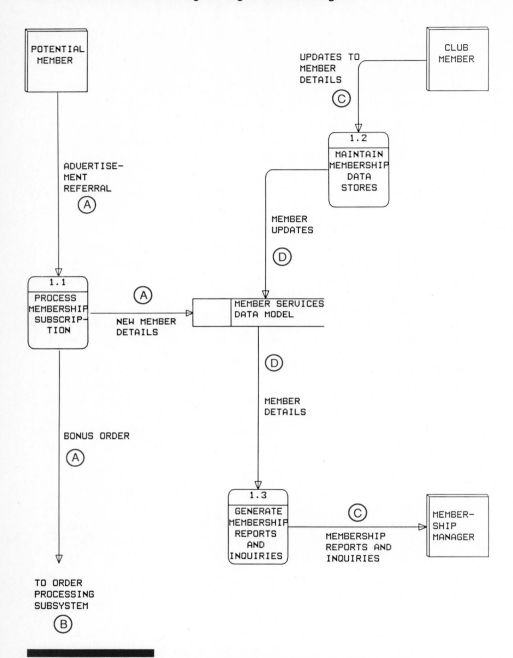

FIGURE 9.10 DFD for the Promotion Subsystem This DFD explodes from process 2 on the systems diagram. It depicts the three principal functions of any subsystem: transaction processing, data maintenance, and reporting (both management reporting and decision support).

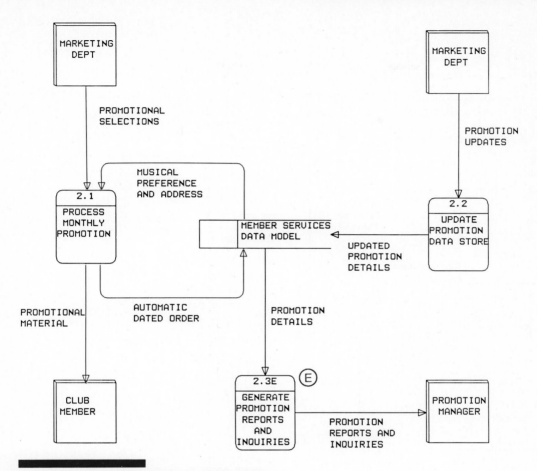

FIGURE 9.11   DFD for the Promotion Subsystem

(again, Figure 9.8). Do that! As you can see, the flow is going to the ORDER-PROCESSING SUBSYSTEM. So that end-users aren't forced to "go back" to other diagrams, we like to add a text label, in this case TO ORDER-PROCESSING SUBSYSTEM, at the end of the interface data flow. This improves readability.

ⓒ Some new net data flows appear on the diagram. They were less common or important, and therefore were not included on the higher-level context and systems diagrams. These new flows reveal new, minor details in the process model. Some of these data flows may be exploded to multiple data flows on the lower levels. For instance, MEMBERSHIP REPORTS AND INQUIRIES is a consolidated flow that represents several different types of reports and inquiries, details about which would unnecessarily clutter this middle-level diagram. This keeps such diagrams more readable.

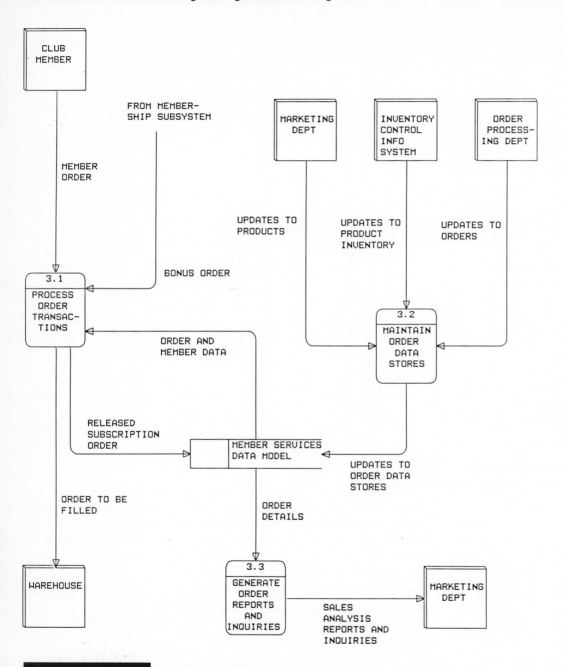

FIGURE 9.12 DFD for the Order-Processing Subsystem

Ⓓ Earlier we suggested that most data flow names should not be plural. An exception is those data flows coming from or going to data stores. A plural name such as MEMBER DETAILS is more descriptive in this case since a single occurrence of the flow contains numerous details about a single member.

The plural names on the output data flows are needed since the data flow is actually a consolidation of numerous distinct outputs that will be identified individually at lower levels of the DFDs (at which time the names will be made singular).

Figure 9.11 is the explosion of the PROMOTION SUBSYSTEM process (process 2) on the systems diagram (Figure 9.8). Once again, we have included the three processes that were suggested on the original decomposition diagram outline. And once again, the data flows for the parent, subsystems process (process 2 in Figure 9.8) have been carried down to this DFD. Finally, and again, a few new data flows were added. We call your attention to the following new item of interest:

Ⓔ The *E* that was added to the identification number for process 2.3 indicates our intention of eventually exploding this process into a more detailed DFD. Recall from our decomposition diagram that our end-users had not yet factored that process into separate reports and inquiries. At whatever time we actually explode the process, we should remove the *E*.

Figure 9.12 is the explosion of the ORDER-PROCESSING SUBSYSTEM process on the systems diagram (Figure 9.8). Once again, we have included the three processes that were suggested on the original decomposition diagram outline. And once again, the data flows for the parent, subsystems process (process 3 in Figure 9.8) have been carried down to this DFD. Finally, we have again added several new data flows to reveal additional detail. There is one new comment on this DFD:

Ⓕ The BONUS ORDER is another example of an *interface data flow*. It was necessary in order to balance the DFD with its parent process, process 3, on the systems diagram (Figure 9.8). We have again added the label to indicate which subsystem it comes from, which can be determined by looking at the parent diagram (Figure 9.8).

Also note that this *input* interface data flow matches the *output* interface data flow that was highlighted in Figure 9.10. Interface data flows will always occur in matching pairs on separate DFDs.

Step 5: Draw Lower- and Primitive-Level Data Flow Diagrams We now complete our leveled set of DFDs by drawing those diagrams that show detailed processing requirements for the system. These are called lower- and primitive-level DFDs.

Let's start with the membership subsystem. Look back to the decomposition diagram. Notice that this subsystem must only support one transaction, MEMBERSHIP SUBSCRIPTIONs. Membership subscription processing's role in the total membership subsystem was depicted in Figure 9.10. It is now time to explode the PROCESS MEMBERSHIP SUBSCRIPTION process in Figure 9.10 to show detailed, logical process requirements. Figure 9.13 is that diagram. We call your attention to the following details:

(A) None of the processes on this diagram is exploded to more detailed DFDs. So that the reader won't look for additional diagrams, the letter *P* has been added to the identification number for each process. The *P* stands for *primitive process*, defined as a process that does not explode to another DFD (although we will further describe the process with other tools in Chapter 11).

(B) The MEMBERSHIP APPLICATION data flow from the parent process does not appear anywhere on this diagram. This may lead you to conclude that the DFD doesn't balance with its parent. This is not true. What we did was to explode the data flow MEMBERSHIP APPLICATION into two distinct data flows, ADVERTISEMENT REFERRAL and CLUB MEMBER REFERRAL. In this DFD, they happen to go to the same initial process. On other DFDs, exploded flows may go to or come from separate processes on the lower-level DFD.

(C) For the first time, we are recognizing distinct data stores (subsets of the total data model). Each store corresponds to a data entity on the entity relationship data model from Chapter 8.

(D) PENDING MEMBERSHIP ORDERS is not a data entity in our data model. Rather, it is a buffer data store to hold subscription orders until the membership application is approved (or rejected). Order data merely resides there until other processes on the DFD have completed their work.

(E) A few new data flows to/from external and internal agents were introduced at this level. Primitive levels should include all data flows, no matter how trivial or uncommon.

(F) All membership applications include subscription orders that give new members and referring members records, tapes, or discs for a super-discounted price—for example, 12 records for 25 cents. These SUBSCRIPTION ORDERS were screened out of the application before the approval process since the order data is irrelevant to the approval process. This is an important concept of logical data flow diagrams, described as follows:

A process should only receive that data that it truly needs to perform its named purpose. This is called decoupling the process from excessive data.

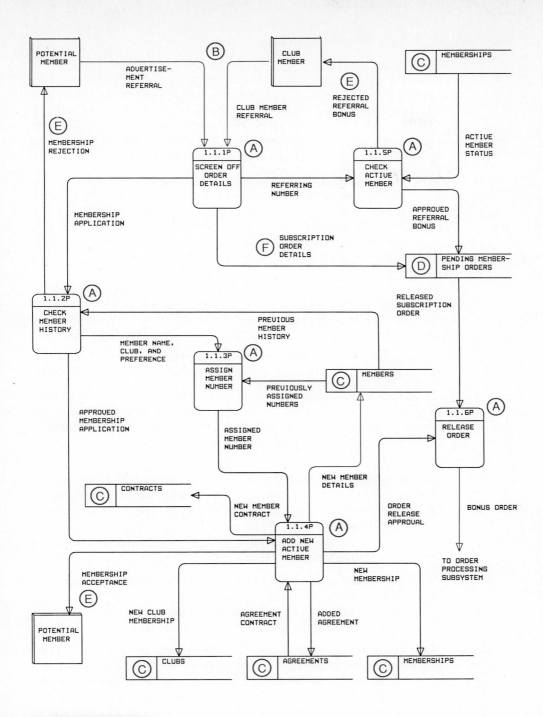

FIGURE 9.13 DFD for Membership Subscription Processing   This is a detailed DFD for transaction processing of a single transaction.

If you think about it, this makes a lot of sense. If we eventually write a program for this process, that program should not receive data that it doesn't need. Why? Because it becomes too dependent on other programs, especially if those other programs necessitate a change to the extraneous data. Also, programs that input unneeded data are not as portable; that is, you cannot as easily reuse the code in other programs.

Now, let's diagram the data maintenance activities for the membership subsystem. Look again at the decomposition diagram, our outline. Notice that the membership subsystem is custodian of three data stores, MEMBERS, CLUBS, and AGREEMENTS. We need to draw detailed DFDs for this data maintenance. Figure 9.14 shows the three data maintenance functions from the decomposition diagram. We did not explode the data store because this is not yet a primitive-level DFD.

Figures 9.15 through 9.17 show the detailed processing requirements for membership data maintenance. Notice in Figure 9.15 that a single operation, such as ADD, can be triggered by several data flows from several sources. We call your attention to the following details, demonstrated in Figure 9.15:

(A) Once again, detailed data stores were deferred until this diagram, the primitive-level DFD. Thus, the preceding levels were kept more readable.

(B) Every data maintenance function should allow end-users to add, change, and delete records in their respective data stores. Keep in mind, as we progress toward design and implementation, that this DFD is usually conceived of as a single program.

(C) Why would we want to include an ADD facility when we already included a MEMBERSHIP SUBSCRIPTION–processing DFD earlier? All systems should provide for routine data maintenance. What, for example, would you do if a member record were accidentally deleted? It seems silly to execute the membership subscription process to correct such a mistake.

This concept should be familiar to those of you who have written a classic file maintenance computer program.

(D) Note that CHANGE OF NAME OR ADDRESS occurs as a single data flow. This logical data flow accommodates the many ways that the data flow may be implemented—for example, by postal change of address card, phone call, letter, and the like. *Logically*, it is one data flow.

Next, let's complete the membership subsystem by drawing the DFD for management reporting and decision support. From the decomposition diagram, we can identify processes that will generate specific reports and inquiry responses. The DFD for these processes is illustrated in Figure 9.18. Pay particular attention to the following details:

(A) Each process draws on those data stores that contain needed details for the report or inquiry response. Since this is a primitive-level DFD, the specific data stores were used in lieu of the information model.

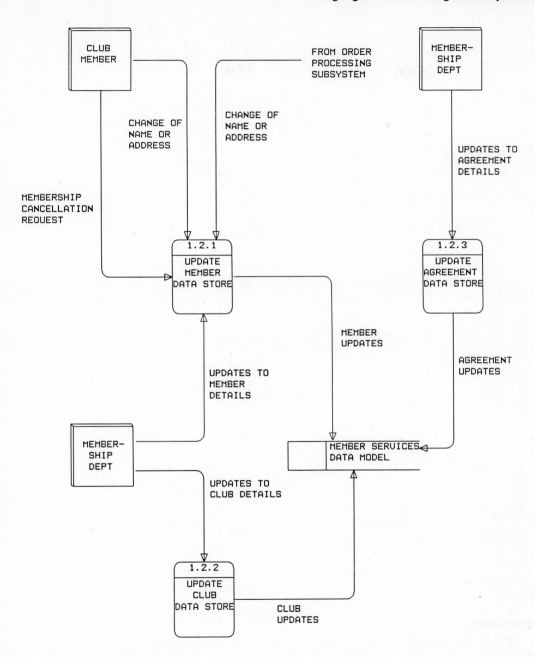

FIGURE 9.14 DFD for Membership Data Maintenance All information systems must provide for routine maintenance of all data stores. This DFD indicates that the membership subsystem is responsible for maintaining three data stores.

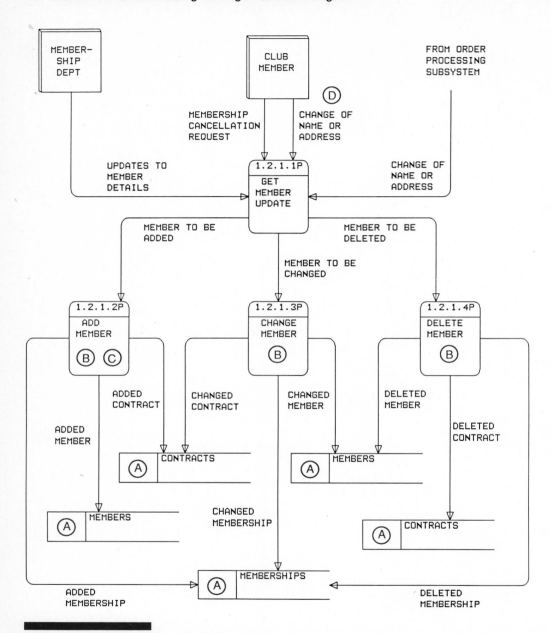

FIGURE 9.15   DFD for the MEMBERS Data Store Maintenance   This DFD shows detailed processing for routine maintenance: adding, deleting, and changing records to/from/in the MEMBERS data store. All data maintenance DFDs are very similar.

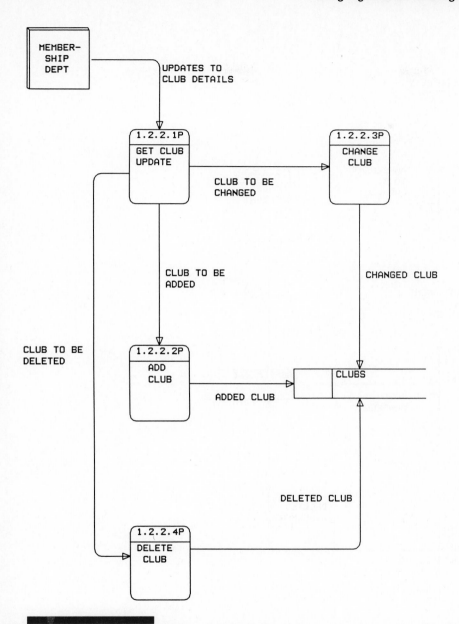

FIGURE 9.16   DFD for CLUBS Data Store Maintenance   This DFD shows routine maintenance requirements for the CLUBS data store.

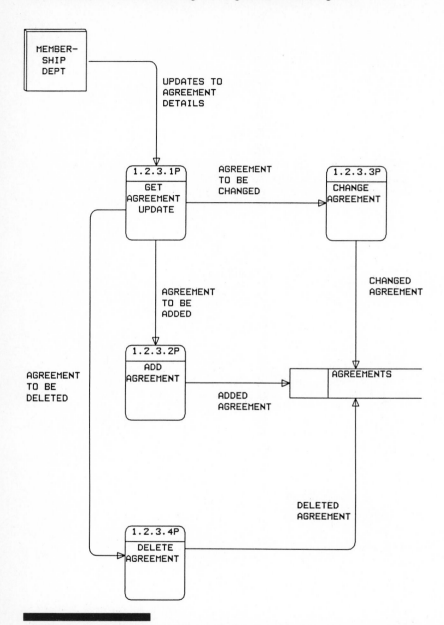

FIGURE 9.17 DFD for AGREEMENTS Data Store Maintenance This DFD shows routine maintenance for the AGREEMENTS data store.

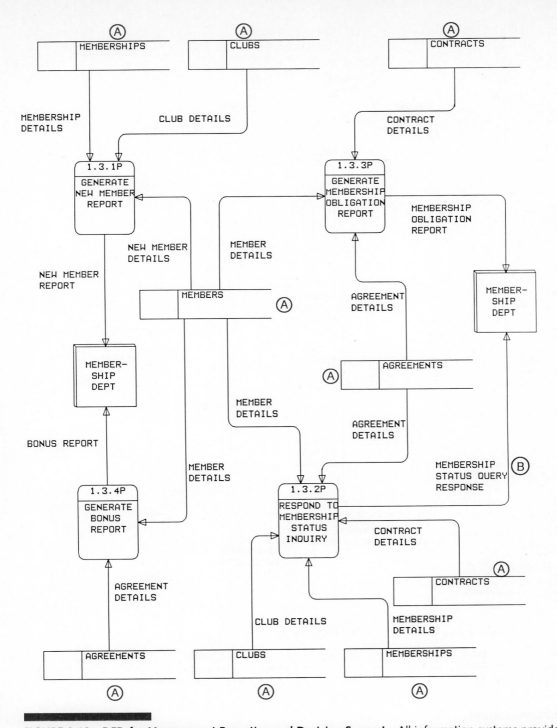

FIGURE 9.18 DFD for Management Reporting and Decision Support All information systems provide opportunities for using data stores to present management reports and decision support responses. The DFDs for such reporting in the other subsystems will be very similar to this diagram. **307**

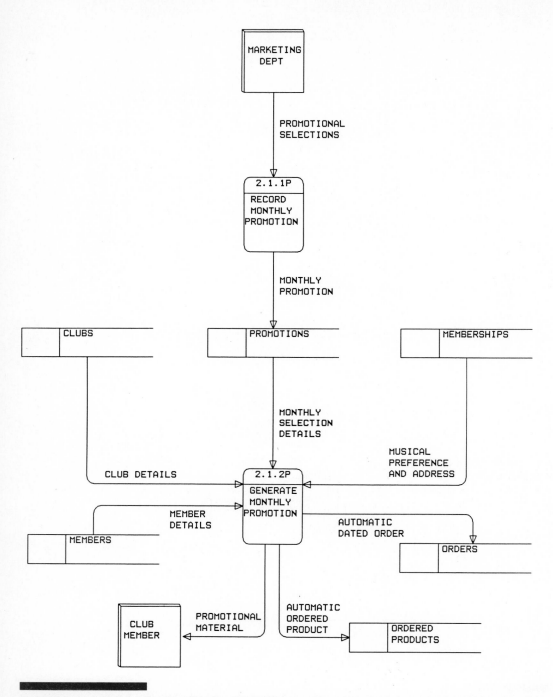

FIGURE 9.19   DFD for the Monthly Promotion Transaction   This DFD demonstrates requirements for generating the monthly promotion and accompanying automatic, dated order.

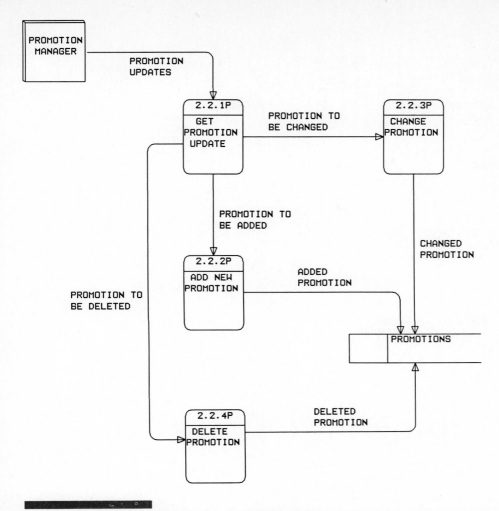

FIGURE 9.20 DFD for Maintaining the PROMOTIONS Data Store This DFD demonstrates the requirement for routine maintenance to be performed on the PROMOTIONS data store.

Ⓑ We realize that all inquiry responses must have been triggered by an inquiry request data flow. However, one of the fundamental guidelines of DFDs is that they only show *net* data flow. Otherwise, inquiry processes would become cluttered with obvious, excessive data flows. In this case, the net data flow is the response itself.

Detailed DFDs for the promotion subsystem are illustrated in Figures 9.19 and 9.20. Like the membership subsystem, the promotion subsystem had only one transaction, which is documented by the primitive DFD in Figure 9.19. Unlike the membership subsystem, the promotion subsystem maintains only a single file, documented by the primitive DFD in Figure 9.20. As has been the case with the entire leveled set of DFDs, these diagrams follow the outline

309

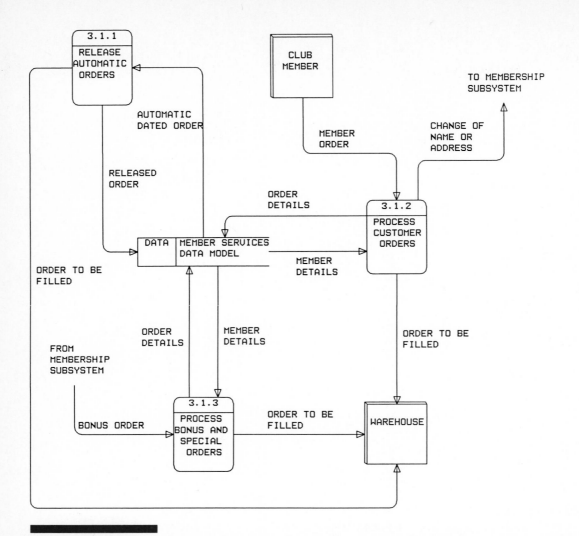

FIGURE 9.21   DFD for Order-Processing Transactions   This DFD shows all of the transactions to which the order-processing subsystem must respond. Each process is exploded in the subsequent figures.

that was established in the decomposition diagram. Each diagram balances with the parent process from which it was exploded.

Finally, in the interest of completeness, detailed DFDs for the order-processing system are presented in Figures 9.21 through 9.31. This subsystem differs from the others in that it supports multiple transactions, shown together in Figure 9.21. The primitive DFDs for those transactions are provided in Figures 9.22 through 9.24. Figures 9.25 through 9.30 address the maintenance of data stores for which the order-processing system has custodial responsibility. Finally, Figure 9.31 depicts the report and inquiry requirements for the subsystem.

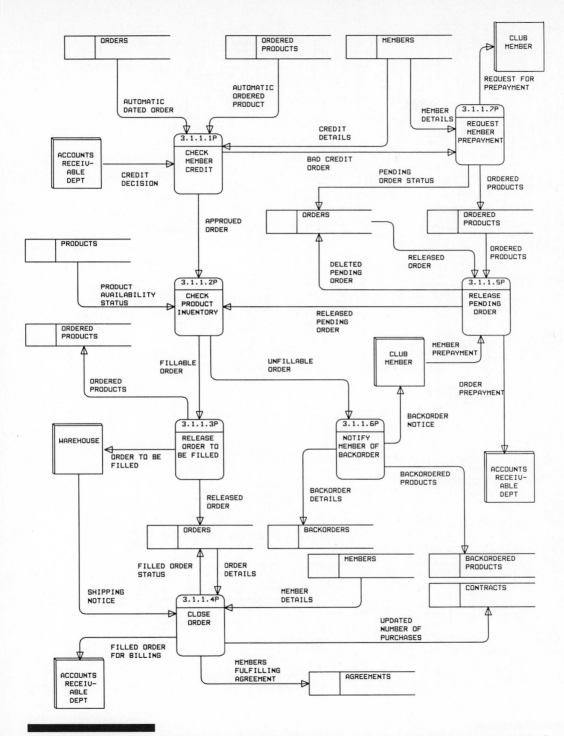

FIGURE 9.22   DFD for Automatically Releasing Dated Orders   This DFD documents requirements for automatically releasing promotion orders for which the customer has not replied by the indicated date.

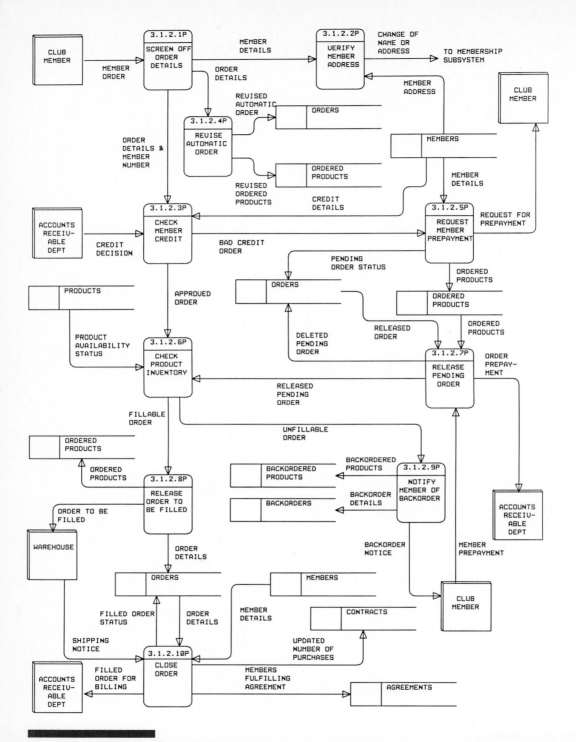

FIGURE 9.23 DFD for Processing Customer Responses to Promotion Orders This DFD documents requirements for amending promotion orders according to customers' responses to the promotion and receipt of a catalog.

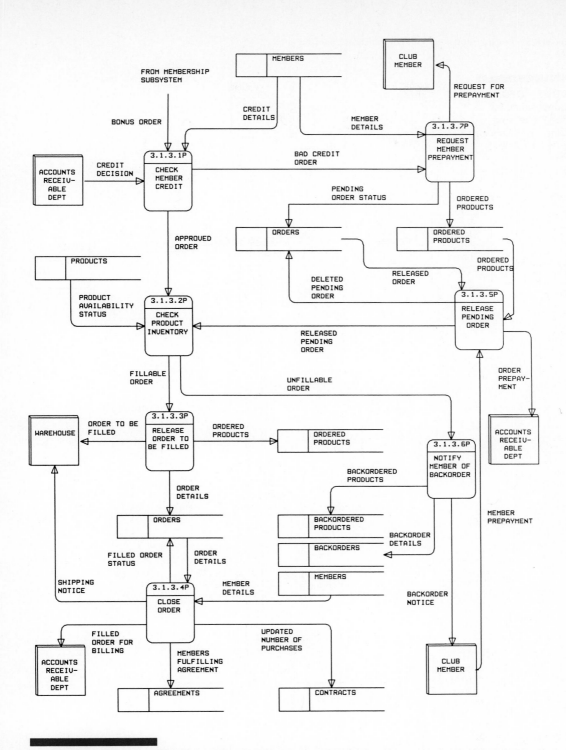

FIGURE 9.24 DFD for Processing Bonus and Special Orders This DFD documents requirements for processing special orders for merchandise or bonus orders for titles.

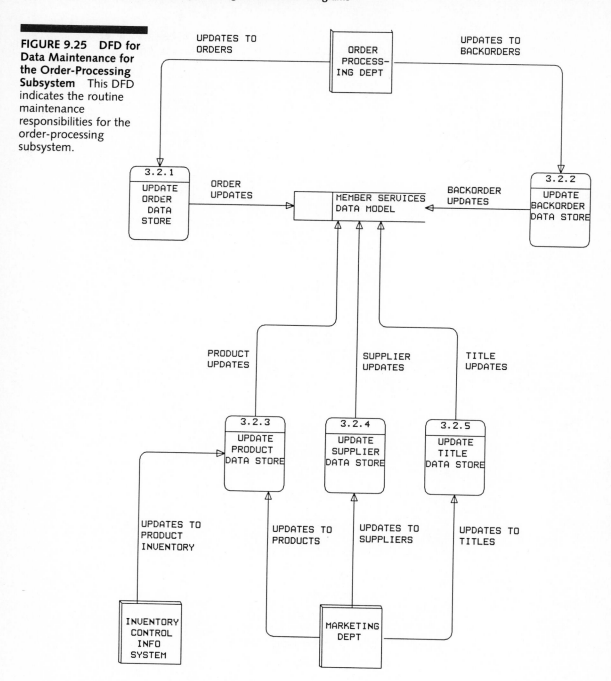

FIGURE 9.25 DFD for Data Maintenance for the Order-Processing Subsystem This DFD indicates the routine maintenance responsibilities for the order-processing subsystem.

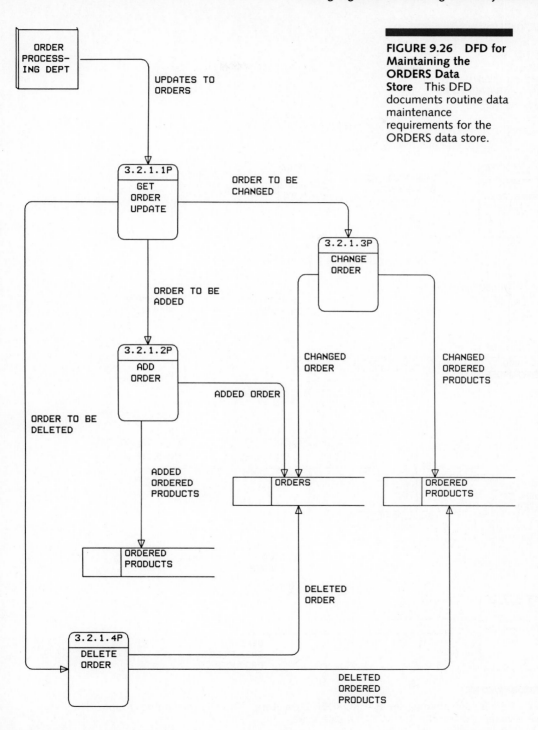

FIGURE 9.26 DFD for Maintaining the ORDERS Data Store This DFD documents routine data maintenance requirements for the ORDERS data store.

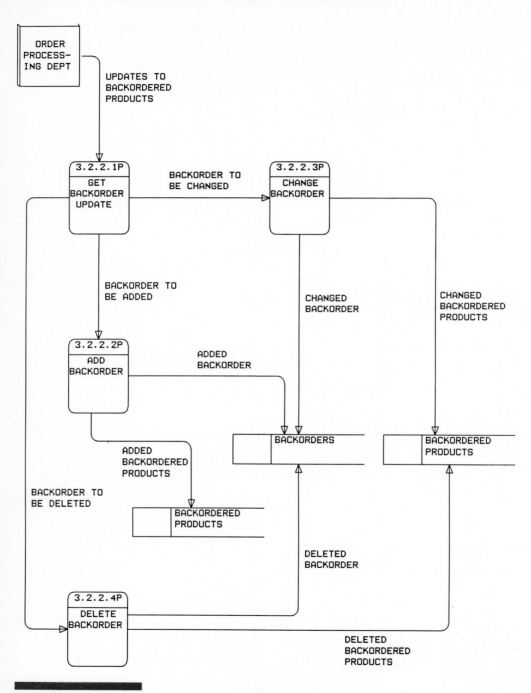

FIGURE 9.27 DFD for Maintaining the BACKORDER Data Store This DFD documents routine data maintenance requirements for the BACKORDER data store.

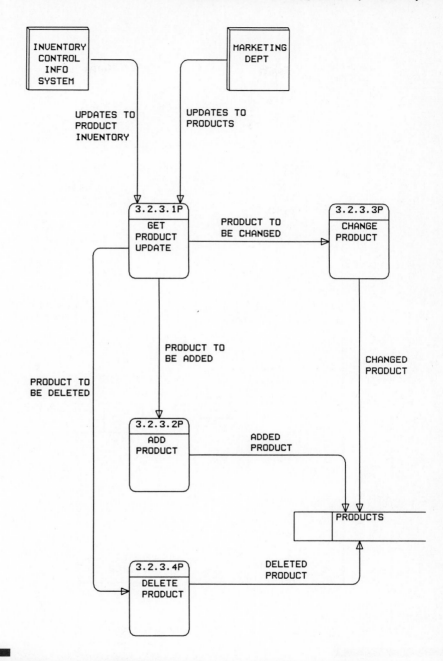

FIGURE 9.28 DFD for Maintaining the PRODUCTS Data Store This DFD documents the routine data maintenance requirements for the PRODUCTS data store.

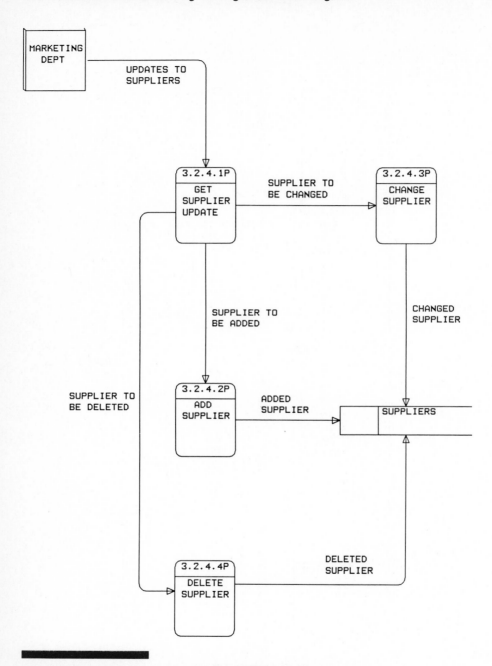

FIGURE 9.29   DFD for Maintaining SUPPLIERS Data Store

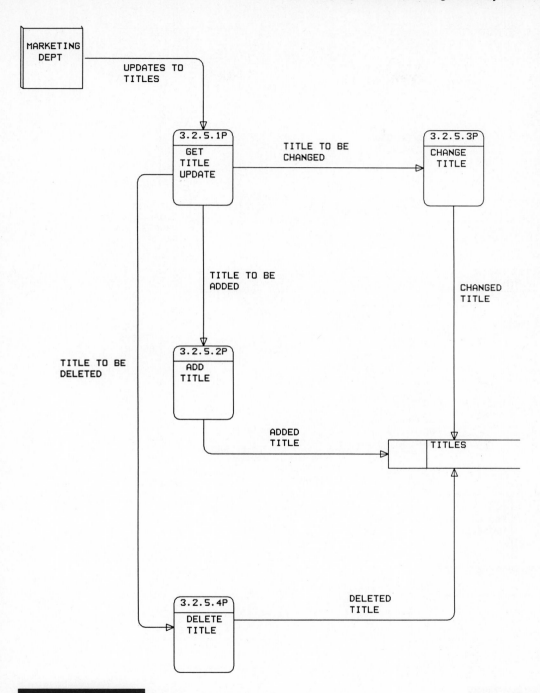

FIGURE 9.30   DFD for Maintaining TITLES Data Store

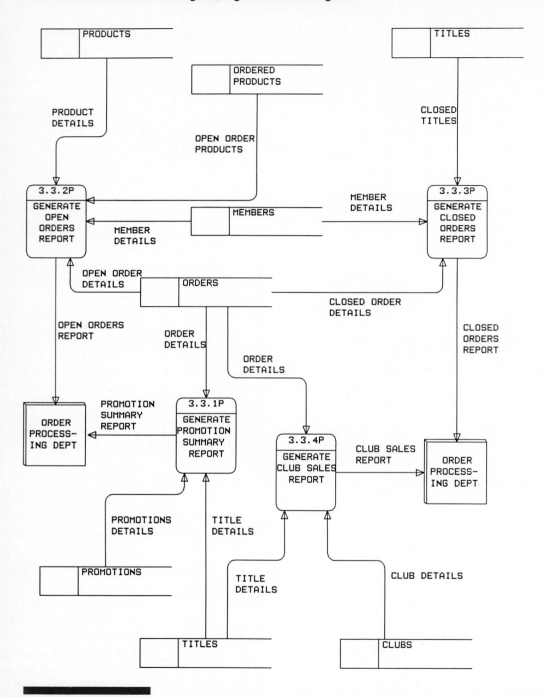

FIGURE 9.31 DFD for Generating Management Reports and Supporting Decisions

Data Flow Diagrams for Subsequent Systems Analysis and Design

We have completed our leveled set of DFDs. By looking at different DFDs in the set, you and your end-users will be able to study the overall system, detailed pieces of the system, and anything in between. Few tools can boast that level of communication.

Data flow diagrams have value throughout the life cycle. Once requirements are approved, we must evaluate alternative ways to implement processes, data stores, and data flows. Gradually, as implementation decisions are made, we will add physical, implementation details to the DFDs. Consequently, we will be able to demonstrate the transition from requirements to reality—an important step for both end-users, who need to see *how* the new system will work, and programmers, who must implement the system according to the preferred methods demonstrated in the DFDs.

Summary

A logical data flow diagram (DFD) is a tool for drawing a model or picture of an information system's processing requirements. It illustrates the flow of data and work through a system. There are only four symbols that can appear on a DFD: the process, the external or internal entity, the data store, and the data flow. With these symbols, you can model processing for virtually any information system. Logical DFDs differ from physical DFDs in that the logical DFDs describe the systems requirements independently of any technology or methods that might be used to implement those requirements. This opens the door for creative solutions. Data flow diagrams can effectively model all transaction-processing, management reporting, and decision support functions in a system.

Because most systems are too complex to be depicted on a single DFD, we draw a series of DFDs that depict the system in various levels of detail, from general to explicit. The most general DFD is called the context diagram because it shows the scope or boundary of the system. The next level of detail, the systems diagram, shows key subsystems or functions and their interfaces. Middle-level DFDs show still greater detail about subsystems and/or subfunctions. Finally, primitive levels show explicit data flows and processes for a small, manageable piece of the system.

To intelligently plan the levels of DFDs, a decomposition diagram is used to functionally decompose a system. The chapter presented useful strategies that are applicable to virtually all information systems.

THE NEXT GENERATION

Transformation Graphs: An Extension of DFDs for Real-Time Systems Requirements

Data flow diagrams are the most popular modeling tool for information systems analysis and design. But the tool doesn't work nearly as well when we are analyzing and designing real-time systems.

Real-time systems are systems that have a more critical time element. Information systems tend to be oriented toward processing data into information. Real-time systems may overlap with this function; however, they are primarily intended to exert control over their environment.

Examples of real-time systems are plentiful: air traffic control, computer-assisted manufacturing, robotics, computer-controlled or computer-assisted appliances, temperature/humidity control, and missile tracking. In all real-time systems, the system monitors the environment and then either controls devices that change the environment or helps humans effect the change.

Data flow diagrams are of limited value in real-time applications since they show only data flows and processes that handle data and information . . . until recently!

Transformation graphs (see Figure 9.32) are a superset of data flow diagrams. They show all data flows, data stores, data processes, and external entities, just as data flow diagrams do. Additionally, they show control processes that respond mainly to conditions instead of just data. They also show control flows. Control flows are either sensed conditions (for example, temperature, time, altitude, and the like) or response stimuli (for example, a signal to turn a furnace on or off, a signal that is a warning to a human operator, a signal that causes a missile to correct its course, and the like). Finally, control stores contain rules for responding to certain conditions with certain actions.

Transformation graphs are part of an exciting new methodology that extends Structured Systems Analysis to the world of real-time systems. For more information on this methodology, read Stephen Mellor and Paul Ward's book *Structured Development for Real-Time Systems* (1985).

This extension of data flow diagrams and structured methods may eventually replace or merge with DFDs, because by definition, it will already support the entire modeling language required for information systems. And since many companies are now seeking to integrate information systems and real-time systems, separate tools and methods would be counterproductive. The common use of transformation graphs may well be destined. This is already evidenced by the inclusion of transformation graphs in many CASE tools.

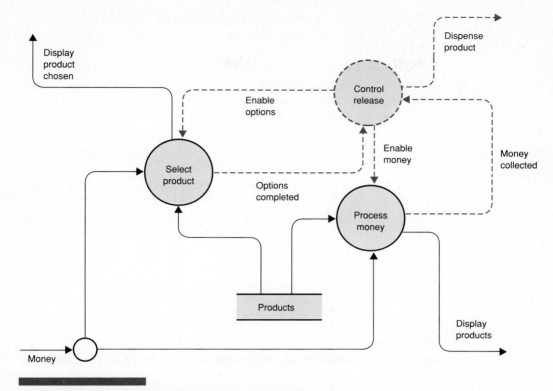

FIGURE 9.32 Transformation Graph for a Vending Machine

Problems and Exercises

1. Explain to an end-user or manager the difference between logical and physical modeling. When and why would you use each?

2. Compare process models and data models. What does each model show? Should you choose between the two modeling strategies? Why or why not?

3. A manager who has noted your use of logical data flow diagrams to document a proposed system's requirements has expressed some concern because of the lack of details that demonstrate the computer's role in the system. Defend your use of logical DFDs. Concisely explain the symbolism and how to read a DFD. (*Note:* The answer to this exercise should be a standard component in any report that will include DFDs. You cannot be certain that the person who reads this report will be familiar with the tool.)

4. Explain why you should exclude implementation details when drawing a logical DFD. Can you think of any circumstances in which physical details might be useful?

5. Explain why a systems analyst might want to draw logical models of an automated portion of an existing information system rather than simply accepting the existing technical information systems documentation, such as systems flowcharts and program flowcharts.

6. Draw a logical data flow diagram for the IRS minicase that opened Chapter 7. How does your DFD compare with the picture you drew in response to the challenge that concluded that minicase?

7. Draw a logical data flow diagram to document the flow of data in your school's course registration and scheduling system.

8. Draw a logical data flow diagram for some day-to-day "system" that you use or observe in use—for instance, your morning routine; making your favorite meal, including appetizer, entree, side dishes, and dessert; constructing something from scratch.

9. Draw a logical data flow diagram for the inventory control system that was described in the minicase that opened Chapter 2.

Minicases

1. Draw a logical data flow diagram for Minicase 2 at the end of Chapter 7.
2. Draw a logical DFD for Minicase 4 at the end of Chapter 7.

Annotated References and Suggested Readings

DeMarco, Tom. *Structured Analysis and System Specification.* Englewood Cliffs, N.J.: Prentice-Hall, 1978. This is the classic book on the Structured Systems Analysis methodology, which is built heavily around the use of data flow diagrams.

Gane, Chris, and Trish Sarson. *Structured Systems Analysis: Tools and Techniques.* Englewood Cliffs, N.J.: Prentice-Hall, 1979. An early Structured Analysis methodology book. Not as thorough as DeMarco but relatively easy to read and grasp. Their DFD symbolism is identical to ours. A methodology called *Stradis* has been developed around this approach. *Stradis* is marketed through a St. Louis–based company called Improved Systems Technologies.

Keller, Robert. *The Practice of Structured Analysis: Exploding Myths.* New York: Yourdon Press, 1983. A nice, concise overview of the first-generation Structured Analysis methodology.

McMenamin, Stephen M., and John F. Palmer. *Essential Systems Analysis.* New York: Yourdon Press, 1984. The most thorough reference to date on logical data flow diagrams. The book also hints at ways to integrate process and information modeling. We recommend that you read DeMarco before you read McMenamin and Palmer.

Mellor, Stephen, and Paul Ward. *Structured Development for Real-Time Systems.* New York: Yourdon Press, 1985.

Ward, Paul. *Systems Development Without Pain.* New York: Yourdon Press, 1985. This book provided our first insight into the possibilities of merging the process modeling and information modeling approaches.

Yourdon, Edward. "Whatever Happened to Structured Analysis?" *Datamation*, June 1, 1986. This article provides a first glimpse at the evolution of Structured Analysis as it will appear in forthcoming works from Yourdon Press, the largest library of titles on Structured Analysis. The article hints at the following: (1) the elimination of current systems modeling to overcome analysis paralysis, (2) the merger of Structured Analysis and information modeling, and (3) the merger of Structured Analysis techniques for information systems and real-time systems.

Yourdon, Edward. *Modern Structured Analysis.* Englewood Cliffs, N.J.: Yourdon Press, 1989. This is the long-awaited update to DeMarco's classic book. It includes all the suggested updates to the structured analysis methodology.

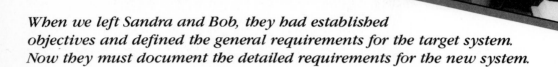

SOUNDSTAGE

Requirements Analysis for the New System (Part II)

When we left Sandra and Bob, they had established objectives and defined the general requirements for the target system. Now they must document the detailed requirements for the new system.

EPISODE 5   Detailed Requirements for the New System

Two weeks later, Sandra and Bob had completed documenting the general requirements for the new system with entity relationship diagrams and logical data flow diagrams. Sandy was starting to work on defining and evaluating alternative solutions (which will be the topic of our next episode). Meanwhile, Bob was working on the detailed systems requirements. We join Bob as he discusses some of those requirements with Ann Martinelli, the Membership Director.

"Well, that should finish the membership subsystems processing requirements. Can we work on the automatic orders now? If you need to get back to work we can do this later."

"That's okay," Ann responded. "I have a few more minutes."

Bob continued, "First, let's define these data flows. Where did I put those definitions? Ah, here they are (see Figure E5.1). We just defined AUTOMATIC DATED ORDER and AUTOMATIC ORDERED PRODUCT as outputs of the GENERATE MONTHLY PROMOTION process, but they are also inputs to this CHECK MEMBER CREDIT process. We need to verify the CREDIT DETAILS and CREDIT DECISION data flows."

"CREDIT DETAILS from the MEMBERS file is just the members DATE OF FIRST MEMBERSHIP and BALANCE PAST DUE," Ann offered. "And the CREDIT DECISION from the A/R Department is simply an approved or disapproved decision and the authorizing person."

Bob continued, "You can see from this diagram (Figure E5.2) that when combined with CREDIT DETAILS from the MEMBERS file and possibly a CREDIT DECISION from the Accounts Receivable Department, the output of this procedure is either an APPROVED ORDER or a BAD CREDIT ORDER. Now can we discuss the procedure itself?"

Ann replied, "I think I'm beginning to understand. First we calculate and verify the TOTAL ORDER PRICE. Then, if the PREPAID AMOUNT is less, we check the member's CREDIT DETAILS from the MEMBERS file. For automatic orders, the amount of prepayment is usually zero. If the member is relatively new, we sometimes need a CREDIT DECISION from the Accounts Receivable Department."

"Is that all?" Bob asked as he wrote a CHECK MEMBER CREDIT procedure description (see Figure E5.3).

"Well, not really," Ann answered. "If the member's credit is bad, we mail him or her a prepayment request and hold the order until we receive payment. If the order was approved, we check the product inventory."

"But both of those are separate processes that we will define later," Bob added. He then showed Ann the procedure description and she verified its correctness.

"I think we should stop here for today. I need to get back to work," Ann said.

Episode 5, continued ▶

FIGURE E5.1 Typical Contents of a Data Flow

An AUTOMATIC DATED ORDER is a dated order created by a promotion that is automatically filled and shipped to the club member.

An occurrence of an AUTOMATIC DATED ORDER is uniquely identified by the data element ORDER NUMBER.

An AUTOMATIC DATED ORDER consists of the following data elements:

ORDER NUMBER   PREPAID AMOUNT

ORDER DATE   TOTAL ORDER PRICE

ORDER STATUS

An AUTOMATIC ORDERED PRODUCT is an ordered product created by a promotion that is automatically shipped to the club member.

An occurrence of an AUTOMATIC ORDERED PRODUCT is uniquely identified by the data elements PRODUCT NUMBER and MEDIUM CODE.

An AUTOMATIC ORDERED PRODUCT consists of the following data elements:

PRODUCT NUMBER   QUANTITY SHIPPED

MEDIUM CODE   PRODUCT PRICE

QUANTITY ORDERED

FIGURE E5.2 CHECK MEMBER CREDIT Data Flow Diagram

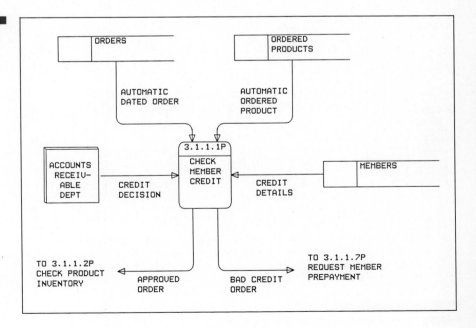

"Thanks for your help," Bob responded.

(This process would continue until the contents of all data flows on all the data flow diagrams had been defined and verified and the procedures for each of the processes had been documented using Structured English.)

Where Do We Go from Here?

This episode reinforced the use of data flow diagrams (discussed in Chapters 7 and 9) for outlining the *general* requirements of a new information system. The case also introduced two new tools used to specify the *detailed* requirements of a system, a data dictionary and Structured English. Both of these tools are very popular for specifying detailed systems requirements.

Chapter 10 introduces you to the concept and use of a project data dictionary. The data dictionary is used to document the *contents* of data flows and data stores depicted on data flow diagrams. Chapter 11 introduces you to tools and techniques for documenting policies and procedures, specifi-

cally Structured English and decision tables. These tools specify how processes on a data flow diagram accomplish their work.

This episode also emphasized the need for analysts to develop sound communications skills. Sandra and Bob are spending considerable amounts of time talking with the system's end-users. Systems analysts don't spend all their time behind a desk. Once again, we direct you to Part Four, Module C, to learn more about making presentations and conducting walkthroughs.

```
For each AUTOMATIC DATED ORDER, do the following:

    For each ordered product do the following:

        Calculate the TOTAL ORDER PRICE using the following formula:

            TOTAL ORDER PRICE = TOTAL ORDER PRICE +
            (PRODUCT PRICE X QUANTITY ORDERED)

        If the PREPAID AMOUNT is less than the TOTAL ORDER PRICE
        then:

            If the DATE OF FIRST MEMBERSHIP is more than six
            months ago and the BALANCE PAST DUE is equal to zero
            or the CREDIT DECISION from the Accounts Receivable
            Department is approved do the following:

                Approve the order and forward it to be checked for
                product availability

            Otherwise (DATE OF FIRST MEMBERSHIP is less than six
            months ago or the BALANCE PAST DUE is greater than
            zero or the CREDIT DECISION from the Accounts
            Receivable Department is disapproved) do the following:

                Disapprove the order and forward it to for
                prepayment request processing
```

FIGURE E5.3   Typical Procedure Description

CHAPTER TEN

Defining Logical Data and

Information Requirements

in a Project Dictionary

Americana Plastics

As the lead analyst on the cost accounting information systems project, Angela had apparently uncovered a terminology problem in the system. Management seemed to be using different definitions of part number.

Americana Plastics is a manufacturer of custom-engineered plastic products that become components of its customers' product lines. Each part is carefully engineered to the customer's specifications of size, shape, durability, temperature resistance, and the like. Because Americana products are custom built for individual customers and are subject to that customer's demand, parts are manufactured and shipped only after they have been ordered.

The cost accounting project had been designed to serve two groups of end-users: the manufacturing group and the accounting group. While working with the accounting group, Angela had learned that all parts sold by Americana were uniquely identified by a six-character part number. She had designed the entire new system using the six-character code as defined by the accounting group. After defining a number of manufacturing cost control analysis reports, she encountered a problem.

"No!" insisted the manufacturing group. "What they are calling a part number, although unique, is actually a mold number. Mold numbers identify the basic molds used to manufacture parts. We identify the actual part by a combination of the mold number and two process codes, the manufacturing method code and the insert codes. The part process method code is a one-letter code that tells us whether the plastic parts are to be formed using heat (H), cold (C), or pressure (P). The insert codes identify slugs that can be inserted into the basic mold to form slightly modified versions of the plastic part to be manufactured."

"Is this really a problem?" Angela asked. "After all, the six-character basic code is unique, so can't we go ahead and use that code for all reports? I'll even design two sets of reports, one set using the heading 'Part Number' and the other set using 'Mold Number.' "

"No," the manufacturing group countered. "We need those additional codes to determine which processing methods and insert requirements are responsible for cost overruns or inefficiencies. How will we be able to use the information to effect changes to our manufacturing methods without the extra codes? In fact, we'd like to suggest alternative structures for the manufacturing reports to make them more useful for pinpointing detailed problems. We'd like to see alternative report structures organized around the different processing methods and inserts as well as the basic molds."

Angela was perplexed! How will she be able to keep track of details when the terminology is not consistent? Each report will have to be redefined to use the terminology of a specific end-user.

Discussion

1. How would you suggest that an analyst keep track of the special terminology characteristic of most business applications?

2. Could Angela have done something to avoid wasting the time she spent designing the reports?

3. What would you have done differently? If you had tried to establish terminology and content before format, how would you have communicated your understanding (or lack thereof) back to the two end-user groups?

What Will You Learn in This Chapter?

This is the first of two chapters dealing with the *project dictionary* as a tool for documenting systems that are being developed. In this chapter, you will learn how to begin documenting details about the data and process models you learned to draw in Chapters 8 and 9. You will know that you have mastered the use of the project dictionary as a systems analysis tool when you can

1. Describe the need for a project dictionary, its contents, and its value as a documentation tool.

2. Define the contents of data entities, data flows, and data stores in terms of restricted data structures that consist of data elements.

3. Create complete project dictionary entries for data entities, data flows, and data stores. These entries should include pertinent facts about terminology, properties, and content.

4. Create complete project dictionary entries for data elements and codes for data elements. These entries should include pertinent facts about terminology, properties, and values (ranges).

Suppose you are reviewing the entity relationship diagrams (ERDs) or logical data flow diagrams (DFDs) for a new information system. These diagrams model the essential business requirements for the new system, but only in general terms. Details still have to be defined. Questions such as these may arise:

What data are we putting in that invoices data store?

Just exactly what information is needed on that overtime analysis report?

Did we remember to include the new disputed amount field in the customer statement?

These are questions that can easily be answered in a project dictionary. In this chapter, we examine the project dictionary and its purpose, content, and use.

What Is a Project Dictionary?

A **project dictionary** is a catalog of requirements and specifications for a new information system. A sample entry, in this case for a data flow, is shown in Figure 10.1. The tool helps the systems analyst keep track of the enormous volume of details that is part of every system, even small ones. Using a project dictionary, the analyst minimizes the chance of becoming overwhelmed by these details.

First, we should acknowledge that we have purposely avoided using a popular synonym, *data dictionary*, in place of the term *project dictionary*. The term data dictionary has a database connotation generally accepted in the computing industry. Firms frequently maintain an integrated data dictionary that stores facts about the files and databases for all systems that are currently being used. Consequently, we prefer the term project dictionary to denote those systems that are in a state of systems development. Once a system is developed and placed into operation, its details may be transferred from the project dictionary to the corporate data dictionary.

What is the purpose of the project dictionary? What facts are recorded in the project dictionary? How is the project dictionary organized?

The Purpose and Content of the Project Dictionary

Why do you need a project dictionary? During the definition phase, the analyst is defining the essential requirements for the new information system. No matter what your methods are, you will likely generate a wish list of

- data to be captured and stored in the system
- inputs to the system
- outputs to be generated by the system

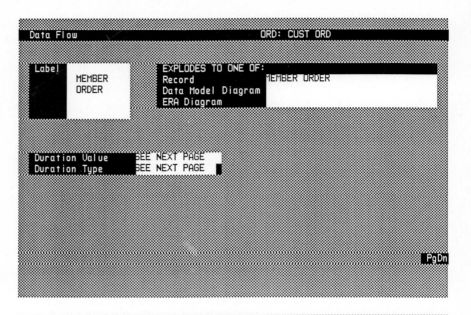

FIGURE 10.1 Sample Project Dictionary Entry A sample entry in a project dictionary. This sample describes a data flow.

If you drew information and process models of the system, all of these items would appear on that diagram as either entity types, data flows, or data stores.

A project dictionary expands on the pictorial models. For instance, given a data flow, let's say MEMBERSHIP SUBSCRIPTION, we need to define the content of a typical occurrence of that data flow. Or, given a data store or entity type (recall that they both describe stored data) such as PROMO-TION(S), we need to define the contents of a single record in that data store or entity type. The project dictionary provides a vehicle for recording these definitions.

As a systems analysis tool, a project dictionary should capture the detailed requirements for every input, output, and data store. These requirements should be taken down to the data element level. Ideally, the dictionary should also include detailed requirements for processing inputs, outputs, and data stores; however, that will be covered in the next chapter. For the time being, let's focus on data and information.

Consistent with our ongoing systems analysis theme, we want to focus on the **logical** data and information requirements, independent of current or possible physical implementations of those requirements. The suggested approach presents the contents of logical data flows and data stores (from process models) and entity types (from data models) in terms of *what* data elements are needed, not in terms of *how* they are currently or possibly formatted. Once this description has been completed, you can sit down with the end-users and discuss that content objectively.

Organization of the Project Dictionary

A project dictionary should be organized for ease of use to allow convenient reference to definitions that you or the end-user need. Furthermore, the project dictionary can be keyed to the names and terms originally docu-mented in the systems models. The ERDs and DFDs from Chapters 8 and 9 provide an outstanding framework for your project dictionary. They are nor-mally included in the project dictionary to serve as a table of contents into the more detailed specifications (Figure 10.2).

As the figure suggests, detailed requirements and specifications would be placed after the models. Ideally, they would be organized by *type*—for instance, data flow, data store, data element, or the like. Consequently, you can easily examine details pertaining to all data flows, data stores, and so on.

A project dictionary for any project, even a small one, can get very large. Size is frequently cited as a major disadvantage of the project dictionary. This is only a disadvantage if you make the mistake of *dumping* the entire project dictionary into the laps of the end-user and request their verification. Nobody's going to read such an imposing document. When you want the end-user to verify a project dictionary, you should present it in pieces. Extract the desired dictionary definitions and organize them so the reader doesn't have to shuffle back and forth through the pages.

FIGURE 10.2 The Organization of a Project Dictionary Although many project dictionaries are maintained on computers, the contents are ultimately output for reference. This book demonstrates the organization of a typical dictionary.

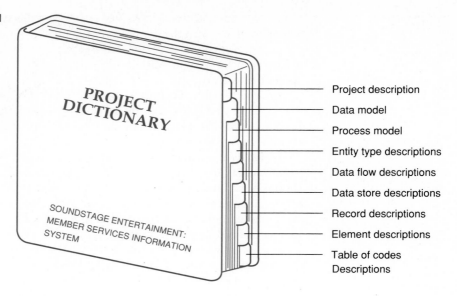

PROJECT DICTIONARY

SOUNDSTAGE ENTERTAINMENT: MEMBER SERVICES INFORMATION SYSTEM

Project description

Data model

Process model

Entity type descriptions

Data flow descriptions

Data store descriptions

Record descriptions

Element descriptions

Table of codes Descriptions

Project Dictionary Conventions and Implementation

There are numerous notations for documenting logical data and information requirements. In this section, we will examine two popular notations. We will also examine different techniques for implementing the actual project dictionary.

Defining Data and Information Structure in the Project Dictionary

Recall that we are defining data and information requirements for inputs, outputs, and data stores (or entity types stored in data stores). These components consist of data elements. By themselves, data elements have little value. However, they combine to form data structures that do have meaning. **Data structures** are a specific arrangement of data elements that define one occurrence of an input, output, or data store.

Inputs, outputs, and data stores can be defined in terms of the following types of data structures:

1. A *sequence* of data elements or group of data elements that occur one after the other

2. The *selection* of one or more data elements from a set of data elements

3. The *repetition* of an individual data element or group of data elements

-English Notation-

A WAGE AND TAX STATEMENT consists of
the following data elements:
    SOCIAL SECURITY NUMBER
    EMPLOYEE NAME
    EMPLOYEE ADDRESS
    EMPLOYER NAME
    EMPLOYER ADDRESS
    WAGES, TIPS, AND COMPENSATION
    FEDERAL TAX WITHHELD
    STATE TAX WITHHELD
    FICA TAX WITHHELD

-Algebraic Notation-

WAGE AND TAX STATEMENT =
    SOCIAL SECURITY NUMBER +
    EMPLOYEE NAME +
    EMPLOYEE ADDRESS +
    EMPLOYER NAME +
    EMPLOYER ADDRESS +
    WAGES, TIPS, COMPENSATION +
    FEDERAL TAX WITHHELD +
    STATE TAX WITHHELD +
    FICA TAX WITHHELD +

FIGURE 10.3 The Sequence Data Structure The sequence data structure consists of a group of serial data elements and/or groups.

Two notations are presented. The most common is the boolean algebraic notation suggested by DeMarco and supported by numerous authors. We also present the form that we prefer—an English derivative of the algebraic notation, which is especially useful for verification by end-users who may be intimidated by the algebraic notation. As we go through the examples, we will present the English notation first since it helps explain the algebraic notation. Your instructor may suggest that you ignore the algebraic notation altogether.

The definition of the data structure for any data flow or data store should begin with the English statement

A WAGE AND TAX STATEMENT consists of the following data elements:

or its algebraic equivalent

WAGE AND TAX STATEMENT =

Note that the equals sign is not used in its arithmetic sense. Here it means "consists of," "is composed of," or "contains."

The Sequence Data Structure The sequence of data elements that define a data flow, data store, or entity type would be documented as illustrated in Figure 10.3. Notice, in the English notation, that we indented the data element names to improve readability. Each of the data elements is *required* to assume a value for every occurrence of WAGE AND TAX STATEMENT. In the algebraic notation, the plus sign is interpreted as the word *and*.

-English Notation-

An ORDER consists of the following data elements:
    ORDER DATE

    Only one of the following elements:
        SOCIAL SECURITY NUMBER
        CUSTOMER ACCOUNT NUMBER

-Algebraic Notation-

ORDER =
    ORDER DATE +

    [SOCIAL SECURITY NUMBER
        CUSTOMER ACCOUNT NUMBER   ]

FIGURE 10.4   The Exclusive OR Selection Data Structure   The exclusive or selection data structure describes a group of data elements from which one and only one may assume a value for the overall data structure.

The Selection Data Structure   The selection construct allows you to show situations where given any single occurrence of a data flow, data store, or entity type, one of the following is true:

- *One and only one* data element from a list of data elements will assume a value (often called *exclusive or*).

- One or more data elements in a list will assume a value (often called *inclusive or*).

There must be at least two data elements in the list from which you choose. Some examples will make the selection construct clear.

Figure 10.4 demonstrates the *exclusive or* construct. Notice how the sequence and selection constructs combine to define the data structure. An order can be placed either by an individual or by a company, but not both. Thus, any given occurrence of an order will consist of either an ORDER DATE and a SOCIAL SECURITY NUMBER *or* an ORDER DATE and a CUSTOMER ACCOUNT NUMBER. In the algebraic notation, the selection elements are set off by the square brackets and separated by commas.

Figure 10.5 demonstrates the *inclusive or* selection construct. Both notations suggest that a TRAVEL EXPENSE VOUCHER contains *one or more* of the listed expenses. TOTAL EXPENSES always occurs. In the algebraic notation the elements from which you select are placed between the angled brackets and separated by commas.

The data structure in Figure 10.6(a) is *incorrect* because STUDENT CLASSIFICATION is a data element, whereas FRESHMAN, SOPHOMORE, JUNIOR, SENIOR are values of the data element CLASS, not data elements themselves. The correct data structure is shown in Figure 10.6(b). Later in this chapter, you'll learn where to document values and value ranges for data elements in the project dictionary.

-English Notation-

A TRAVEL EXPENSE VOUCHER consists of
the following data elements:

      EMPLOYEE ID NUMBER
      EMPLOYEE NAME
      DATE TRIP STARTED
      DATE TRIP COMPLETED
      PURPOSE OF TRIP
      MILES TRAVELED
      MILEAGE CHARGE

One or more of the following
elements:

      AIR TRAVEL EXPENSE
      TAXI FARE EXPENSE
      REGISTRATION FEES
      LODGING EXPENSES
      MEAL EXPENSES
TOTAL EXPENSES

-Algebraic Notation-

TRAVEL EXPENSE VOUCHER =
    EMPLOYEE ID NUMBER +
    EMPLOYEE NAME +
    DATE TRIP STARTED +
    DATE TRIP COMPLETED +
    PURPOSE OF TRIP +
    MILES TRAVELED +
    MILEAGE CHARGE +
    <   AIR TRAVEL EXPENSE,
         TAXI FARE EXPENSE,
         REGISTRATION FEES,
         LODGING EXPENSES,
         MEAL EXPENSES   > +
    TOTAL EXPENSES

FIGURE 10.5   The Inclusive OR Selection Data Structure   The inclusive or selection data structure describes a
group of data elements from which one or more may assume a value for the overall data structure.

The Repetition Data Structure   The repetition construct is used to set off a
data element or group of data elements that will repeat a specified number
of times for a single occurrence of the data flow or data store. The English
and algebraic notations are documented as

M to N occurrences of each of the following elements:

and

M {list of elements or groups of elements} N

respectively. M is the minimum number of occurrences of the repeating group
and N is the maximum number of occurrences of the data structure. If
$M = 0$, the entire group occurs optionally. In any case, the entire group of
data elements repeats as a group. If the number of repetitions is indefinite,
you may write

One or more occurrences of:

for the English notation or simply not assign a numeric value to N in the
algebraic notation.

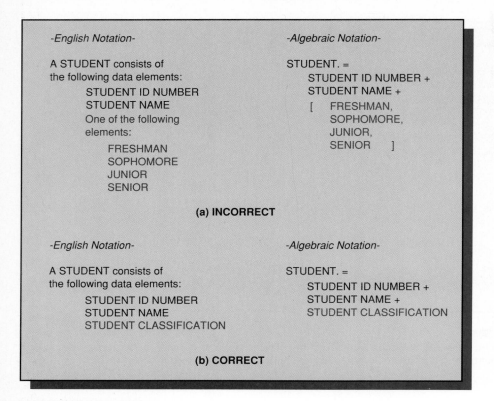

-English Notation-

A STUDENT consists of
the following data elements:

      STUDENT ID NUMBER
      STUDENT NAME
    One of the following
    elements:

        FRESHMAN
        SOPHOMORE
        JUNIOR
        SENIOR

-Algebraic Notation-

STUDENT. =
    STUDENT ID NUMBER +
    STUDENT NAME +
    [FRESHMAN,
        SOPHOMORE,
        JUNIOR,
        SENIOR   ]

(a) INCORRECT

-English Notation-

A STUDENT consists of
the following data elements:

      STUDENT ID NUMBER
      STUDENT NAME
      STUDENT CLASSIFICATION

-Algebraic Notation-

STUDENT. =
    STUDENT ID NUMBER +
    STUDENT NAME +
    STUDENT CLASSIFICATION

(b) CORRECT

FIGURE 10.6   A Common Misuse of the Selection Data Structure   Figure 10.6(a) demonstrates a common mistake: using the section structure to document values of data elements instead of data elements themselves. Figure 10.6(b) fixes the mistake.

The contents of a group may contain a sequence of data elements, a selection construct or constructs, or even additional (sometimes called *nested*) repetition constructs. Let's expand on our ORDER example to demonstrate the repetition construct. Figure 10.7 shows both notations. Note that a sequence of related data elements was defined as the repeating group. Let's look at one more example in Figure 10.8. There are four weeks in a pay period. That explains the first repeating group. For any or all of those weeks, an employee will have from one to seven values of hours worked—one value for each day worked. This is a **nested repeating group**, one repeating group inside another. For the algebraic notation, the inner bracket occurs one to seven times (for days of the week) for each occurrence of pay period, which also happens one to four times.

-English Notation-

An ORDER consists of the following
data elements:

    ORDER NUMBER
    ORDER DATE
    Only one of the following elements:
        SOCIAL SECURITY NUMBER
        CUSTOMER ACCOUNT NUMBER
    SHIPPING ADDRESS
    1 to 20 occurrences of:
        PRODUCT NUMBER
        PRODUCT DESCRIPTION
        QUANTITY ORDERED
        PRODUCT PRICE
        EXTENDED PRICE
    TOTAL ORDER COST

-Algebraic Notation-

ORDER =
    ORDER DATE +
    [SOCIAL SECURITY NUMBER,
        CUSTOMER ACCOUNT NUMBER   ] +
    SHIPPING ADDRESS +
    1 { PRODUCT NUMBER +
        PRODUCT DESCRIPTION +
        QUANTITY ORDERED +
        PRODUCT PRICE +
        EXTENDED PRICE   } 20 +
    TOTAL ORDER COST

FIGURE 10.7 The Repetition Data Structure The repetition data structure consists of a group of data elements that repeat as a group for each occurrence of the main data structure.

-English Notation-

A TIME WORKED RECORD consists of the
following data elements:

    EMPLOYEE IDENTIFICATION NUMBER
    EMPLOYEE NAME
    1 to 4 occurrences of the following:
        START DATE FOR PAY PERIOD
        END DATE FOR PAY PERIOD
        1 to 7 occurrences of the following:
            HOURS WORKED

-Algebraic Notation-

TIME WORKED RECORD =
    EMPLOYEE IDENTIFICATION NUMBER +
    EMPLOYEE NAME +
    1{   START DATE FOR PAY PERIOD
        END DATE FOR PAY PERIOD
        1{   HOURS WORKED   }7   }4

FIGURE 10.8 A Nested Repeating Data Structure A nested repeating data structure contains a group of data elements that repeat inside another group, which itself repeats.

-English Notation-

A CLAIM consists of the following
data elements:

      POLICY NUMBER
      POLICYHOLDER NAME
      POLICYHOLDER ADDRESS
      SPOUSE NAME (optional)
      0 to 15 occurrences of:
            DEPENDENT NAME
            RELATIONSHIP
      CLAIMANT NAME
      1 or more of the following:
            EXPENSE DESCRIPTION
            NAME OR FIRM PROVIDING SERVICE
            TOTAL CHARGE FOR SERVICE

-Algebraic Notation-

CLAIM =
    POLICY NUMBER +
    POLICYHOLDER NAME +
    POLICYHOLDER ADDRESS +
    (SPOUSE NAME   ) +
    0   { DEPENDENT NAME +
          RELATIONSHIP   }   15
    CLAIMANT NAME +
    1   {   EXPENSE DESCRIPTION +
              NAME OR FIRM PROVIDING
              SERVICE +
              TOTAL CHARGE FOR SERVICE }   M

FIGURE 10.9   Optional Data Elements in a Data Structure   This figure demonstrates how to indicate that an element does not have to take on a value. The example also demonstrates how to show that a repeating group does not have to occur for an occurrence of the entire data structure.

A Notation for Optional Data Elements   Sometimes certain data elements or groups of data elements optionally take on values for occurrences of data flows. In Figure 10.9, for any occurrence of CLAIM, the element SPOUSE NAME may or may not take on a value. In the algebraic notation, parentheses set off the optional data element.

Note that we still used the repeating construct for the optional group of data elements. But we only have to provide dependent relationship and age if the claim is for a dependent over a certain age. We also provided a mandatory repeating group to reinforce the nonoptional repeating group concept.

For the next example, let's study only the English notation. Let's consider the difference between optionally, the following:

    OVERAGE DEPENDENT AGE
    OVERAGE DEPENDENT RELATIONSHIP

and

    OVERAGE DEPENDENT AGE (optional)
    OVERAGE DEPENDENT RELATIONSHIP (optional)

The former suggests that both elements occur together or neither element

-English Notation-

A WAGE AND TAX STATEMENT consists of
the following data elements:

    SOCIAL SECURITY NUMBER
    EMPLOYEE NAME
    EMPLOYEE ADDRESS which is defined
        separately as ADDRESS
    EMPLOYER NAME
    EMPLOYER ADDRESS which is defined
        separately as ADDRESS
    WAGES, TIPS, COMPENSATION
    FEDERAL TAX WITHHELD
    STATE TAX WITHHELD
    FICA TAX WITHHELD

-Algebraic Notation-

WAGE AND TAX STATEMENT =
    SOCIAL SECURITY NUMBER +
    EMPLOYEE NAME +
    EMPLOYEE ADDRESS=ADDRESS +
    EMPLOYER NAME +
    EMPLOYER ADDRESS=ADDRESS +
    WAGES, TIPS, COMPENSATION +
    FEDERAL TAX WITHHELD +
    STATE TAX WITHHELD +
    FICA TAX WITHHELD

-English Notation-

ADDRESS consists of the following data elements:
    One or both of the following:
        STREET ADDRESS
        POST OFFICE BOX NUMBER
    CITY
    STATE
    ZIP CODE

-Algebraic Notation-

ADDRESS =
    <   STREET ADDRESS,
          POST OFFICE BOX NUMBER > +
    CITY +
    STATE +
    ZIP CODE

FIGURE 10.10 Common Data Structures It is useful to define reusable data structures that can be referenced from within other data structures. Thus, we avoid reinventing the wheel when it comes to common structures like ADDRESS.

occurs. It's all or nothing. The second suggests that either element may occur independently of the other.

Groups of Elements Groups of data elements that always occur together can be given a group name. The group name would then be documented as a separate data structure. This approach is demonstrated in Figure 10.10. This approach allows many data structures to refer to a common substructure. Thus, instead of having to separately define data elements for employee address, customer address, supplier address, and the like, we need only set each of those addresses equal to the common definition of ADDRESS.

Computer-Assisted Systems Engineering (CASE) and Other Alternatives for Implementing the Project Dictionary

Computer-Assisted Systems Engineering (CASE) was introduced in Chapter 5 as an enabling, emerging technology for systems analysis and design. All CASE products are built around a project dictionary (sometimes called an *encyclopedia*). The project dictionary is usually fragmented into project dictionary subsets. The project dictionary catalogs all systems models and details about the objects contained on those models. The analyst can easily call up a model, such as a data flow diagram, then select a data flow from that model and instantly call up the data structure for that data flow.

CASE products also make it possible to look at subsets of the project dictionary, such as all data flows for a particular subsystem or end-user. In addition, CASE products provide numerous analytical tools to check for consistency and completeness in the project dictionary. Many CASE tools can import data from and export data to other project dictionaries. CASE is becoming the tool of choice for managing the overwhelming detail that makes all but the smallest systems projects difficult.

Alternatively, if CASE is not available, there are several ways to implement a project dictionary. One of the more popular options is to print special forms (recall Figure 10.1) for documenting entity types, data flows, data stores, data elements, and so forth. The advantage of forms is that the analyst is provided with standards for what to include in the dictionary.

Another technique is to implement a simple computerized project dictionary using a word processor or line editor on a computer. This technique offers the advantage of easier editing; however, the disadvantage is that it can be difficult to get listings of subsets of the dictionary—it's all or nothing.

In the remaining section of this chapter, we will demonstrate how to enter complete descriptions of our information and process models into the project dictionary. All of the dictionary entries in this chapter were created with a customized version of a popular CASE tool, Excelerator. Only the color screens and annotations were added by an artist. Each CASE tool has its own way of dealing with data structures. Excelerator is no different. We will explain its idiosyncrasies in the examples.

Using a Project Dictionary for Systems Analysis

The project dictionary is a central tool for systems analysis. It is especially integral to the use of the popular Structured Analysis methodology. Let's place that methodology in the context of the life cycle and describe how the project dictionary would be developed.

Modern Structured Analysis Using a Project Dictionary

In Chapters 8 and 9, you learned that modern Structured Analysis is built around a combination of information modeling (via entity relationship diagrams) and process modeling (via data flow diagrams). Also recall from earlier chapters that systems analysis consists of four phases: survey, study, definition, and selection.

During the **survey** and **study phases**, you primarily define problems and opportunities within the context of the system's environment. As you study the current system, you discover that it is couched in some business vocabulary that consists of terminology and acronyms. Thus, the study phase represents an ideal time to start your dictionary. Recall that DFDs are not essential to the study phase for modern Structured Analysis. However, the study phase can still benefit from a *glossary* of data flows and data stores with appropriate business definitions. This study phase strategy was discussed in Chapter 8 as a preface to information modeling.

The **definition phase** looks at requirements independently of the ways that we will implement those requirements (logical modeling). This is the phase during which the project dictionary expands most rapidly. Information models, process models, and any other graphic models would normally be placed at the beginning of the dictionary to serve as a table of contents for the detailed requirements.

While a picture *is* worth a thousand words, the models just mentioned do not contain sufficient detail to fully document requirements. Much in the manner that we exploded entity types to reveal subtypes and data flow diagrams to reveal more detailed DFDs, we explode or *describe* the objects on the models to the project dictionary. In this chapter, we have focused on describing data-oriented objects to the dictionary. In the next chapter we will show how to describe process-oriented objects to the dictionary.

Moving on, the **selection phase** identifies and analyzes alternative physical solutions that will fulfill defined requirements. Final decisions on how data flows, data stores, entity types, and processes will be added to the dictionary. Once again, you should notice how the phases gradually move from logical perspectives (survey, study, and definition phases) to the beginnings of physical modeling (selection phase). From there, the design and implementation phases will add additional details to the physical models.

How to Describe Information and Process Models in the Dictionary

We've learned how to define data structures for logical data flows, data stores, and entity types. Our goal, however, is to define all logical (nonphysical) facts

FIGURE 10.11 Project Dictionary Entry for an Entity Type from an Information Model
This is a typical dictionary description for an entity type. The Explosion attribute points to a separate description for a Record.

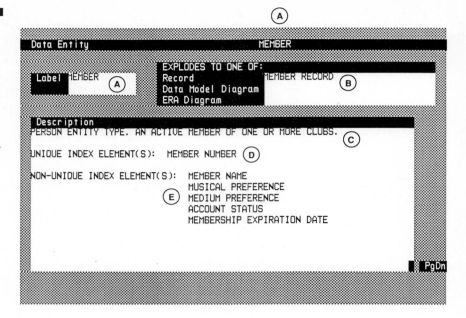

about these components. In this section, you will learn how to write complete logical project dictionary entries.

These entries have been gleaned from the logical models for the Entertainment Club case study. We are not presenting this section step by step, since you would normally begin recording the dictionary entries as you drew the information and process models. In other words, the step "Record details in the project dictionary" would immediately follow the last step of information modeling (Chapter 8) and process modeling (Chapter 9).

Entity-Type Entries in a Project Dictionary   Entity types were the only data-oriented object on a data model (entity relationship diagram). A sample dictionary screen or form for an entity type is shown in Figure 10.11. We call your attention to the following:

(A) Every entity type should have a name and label, which are normally the same thing. Naming conventions were suggested in Chapter 8.

(B) The contents of every entity type should be described as a record. The record description, to be described shortly, documents the data structure that was covered earlier in the chapter. Why record the data structure separately from the entity description? Because by separating the record, that record can be reused (which makes sense for entity types, since that data is usually stored for use by several processes). As a general rule, we give the record the same name as the entity type.

Ⓒ Every entity type should contain a brief definition to establish vocabulary.

Ⓓ Every entity type's unique identifier elements (covered in Chapter 8) should be recorded in the dictionary. The "Opr" column allows you to specify AND (for concatenated identifiers) and OR (for candidate identifiers).

Ⓔ Optionally, you may want to record nonunique subsetting elements. These elements divide all occurrences of the entity types into useful subsets. For instance, the values for the element GENDER might divide all occurrences of the MEMBER entity type into two subsets: male members and female members.

Data Flow Entries in a Project Dictionary   Data flows are objects on data flow diagrams. Figure 10.12 demonstrates a screen or form for describing a data flow to the dictionary. You should note the following:

Ⓐ Every data flow should be named. If desired, labels can be used to give longer, more descriptive names that will appear on the data flow diagrams.

Ⓑ The contents of data flows should be described as records. A record description, described shortly, documents the data structure that was covered earlier in the chapter. Why record the data structure separately from the data flow description? Because by separating the record, that record can be reused for other data flows that are identical, though perhaps with different names. As a general rule, we give the record the same name as the data flow except when its contents correspond to an entity type (a data flow from the data store for that entity type).

Ⓒ Every data flow should contain a brief definition to establish vocabulary.

Ⓓ Data flows are triggered by business events or business cycles (time). The event/time that triggers a single occurrence of the data flow should be described.

Ⓔ The average, peak, and valley volumes of the data flow should be recorded. Ultimately, any physical implementation of the data flow must accommodate this logical volume.

Data Store Entries in a Project Dictionary   Data stores are also easy to describe using a data dictionary. A typical data store entry is illustrated in Figure 10.13. Notice the similarity to an entity type. This makes sense since an entity type

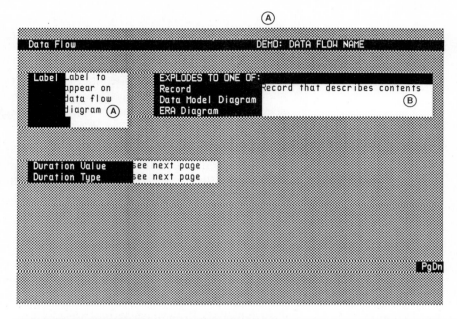

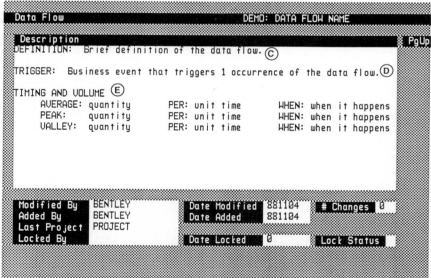

FIGURE 10.12 Project Dictionary Entry for a Data Flow from a Process Model This is a typical dictionary description for a data flow. The Explosion attribute points to a separate description for a Record.

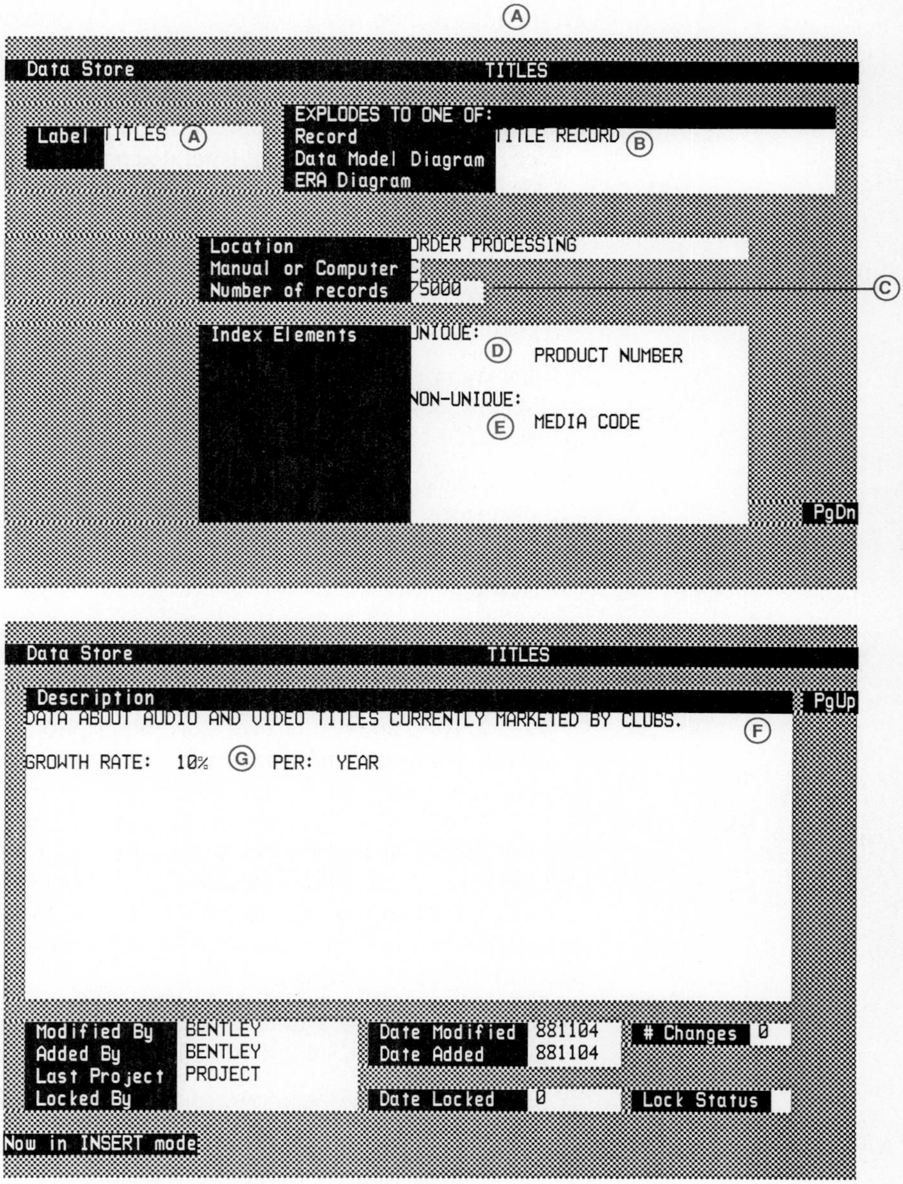

Ⓐ

```
Data Store                                      TITLES

                          EXPLODES TO ONE OF:
   Label  TITLES  Ⓐ       Record              TITLE RECORD  Ⓑ
                          Data Model Diagram
                          ERA Diagram


                   Location          ORDER PROCESSING
                   Manual or Computer C
                   Number of records  75000                              ──Ⓒ

                   Index Elements    UNIQUE:
                                      Ⓓ   PRODUCT NUMBER


                                     NON-UNIQUE:
                                      Ⓔ   MEDIA CODE

                                                                  PgDn
```

```
Data Store                                      TITLES

   Description                                                   PgUp
   DATA ABOUT AUDIO AND VIDEO TITLES CURRENTLY MARKETED BY CLUBS.
                                                                Ⓕ
   GROWTH RATE:  10%  Ⓖ  PER:  YEAR




   Modified By    BENTLEY        Date Modified  881104    # Changes  0
   Added By       BENTLEY        Date Added     881104
   Last Project   PROJECT
   Locked By                     Date Locked    0         Lock Status
 Now in INSERT mode
```

**FIGURE 10.13 Project Dictionary Entry for a Data Store from a
Process Model** This is a typical dictionary description for a data store.
The Explosion attribute points to a separate description for a Record.

on the information model corresponds to all occurrences of that entity type in a data store on the process model (primitive level). The following annotation will help to clarify the figure:

Ⓐ Every data store should have a unique name. Naming conventions were covered in Chapter 9. The label that appears on the DFD is usually identical to the name.

Ⓑ Like entity types and data flows, the content or data structure will be described separately as a record. In this case, the record name should be identical to that of the entity type since the data store corresponds to all occurrences of that entity type.

Ⓒ For every primitive data store—one that corresponds to an entity type—specify how many occurrences (or records) will be stored in the data store.

Ⓓ What data elements uniquely identify one occurrence (or record) in the data store?

Ⓔ What data elements uniquely identify a subset of occurrences in the data store?

Ⓕ Every data store should include a brief definition to establish vocabulary.

Ⓖ What is the anticipated growth rate of the data store?

Figure 10.14 shows the data store description for a high-level data store that explodes to our information model. Notice that many of the attributes for this data store are not applicable.

Record Entries in the Project Dictionary   So far, all of the objects we've demonstrated explode to a **record**. Records describe the data structures that were covered earlier in the chapter. Figure 10.15 illustrates a simple record description in the dictionary. This record corresponds to an entity type or data store. Note the following:

Ⓐ Every record has a unique name. Recall that the named record (data structure) may be shared by many data flows, data stores, and entity types.

Ⓑ Some records may be known to end-users by names other than the logical names we've given the records. Aliases should be recorded to provide reference.

Ⓒ The Normalized attribute happens to be specific to the CASE product we used to demonstrate this entry. It is marked Y for yes or N for no to indicate whether the record should be in third normal form. The CASE product includes automated analysis tools to verify or dispute the recorded claim.

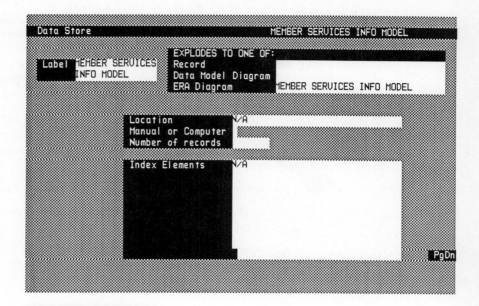

FIGURE 10.14 Project Dictionary Entry for a Data Store (Process Model) That Explodes to an Entity Relationship Diagram (Information Model)

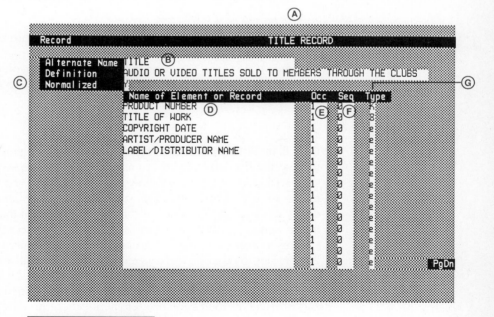

FIGURE 10.15 Project Dictionary Entry for a Record This is a typical dictionary description for a record. A record describes a data structure.

Ⓓ The data structure is recorded in terms of data element names and/or data group names. A data group is called a subrecord. A subrecord would be described in a separate record description screen or form. This allows common subrecords like address to be used for multiple group names (such as CUSTOMER ADDRESS, EMPLOYEE ADDRESS, VENDOR ADDRESS, and so forth).

Ⓔ The attribute *Occ* means maximum number of occurrences of the element or group for a single occurrence of the record. Because the entity types were normalized (Chapter 8), every data element occurs only once for each occurrence of this entity type.

Ⓕ The attribute *Seq* is not important to this discussion.

Ⓖ The attribute *Type* describes the type of element or group that is being described. For elements, type can be

K   meaning   identifier element (key)
E   meaning   non-identifier element

For groups or subrecords

R   meaning   subrecord

If type is *R*, there must be another record description in the project dictionary to describe the subrecord's data structure.

The data structure for a data flow is frequently more complex since it may contain repeating groups of elements. Such a data structure is demonstrated in Figure 10.16 (top). The basic difference is the *Occ* column, which indicates that the group of elements in Figure 10.16 (bottom) occurs more than once for a single occurrence of the data flow.

Data Element Entries in the Project Dictionary   Preparing data dictionaries to describe data elements is just as important as preparing data dictionaries to describe entity types, data flows, data stores, and records. Why? Data elements are not composed of data structures or other data elements. But data elements are assigned values, and it is very important that we learn the legitimate values for data elements so we can design controls in information systems to guarantee that occurrences of data elements are valid. We can use dictionaries to describe the data elements and to verify our understanding of the data element with the end-user. A complete data element entry in the dictionary is illustrated in Figure 10.17. It is more difficult to separate logical and physical attributes for data elements; therefore, they are combined on this form/screen. We will discuss only the pertinent attributes at this time:

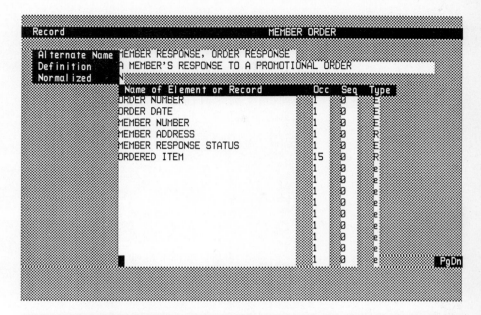

```
Record                                           MEMBER ORDER

Alternate Name MEMBER RESPONSE, ORDER RESPONSE
Definition     A MEMBER'S RESPONSE TO A PROMOTIONAL ORDER
Normalized     N
                   Name of Element or Record       Occ  Seq  Type
                   ORDER NUMBER                      1    0    E
                   ORDER DATE                        1    0    E
                   MEMBER NUMBER                     1    0    E
                   MEMBER ADDRESS                    1    0    R
                   MEMBER RESPONSE STATUS            1    0    E
                   ORDERED ITEM                     15    0    R
                                                     1    0    e
                                                     1    0    e
                                                     1    0    e
                                                     1    0    e
                                                     1    0    e
                                                     1    0    e
                                                     1    0    e
                                                     1    0    e  PgDn
```

```
Record                                           ORDERED ITEM

Alternate Name ORDERED PRODUCT, ORDERED TITLE
Definition     AN ORDERED TITLE/PRODUCT ON A MEMBER ORDER
Normalized     N
                   Name of Element or Record       Occ  Seq  Type
                   PRODUCT NUMBER                    1    0    E
                   MEDIA CODE                        1    0    E
                   CREDIT                            1    0    E
                   QUANTITY ORDERED                  1    0    E
                   QUANTITY SHIPPED                  1    0    E
                   PRODUCT PRICE                     1    0    E
                   EXTENDED ORDERED PRODUCT PRICE    1    0    E
                                                     1    0    e
                                                     1    0    e
                                                     1    0    e
                                                     1    0    e
                                                     1    0    e
                                                     1    0    e
                                                     1    0    e  PgDn
```

FIGURE 10.16   Project Dictionary Entry for a Record That Contains Repeating Groups (Subrecords)   This is a typical dictionary description for a record that contains repeating groups. The repeating group is specified first in the *Occ* column (top). The elements and groups that make up the group are specified in the bottom figure.

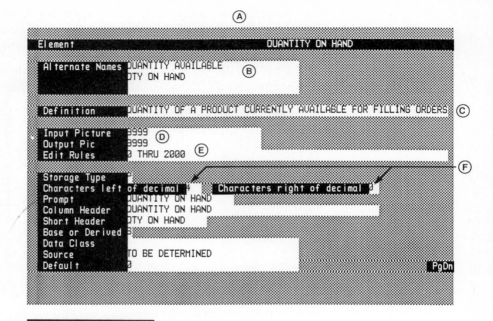

FIGURE 10.17 Project Dictionary Entry for a Data Element

Ⓐ Every data element should have a logical, descriptive name. Naming conventions for data elements were described in Chapter 8.

Ⓑ Data elements frequently have business aliases, synonyms, or acronyms. They should be recorded as a frame of reference for end-users and analysts.

Ⓒ Every data element should include a brief definition to establish vocabulary.

Ⓓ Data elements may be constrained by certain business formatting restrictions. The CASE product we used required that the format be specified as a COBOL-like PICTURE clause.

Ⓔ Legitimate data element values must be specified. Values fall into three categories:

  • *indefinite values*, meaning that the element can assume a virtually limitless number of values. In this case, the *Value Range* attribute is left blank.

- *value ranges*, usually numeric, meaning that the element can assume any value in the range of values—for example, 3.35 through 6.50. Such a range is recorded in the *Value Range* attribute.

- *finite values*, meaning that the element may assume any one of a limited set of well-defined values. These values can also be listed in the *Value Range* attribute. Sometimes the set of values is a set of **codes**, each of which carries a meaning. In this case, the attribute *Value Range* should be specified as FROM <name of a table of codes>. Tables of codes are described separately in the dictionary.

(F) Decimal positioning should be specified for all numeric data elements.

Table of Codes Entries in the Project Dictionary and Coding Techniques A **code** is a group of characters and/or digits that identify and describe something in the business system. Codes are frequently used to describe customers, products, materials, or events. The use of codes is popular for several reasons. First, codes can often be used for quick and easy identification of people, objects, and events. For instance, a product coded A-57-G may mean "product number 57, gallon can, in warehouse zone A." Second, codes usually condense numerous facts into concise format. And finally, the concise format usually reduces data storage space requirements.

Systems analysts are frequently charged with analyzing and defining coding schemes for information systems. Let's examine some of the more common coding schemes. Then we'll show you how to describe them to the project dictionary.

Sequential and **serial codes** are quite similar. Both number items with consecutive numbers—for example, 1,2,3 . . . n. Sequential numbers are typically assigned to a set of items, such as customers, that have been previously ordered (alphabetically, for example). Although the scheme is simple, new items cannot easily be inserted without disrupting the original ordering or changing many assigned numbers.

Serial numbers are assigned according to *when* new items are first identified. The first item identified is numbered 1, the second item is numbered 2, and so forth. Although serial coding also offers simplicity and a nearly infinite number of occurrences, the code has little information value.

Sequential and serial coding are frequently used as components in the more complicated coding schemas described in the following sections.

Block codes are a variation on sequential coding. A set of sequential or serial codes is divided into *blocks* that classify items into specific classes. For instance, a block code could be defined for customers as follows:

1000 through 4999 Record/Tape Club Customers
5000 through 6999 Compact Disc Club Members
7000 through 7999 Video Club Members

THE NEXT GENERATION

Integrated Data Dictionaries: An Idea Whose Time Has Come

In many businesses, data has become their most precious resource—and their most uncontrolled resource! As data files and databases proliferate, data administration (or the lack thereof) becomes a serious problem. Just how serious can the problem get? Consider a law that mandates a nine-digit zip code (or larger). Do you have any idea how difficult it is to find every computer program and file that has a zip code field in it? Why so difficult? Over the years, programmers and analysts came and went. Each introduced a bit of *uniqueness* into the systems and computer program library. How many unique field names exist for zip code? Let's see—there's ZIPCODE, ZIP.CODE, ZIP, ZP, Z, ZC, ZCODE, ZCOD, ZIPCD—well, you get

the idea. Where do you look for the fields? And how do you look for them? Clearly, there's got to be a better way of keeping track.

And there is! A formal data administration function has developed in some businesses. That's *data administration*—not *database administration*. The idea is to create and maintain a data dictionary for the business as a whole. That dictionary keeps track of where data elements are used and stored— every file, database, computer program, report, document, and so forth. The aliases, descriptions, formats, limitations, and properties of every data element are recorded in a centralized data dictionary. As new systems are developed and old systems are maintained, analysts and programmers are urged to use

the dictionary to curb the proliferation of aliases and to minimize the creation of new data elements. In other words, we are taking the project data dictionary concept and increasing its perspective to include the business as a whole!

The only way to implement an enterprise-wide data dictionary is to use one of the commercially available data dictionary software packages. Automated data dictionaries aren't new. They've been with us almost since the advent of database management systems software. In fact, you usually get a data dictionary package when you buy a database management system (database-independent products are also available). These automated data dictionaries were originally conceived to

Every customer would be assigned a serial or sequential number within its proper block classification. The number for any given customer identifies the customer type.

Alphabetic codes use different combinations of letters and/or numbers to describe items. Alphabetic codes have greater information value than any of the previously described codes. Many such codes have been standardized—for example, the two-letter system of abbreviations for states, in which TX = Texas, for example. Most alphabetic codes are abbreviations; therefore, with a little practice, such codes can easily be interpreted.

THE NEXT GENERATION

Integrated Data Dictionaries: An Idea Whose Time Has Come, Continued

help the database professionals keep track of the data elements and records in databases. But they are being increasingly used to catalog all the data flows, stores, and elements in a business.

What can an automated data dictionary do for you? Data dictionaries may allow you to do the following:

- Get full listings of all known facts about all or specific subsets (such as a group of related applications, a single related application, and/or a single program) of data flows, stores, elements, and so forth.

- Get partial listings (such as name and description only) for data flows, stores, elements, and so forth.

- Reorganize facts into a convenient format. For example, give full facts for all data elements that are part of specific data flows.

- Find all data flows, data stores, or data elements that share a certain property (such as the same unit of measure or the same length).

- Answer queries about data. For instance, LIST ALL KNOWN ALIASES FOR ZIP CODE. And then, LIST ALL FILES IN WHICH ZIP CODE (including its aliases) IS STORED. And finally, LIST ALL PROGRAMS THAT USE THE FIELD ZIP CODE (again, including all its aliases).

Where does the systems analyst fit into this picture? The analyst's system data dictionary can be a subset of the business data dictionary.

One issue that will have to be addressed is the interface between CASE dictionaries (for systems being developed) and production data dictionaries. Few corporate data administrators are willing to allow that automatic transfer from CASE to occur without some quality check taking place.

Thus, the systems analyst of the future may not have to worry about proliferating the creation of duplicate data elements and data stores. Indeed, the future systems analyst may not be allowed to operate in isolation from other analysts and information systems!

Group codes are the most powerful of the coding schemes because they convey much more information content than the other coding schemes, which they frequently utilize. Each position or group of positions in the code describes some pertinent characteristic of the item being coded. Thus, the code number tells the reader a great deal about the item itself. There are two common types of group codes: significant position codes and hierarchical codes.

For **significant position codes**, each digit or group of digits describes a measurable or identifiable characteristic of the item. Significant digit codes are frequently used to code inventory items.

For example, the following code might be defined for a record/tape/compact disc product:

first digit	Product Classification:
	R = Rock
	C = Classical
	J = Jazz
	W = Country/Western
	E = Easy Listening
	O = Other
second and third digit	Year of Issue
fourth through seventh digit	Sequential Assigned No.
eighth digit	Media Code
	R = Record
	C = Cassette
	D = Compact Disc
	T = Digital Audiotape
	V = VHS Videotape
	B = Beta Videotape
	8 = 8mm Videotape

As was the case with alphabetic codes, significant digit codes are frequently standardized for some industries (zip codes, light bulb codes, tire codes).

Hierarchical codes provide a top-down interpretation for an item. Every item coded is factored into groups, subgroups, and so forth. Each group and subgroup can be coded such that the codes identify specific groups and subgroups. For instance, we could code—or partially code—all inventory items by warehouse location as follows:

first digit	Warehouse Zone (A, B, C, D, or E)
second digit	Section in Zone (1–5)
third and fourth digits	Aisle in Section (1–20)
fifth digit	Shelf Number (A–M)

The five coding schemes discussed can be combined in any arrangement desired to achieve business goals. Codes are intended to make the handling of data easier. Whenever you encounter a data element, which can only assume a finite set of values, a coding scheme may have to be studied, defined, or expanded. Codes are, first and foremost, a business tool. Therefore, you should consider the following business issues when analyzing or proposing codes:

1. *Codes should be expandable.* To accommodate natural growth, a code should allow for additional entries.

2. *Codes should be unique.* Each occurrence should define one, and only one, occurrence of data.

3. *Size of codes is important.* A code should be large enough to describe relevant characteristics but small enough to be easily read and interpreted by people.

4. *Codes should be convenient.* A new occurrence of a code should be easy to construct and interpret. A computer should *not* be required.

Once a code has been defined, it can be entered into the project dictionary. Figure 10.18 demonstrates one such entry. It is self-explanatory.

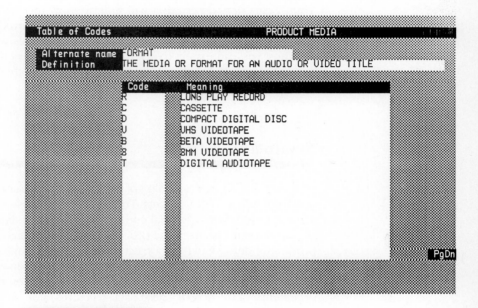

FIGURE 10.18   Project Dictionary Entry for a Table of Codes

The Project Dictionary as a Foundation Tool for Systems Analysis and Design

Like entity relationship diagrams and data flow diagrams, the project dictionary is a foundation tool for systems analysis and design. You've learned that as systems analysis tools, your ERDs and DFDs are catalogued into the dictionary. All of the objects on those graphic models are then described to the dictionary to complete the logical requirements specification.

Looking ahead to systems design and implementation, the project dictionary will gradually evolve to include implementation decisions and details. Thus a data flow description will indicate how the data flow will be implemented—for example, as a printed report or displayed terminal screens—and will include details about format, layout, controls, and the like.

Thus, you can expect to see the project dictionary throughout this book as you further study systems analysis and design methods.

Summary

A project dictionary is a catalog of details for a new information system's data and information requirements. The principal purpose of the dictionary is to record details about essential entity types (from the data model), data flows (inputs and outputs from the process model), and data stores (also from the process model). All of these objects consist of data elements, the smallest unit of data or information that has any meaning to an end-user. Data elements are arranged in specific patterns, called data structures, that describe a single occurrence of the object. These data structures, along with other logical attributes, are recorded in the dictionary.

The content of any occurrence of a data object can be documented as combinations of three data structures: a *sequence* of data elements, the *selection* from a list of data elements, and the *repetition* of one or more data elements.

A project dictionary is normally initiated during the study phase of the systems development life cycle; however, it expands most rapidly during the definition phase, when end-user requirements are the central focus. It remains the principal source of facts throughout the remainder of the life cycle.

Project dictionaries can be implemented manually, with forms, or automatically, using CASE tools. Dictionary entries should be defined for entity types, data flows, data stores, records, data elements, and tables of codes. All entries should specify important logical attributes for that object type and should be cross-referenced with one another.

Problems and Exercises

1. Why is a project dictionary a valuable systems analysis tool? What are the possible consequences of not creating a project dictionary during systems analysis?

2. Can you think of any specific times that a project dictionary might have been helpful when you were writing a computer program? Can you think of a situation in which you misinterpreted a computer program requirement because you didn't know something that could have been recorded in a project dictionary?

3. Dig out your last computer program. Prepare project dictionary entries for the following:
   a. Inputs (data flows)
   b. Outputs (data flows)
   c. Files or database (data store)
   d. All variables or fields (data elements)

4. Describe the relationship between a project dictionary and data and process models.

5. You have compiled a complete project dictionary for your new inventory control system. It is time to verify the contents of three summary reports specified for your end-users. Each data flow (report), record (data structure), and the data elements should be reviewed. Unfortunately, the entire dictionary is 373 pages long. You can't legitimately mark the relevant pages and "thumb" back and forth between pages during your review. How should the report specifications be presented?

6. Why shouldn't a project dictionary be organized alphabetically independent of type—for example, data flow, data store, data element, and so on—like a traditional dictionary such as *Webster's?*

7. Using the English data structure notation in this chapter, create a project dictionary entry for the following:
   a. Your driver's license
   b. Your course registration form
   c. Your class schedule
   d. IRS Form 1040 (any version)
   e. An account statement and invoice for a credit card
   f. Your telephone, electric, or gas bill
   g. An order form in a catalog
   h. An application for anything (for example, insurance, housing)
   i. A retail store catalog
   j. A typical real estate listing
   k. A computer printout from a business office or computer course
   l. A catalog that describes the classes to be offered next semester
   m. Your checkbook
   n. Your bank statement

8. Repeat Exercise 7 using the algebraic data structure notation.

9. Select one of the data structures you developed in Exercise 7 or 8 and complete a set of data element dictionary entries for each element appearing in the project dictionary.

10. During the study phase of systems analysis, the analyst must gather facts concerning both the manual and automated portions of the system. Why would it be desirable for a systems analyst to obtain samples of the existing computer files and computer-generated outputs? What value would project dictionary entries for computer files and computer-generated outputs have during systems analysis?

11. Visit a local business or school office. Ask for samples of five business forms, logsheets, or regular reports. Prepare complete project dictionary entries for each sample. If possible, review your entries with the end-users. Did they find your entries easy to read and understand? Can they think of additional data elements that would make their job easier? Add these elements to your project dictionary entries.

12. Find an example of a business code, possibly on the forms from Problem 11. Make a data element project dictionary entry for that code. Analyze that code according to the guidelines presented in this chapter. Can you suggest a better coding scheme?

Projects and Minicases

1. Create a simple project dictionary using your local word processor or line editor. Try to implement standard forms that can be read into the dictionary (to initiate new entity types, data flows, data stores, records, data elements, and table of codes entries).

2. Through information systems trade journals, research a commercial CASE product. Evaluate that package's *project* dictionary. Can you define both information and process models? Can you describe data structures to the dictionary? Can you describe individual data elements to the dictionary? What types of analytical reports can be generated from the dictionary?

3. The District Sales Manager for Grayson Industries, Inc., has requested an improved sales analysis report. He envisions a new report that might appear as follows:

Grayson Industries' sales regions cover four territories, each divided into two to four districts. There are 45 salespersons, and each salesperson is assigned to one and only one district. Prepare a project dictionary entry to describe the content of the envisioned sale analysis report. Why would developing a project dictionary be beneficial even though the manager already has an idea of what he wants? What additional information concerning this report should be included in the project dictionary before the layout of the final report is designed?

4. Identify and select a system from which you can gather three to five related forms. Discuss the forms with somebody who is knowledgeable about them. Then, prepare a logical project dictionary for the consolidated forms. The forms correspond to data flows. Define contents as records. Be sure to define all data elements. No flow, record, or element should be redundantly stored, even if it exists redundantly on the forms under separate names. Be sure to use logical, implementation-independent naming conventions. Review your logical dictionary with whomever you originally discussed the forms with. Try to evaluate how well the dictionary describes the essence of the forms.

Annotated References and Suggested Readings

DeMarco, Tom. *Structured Analysis and System Specification*. Englewood Cliffs, N.J.: Prentice-Hall, 1978. Chapters 11 through 14 present the most comprehensive treatment of the project dictionary that we've seen so far. DeMarco uses an algebraic notation for specifying data structures.

Gane, Chris, and Trish Sarson. *Structured Systems Analysis: Tools and Techniques*. Englewood Cliffs, N.J.: Prentice-Hall, 1979. Another classic reference on the project dictionary (Chapter 4 covers automated data dictionaries in greater depth).

CHAPTER ELEVEN

Defining Logical Policies

and Procedures in a

Project Dictionary

Granger's Restaurant Supply

We'll try a little different minicase approach for this chapter. First, let's get away from computer-related minicases. Second, let's start with the challenge:

Solve the dilemma described in the minicase. Are you willing to gamble your course grade that you can correctly solve the problem?

Make the following assumptions: (1) none of the characters in the story cheated, and (2) all of the characters are reasonably intelligent—that is, they would have guessed any obvious answer. And now the story.

Joe, Gordon, and Susan own Granger's Restaurant Supply. They are in dire financial straits. They don't have enough money to meet their debts. They are $250,000 short and cannot get a bank loan because of their poor credit rating. Among them, they can only collect $50,000.

They have decided on a drastic and risky solution to their problem. They will go to Atlantic City and try to gamble their $50,000 into enough money to cover their debts and save the company . . . or lose it! There is one problem. They are lousy gamblers! Within one short hour, they lose the entire $50,000. As they leave the casino, they run into the president of Premier Restaurant & Supply, Inc., their fiercest competitor. He has been trying, unsuccessfully, to buy Granger's for some time.

The unlucky trio offer Granger's to the greedy competitor for a bargain basement price. However, their rival, sensing an opportunity to get the business for absolutely nothing, offers the following proposition:

"I have five poker chips in my pocket: three blue and two white. I propose to blindfold each of you and then give you each a chip. One by one, I will remove your blindfolds. You will be permitted to see the chip in your colleagues' hands; however, you must keep your own chip concealed in your closed palm. If any one of you can tell me the color of your own chip, then I will give you $1 million cash, more than enough to ensure the financial future of your business. Each of you has the option of guessing or not guessing. However, if any one of you guesses wrong, you must give me your company, free and clear. Is it a deal?"

The partners have little choice and no other reasonable hope, so they accept the challenge. The competitor then shows them the five chips—three blue and two white—and chuckles as he places the blindfolds in place and gives each person one chip. He returns the two unused chips to his pocket.

The blindfold is removed from Joe, the eldest businessman and a world-class chess master. He looks at his partners' chips but, despite his logical mind, cannot determine the color of his own chip. He responds, "I just cannot give an answer. It's too risky. I'm better off giving my partners the opportunity for a better guess."

The blindfold is removed from Gordon, a graduate of a prestigious business school. After looking at the chips of his two partners, he too is unable to guess the color of his own chip. He passes the opportunity to Susan.

The competitor grins as he starts to remove the blindfold from Susan. He doesn't give her any more of a chance than he gave Joe or Gordon.

Susan interrupts confidently, "You can leave my blindfold on. How about double or nothing!" The competitor laughs aloud, "It's your funeral!"

Susan replies, "I'll take that $2 million! I know from the answers of my colleagues that my chip is _____ ." She is correct, and the winnings save Granger's from financial ruin.

What color was Susan's chip? Can you prove it without any doubt?

What Will You Learn in This Chapter?

This is the second of two chapters dealing with the *project dictionary* as a tool for documenting systems being developed. In this chapter, you will learn about tools and techniques for specifying processing requirements. You will have mastered process requirements specification techniques when you can

1. Differentiate between a policy and a procedure.

2. Describe some of the typical problems encountered in documenting procedures, and explain the ambiguities of ordinary English as a policy and procedure specification tool.

3. Construct a decision table to describe a policy in terms of conditions and actions to be taken under various combinations of those conditions.

4. Use Structured English to write procedure specifications.

In the last chapter, we introduced the project dictionary as a tool for recording details that are omitted from information and process models. We focused on details about data-oriented objects in the models, namely, entity types (from information models) and data flows and data stores (from process models). We intentionally omitted details about an equally important component of the system—processes (from the process models). Processes describe business policies and procedures that must eventually be implemented, either manually or through computer programs.

In this chapter, we will study two tools for specifying business policies and procedures: decision tables and Structured English. Both tools document data flow diagram processes into the project dictionary. Both tools provide us with an alternative that avoids the natural ambiguity of the English language.

Policies, Procedures, and the Project Dictionary

Processes on DFDs represent required tasks performed by the system. These tasks are performed according to business policies and procedures. What are policies and procedures? Why are they difficult to specify? How should they be specified?

What Is the Business Policy or Procedure?

A **policy** is a set of rules that govern some task or function in the business. In most firms, policies are the basis for decision making. For instance, most companies have a credit policy for determining whether to accept or reject an order. A credit card company must bill cardholders according to policies that adhere to restrictions imposed by state and federal governments—concerning maximum interest rates and minimum payments, for instance. Policies consist of rules that can often be translated into computer programs *if* the systems analyst can accurately convey those rules to the computer programmer.

What are procedures? To the programmer, **procedures** may represent the executable instructions in a computer program. But to the end-user, **procedures** are step-by-step instructions for accomplishing a task or tasks. And in many businesses, well-defined procedures are sorely missed by the end-users and management. Procedures put those policies into action. For instance, most companies have a policy on vacations, leaves, sick days, and the like. Those policies are implemented by procedures that define how to call in sick, request and approve vacations, and so forth. The procedures typically indicate how policies are to be documented.

Many of you have come from a programming course or background. As computer programmers, your job is to translate business requirements into syntactically correct code. Unfortunately, the language and idiosyncrasies of the business world and the programming world are vastly different. The language of computer programming is extremely precise, much more so than natural English. For this reason, many programmers spend more time debugging the business requirements than they do their computer programs! We would like to suggest that the programmer should not have to interpret or clarify business requirements. That is the analyst's job!

The Problems with Procedures

When was the last time you received a programming assignment that you didn't spend class or office time clarifying? In defense of analysts and teachers, the complexity of English and other natural languages can make it difficult to convert business policies and procedures into computer programs. Manual procedures are no easier to write. Indeed, you may have experienced frustration in completing your tax return. Why? The manual procedures that describe the process are difficult to communicate. Let's examine some of the problems associated with policy and procedure specification. We must overcome these problems.

General Criticisms of Procedures   Leslie Matthies is a writer with some unique insight into the often-ignored art of procedure writing. In *The New Playscript Procedure* (1977), Matthies describes several problems encountered with typical procedures. We'd like to paraphrase a few of those problems that are pertinent to policy and procedure specification.

1. Simply writing a procedure may not be enough. Not only do many of us not write well, but we also tend not to question our writing abilities. It's important to write clearly and accurately.
2. Most of us are *too* educated! It's often difficult for a highly educated person to communicate with an audience that may not have had the same educational opportunities. For example, the average college graduate (including most analysts) has a working vocabulary of 10,000 to 20,000 words; on the other hand, the average non-college graduate has a working vocabulary of around 5,000 words. This underscores the importance of the first law of good writing—*know your audience*—since a procedure has little value if it cannot be interpreted by those who will perform it.
3. A related problem deals with *jargon*. Often, we allow the jargon of computing to dominate our procedures. The computer industry constantly invents terms and acronyms to describe its products and discipline.

Problems with Ordinary English   Let's examine some of the problems that arise when the English language is used to specify procedures.

1. Statements tend to have excessive or confusing scope. How would you carry out this procedure: "If customers walk in the door and they do not want to withdraw money from their account or deposit money to their account or make a loan payment, send them to the trust department." Does this mean that the only time you should not send the customer to the trust department is when he or she wishes to do all three of the transactions? Or does it mean that if a customer does not wish to perform at least one of the three transactions, that customer should not be sent to the trust department? The scope of the procedure is not clear.

2. Compound sentences (two complete sentences connected by *and*, *but*, or other conjunctions) can be another serious problem. Consider the following example—a procedure describing how to replace an electrical outlet: "Remove the screws that hold the outlet cover to the wall. Remove the outlet cover. Disconnect each wire from the plug, but first make sure the power to the outlet has been turned off." Did you catch the compound sentence structure in the last instruction? An unwary person might try to disconnect the wires prior to turning off the power!

3. Multiple definitions associated with many words are another problem. An example of this problem was seen in the Americana Plastics case that began Chapter 10, where part number did not mean the same thing to each of the end-users!

4. Undefined adjectives confuse readers. Each semester, we receive several "Good Student Driver" discount forms from our students. They want to know if the student is in *good* standing. What's *good*? They sometimes define good as "the upper 10 percent of his/her class." Ten percent of what class? The entire university's class? The class of Computer Information Systems majors?

5. Conditional instructions sometimes present difficult problems. These instructions determine whether or not certain steps are performed. The statement of the conditions can be part of the problem. If we state that "all applicants under the age of 19 must secure parental permission," do we really mean less than 19 or less than or equal to 19? Although programmers are familiar with this problem, end-users and analysts frequently forget to carefully specify value ranges.

   This problem is further complicated by combinations of conditions. For example, credit approval may be a function of several conditions: credit rating, credit ceiling, annual dollar sales for that customer, and payment history. Different combinations of these factors result in different decisions, such as accept order on credit, reject order on credit and require full prepayment, or reject order until a down payment is received. As the number of conditions and possible combinations increases, the procedure becomes more and more tedious and difficult to write.

Fortunately, there are ways to formalize the specification of policies and procedures. These techniques address the problems we've stated.

Policies and Procedures in the Project Dictionary

The **project dictionary** was introduced in Chapter 10 as a catalog of requirements and specifications for a new information system. Chapter 10 demonstrated the descriptions of entity types, data flows, and data stores into the dictionary. These entries expand on the pictorial models.

Now we extend our entries in the project dictionary to include processes. Processes were objects contained only on process models (data flow diagrams or DFDs). Specifically, we are interested in documenting those primitive processes that do not explode to more detailed DFDs.

Consistent with our ongoing systems analysis theme, we want to document **logical** processing requirements, independent of current or possible physical implementations—especially programming languages and logic dependent on such languages. Such an approach allows us to more easily verify and evaluate policies and procedures with end-users who may feel uncomfortable with computer-like logic and tools.

Process Specification Conventions and Dictionary Implementation

Structured systems analysis provides two common tools for documenting primitive process details: *decision tables* and *Structured English*. In this section, we describe these tools and the methods for implementing them in the project dictionary.

Decision Tables: A Policy Specification Tool

Decision tables, unfortunately, don't get enough respect! People who are unfamiliar with the tool tend to avoid them. But decision tables are very useful for specifying complex policies and decision-making rules. Figure 11.1 illustrates the three components of a standard decision table.

- **Condition stubs** describe the conditions or factors that will affect the decision or policy.
- **Action stubs** describe, in the form of statements, the possible policy actions or decisions.
- **Rules** describe which actions are to be taken under a specific combination of conditions.

CHECK CASHING IDENTIFICATION CARD

Upon presentation person named hereon is entitled to cash personal checks up to $75.00 and payroll checks of accredited companies at Save Super Markets. Card is issued in accordance with terms and conditions of application, remains property of Save Super Markets, Inc., and shall be returned upon request.

Charles C. Parker, Jr

SIGNATURE
ISSUED BY
SAVE SUPER MARKETS, INC.

(a)

Check Cashing Policy

		Rules				
		1	2	3	4	5
Conditions	Type of check	1	2	1	2	2
	Check amount less than or equal to $75	Y	Y	N	N	–
	Company accredited by store	–	Y	–	Y	N
Actions	Cash check	X	X			
	Refuse check			X	X	X

(b)

FIGURE 11.1   A Decision Table for Specifying a Store's Check Cashing Policy   Decision tables offer a number of advantages over ordinary English. Although the narrative equivalent of the decision table is also short and concise, the decision table clearly ensures that the policy is complete and without contradictions.

Figure 11.1(a) depicts a check cashing policy that appears on the back of a check cashing card for a grocery store. In Figure 11.1(b), this same policy has been defined with a decision table. Three conditions affect the check-cashing decision: the type of check (1 = Personal, 2 = Payroll), whether the amount of the check exceeds the maximum limit (Y = Yes, N = No), and whether the company is accredited by the store (Y = Yes, N = No). The actions (decisions) are either to cash the check or refuse to cash the check. Notice that each combination of conditions defines a rule that results in an action, denoted by an X. Finally, note that rules 1, 3, and 5 contain a "—" entry for certain conditions. This means that the condition is *irrelevant* for these rules.

Decision tables offer a number of advantages over ordinary English. They use a standard format and handle combinations of conditions in a very concise manner. The English equivalent of a decision table would be much more difficult to write and read, because each combination of conditions would have to be described. Decision tables also provide techniques for identifying policy incompleteness and contradictions. In other words, if a combination of conditions has no actions, the policy is incomplete.

Let's learn how to construct decision tables. But first, let's end the suspense! What color chip did Susan have?

FIGURE 11.2 A Decision Table for Solving the Poker Chip Problem This is a first-draft decision table for solving the chapter's opening minicase.

Process Name		Rules							
		1	2	3	4	5	6	7	8
Conditions	Joe	W	W	W	W	B	B	B	B
	Gordon	W	W	B	B	W	W	B	B
	Susan	W	B	W	B	W	B	W	B
Actions	Impossible—only two white chips	X							
	Joe would have guessed					X			
	Gordon would have guessed			X					

Solving the Poker Chip Problem in Our Minicase   Figure 11.2 sets up a decision table for solving our poker chip problem. All possible combinations of poker chips are recorded as rules. The actions represent possible outcomes of the problem. Note that the first rule results in the action "Impossible" because we know that there were only two white chips. We can also eliminate rules 3 and 5 because, if either Joe or Gordon had seen two white chips, he would have known his own chip was blue. But that didn't happen. We cannot eliminate rule 2 because Susan requested that her blindfold not be removed.

Examine the remaining rules for the third condition stub carefully. Do you see that only one of the remaining rules (rule 7) results in Susan having a white chip? She solved the problem by concluding from her colleagues' responses that she didn't have a white chip. Here's how to eliminate rule 7.

If rule 7 had been true, Gordon would have known that his chip was blue, and the business would have been saved *before* Susan was asked to guess. Why? Because Joe couldn't answer, Gordon knew that Joe did not see two white chips. Joe had to see either two blue chips or a white and a blue chip. Now, remove the blindfold from Gordon *and* assume that rule 7 is true. If rule 7 is true, then Gordon is looking at a white chip in Susan's hand. Therefore, his own chip could not possibly be white because, if Joe had seen a

FIGURE 11.3 The Final Solution to the Poker Chip Problem
Through the process of elimination, Susan was able to deduce the color of her chip.

Process Name			Rules							
		1	2	3	4	5	6	7	8
C o n d i t i o n s	Joe	W	W	W	W	B	B	B	B
	Gordon	W	W	B	B	W	W	B	B
	Susan	W	(B)	W	(B)	W	(B)	W	(B)
A c t i o n s	Impossible—only two white chips	X							
	Joe would have guessed					X			
	Gordon would have guessed			X					
	Susan knows her chip is blue		↑		↑		↑		↑

white chip in both other people's hands, he would have known that his was blue. Thus, we can eliminate rule 7 because Gordon, as an intelligent person, would have known his chip was blue.

Looking at the revised decision table in Figure 11.3, we see that all the remaining rules result in Susan's chip being blue. Therefore, she didn't need to see her colleagues' chips. All she was required to do was recognize that her chip was blue—and all rules resulting in a white chip had been eliminated by her colleagues' answers!

Building a Decision Table   Let's begin with a policy statement.

A local credit union offers two types of savings accounts, regular rate and split rate. The regular rate account pays dividends on the account balance at the end of each quarter—funds withdrawn during the quarter earn no dividends. There is no minimum balance on the regular rate account. Regular rate accounts may be insured. Insured accounts pay 5.75 percent annual interest. Uninsured regular rate accounts pay 6.00 percent annual interest.

For split rate accounts, dividends are paid monthly on the average daily balance for that month. Daily balances go up and down according to deposits and withdrawals. The average daily balance is determined by adding each day's closing balance and dividing this sum by the number of days in the month. If the average daily balance is less than $25, then no dividend is paid. Otherwise, if the average daily balance is $25 or more, 6 percent per annum is paid on the first $500, 6.5 percent on the next $1,500, and 7 percent on funds over $2,000. There is no insurance on split rate accounts.

Let's construct a decision table for the above example. Recall the simple decision table presented in Figure 11.1. That table contained simple condition stubs that could only assume one of two values, *yes* and *no*. Many conditions, like those presented in the above policy, may assume more than two possible values. The rules for constructing all decision tables are identical. Let's learn the procedure.

1. Identify the conditions and values. Identify the data element each condition tests and all of the values that these data elements can assume. For our example:

DATA ELEMENTS OR CONDITIONS	VALUES
Account Type	R = Regular
	S = Split Rate
Insurance?	Y = Yes
	N = No
Balance Dropped Below $25 During Month?	Y = Yes
	N = No
Average Daily Balance	1 = 0.00–24.99
	2 = 25.00–500.00
	3 = 500.01–2,000.00
	4 = more than 2,000.00

   Note that we have yes and no conditions as well as multivalue conditions.

2. Determine the maximum number of rules. The maximum number of rules in a decision table is calculated by multiplying the number of values for each condition data element by each other. For example:

Condition 1 offers two values	2
Condition 2 offers two values	× 2
Condition 3 offers two values	× 2
Condition 4 offers four values	× 4
Number of rules in decision table	32

   (calculated as 2 × 2 = 4 × 2 = 8 × 4 = 32)

THE NEXT GENERATION

Will Expert Systems Create New Demand for Decision Tables?

Decision tables were, at one time, a favorite tool of systems analysts, who used them to communicate decision logic to programmers. Also, some researchers spent considerable time trying to fashion automatic code generators that would work from decision tables. However, their period of great popularity was brief. While still a useful tool, the decision table has generally taken a back seat to other policy and procedure-specification tools.

Expert systems represent an application area that is only beginning to be explored by users. Essentially, the concept extends the simple notion of storing data to include the storage of knowledge—the knowledge used by experts to either identify decision-making opportunities or help make the decisions (or possibly make the decisions unilaterally!).

Consider the decision-making process used by an expert in any discipline. Decisions are based on the expert's understanding of the conditions and the possible actions that can be taken under different conditions (or combinations of conditions). Sound familiar? Decision tables document a similar process.

Conditions can generally be mapped to the business data that describes those conditions.

Increasingly, that data is being captured in various computer files and databases. Actions can be gleaned from the experts and similarly stored (in what has come to be known as a *knowledge base*). Rules for defining the relationships between conditions and actions must also be stored in a knowledge base. Decision tables are beginning to look like a tool for concisely defining expert system requirements.

In this book, we've been describing the impact of computer-assisted systems engineering (CASE) tools for building systems. Is CASE influencing expert systems today? Not really. However, one could argue that decision tables (in some form) could easily be incorporated into CASE products. Would that be sufficient? Not really.

A special professional called a *knowledge engineer* has been suggested as a necessary facilitator of expert systems. A knowledge engineer is specially educated or trained to identify and record an expert's knowledge, which is to be emulated or supported by the expert system. The problem is that properly trained knowledge engineers are, and will be, scarce.

But a solution is possible. Systems analysts are becoming more plentiful. They are generally well schooled or experienced in fundamental fact-finding techniques. Unfortunately, they are not well schooled in knowledge acquisition techniques. But what if they had a CASE tool that made up for the lack of knowledge acquisition expertise?

That's right! An expert system for building expert systems! Incorporate the expert system into the basic CASE product. Put that CASE product in a laptop computer so that the analyst has the expert system in those meetings and interviews with experts in which knowledge acquisition is the objective. As the data is elicited from the expert and entered into the CASE tool, it is reformatted into decision tables—a tool familiar to analysts and programmers who must eventually implement the business expert system. Specifications can be easily analyzed against the CASE dictionary for completeness and consistency.

Thus, a tool that has lost favor could enjoy a renaissance in the next generation.

Dividend Rate

	Process Name	Rules																																
		1	2	3	4	5	6	7	8	9	10	11	12	13	14	15	16	17	18	19	20	21	22	23	24	25	26	27	28	29	30	31	32
C o n d i t i o n s	Account type	R	S	R	S	R	S	R	S	R	S	R	S	R	S	R	S	R	S	R	S	R	S	R	S	R	S	R	S	R	S	R	S	(a)
	Insurance	Y	Y	N	N	Y	Y	N	N	Y	Y	N	N	Y	Y	N	N	Y	Y	N	N	Y	Y	N	N	Y	Y	N	N	Y	Y	N	N	(b)
	Balance dropped below $25 during month	Y	Y	Y	Y	N	N	N	N	Y	Y	Y	Y	N	N	N	N	Y	Y	Y	Y	N	N	N	N	Y	Y	Y	Y	N	N	N	N	(c)
	Average daily balance	1	1	1	1	1	1	1	1	2	2	2	2	2	2	2	2	3	3	3	3	3	3	3	3	4	4	4	4	4	4	4	4	(d)
	Pay no dividend																																
A c t i o n s	5.750% ÷ 4 quarterly dividend on entire balance																																
	6.000% ÷ 4 quarterly dividend on entire balance																																
	6.000% ÷ 12 monthly dividend on balance up to $500																																
	6.500% ÷ 12 monthly dividend on balance between $500.01 and $2,000																																
	7.000% ÷ 12 monthly dividend on balance over $2,000																																

FIGURE 11.4 Entering All Possible Rules into a Decision Table This decision table identifies all possible combinations of conditions. Each condition defines a rule that must be verified or eliminated as impossible. The circled letters correspond to steps described in the text to determine all possible conditions.

3. Identify the possible actions. Identify each independent action to be taken for the decision or policy. For our example:

> Pay no dividend
> Pay 5.750%/4 quarterly dividend on entire balance
> Pay 6.000%/4 quarterly dividend on entire balance
> Pay 6.000%/12 monthly dividend on balance up to $500
> Pay 6.500%/12 monthly dividend on balance between $500.01 and $2,000
> Pay 7.000%/12 monthly dividend on balance over $2,000.00

4. Enter all possible rules. Record the conditions and actions in their respective places in the decision table (see Figure 11.4). All possible rules can easily be identified by completing the following steps:

Ⓐ For the first condition, alternate its possible values.

Ⓑ Note the size of the pattern that repeats in step (a) (in our example, two values). Cover each pattern of two values (R S) in the previous condition with the values of the second condition, repeating as necessary until the row is filled.

Ⓒ Again, note the size of the pattern that repeats in the second condition (this time, four values: Y Y N N). Cover each pattern of four values with the values of the next condition, repeating as necessary until the row is filled.

Ⓓ Once again, note the size of the pattern that repeats in the third condition (this time, eight values: Y Y Y Y N N N N). Cover each pattern of eight values with the values of the next condition, repeating as necessary until the row is filled.

This simple process defines all possible combinations of conditions for any decision table. We could have continued the process for any number of conditions.

5. Define the actions for each rule. Determine which action(s) are appropriate for each rule and mark them with an X. In the event that certain rules are impossible (cannot happen), add an action stub *Impossible* and mark the rules with an X. A question mark denotes an action for a rule is unknown. It reminds you to check with your end-users to learn how this rule should be handled. Figure 11.5 illustrates the actions for our rules.

6. Verify the policy. Your completed decision table should be reviewed with your end-users. Resolve any rules for which the actions are not specified. Verify that rules specified as *impossible* cannot occur. Resolve apparent contradictions, such as one rule with two possible interest rates covering a single balance. Finally, verify that each rule's actions are correct.

Process Name	Dividend Rate	Rules																														
		1	2	3	4	5	6	7	8	9	10	11	12	13	14	15	16	17	18	19	20	21	22	23	24	25	26	27	28	29	30	31	32
Conditions																																	
Account type		R	S	R	S	R	S	R	S	R	S	R	S	R	S	R	S	R	S	R	S	R	S	R	S	R	S	R	S	R	S	R	S
Insurance		Y	Y	N	N	Y	Y	N	N	Y	Y	N	N	Y	Y	N	N	Y	Y	N	N	Y	Y	N	N	Y	Y	N	N	Y	Y	N	N
Balance dropped below $25 during month		Y	Y	Y	Y	N	N	N	N	Y	Y	Y	Y	N	N	N	N	Y	Y	Y	Y	N	N	N	N	Y	Y	Y	Y	N	N	N	N
Average daily balance		1	1	1	1	1	1	1	1	2	2	2	2	2	2	2	2	3	3	3	3	3	3	3	3	4	4	4	4	4	4	4	4
Actions																																	
Pay no dividend		?							X																	X				X				
5.750% ÷ 4 quarterly dividend on entire balance		X	?	X		X								X								X				X							
6.000% ÷ 4 quarterly dividend on entire balance		?		X				X				X				X				X				X				X				X	
6.000% ÷ 12 monthly dividend on balance up to $500		?															X								X								
6.500% ÷ 12 monthly dividend on balance between $500.01 and $2,000		?															X								X								
7.000% ÷ 12 monthly dividend on balance over $2,000		?					X					X															X				X		
Impossible (no insurance for split rate)		?			X				X				X				X				X				X				X				X

FIGURE 11.5 Defining the Actions for Each Rule in a Decision Table This decision resulted in some rules (combinations of conditions) that are impossible. For other rules, we are uncertain as to what action should be performed.

7. Simplify the decision table. At this point, our decision table is both complete and correct. Still, 32 rules can be a bit overwhelming. We can simplify the decision table by eliminating and consolidating certain rules. (*Note:* This step should never be done until step 6 is complete!) The technique is described as follows:

a. Eliminate impossible rules.

b. Look for indifferent conditions. An **indifferent condition** is a condition whose values do not affect the decision and always result in the same action. These rules can be consolidated into a single rule. The technique is described as follows:

   1. Find a *set* (pair, trio, and so on) of rules for which

     • The actions are *identical.*

     • The condition values are the same except for *one and only one* condition or factor.

     • *All* possible values of an entry for a given condition must become indifferent before the rules can be collapsed. (This is important, especially in extended entry tables.)

   2. Consolidate that set of rules into a single rule, replacing the value of the indifferent condition with a minus sign—the *indifference symbol.*

This technique should be repeated as often as sets of rules satisfy the criteria in step (b). *But be careful! Never consolidate rules based on conditions that have already been identified as indifferent!*

Before we simplify our dividend table, let's look at an easier example. In Figure 11.6(a), notice that rules 1, 5, and 9 result in the same action. Also note that all condition values for the three rules are the same except for the third condition. In addition, observe that the three rules cover all possible values for the third condition. Do you see that the third condition's value does not affect the action? That condition is indifferent. We can consolidate the three rules into a single rule, as in Figure 11.6(b). We recorded the original rule numbers above the consolidated rules for your convenience. Also note that we consolidated rules 11 and 12 based on indifference for the first condition stub.

You may have been tempted to consolidate rules 3 and 7. But notice that the third condition hasn't been satisfied: only two (L and M) of the three values (L, M, and H) of the alleged indifferent condition are covered! Therefore, that condition is not indifferent.

Figure 11.7 depicts the final simplified decision table for the credit union's dividend policy. Convince yourself that our simplifications are valid. This condensed decision table is much easier to read than its equivalent in Figure 11.5.

FIGURE 11.6
Simplifying a Decision Table Most decision tables can be simplified by collapsing combinations of certain rules into single rules. For teaching purposes, we placed original rule numbers above the collapsed rules. In practice, the rules would be renumbered from 1 to 9 (for the nine remaining rules).

		1	2	3	4	5	6	7	8	9	10	11	12
C o n d i t i o n s	Condition 1	Y	N	Y	N	Y	N	Y	N	Y	N	Y	N
	Condition 2	Y	Y	N	N	Y	Y	N	N	Y	Y	N	N
	Condition 3	L	L	L	L	M	M	M	M	H	H	H	H
A c t i o n s	Action 1	X				X				X			
	Action 2		X							X	X	X	X
	Action 3				X		X						
	Action 4			X				X					

(a)

		1 5, 9	2	3	4	6	7	8	10	11, 12			
C o n d i t i o n s	Condition 1	Y	N	Y	N	N	Y	N	N	–			
	Condition 2	Y	Y	N	N	Y	N	N	Y	N			
	Condition 3	–	L	L	L	M	M	M	M	H			
A c t i o n s	Action 1	X											
	Action 2		X					X	X	X			
	Action 3				X	X							
	Action 4			X			X						

(b)

Structured English: A Procedure Specification Tool

Whereas decision tables are particularly effective for describing policies, Structured English is a tool for describing procedures. Structured English is based on the principles of structured programming. An example of Structured English is illustrated in Figure 11.8.

**FIGURE 11.7
Simplified Decision
Table for the Credit
Union Dividend
Policy** Notice that
conditions 3 and 4 are
indifferent to regular
accounts. Thus, more
than one condition can
be indifferent to a single
action.

Process Name		Dividend Rate Rules						
		1	2	3	4	5	6	7
C o n d i t i o n s	Account Type	R	R	S	S	S	S	S
	Insurance	Y	N	–	–	N	N	N
	Balance dropped below $25 during month	–	–	Y	N	N	N	N
	Average daily balance	–	–	–	1	2	3	4
A c t i o n s	Pay no dividend			X	X			
	5.750% ÷ 4 quarterly dividend on entire balance	X						
	6.000% ÷ 4 quarterly dividend on entire balance		X					
	6.000% ÷ 12 monthly dividend on balance up to $500					X	X	X
	6.500% ÷ 12 monthly dividend on balance between $500.01 and $2,000						X	X
	7000% ÷ 12 monthly dividend on balance over $2,000							X

Account type: R = regular
S = split rate

Insurance: Y = yes
N = no

Balance dropped below
$25 during month: Y = yes
N = no

Average daily balance: 1 = daily balance of 0.00-24.99
2 = daily balance of 25.00-500.00
3 = daily balance of 500.01-2,000.00
4 = daily balance more than 2,000.00

You may notice some similarity between Structured English and computer program pseudocode. **Pseudocode** is a tool to define detailed program algorithms or logic prior to coding. But pseudocode often tends to take on a programming accent that makes it unsuitable for nonprogrammers. For instance, array and variable initialization, opening and closing files, and read/write operations are often included in pseudocode. Also, most programmers tend to accent their pseudocode with the syntax of a computer programming language (for example, BASIC, COBOL, FORTRAN), usually the first such language they learned. Otherwise, the two tools are essentially equivalent.

**FIGURE 11.8 A
Sample Structured
English Description of
a Business Procedure**

For each LOAN ACCOUNT NUMBER in the LOAN ACCOUNT
FILE do the following steps:
If the AMOUNT PAST DUE is greater than $0.00 then:
While there are LOAN ACCOUNT NUMBERs for the
CUSTOMER NAME do the following steps:

Sum the OUTSTANDING LOAN BALANCEs.
Sum the MINIMUM PAYMENTs.
Sum the PAST DUE AMOUNTs.

Report the CUSTOMER NAME, LOAN ACCOUNTs on
OVERDUE CUSTOMER LOAN ANALYSIS.

Structured English borrows the logical constructs of structured pseudo-code but restricts the use of nouns, verbs, adjectives, adverbs, and computer jargon to make the specification easier to read for end-users. Let's learn how to write Structured English.

Structured English Guidelines and Syntax   Structured English is the marriage of the English language with the syntax of structured programming. The following restrictions are placed on the use of English within Structured English:

1. Only strong, imperative verbs may be used.

2. Only nouns and terms defined in the project dictionary may be used. These nouns may include names of data flows, data stores, entity types, records, data elements, and tables of code (and also decision tables!).

3. Compound sentences should be avoided.

4. Undefined adjectives and adverbs (*good,* for instance) are not permitted unless clearly defined in the project dictionary as value ranges for data element descriptions.

5. Avoid language that destroys the natural flow. Examples include *go to* . . . , *do* . . . , and *perform*, all of which are excessively physical, programming verbs.

6. A limited set of logic or flow constructs must be used. These constructs are familiar to those who practice structured programming. The three valid constructs are
   a. A sequence of single declarative statements
   b. The selection of one or more declarative statements based on a decision (the if-then-else or decision construct)
   c. The repetition of one or more declarative statements (the looping construct)

The best way to learn Structured English is to study some examples. We won't explain the policies involved because, if that were required, the value of Structured English would be highly questionable. You will likely develop your own Structured English style. We try to write simple but complete sentences. To as great an extent as possible, we want to eliminate the rigid style of computer programming while maintaining the syntax.

Sequential instructions are simple, declarative statements that follow one another. Declarative statements should begin with a strong action verb that describes exactly what should be done in that step of the procedure. Avoid vague, meaningless verbs such as *process*, *handle*, and *perform*. Also avoid computer programming language verbs such as *move*, *open*, or *close*. Many declarative statements are arithmetic, specifying how to calculate data elements such as GROSS PAY, FEDERAL TAX WITHHELD, and NET PAY. The statement should always specify the formula to be used. For example:

Calculate <insert data element> using the formula:
<insert formula>

All data elements should be defined in the project dictionary.

The following statements are valid sequential instructions:

Find the MEMBER ACCOUNT using the MEMBER ACCOUNT NUMBER.
Compute the new ACCOUNT BALANCE using formula:
ACCOUNT BALANCE = ACCOUNT BALANCE + ADJUSTMENT AMOUNT.
Record the new ACCOUNT BALANCE in the MEMBER ACCOUNT data store.
Write the new ACCOUNT BALANCE on the CUSTOMER STATEMENT.

Note that we capitalized the names of data flows, data stores, data elements, and other nouns recorded elsewhere in the project dictionary.

The **decision construct** of Structured English allows you to place branching instructions within a sequence of instructions. The following formats for the decision construct are permitted.

- If <insert condition> then:
    <insert instruction(s)>
  Otherwise: (not condition)
    <insert instruction(s)>.

- Select the appropriate case:
  Case 1: <insert condition value 1>
        <insert instruction(s)>
  Case 2: <insert condition value 2>
        <insert instruction(s)>

    .

    .

    .

  Case n: <insert condition value n>
        <insert instruction(s)>

FIGURE 11.9 The Case and If-Then-Otherwise Constructs in Structured English

Find the MATERIAL NUMBER in the INVENTORY FILE.

Select the appropriate case:

Case 1: MATERIAL CLASS = 'stock,' then:

If the QUANTITY ON HAND is greater than or equal to the QUANTITY REQUISITIONED then:

Calculate new QUANTITY ON HAND using the formula:

QUANTITY ON HAND - QUANTITY REQUISITIONED.

Record QUANTITY ON HAND in the INVENTORY FILE. Issue a STORES TICKET.

Otherwise (QUANTITY ON HAND is not greater than the QUANTITY REQUISITIONED) then:

Issue a STORES STOCKOUT TICKET.

Case 2: MATERIAL CLASS = 'seasonal,' then:

Calculate QUANTITY NEEDED using the formula:

REQUISITION QUANTITY x SEASONAL ADJUST RATE.

Issue a PURCHASE REQUISITION.

Case 3: MATERIAL CLASS = 'requisition,' then:

Issue a PURCHASE REQUISITION.

Use the first format for conditions that can only assume two values—for example, *yes* and *no*, *male* and *female*. Use the second format whenever the condition can assume more than two values—for example, *freshman*, *sophomore*, *junior*, and *senior.* The instructions within the decision construct include one or more statements.

Figure 11.9 demonstrates the *case* and *if-then-otherwise* constructs. The instructions for each case are *indented* and *blocked* for readability. The constructs are *nested*, one within another. Notice how we *indent* and *block* the instructions to improve readability. We also recorded the negative of the condition in parentheses with the *otherwise* statements, again to enhance readability.

The **repetition** or **looping construct** allows us to specify that a sequence of instructions is to be repeated until some condition or desired result is satisfied. The following formats are permitted:

FIGURE 11.10   The Repetition Construct of Structured English

For each CUSTOMER NUMBER in the CUSTOMER ACCOUNT file, do the following:

> Repeat the following steps for each ACCOUNT NUMBER:
>
>> For each ACCOUNT TRANSACTION for the ACCOUNT NUMBER, do the following:
>>
>>> Report each ACCOUNT TRANSACTION.
>>> Sum the following account totals:
>>>> NUMBER OF DEBIT TRANSACTIONS
>>>> NUMBER OF CREDIT TRANSACTIONS
>>>> TOTAL OF DEBIT TRANSACTIONS
>>>> TOTAL OF CREDIT TRANSACTIONS
>>>> ACCOUNT EXPENSES
>>
>> Report the account totals for the ACCOUNT NUMBER.
>
> Until there are no more ACCOUNT NUMBERs for the CUSTOMER NUMBER.

- Do the following <insert some number> times:
    <insert instruction(s)>
- Repeat the following steps:
    <insert instruction(s)>
  Until <insert condition> is satisfied.
- While <insert condition>, do the following steps:
    <insert condition(s)>
- For each <insert condition>, do the following steps:
    <insert instruction(s)>

The first format allows you to specify that certain steps be performed a specific number of times. Do you see the difference between the second and third formats? The second format requires that the instructions be executed at least one time. The third format specifies that the instructions might not be executed at all, if the condition is not initially satisfied.

Figure 11.10 demonstrates the repetition construct for the procedure that produces a consolidated bank statement for each customer. The first repetition construct, *for each* . . . , drives the entire procedure. Because a customer can have numerous accounts but must have at least one account, we use the *repeat-until* construct for an inner loop of the procedure.

Decision Table or Structured English?

This is the first chapter in which we've offered two tools for specifying one object, a process. Which tool should be used? It should not surprise you that decision tables and Structured English can complement one another to describe a single process. Recall that procedures often implement policies. Why then can't we describe all procedures using Structured English and reference appropriate decision tables to describe any policies implemented by the procedures? The answer, of course, is that we can.

Computer-Assisted Systems Engineering for Process Specification

Computer-Assisted Systems Engineering (CASE) has been a continuing theme throughout this book. Virtually all CASE products include specification tools for primitive-level processes. These specifications are recorded in the project dictionary.

CASE specifications for processes vary greatly from one product to another. Some products impose more structure than others. For instance, Figure 11.11 illustrates a CASE-based tool called *action diagrams* that, among other things, can implement Structured English. All of the examples in the next section were created using a customized CASE product called Excelerator.

Using a Project Dictionary for Systems Analysis (Continued from Chapter 10)

As described in Chapter 10, a project dictionary is essential to the Structured Systems Analysis methodology. Let's place the methodology in the context of the life cycle and describe when process descriptions would be written. Then we'll demonstrate some sample dictionary entries.

Modern Structured Analysis Using a Project Dictionary

In Chapter 5, we introduced a popular methodology called Structured Analysis. Structured English was first introduced as part of the methodology. Decision tables, a classic tool that predates Structured English, were suggested as

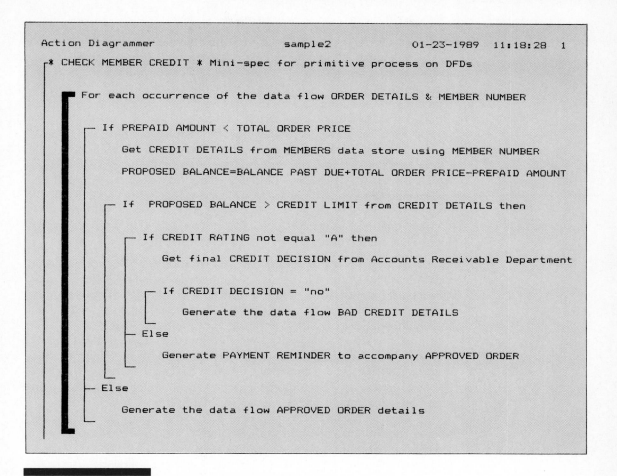

```
Action Diagrammer                    sample2              01-23-1989  11:18:28  1
* CHECK MEMBER CREDIT * Mini-spec for primitive process on DFDs

   For each occurrence of the data flow ORDER DETAILS & MEMBER NUMBER

      If PREPAID AMOUNT < TOTAL ORDER PRICE

         Get CREDIT DETAILS from MEMBERS data store using MEMBER NUMBER

         PROPOSED BALANCE=BALANCE PAST DUE+TOTAL ORDER PRICE-PREPAID AMOUNT

         If  PROPOSED BALANCE > CREDIT LIMIT from CREDIT DETAILS then

            If CREDIT RATING not equal "A" then

               Get final CREDIT DECISION from Accounts Receivable Department

               If CREDIT DECISION = "no"

                  Generate the data flow BAD CREDIT DETAILS
            Else

               Generate PAYMENT REMINDER to accompany APPROVED ORDER

      Else

         Generate the data flow APPROVED ORDER details
```

FIGURE 11.11   Sample Dictionary Description for a Procedure

an alternative or complementary tool. Let's place process modeling for Structured Analysis into the context of systems analysis and the life cycle.

Recall the four phases of systems analysis: survey, study, definition, and selection. During the **survey** and **study** phases, you first encounter existing policies and procedures in the system. These should be evaluated for consistency and completeness as well as for other problems and opportunities. Decision tables and Structured English are not essential to complete the survey and study phases. You may, however, find it useful to translate existing policies and procedures into decision tables and Structured English so that they might more easily be verified and analyzed.

The **definition phase** identifies end-user requirements independent of implementation alternatives. Structured Analysis uses this phase to model logical, implementation-independent requirements. One of those models consists of data flow diagrams (DFDs). In Chapter 10, you learned how data flows and data stores on DFDs are further described in the dictionary. The primitive processes on DFDs should similarly be described in the dictionary using decision tables and Structured English.

The **selection** phase determines how processes will ultimately be implemented—computer programs, manual procedures, and other machine operations. As you move into the design phases, detailed process specifications can greatly reduce the amount of time required to write programs and procedure manuals.

How to Describe Process Models in the Dictionary

Process models (data flow diagrams) were introduced in Chapter 9 as an effective requirements modeling tool. The models show the essential processes, their inputs, outputs, and data store interactions. In Chapter 10, we described how to further document data flows and data stores. In this section we'll describe how to complete the specification of primitive processes.

Primitive processes—those that are not exploded into more detailed DFDs—should be exploded into project dictionary descriptions that include Structured English and, where applicable, decision tables.

These entries have been taken from the logical models for the Entertainment Club case study. The entries were created using a CASE product.

Process Description Entries in the Dictionary   A sample dictionary screen or form for a primitive process is called a Primitive Process Procedure, demonstrated in Figure 11.12. We call your attention to the following:

Ⓐ Every primitive process procedure should have a name that corresponds to the name of the process on the data flow diagram.

Ⓑ A primitive process may be further described by a decision table that is named here and referenced from within the procedure (see below).

Ⓒ The procedure statement is expressed in Structured English.

Decision Table in the Dictionary   Decision tables are always referenced from within a Primitive Process Procedure (see above). The decision table itself is described separately in the dictionary, as shown in Figure 11.13. Note the following (page 388):

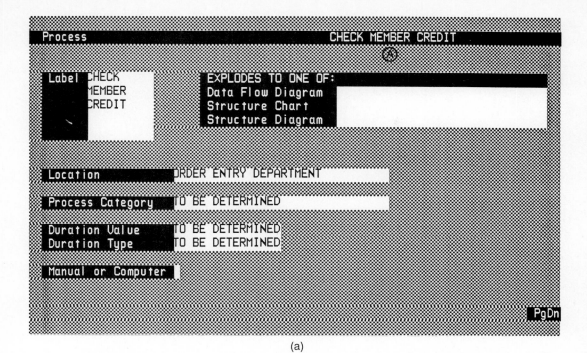

(a)

Process   CHECK MEMBER CREDIT

Ⓐ

Label CHECK
       MEMBER
       CREDIT

EXPLODES TO ONE OF:
Data Flow Diagram
Structure Chart
Structure Diagram

Location   ORDER ENTRY DEPARTMENT

Process Category   TO BE DETERMINED

Duration Value   TO BE DETERMINED
Duration Type   TO BE DETERMINED

Manual or Computer

PgDn

(b)

Process   CHECK MEMBER CREDIT

Description   PgUp

For each ORDER DETAILS & MEMBER NUMBER, do the following steps:
Ⓒ   If PREPAID AMOUNT is less than the TOTAL ORDER PRICE then:
   (1)   Search MEMBERS for the CREDIT DETAILS using the MEMBER NUMBER.
   (2)   Calculate the PROPOSED BALANCE using the formula:
          PROPOSED BALANCE = BALANCE PAST DUE + TOTAL ORDER PRICE –
                        PREPAID AMOUNT
   (3)   Select the CREDIT DECISION using the CREDIT DETAILS and
        CREDIT decision table. Ⓑ
   (4)   If action is "approve credit" then:
        Report the APPROVED ORDER details.
        Otherwise:
        Report the BAD CREDIT ORDER details.

Modified By	BENTLEY	Date Modified	890120	# Changes 0
Added By	BENTLEY	Date Added	890120	
Last Project	PROJECT			
Locked By		Date Locked	0	Lock Status

FIGURE 11.12 Reference to a Decision Table from Within Structured English Note the use of a numbering convention within the procedure. Some analysts feel this notation eliminates the programming look and feel of Structured English.

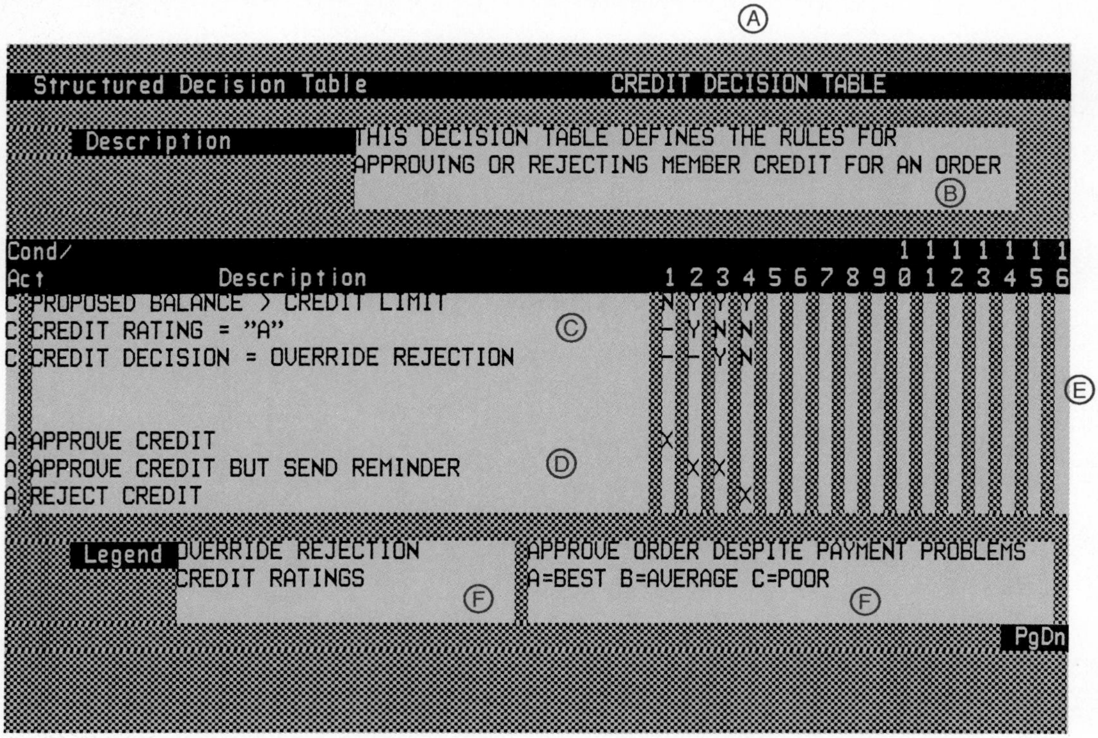

Ⓐ

Structured Decision Table   CREDIT DECISION TABLE

Description   THIS DECISION TABLE DEFINES THE RULES FOR
                   APPROVING OR REJECTING MEMBER CREDIT FOR AN ORDER

Ⓑ

Cond/ Act	Description		1	2	3	4	5	6	7	8	9	1 0	1 1	1 2	1 3	1 4	1 5	1 6
C	PROPOSED BALANCE > CREDIT LIMIT		N	Y	Y	Y												
C	CREDIT RATING = "A"	Ⓒ	-	Y	N	N												
C	CREDIT DECISION = OVERRIDE REJECTION		-	-	Y	N												
A	APPROVE CREDIT		X															
A	APPROVE CREDIT BUT SEND REMINDER	Ⓓ		X	X													
A	REJECT CREDIT					X												

Ⓔ

Legend   OVERRIDE REJECTION   APPROVE ORDER DESPITE PAYMENT PROBLEMS
          CREDIT RATINGS   A=BEST B=AVERAGE C=POOR

Ⓕ   Ⓕ

PgDn

FIGURE 11.13 Sample Dictionary Description for a Decision Table

Ⓐ Every decision table should have a descriptive, logical name.

Ⓑ Every decision table should have a definition that suggests its purpose, if the name is not self-explanatory.

Ⓒ Conditions should be expressed in terms of data elements that are described elsewhere in the dictionary (see Chapter 10). The CASE product we used allows you to enter and print many more conditions than this screen can show at one time.

Ⓓ Actions are recorded in the usual location. The CASE product we used allows you to enter and print many more actions than this screen can show at one time.

Ⓔ Only rules that result from simplification of the decision table are entered. Rules are defined in terms of combinations of conditions that result in specified actions. The CASE product we used allows you to enter and print many more rules than this screen can show at one time.

Ⓕ The legend area is used to describe the meanings of the values that can be assumed by condition data elements. This is necessary because the CASE product restricts values of the condition data elements to a single letter or digit.

Summary

In addition to documenting details about data flows, data stores, and entity types (Chapter 10), analysts must describe processes—from data flow diagrams—to the dictionary. Processes are documented as policies and procedures. Policies specify management decisions and rules. Procedures execute those policies as well as the tasks that support day-to-day business operations and management. Although most existing procedures are specified in ordinary English, we have learned that most such procedures are often unclear and incomplete.

Fortunately, two tools can help us specify policies and procedures more effectively. Decision tables are particularly effective for policies and for procedures that contain complex combinations of decisions. Structured English is a procedure specification tool that overcomes many of the limitations and ambiguities of ordinary English. Together, these two tools complement data flow diagrams and data dictionaries. Furthermore, these tools help us effectively communicate business requirements to computer programmers and end-users alike!

Policy and procedure specifications are entered into the project dictionary that was first introduced in Chapter 10. They explode out of primitive processes on data flow diagrams, those processes which do not explode into more detailed DFDs.

This chapter concludes Part Two. It is important for you to realize that the tools presented in Part Two have their greatest value in the systems analysis phases. In Part Three, we will demonstrate some of these uses and emphasize how the tools used in the analysis phases can complement and support the tools used in the design phases.

Problems and Exercises

1. Explain the difference between a policy and a procedure. Give an example of a policy and the procedure to implement or administer that policy.

2. Obtain a formal statement of a policy and procedure from a local company (or from home, such as a policy for a credit card). Evaluate the policy and procedure statement in terms of the common specification problems identified earlier in this chapter.

3. Reconstruct the policy and procedure used in Exercise 2 with the tools you learned in this chapter. Simplify the decision table.

4. Simplify the following decision table:

Rules

	1	2	3	4	5	6	7	8	9	10	11	12
condition 1	Y	N	Y	N	Y	N	Y	N	Y	N	Y	N
condition 2	Y	Y	N	N	Y	Y	N	N	Y	Y	N	N
condition 3	A	A	A	A	B	B	B	B	C	C	C	C
action 1			X				X				X	
action 2	X				X				X			
action 3		X				X						
action 4				X				X		X		X

5. Were you able to combine rules 2 and 6 in problem 4? Why?

6. Write a mini spec (Structured English) for balancing your checkbook.

7. Produce a mini spec to describe how to prepare your favorite recipe, tune a car, or perform some other familiar task. Ask a novice to perform the task, working from your specification.

Projects and Minicases

1. Obtain a copy of your last programming assignment. Using Structured English, write procedure specifications to clearly communicate the procedures that you were asked to implement. Did you avoid unnecessary details? Does your Structured English take on an unnecessary programming accent? Does your Structured English accurately communicate the processing procedures that are to be implemented as a computer program? Did you avoid programming-dependent statements, such as OPEN and PERFORM? Would you have appreciated such a description when you were originally given the assignment?

2. Prepare a decision table that accurately reflects the following course grading policy: A student may receive a final course grade of A, B, C, D, or F. In deriving the student's final course grade, the instructor first determines an *initial* or tentative grade for the student. The initial course grade is determined in the following manner.

A student who has scored a total of no lower than 90 percent on the first three assignments and exams and received a score no lower than 70 percent on the fourth assignment will receive an initial grade of A for the course. A student who has scored a total lower than 90 percent but no lower than 80 percent on the first three assignments and exams and received a score no lower than 70 percent on the fourth assignment will receive an initial grade of B for the course. A student who has scored a total lower than 80 percent but no lower than 70 percent on the first three assignments and exams and received a score no lower than 70 percent on the fourth assignment will receive an initial grade of C for the course. A student who has scored a total lower than 70 percent but no lower than 60 percent on the first three assignments and exams and received a score no lower than 70 percent on the fourth assignment will receive an initial grade of D for the course. A student who has scored a total lower than 60 percent on the first three assignments and exams, *or* received a score lower than 70 percent on the fourth assignment, will receive an initial and final grade of F for the course. Once the instructor has determined the initial course grade for the student, the *final* course grade will be determined. The student's final course grade will be the same as his or her initial course grade if no more than three class periods during the semester were missed. Otherwise, the student's final course grade will be one letter grade lower than his or her initial course grade (for example, an A will become a B).

Are there any conditions for which there was no action specified for the instructor to take? If so, what would you do to correct the problem? Can your decision table be simplified? If so, simplify it.

Annotated References and Suggested Readings

Copi, I. R. *Introduction to Logic.* New York. Macmillan, 1972. Copi provides a number of problem-solving illustrations and exercises that aid in the study of logic. The poker chip problem in our chapter case was adapted from one of Copi's reasoning exercises.

DeMarco, Tom. *Structured Analysis and System Specification.* Englewood Cliffs, N.J.: Prentice-Hall, 1978. This classic book introduced Structured English.

Gane, Chris, and Trish Sarson. *Structured Systems Analysis: Tools and Techniques.* Englewood Cliffs, N.J.: Prentice-Hall, 1979. Gane and Sarson include an entire chapter on defining process logic. In addition to discussing decision tables and Structured English, this chapter explains how decision trees and pseudocode can be used to define process logic.

Gildersleeve, T. R. *Successful Data Processing Systems Analysis.* Englewood Cliffs, N.J.: Prentice-Hall, 1978. The first edition of this book includes an entire chapter on the construction of decision tables. Gildersleeve does an excellent job of demonstrating how narrative process descriptions can be translated into condition and action entries in decision tables. Unfortunately, the chapter was deleted from the second edition.

Martin, James, and Carma McClure. *Action Diagrams: Towards Clearly Specified Programs.* Englewood Cliffs, N.J.: Prentice-Hall, 1986. This book describes a formal grammar of Structured English that encourages the natural progression of a process (program) from Structured English to code. Action Diagrams are supported directly through CASE tools sold by Knowledgeware, Inc.

Matthies, Leslie H. *The New Playscript Procedure.* Stamford, Conn.: Office Publications, 1977.

PART THREE

Systems Design and Implementation: Tools and Techniques

12
How to Design an Information System: Traditional and Prototyping Approaches

13
Designing Conventional Computer Files and Controls

14
Designing Modern Computer Databases

15
Designing and Prototyping Computer Outputs and Controls

You may have been wondering if we were ever going to discuss the design of computer-specific specifications for an information system. We all know that the computer elements of a system must be designed sooner or later. Parts One and Two deliberately avoided concerns about computer elements; if the problem and requirements are not well understood, the best-designed computer elements will not succeed! The basis for a good system design is a good system analysis. In Part Three, we deal exclusively with *computer*-based system design.

Like systems analysis, systems design is an essential activity in the system development life cycle. Although systems design is usually performed with somewhat greater rigor and standards than systems analysis, systems design is still often shortchanged. The fault does not lie entirely with the systems analyst. Users frequently become impaitent with both the analysis and the design activities. The prevalent attitude is "You are a computer person. Computer people write programs. Programs are what make the new system go. So why aren't you programming?" The initial result? The project moves prematurely into the implementation phases. The net result? The system takes longer to implement, costs more to implement, contains more errors and omissions, is more difficult to work with, and is very costly to maintain and enhance. Whose reputation is on the line? Yours, of course. After all, you (the analyst) worked most closely with the users and management.

We, as computer professionals, must also assume some of the blame. For years, we have forcefed our technical terminology and documentation to end-users. Like castor oil, it's left a bad taste. Better tools and techniques are available, but we've abused them. We've worried too much about the computer and not enough about the person who will use the computer-based system. The result? Many users don't want out help, and they don't necessarily want to help us do a better job. Meanwhile, the tools and techniques of systems design get an undeserved bad rap. We must try to correct this situation.

The purpose of this unit is to introduce you to systems design, the process tools, and techniques. Here's an overview of the chapters you will study.

16

Designing and Prototyping Computer Inputs and Controls

17

Designing and Prototyping the User Interface and On-Line Terminal Dialogues

18

Designing Computer-Based Methods, Procedures, and Controls

19

Designing Structured Programs

20

How to Implement and Evaluate a new Information System

Chapter 12 introduces the **process of systems design.** It also introduces you to the notion of buying software as opposed to building it. And, most important, it introduces you to a technique called *prototyping*. Prototyping addresses two problems: getting the users involved and dealing with user impatience with the process of design.

Chapter 13 covers the design of **conventional computer files,** like those you've learned to use in programming courses. Now you'll learn about some of the preprogramming file decisions that influence the choice of different files.

Chapter 14 is new to this edition. **Database** has become a technology of choice and a viable alternative to conventional files. Since you may not have much programming experience with databases, we'll teach you some fundamentals. Then we'll show you the basics of designing different types of databases.

Chapter 15 discusses the most visible component of any computerized application—its **outputs.** You learn how to intelligently design outputs and communicate the specifications to programmers.

The flip side of outputs is inputs. Chapter 16 presents **input design.** Input design not only includes the actual computer inputs, but also the forms used to originally record new data so that it can be input.

The trend toward on-line information systems necessitates Chapter 17, the design of the **on-line user interface** (sometimes called *terminal dialogue*). Nowhere is human engineering more important than in designing on-line interfaces for end-users.

Chapter 18 starts to bring all the components of the earlier chapters together. You'll learn how **computer procedures** are designed and documented. Procedures determine the flow of inputs to processes to outputs and the role of files and databases in that flow. The process is similar to data flow diagramming; however, it is complicated by the addition of computer details.

Chapter 19 teaches you how to **package the system,** input, output, interface, file, and database specifications for the programmer who must construct the system. Additionally, you will learn how programs should be designed to reduce costs of implementation and maintenace.

Chapter 20 is not really about systems design. However, because the systems analyst assumes an important role during **systems implementation,** we thought it useful to present systems implementation in much the same manner that we introduced systems analysis (Chapter 6) and design (Chapter 12).

It should be noted that Part Four presents a number of complementary topics for systems design. These include project management, fact-finding, interpersonal communications, and cost/benefit analysis. The material can and should be integrated into your study of Part Three.

CHAPTER TWELVE

How to Design an Information System: Traditional and Prototyping Approaches

Schuster and Petrie, Inc.

Keith Stallard is a relatively new programmer/analyst at Schuster and Petrie, Inc. Having spent two years as a programmer, he was promoted one year ago. The Information Services Division of S & P requires twice-a-year job performance reviews. Tim Hayes, Associate Director of Financial Systems, has scheduled a job performance review with Keith. Let's listen in.

"Well, Keith, do you still want this programmer/analyst job?" asked Tim. "You've had about six months to get used to your new responsibilities."

"More than ever," answered Keith. "Now that I've had a taste of systems work, I know it's right for me. I assume this meeting will determine if I'm making progress. How am I doing?"

"You're right. I've discussed your performance on the job cost accounting system project with both your supervisor and your key user contact," replied Tim. "Your design specification document was quite impressive. But I have to ask you, where did you learn to complete such thorough specifications? The implementation appears to be moving along more smoothly than expected, largely because of your specifications."

Keith smiled and explained, "I did a lot of reading in systems analysis and design textbooks at the college library. I also queried both users and programmers about problems with typical specifications. My own experience as a programmer has influenced my specifications. But to be honest, I was really embarrassed by my performance on the account aging project. That's why I did all those things!"

Tim interrupted, "I don't understand. We gave you acceptable ratings. The account aging project was a little off schedule, but that's the only problem I recall."

Keith answered, "It was a little more complicated than that. Bill was supervising me since it was my first experience with systems design [*Note:* Bill had done the systems analysis]. But Bill had to be called off the project to fix a major flaw in another system. I kept working on the design and passed the design document to Rita [a programmer]. Then lightning struck for a second time. Rita had to go into the hospital and I had to assume her programming responsibilities. It was the first time I ever had to cut code from my own specifications. There were so many details, and I hadn't documented all of them. Surprisingly, I couldn't even remember all of the thought process that went into my own specifications. Now I know how the maintenance programmers feel!"

Keith continued, "But eventually I got the system up and running—only to find out that some of the reports were not acceptable to my users. The content was there. The format was wrong. I had to take certain liberties with the format. In my school days, that was what we did with all programming assignments. I just didn't appreciate the importance of user involvement in the design process. I assumed that systems analysis took care of all the user issues."

"And then, the Internal Audit Department got ahold of my design specifications. They didn't like them at all! There weren't enough internal controls to satisfy their standards. By that time, I had half the programs written and tested. I had to redesign many system components and rewrite several affected programs. To make a long story short, I never want to go through that kind of design experience again. So I learned about systems design."

Tim asked, "And you still want to be an analyst? After all that?"

"Yes," answered Keith. "Despite the problems, I found the work to be so much more satisfying than programming. I knew it wasn't going to be easy. But it was enjoyable."

"Well," said Tim. "We've discussed your strengths. But we do need to work on a few things. First, as you know, most of our older systems are being converted to on-line systems using databases. I'm sending you to a one-week intensive course at IBM. There you will learn about the new DB2 database management system. I also want you to go through our user interface course the next time it's offered. I don't know if you've heard that the on-line interface on your accounts aging system hasn't lived up to expectations."

"You also need to work on your writing and speaking skills. The report you did to sell the new job costing system wasn't well organized, was too wordy, and contained numerous grammatical errors and typos. You were lucky that Bill got it before your users. You might have lost the sale. And your presentation of that system to management could have gone a little smoother. Public speaking is tough. I realize that! But you seemed unconfident. You had a good design! But if you don't seem confident and comfortable with your design, how will management feel about it? We did get it through, though. But your communications skills need improvement . . . especially if you want my job in the future. You have that potential, Keith. Don't waste it!"

"I understand," responded Keith. "And I appreciate your honesty. I've suspected the problem. I guess I never took those English and communications courses seriously. I'll get enrolled in some evening continuing education courses for the next term. I'm not going to let poor communications skills get in the way of my future."

Tim leaned back in his chair and pondered for a moment. "Let's get to the bottom line, Keith. You've shown better than average progress in your new assignment. That's why, effective next month, you'll see a little increase in your paycheck. If you keep up the good work and improve in the areas we've outlined, I'm certain you'll be promoted to systems analyst within two years. Now, let's talk about the design specification some more. Do you think we could teach our other analysts to do that? We could sure use some new standards to. . . ."

Discussion

1. Thinking back to your programming courses (or experiences), what are some of the problems you've had responding to programming assignments (the clasroom equivalent to a systems design)?

2. What did Keith learn about working from his own specifications?

3. Would you be able to easily change a design requirement in a program that you wrote one or two years ago? What implication would design documentation have on your answer?

4. As a systems analyst working on the design phases of a project, what types of people did Keith have to communicate with? Why does communication become tougher during systems design than during systems analysis?

What Will You Learn in This Chapter?

In this chapter, you will learn more about the **systems design process** of the systems development life cycle. This process consists of two phases: *acquisition* and *physical design.* You will know that you understand the complete systems design process when you can

1. Define the *systems design process* in terms of the acquisition and physical design phases of the life cycle.

2. Describe the acquisition and physical design phases in terms of
   a. Purpose and objectives.
   b. Tasks and activities that must or may be performed.
   c. Sequence or overlap between tasks and activities.
   d. Techniques used.
   e. Skills you must master to perform the phase properly.

3. Describe traditional and prototyping approaches to physical systems design.

4. Describe the continuing roles of fact-finding, interpersonal communications, and cost-benefit analysis in systems design.

What is *systems design*? You already know what a system is, so the key term here is *design*. For our purposes, design is the skillful planning of the computer elements for an improved information system. Whereas systems analysis primarily focused on the logical, implementation-*independent* aspects of a system (the requirements), systems design deals with the physical, implementation-*dependent* aspects of a system (the system's *technical specifications*).

Putting the Design Process in Perspective

Most of us place too restrictive a definition on the process of design. We envision ourselves drawing blueprints of the computer-based systems to be programmed and developed by ourselves or our own programmers. Thus, we design inputs, outputs, files, databases, and other computer components. Recruiters of computer-educated graduates refer to this restrictive definition as the "not invented here" syndrome. In reality, many companies purchase more software than they write in-house. That shouldn't surprise you. Why reinvent the wheel? Many systems are sufficiently generic that computer vendors have written adequate—but rarely, if ever, perfect—software packages that can be bought and possibly modified to fulfill end-user requirements.

Consequently, we must expand our definition of the *design process* to include the acquisition or purchase of computer software and hardware, as well as the more traditional physical design of computer-based components. That definition is illustrated in Figure 12.1. For each of the highlighted phases, we will study *purpose*, *objectives*, and *tasks* that should be performed and important *skills* to be mastered.

Which happens first, acquisition or physical design? Recall that most of the phases of systems analysis covered in Chapter 6 focused on problem solving at the logical, implementation-independent level. However, the last phase of analysis, *selection*, began to make the transition to physical issues, how to implement the logical requirements for the system. During the selection phase, we studied alternative physical, implementation solutions. A key consideration was *make or buy*—that is, whether to make the system in-house versus buying it from someone who had already made it. That decision determines the sequence of the acquisition and physical design phases.

Still, in reality, many projects will require both the acquisition and physical design phases. For example, if we decided to design and build the system in-house, we may still need to buy computers, peripheral devices such as terminals, or software tools to support the system to be designed. Likewise, if

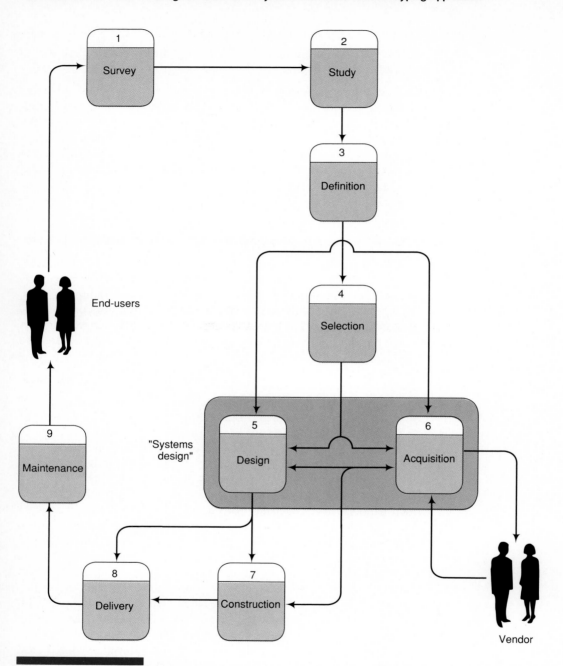

FIGURE 12.1 The Systems Design Process The systems design process is defined as the acquisition and physical design phases of your systems development life cycle.

we decided to purchase a software package, we may need to design additions and/or modifications to the chosen package. Furthermore, most packages must be integrated into a total environment that includes other software packages, possibly from other computer vendors and/or built in-house. Those interfaces between systems must be designed!

We have deliberately chosen to present acquisition before physical design. Why? In most cases, we find that computer hardware and software constrain the physical design of a computer-based system. For example, a database management system (software) will constrain the design of a database. Similarly, a terminal's display screen and attributes will limit the design of screens and dialogues. Furthermore, the acquisition of software packages presents new design issues concerning integration, as we have just mentioned. Still, acquisition and physical design are often done concurrently. Since the acquisition (and subsequent delivery) tasks can be prolonged, much of the physical design can be completed prior to delivery of the software and hardware.

How to Acquire Software and Computer Equipment for a New Information System

The acquisition of software and hardware (computer equipment) is not necessary for *all* new systems. On the other hand, when new software or hardware is needed, the selection of appropriate products is often difficult. Decisions are complicated by technical, economic, and political considerations. A poor decision can ruin an otherwise successful analysis and design. The systems analyst is becoming increasingly involved in the acquisition of software packages, peripherals, and computers to support specific applications being developed by that analyst.

Purpose and Objectives of the Acquisition Phase

The purpose of the acquisition phase is to choose and obtain the *specific* software and hardware products to support an approved information system. The following objectives are addressed.

End-User and Technical Staff Must Be Involved in the Acquisition Phase   In our not-too-distant past, end-users seldom wanted to be involved in the acquisition phase. But times have changed. End-users are more computer literate. Microcomputers and intelligent terminals have found their way into most business offices. Consequently, end-users and management have become more interested in the choice of computer technology.

Our concern for the end-users goes beyond their increased interest in the technology. Issues such as ergonomics and human engineering have become important factors that only the end-users can address. **Ergonomics** refers to the physical orientation of the person-machine environment. For instance, the tilt of a terminal screen, the color and glare of a display, and the slope and layout of a keyboard are all ergonomic issues. **Human engineering** refers to the friendliness and ease of use of a software package, manual, computer dialogue, or report.

Additionally, most technology decisions involve sizable capital outlays that must be approved by end-users. Management typically defers the research and recommendations to information systems professionals, but the final decision is economic and is, therefore, made by management. This underscores an important characteristic of the acquisition phase—that analysts rarely have "carte blanche" authority in software and hardware selection. Possible participants are listed in the margin.

Address Known Business Issues and Directions in the Acquisition Phase   The business dimension should drive the technology selection decision, not vice versa. It is the analyst's responsibility to ensure that the organization's purpose, goals, objectives, and policies are best served by the technology that is selected. Implicit in the term *mission* is the implication of *future*. Especially when evaluating software packages, it is important that the packages selected not only fulfill current requirements, but be adaptable to tomorrow's needs!

To achieve this objective, most computer professionals must inoculate themselves against "bells and whistles" disease. We must confess our infatuation with the marvelous technology that becomes available with each passing day (the same might be said of today's more computer-literate end-users). We must recognize as frivolous any technology that does not cost-effectively address the business. To this end, we need to be wary of placing too much emphasis on technical specifications and features.

Match Technology to Functional Requirements for the Information System   Some consultants believe that most software should be purchased, not built. They argue that most applications can be supported through packaged software written by software companies. Still, functional requirements should drive decisions to purchase software and hardware. If you don't take time to understand and document requirements, you frequently pick inadequate software and hardware. Consequently, the acquisition phase may actually be more dependent, not less, on the requirements specification techniques you learned in Chapters 6 through 10. The model for functional requirements is reproduced in the margin.

Evaluate and Select Appropriate Systems Components   The acquisition phase is primarily concerned with the software and hardware components of your IPO model. The analyst must evaluate how well alternative products complement the other existing hardware and software as well as the other compo-

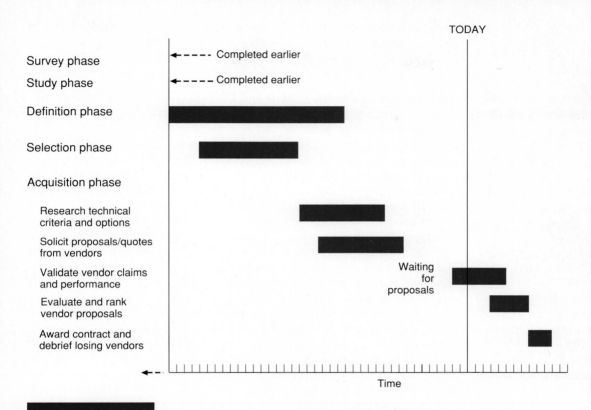

FIGURE 12.2 Acquisition Phase Tasks The acquisition phase can be broken down into the tasks suggested in this diagram.

nents of the model. For instance, a specific software package may require a specific operating system, memory requirements, and so on. A software package may also dictate new methods and procedures for exploiting its full potential. For any product, you must ask if that hardware and software configuration is suitable to the other components in the system.

How to Complete the Acquisition Phase

Now that you know *what* we want to accomplish in the acquisition phase, let's discuss *how* you do that phase. Although there is no way to guarantee that the best hardware or software will be selected, there are ways to minimize the probability of making a bad decision. Figure 12.2 illustrates the tasks. The horizontal bars depict tasks (the acquisition tasks are highlighted in color). The analysis phases were added to put the acquisition phase into the perspective of the complete life cycle. A vertical line indicates one specific point in time, usually a date. Notice that the tasks are not sequential; they overlap. On any given date, you will likely be working on more than one task. Let's study each acquisition phase task in greater detail.

✓ TECHNICAL CRITERIA

Quality of
  Documentation
Ease of Learning
Ease of Use
Response Time
Throughout
Number of Installed
  Copies
Number of Improved
  Versions Over Time
  (Maturity)
Licensing Arrangements
Training
Maximum File/Database
  Size
Internal Controls

Task: Research Technical Criteria and Options   The first task is to research technical alternatives. This task identifies specifications that are important to the hardware and/or software that is to be selected. The task responds to inputs from the analysis phases: *end-user requirements* (from the definition phase) and the *recommended* (and hopefully approved) *hardware/software requirements* (from the selection phase). These requirements specify the functionality, features, and critical performance parameters. Examples of criteria for hardware and software are listed in the margin.

To complete this task you must do your homework. Do not get your education from a salesperson! Remember, a computer salesperson's goal is the sale. We aren't suggesting that vendor sales representatives are dishonest, but the number one rule of salesmanship is to emphasize your product's strengths and deemphasize its weaknesses. Legally, most courts of law adopt the attitude *caveat emptor*, Latin for "Let the buyer beware."

A little reading in appropriate magazines and journals helps you identify those technical and business issues and specifications that will become important to the selection decision. Examples are listed in the margin (left). Compile a list of the requirements and features for the technology to be selected. Then, determine whether each requirement or feature is absolutely essential, desirable (could be acquired through a third-party vendor or built in-house), or just nice to have (you could live without it).

Useful sources of information for your research include the following:

- *Internal standards* may exist for hardware and software selection. Some companies insist that certain technology will be bought from specific vendors if those vendors offer it. For instance, some companies have standardized on specific brands of microcomputers, terminals, printers, database management systems, network managers, data communications software, spreadsheets, and programming languages. A little homework here can save you a lot of unnecessary research.

- *Information services* are primarily intended to constantly survey the marketplace for new products and advise prospective buyers on what specifications to consider. They also provide information such as the number of installations and general customer satisfaction with the products. Some information services are listed in the margin (left).

✓ INFORMATION SERVICES

Data Pro
EDP Auerbach
*International Computer
  Programs*
The Source

- *Trade newspapers and periodicals* offer articles and experiences on various types of hardware and software that you may be considering. Some examples of these are listed in the margin (top right). Many can be found in school and company libraries. Subscriptions (sometimes free) are also available.

✓ **TRADE**
   PUBLICATIONS

Computerworld
   (weekly)

InfoWorld (weekly)

InformationWeek
   (weekly)

Datamation (monthly)

Computer Decisions
   (monthly)

Infosystems (monthly)

Mini-Micro Systems
   (monthly)

Communications Week
   (weekly)

PC (biweekly)

PC Week (weekly)

PC World (monthly)

Byte (monthly)

Your research should also identify potential vendors that supply the products you will consider. After you've done your homework, initiate contact with these vendors. You will be better equipped to deal with vendor sales pitches after doing your research!

Task: Solicit Proposals (or Quotes) from Vendors   Given the hardware/software specifications, your next task is to solicit proposals from vendors. If your company is committed to buying from a single source (IBM, for example), the task is quite informal. You simply contact the supplier and request price quotations and terms. On the other hand, most decisions offer numerous alternatives. In this situation, good business sense dictates that you use the competitive marketplace to your advantage.

The solicitation task prepares one of two documents: a request for quotations (RFQ) or a request for proposals (RFP). The **request for quotations** is used when you have already decided on the specific product, but that product can be acquired from several distributors. Its primary intent is to solicit specific configurations, prices, maintenance agreements, conditions regarding changes made by buyers, and servicing. The **request for proposals** is used when several different vendors and/or products are candidates and you want to solicit competitive proposals and quotes. RFPs can be thought of as a superset of RFQs. We'll address the RFP for the remainder of this task description.

The quality of an RFP has a significant impact on the quality and completeness of the resulting proposals. A suggested outline for an RFP is presented in Figure 12.3, since an actual RFP is too lengthy to include in this book. Obviously, your ability to write clearly will affect the quality of proposals you get in response to your RFP. Furthermore, you can expect that any RFP will raise additional questions that you will address in meetings and other communications with prospective vendors. Therefore, verbal communications skills will also be tested in this task.

The primary purpose of the RFP is to communicate your requirements and desired features to prospective vendors. Requirements and desired features must be categorized as *mandatory* (must be provided by the vendor), *extremely important* (desired from the vendor but can be obtained in-house or from a third-party vendor), or *desirable* (can be done without). Requirements might also be classified by two alternate criteria: those that satisfy the needs *of the systems* and those that satisfy our needs *from the vendor* (for example, service).

Entire books could be written on the flow and processing of RFPs and RFQs (see references at the end of the chapter). Mechanisms must be implemented to answer vendor questions and control the format of the vendor's subsequent proposals. Often, vendors are invited to a *bidder's meeting* where common questions and issues can be addressed. Ultimately, interested vendors will submit proposals. The remaining tasks address the analysis of proposals.

Many of the skills you developed in Part Two, such as process and data modeling, can be very useful for communicating requirements in the RFP. We

FIGURE 12.3   Request for Proposals (RFP)
This is an outline for a typical request for proposals. The outline for a request for quotations would be similar; however, the RFQ's requirements are more technical and don't allow the vendor as much flexibility to tailor alternatives to the customer's business.

Request for proposals

I. Introduction
   A. Background
   B. Brief summary of needs
   C. Explanation of RFP document
   D. Call for action on part of vendor

II. Standards and instructions
   A. Schedule of events leading to contract
   B. Ground rules that will govern selection decision
      1. Who may talk with whom and when
      2. Who pays for what
      3. Required format for a proposal
      4. Demonstration expectations
      5. Contractual expectations
      6. References expected
      7. Documentation expectations

III. Requirements and features
   A. Hardware
      1. Mandatory requirements, features, and criteria
      2. Essential requirements, features, and criteria
      3. Desirable requirements, features, and criteria
   B. Software
      1. Mandatory requirements, features, and criteria
      2. Essential requirements, features, and criteria
      3. Desirable requirements, features, and criteria
   C. Service
      1. Mandatory requirements
      2. Essential requirements
      3. Desirable requirements

IV. Technical questionnaires

V. Conclusion

have found that vendors are very receptive to these tools because they find it easier to match products and options and package a proposal that is directed toward your needs. Everybody benefits from a clear and complete statement of requirements.

Task: Validate Vendor Claims and Performance   Soon after the RFPs are sent to prospective vendors, you will begin receiving proposals. Because proposals cannot and should not be taken at face value, claims and performance must be validated. This task is performed *independently* for each proposal; proposals are not compared with one another.

Eliminate any proposal that does not meet *all* of your *mandatory* requirements. If you clearly specified your requirements, no vendor should have submitted such a proposal. For proposals that cannot meet one or more *extremely important* requirements, verify that the requirements or features can be fulfilled by some other means. Finally, validate vendor claims and promises.

Claims about mandatory, extremely important, and desirable requirements and features can be validated by completed questionnaires and checklists (included in the RFP) with appropriate vendor-supplied references to user and technical manuals. Promises can only be validated by ensuring that they are written into the contract. Performance is best validated by a demonstration, which is particularly important when you are evaluating software packages. Demonstrations allow you to confirm capabilities, features, and ease of use.

Task: Evaluate and Rank Vendor Proposals   The validated proposals can now be evaluated and ranked. The evaluation and ranking task is, in reality, another cost-benefit analysis performed during systems development. It is highly recommended that the evaluation criteria and scoring system be established *before* the actual evaluation takes place. Why? Because we don't want to bias the criteria and scoring to subconsciously favor any one proposal.

Some methods suggest that requirements be weighted on a point scale. Better approaches use dollars and cents! Monetary systems are easier to defend to management than points. One such technique is to evaluate the proposals on the basis of hard and soft dollars. Hard-dollar costs are the costs you will have to pay to the selected vendor for the equipment or software. Soft-dollar costs are additional costs you will incur if you select a particular vendor (for instance, if you select vendor A, you may incur an additional expense to vendor B in order to overcome a shortcoming of vendor A's proposed system). This approach awards the contract to the vendor who fulfills all essential requirements while offering the lowest total hard-dollar plus soft-dollar penalties for desired features not provided (for a detailed explanation of this method see Isshiki, 1982, or Joslin, 1977).

Task: Award (or Let) Contract and Debrief Losing Vendors   Having ranked the proposals, you usually present your recommendations to management for final approval. Once again, communications skills, especially salesmanship, will be important to your ability to persuade management to follow the recommendations.

Once the final decision is made, a contract must be negotiated with the winning vendor. Certain special conditions and terms may have to be written into the standard contract. Ideally, no computer contract should be signed without the advice of a lawyer. For microcomputers and software, legal advice can be prohibitively expensive (compared to the cost of the products themselves). In this case, the analyst must be careful to read and clarify all licensing agreements. No final decision should be approved without the consent of a qualified accountant or management. Purchasing, leasing, and leasing with a purchase option involve complex tax considerations.

How to Physically Design a Computer-Based Information System

Now we come to a more traditional phase of the systems development life cycle, the design of computer specifications for a new information system. Let's set the stage. From the definition phase we have a requirements statement. From the selection phase, we have a feasible target solution that specifies which aspects of the system to computerize and how they should be automated—for instance, batch, on-line, centralized, distributed. Given this target solution, we must provide the detailed computer specifications that the computer programmers and technicians will need to implement that solution (or modify, enhance, and integrate a selected software package).

Purpose and Objectives of the Physical Design Phase

The purpose or goal of the physical design phase is twofold. First and foremost, the analyst seeks to design a system that both fulfills requirements and will be friendly to its end-users. Human engineering will play a pivotal role during design. Second and still very important, the analyst seeks to present clear and complete specifications to the computer programmers and technicians. To achieve these two goals, we must accomplish the following objectives.

End-Users Should Be Involved in the Physical Design Phase Few people will question the importance of end-user participation during the study, definition, and selection phases. But during the physical design phase? Many analysts go off and design the new system without significant end-user participation. This frequently results in end-user dissatisfaction with the way the system works (even if it does fulfill essential requirements), or in end-users who feel that the system has been forced on them! In such cases, the analyst has failed to appreciate a basic reality—the end-user must live with the system long after the analyst is gone.

End-users should be intimately involved in the physical design phase. As requirements are transformed into computer inputs, outputs, files/databases, and program specifications, the end-user should at least review key components. Ideally, the end-users themselves should assist in the design tasks. Later in this chapter, you'll learn about an important new strategy for involving end-users, *prototyping.*

Ensure Physical Design Consistency with the Business   The analyst's principal concern during the design phase is to ensure that the subsequent design remains consistent with business directions. These directions were identified during the study phase of the life cycle.

Fulfill Current and Projected Functional Requirements   The analyst should have defined functional requirements long before systems design. The analyst should now ensure that the design fulfills those requirements. This phase determines *how* those requirements are fulfilled. Additional requirements may be introduced during systems design because a computer is used. For instance, edit reports for input data and audit trails for transactions are typically added to the basic user-defined requirements to ensure the proper operation of a *computer-based* system. Additionally, performance and security requirements become important when you are designing computer systems.

There is a tendency to adopt a physical design attitude of "get the system running." This should be avoided. Physical designs should constantly be evaluated for flexibility and adaptability. Future requirements should not necessitate the redesign of well-conceived systems.

Design All Information Systems Components   For the physical design phase, the information components dimension is the most visible dimension. We design each of the following components:

- *Data and information.* We specified the content of each data and information flow during the definition phase. We specified the media during the selection phase. Now we need to physically design the style, organization, and format of all inputs and outputs.
- *Data stores.* Once again, content and media have already been defined. During physical design, we must specify format, organization, and access methods for all files and databases to be used in the computer-based system.
- *End-users.* The roles people must play in the new system must be specified. For instance, who will capture and input data? Who will receive outputs?
- *Methods and procedures.* During physical design, the sequence of steps and flow of control through the new system must be specified. The processing methods and intermediate manual procedures must also be clearly documented.
- *Computer equipment.* Although hardware is not selected or designed during the physical design phase, the hardware does constrain the system.

The specific hardware configuration specified during the selection phase must be considered as various other components are designed.

- *Computer programs.* Complete programming specifications must be prepared for every program that must be written or modified.
- *Internal controls.* During physical design, we must specify internal controls to ensure the security and reliability of the system.

Clearly, the physical design phase gets into considerably greater detail than any of the previous phases of the life cycle.

How to Complete the Physical Design Phase

Now we can discuss the specific tasks and documentation for the physical design phase. Figure 12.4 depicts the physical design phase tasks and documentation. Notice how these tasks are completed in parallel with one another (the bars overlap more than for tasks of other phases we covered). Each task represents a unique set of skills that you will learn and master in the forthcoming chapters and modules. One set of skills is common to the entire physical design phase—communications skills! As an analyst, you will be conducting meetings and walkthroughs during or on completion of each task. Some of these meetings and walkthroughs are with end-users. Others are with technical specialists. You need to learn how to communicate with each audience on its own level. No other phase of systems development places the analysts in contact with such diverse audiences.

One issue that recurs during the physical design phase tasks is the importance of internal controls. The complexity of modern computer-based systems, many of which use sophisticated database, data communications, and on-line technology, makes businesses more vulnerable to mistakes, security violations, and disasters. For this reason, internal controls are designed into the system. Some internal controls prevent these problems; others aid in the recovery from them. In this section, we'll only mention the tasks during which the controls are designed, but in subsequent chapters, you'll learn how to design specific controls. We adopted this strategy because design of internal controls should never be regarded as an afterthought of the physical design phase—it is integral to each task.

As we go through these tasks, keep in mind that we are only discussing *what* the analyst does. In the next several chapters, we'll discuss how to do these tasks and demonstrate specific tools and techniques with which to accomplish them.

Task: Design Computer Files and/or Databases Given the data store requirements from the definition phase, we must design the corresponding computer files and/or databases. The physical design of data stores goes far

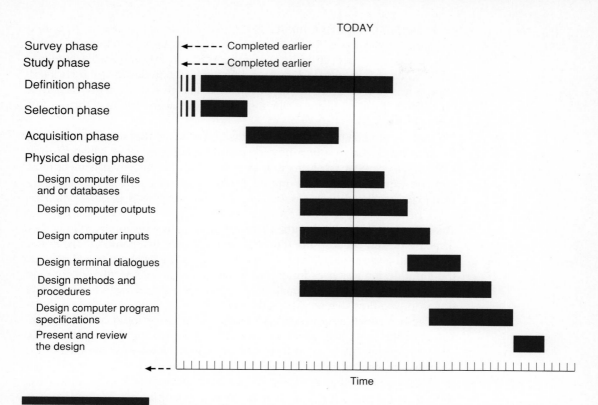

FIGURE 12.4 Physical Design Phase Tasks The physical design phase can be broken down into the tasks illustrated in this diagram.

beyond the simple layout of records. Files/databases are a shared resource. Many programs will typically use them. Future programs may use files and databases in ways not originally envisioned. Consequently, we must be especially attentive to designing files and databases that are adaptable to future requirements and expansion.

We must also analyze how programs will access the data in order to improve performance. You may already be somewhat familiar with various programming data structures (for example sequential, ISAM, VSAM, relative, and linked lists) and their impact on performance and flexibility. These issues affect file/database organization decisions. Other issues to be addressed during both file and database design include record size and storage volume requirements. Finally, because files and databases are shared resources, we must also design internal controls to ensure proper security and disaster recovery techniques, in case data is lost or destroyed.

Task: Design Computer Outputs During the definition phase, we specified the required content and timing for the output data flows (Chapter 9). In this task, we must format each of those outputs. Transaction outputs will frequently be designed as preprinted forms onto which transaction details will be printed. Reports and other outputs are usually printed directly onto paper or displayed on a terminal screen. In any event, the precise format and layout of the outputs must be specified.

Because end-users and managers will have to work with these outputs, we must be careful to solicit their ideas and suggestions, especially regarding format. Management involvement is also necessary because management must approve expenditures on all outputs. Finally, internal controls must be specified to ensure that the outputs are not lost, misrouted, misused, or incomplete.

Task: Design Computer Inputs The design of computer inputs follows file/database design because the files and databases are maintained through the timely and accurate input of data. In addition to file/database specifications, the input-design task uses input requirements that were specified during the definition phase of systems analysis (Chapter 8 identified such inputs as data flows on data flow diagrams).

It is crucial to design the data capture method for inputs. For instance, you may design a form on which data to be input will be initially recorded. "Easy," you say? Think about it! You want to make it easy for the data to be recorded on the form, but you also want to simplify the entry of the data from the form into the computer or onto a computer-readable medium. This is particularly true if the data is to be input by people who are not familiar with the business application (keypunch operators, for example). The layout of the input record as it will be presented to the computer must also be designed.

And that brings us to a second complication. Any time you input data to the system, you can make mistakes. We need to define editing controls to ensure the accuracy of input data. Normally, we will be forced to define additional outputs (called *edit reports* or *edit screens*) to identify input errors. As an additional control to prevent lost or erroneous inputs, you usually design historical reports that commit input transaction processing to paper, where they can be audited and confirmed.

Task: Design the On-Line User Interface This is a task that is omitted from many physical designs. But for on-line systems, the design of the user interface—the dialogue between the end-user and the computer—may well be the most critical design task! Too many on-line systems are difficult to learn and use because they exhibit poor human engineering.

The idea behind user interface design is to build an easy-to-learn-and-use dialogue around the on-line input and output screens that were designed in earlier tasks. This dialogue must take into consideration such factors as terminal familiarity, possible errors and misunderstandings that the end-user may have or encounter, the need for additional instructions or help at certain points in time, and screen content and layout. Essentially, you are trying to anticipate every little error or keystroke that an end-user might make—no matter how improbable. Furthermore, we are trying to make it easy for the end-user to understand what the screen is displaying at any given time. Appropriate attention to this task can save many a late-night wake-up call: "We just crashed your system. We need you to come in and fix it right away!"

Task: Design Methods and Procedures   The general procedures to be used in the new system were approved during the selection phase. For example, we should already know whether the system is on-line or batch, centralized or distributed, and so forth. In this task, we specify exactly *how* the new system will work, answering these questions: Who does what? When? Where? What comes next?

The internal controls from the previous tasks must be integrated into the new systems work flow. This is particularly important for batch systems, because data entry, data editing, and processing tasks must be executed in a specific sequence. The timing of scheduled reports and transaction processing must be clearly specified. Internal controls to ensure the system cannot be deliberately or accidentally abused must be installed. For on-line systems, how will access be controlled? How will transactions be monitored and controlled? For all systems, when will files and databases be backed up? These are all issues to be addressed during the physical design of methods and procedures.

Task: Design Computer Program Specifications   This task packages all of the specifications from the previous tasks into computer program specifications that will guide the computer programmer's activities during the construction phase of the systems development life cycle. But there is more to this task than packaging.

How much more depends on where you draw the line between the systems analyst's and computer programmer's responsibilities (this issue is moot if the analyst and programmer are one and the same person). In addition to packaging, you need to determine the overall program structure. There are numerous strategies for top-down, modular decomposition. They will be surveyed in Chapter 18.

Task: Present and Review the Design   This is a much more elaborate task than the name indicates. Before the design can be presented, you need to prepare two more components:

- *An implementation plan* that presents a proposed schedule for the construction and delivery phases (detailed in Chapter 19).
- *A final cost-benefit analysis* that determines if the design is still feasible.

Given the design specifications and implementation plan, you should be able to make much more refined estimates for the remaining costs!

The final systems design specifications are typically organized into a workbook or technical report. As demonstrated in Figure 12.5, physical design specifications evolve from the logical requirements specifications that were prepared during the systems analysis phases. Thus, the project dictionary that was started during systems analysis will eventually become the design specifications document.

The systems design should be reviewed with all appropriate audiences, which may include the following:

- *End-users.* End-users have already seen and approved the outputs, inputs, and terminal dialogue. The overall work and data flow for the new system should get a final walkthrough and approval.
- *Management.* Management should get a final chance to question the project's feasibility, given the latest cost-benefit estimates.
- *Technical support staff.* Computer center operations management and staff should get a final chance to review the technical specifications to be sure that nothing has been forgotten and so that they can commit computer time to the construction and delivery phases of the project.
- *Audit staff.* Many firms have full-time audit staffs whose job it is to pass judgment on the internal controls in a new system.

As you probably guessed, the results of any of these reviews may necessitate a return to previous tasks in the physical design phase.

Design by Prototyping

Traditionally, physical design has been a paper-and-pencil process. Analysts drew pictures that depicted the layout or structure of outputs, inputs, and files and the flow of dialogue and procedures. This is a time-consuming process that is prone to considerable error and omissions. Frequently, the resulting paper specifications did not prove themselves inadequate, incomplete, or inaccurate until programming started.

Today, analysts are turning in increasing numbers to a modern, engineering-based approach called *prototyping*. A **prototype**, according to the dictionary, is "an original or model on which something is patterned" and/or "a first full-scale and usually functional form of a new type or design of a construction (as an airplane)." Engineers build prototypes of engines, machines, automobiles, and the like, prior to building the actual products. Prototypes allow them to isolate problems in both requirements and designs.

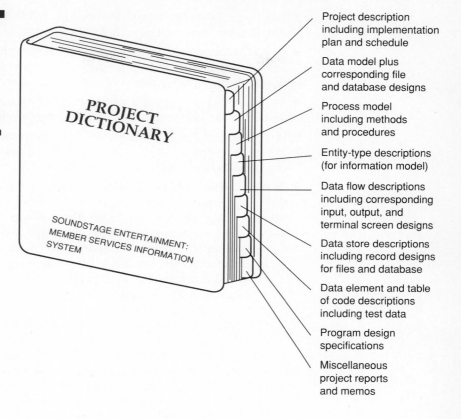

FIGURE 12.5 Format of the Physical Design Specifications This is one possible format for organizing the physical design specifications into a single document. Notice that the physical specifications expand on those logical specifications that were created in the project dictionary during the analysis phases.

Project description including implementation plan and schedule

Data model plus corresponding file and database designs

Process model including methods and procedures

Entity-type descriptions (for information model)

Data flow descriptions including corresponding input, output, and terminal screen designs

Data store descriptions including record designs for files and database

Data element and table of code descriptions including test data

Program design specifications

Miscellaneous project reports and memos

Systems analysts are using powerful prototyping tools and languages to implement this concept. In this section, we examine how prototyping is being used to improve the physical design phase of our life cycle.

The Prototyping Approach: Advantages and Disadvantages

Before we get too engrossed in the how of prototyping, let's discuss the advantages and disadvantages of prototyping. The advantages are as follows:

- Prototyping encourages and requires active end-user participation. This increases end-user morale and support for the project. End-user morale is enhanced because the system appears real to them.

- Iteration and change are a natural consequence of systems development—that is, end-users tend to change their minds. Prototyping better fits this natural situation since it assumes that a prototype evolves, through iteration, into the required system.

- It has often been said that end-users don't fully know their requirements until they see them implemented. If so, prototyping endorses this philosophy.

- Prototypes are an *active*, not *passive*, model that end-users can see, touch, feel, and experience. Indeed, if a picture such as a DFD is worth a thousand words, then a working model of a system is worth a thousand pictures.

- An approved prototype is a working equivalent to a paper design specification—with one exception. Errors can be detected much earlier.

- Prototyping can increase creativity through quicker user feedback that can lead to better solutions. (*Note*: See how the next list for ways creativity can be stifled by prototyping.)

- Prototyping accelerates several phases of the life cycle, possibly bypassing the programmer. In fact, prototyping consolidates parts of phases that normally occur one after the other. These phases include the following:

  —*Definition*. As mentioned above, prototyping can be used to quickly experiment with different requirements. Each prototype can change not only the design, but the actual requirements—until the end-users accept the requirements. In many cases, requirements can be defined more quickly with this approach.

  —*Design*. Screen and report layouts can be very quickly changed until end-users accept their design. Terminal dialogue can be user-tested for friendliness and completeness. Even if the prototype is reconstructed using a traditional language such as COBOL, the prototype serves as a model of how the system must work. In most cases, the design, as developed through prototyping, can be completed faster than one developed with paper and pencil. And because the working prototype has been seen by end-users, it is less likely to be redesigned after it has been implemented in final form.

  —*Construction*. The very act of prototyping requires construction, also known as programming. The analyst programs—if it's even possible to use that term—the prototype. Although many prototypes are eventually discarded in favor of a final system implemented in a traditional language like COBOL, many prototypes are being implemented in the prototyping language, discussed in the next section. This can significantly reduce implementation time and effort.

But prototyping is not without disadvantages, or in this case, pitfalls that should be avoided. Most of the disadvantages can be summed up in one statement: Prototyping encourages ill-advised shortcuts through the life cycle. Fortunately, the pitfalls we are about to describe can all be avoided through proper discipline. The disadvantages are as follows:

- Prototyping encourages a return to the "code, implement, and repair" life cycle that used to dominate information systems. As many companies have learned, systems developed in prototyping languages can present the same maintenance problems that have plagued systems developed in languages like COBOL.

- Prototyping does not negate the need for the survey and study phases. A prototype can just as easily solve the wrong problems and opportunities as a conventionally developed system.

- You cannot completely substitute any prototype for a paper specification. No engineer would prototype an engine without *some* paper design. Yet many information systems professionals try to prototype without a specification. Prototyping should be used to complement, not replace, other methodologies. The level of detail required of the paper design may be reduced, but it is most certainly not eliminated. In the next section, we'll discuss just how much paper design is needed.

- There are numerous design issues not addressed by prototyping. These issues can inadvertently be forgotten if you are not careful.

- Prototyping often leads to premature commitment to a physical design. In other words, the selection phase gets shortchanged.

- When prototyping, the scope and complexity of the system can quickly expand beyond original plans. This can easily get out of control.

- Prototyping can reduce creativity in physical designs. The very nature of any physical implementation—for instance, a prototype of a report—can prevent analysts and end-users from looking for *better* solutions.

- Prototypes often suffer from slower performance than their third-generation language counterparts.

Once again, all of these disadvantages can be overcome through discipline. You need only remind yourself that prototyping does not replace any phase of the life cycle; it merely improves your productivity and quality in several of the phases.

The Technology and Strategy of Prototyping

Given the advantages of prototyping, and realizing that all disadvantages can be overcome, let's examine the technology that makes prototyping possible and the overall strategy for prototyping.

Prototyping Languages and Tools Building prototypes makes so much sense that you may wonder why we didn't always do it. The reason is simple: the technology wasn't available. Traditional languages like COBOL, FORTRAN, BASIC, Pascal, and C (often called *third-generation languages*) don't lend themselves to prototyping. Prototypes must be developed and modified quickly, neither of which is possible with third-generation languages. Consider the prospects of continually modifying the DATA and PROCEDURE divisions of a COBOL program as end-users try to make up their mind what they want and how it should look. Now consider *fourth-generation languages* and *applications generators*.

Fourth-generation languages (4GLs) and applications generators (AGs) are software tools that make building systems a simpler task. At the risk of oversimplifying 4GLs and AGs, they are less procedural than traditional languages. This means that the tools specify more of *what* the system is or should do, and less of *how* to do it. In other words, they are not as dependent on specification of *logic*.

The syntax of a 4GL or AG tends to be more concise and English-like. Many 4GLs and AGs substitute menus and question-answer dialogues for most of the procedural specification common to traditional languages. As an end result, 4GLs and AGs allow analysts and programmers to define and load databases, develop input records, define terminal screens, develop terminal dialogues, and write reports—all within a matter of hours or days instead of the usual weeks and months associated with languages such as BASIC, COBOL, and PL/1! Hence the increased popularity of prototyping as a technique.

Virtually all 4GLs and AGs are built around the technology of a database management system (DBMS). A DBMS helps you organize and store different, but related, collections of data (for instance, CUSTOMERS, ORDERS, and PARTS are different sets of data but related since CUSTOMERS "place" ORDERS "for" PARTS). It is becoming increasingly difficult to distinguish between DBMSs, 4GLs, and AGs. They are inclusive to most database products. In fact, it is increasingly difficult to differentiate between 4GLs and AGs, since one rarely occurs without the other today.

Prototyping technology is widely available in mainframe computer, minicomputer, and microcomputer environments. As just noted, most mainframe DBMSs include a 4GL/AG. Examples include Cullinet's IDMS database with its ADS/O applications generator and ADR's DATACOM with its IDEAL applications generator. Traditional mainframe 4GLs/AGs include Information Builder's FOCUS (arguably the most widely used 4GL) and Mathematica's RAMIS, both of which include their own proprietary DBMS. Other popular prototyping tools are listed in the margin.

Many mainframe prototyping tools have been released as PC versions. For instance, both FOCUS and RAMIS are now available for microcomputers (with compatibility between mainframe- and microcomputer-developed systems). It may actually surprise you that traditional microcomputer database packages are also suitable for prototyping. Most people, for example, think of dBASE IV and R:BASE System V as database packages. In reality, they have evolved into sophisticated 4GLs and AGs that are, in most respects, the equals of their mainframe counterparts.

It should also be noted that many computer-assisted systems engineering (CASE) products now contain limited prototyping tools for designing screens and reports. Some experts believe that CASE will ultimately become the embodiment of both 4GLs and AGs.

✓ **SAMPLE PROTOTYPING TOOLS**

ADS/ON-LINE

APPLICATION FACTORY

DATATRIEVE

FOCUS (also PC/FOCUS)

IDEAL

NATURAL

INTELLECT

MANTIS

NOMAD2

TELON

RAMIS (also PC/RAMIS)

USE.IT

SAS (also PC/SAS)

SPSS

UFO

dBASE IV

R:BASE for DOS and OS/2

KNOWLEDGEMAN 2

Prototyping Strategy   The design-by-prototyping strategy is being used by an increasing number of businesses. Unfortunately, prototyping is also misused by a large number of computer specialists. The method requires a somewhat different approach to the traditional life cycle, but not as radically different as some think. Although it is being presented here as a design strategy, its impact on the entire life cycle should be fully understood.

We've already emphasized that the study phase is still crucial, no matter what design approach is forthcoming. Problems and opportunities must be identified, analyzed, and understood so that objectives may be established for the new system.

Although the definition phase can be simplified in those cases where prototyping will be used, some general requirements should be specified prior to the generation of prototypes. Specifically, since prototyping languages utilize database technology, it makes sense to use a well-conceived data model in building databases. You learned how to build a data model in Chapter 8. You'll learn how to convert it into a database in Chapter 14.

In addition to a data model, process models can be useful for defining those processes that will be designed through prototypes. Where can you save some time in the definition phase? By not defining detailed requirements—for example, records, data elements, policies, and procedures—for the process model. Instead, prototypes will be used to zero in on the detailed requirements for those components . . . but not quite yet.

It is extremely important to complete the selection phase prior to prototyping. Why? To ensure that the target system is the most feasible solution to the end-users' problems and requirements. Perhaps a software package would be better. Or perhaps a microcomputer-based solution would be better than a mainframe solution. The selection phase (covered in Chapter 6) deals with such issues.

Given the most feasible solution, prototyping can begin. Prototypes can be quickly developed using the prototyping technology just described. Prototypes can be built for simple outputs, computer dialogues, key functions, entire subsystems, or even the entire system. Each prototype system is reviewed by end-users and management, who make recommendations about requirements, methods, and formats. The prototype is then corrected, enhanced, or refined to reflect the new requirements. Prototyping technology makes such revisions relatively straightforward. The revision and review process continues until the prototype is accepted. At that point, the end-users are accepting both the requirements and the design that fulfills those requirements.

What happens after design by prototyping? For one thing, design by prototyping doesn't necessarily fulfill all design requirements. For instance, prototypes don't always address important performance issues and storage constraints. Prototypes rarely incorporate internal controls. These must still be specified by the analyst. So what happens once the design is completed?

Design is usually followed by the construction and delivery phases of the life cycle. But haven't we, simply by virtue of the fact that we used prototyping, constructed the system? Not necessarily! A decision must be made on whether to reprogram the system into a more traditional language like COBOL. Why? To improve performance or perhaps to standardize for purposes of maintenance. Many prototypes perform satisfactorily with their small test database; however, performance declines as the database size grows to more realistic and expected levels. Also, a prototype may be developed on a microcomputer for eventual implementation on a mainframe computer. Thus, we see several reasons to possibly reconstruct the system.

Next, let's consider the possibility of implementing the system using the prototype—in other words, letting the prototype evolve into the final system. Even in this situation, the acceptance of the prototype does not signal the completion of the construction phase. Prototype programs must usually be modified to include internal controls. For example, in the interest of speed, prototypes may not include many or any edits for inputs; the end-users can input garbage data. These edits must be added to programs. Also, prototypes rarely create audit trails to track updates to the database or files. This could lead to inability to recover data that is accidentally lost. Such audit trails would have to be added. Thus, we see that the construction phase must still follow design by prototyping.

The delivery phase is clearly not affected by the choice of the traditional or prototyping design strategy. End-users must still be trained. The old system must still be converted to the new system. The final system should still be audited after it is placed into operation.

Rapid Versus Systems Prototyping We've described both the prototyping technology and general strategy. Next, we want to compare and contrast two distinctly different types or styles of prototyping.

Rapid prototyping, our term, is the simplest type of prototyping, albeit a very powerful type. Rapid prototyping allows you to create and test input designs, output designs, terminal dialogues, and simple procedures. You are not building a prototype system. Instead, you are building prototypes of selected components of a system.

The technology of rapid prototyping is unique in that it is not built around a complete applications generator or 4GL. And instead of being built around a DBMS, it is built around a computerized data or project dictionary. This dictionary may be part of a CASE (Computer-Assisted Systems Engineering) product such as Index Technology's Excelerator, or it may be self-contained as part of a dedicated rapid prototyping product such as Pansophic's Telon.

Rapid prototyping proceeds as follows. The prototyping tool is used to create screens or reports. Consider a screen prototype. The analyst can place headings, comments, instructions, and the like anywhere on the screen. The analyst can also define *fields* (or *variables*) to appear on the screen. As fields

THE NEXT GENERATION

Reverse Engineering: Can We Undo the Design Mistakes of the Past?

Systems design is indeed an important process. Today's analysts are generally more careful when designing systems. Structured methods and prototyping have significantly improved the process. On the other hand, systems design does not currently address a major class of problems from our past: billions of lines of program code based on poor designs, poor programming techniques, and a general lack of documentation.

The future looks more promising for these older applications systems. A new technology—reverse engineering—is slowly finding its way out of research laboratories and into commercial use.

Reverse Engineering can be thought of as an extension of the automation theme in Computer-Assisted Systems Engineering (CASE). You might even think of this as reverse CASE. Here's how it works.

Existing program code is reverse compiled into some resemblance to a design specification. The ability to fully achieve this goal is based to some degree on the quality of the code (admittedly poor in many situations), amount of internal program documentation, existing modularization that can be extracted from the code, and a number of artificial intelligent inferences that can be drawn from the code.

Think of the possibilities for the future! Consider, for example, an old, poorly designed system. Problems could manifest themselves in the form of frequent end-user complaints, excessive maintenance, excessive bugs, and so on. We would "decompile" the system's programs into some type of graphic model (like data flow diagrams, although the best model is a question that will have to be addressed by researchers).

The analyst can then get at the model via a CASE tool, revise the model and specifications using better methods and the CASE tool, and possibly recompile the model into a superior collection of programs. But it could become a reality in the next generation of systems work. You should make it a point to read articles that you find about topics like *Reverse Engineering*, *Reverse Compiling*, *Program Optimization*, and *Code Generation* or *Regeneration*.

Too farfetched? Programs are already available to rewrite existing code with structured design and structured programming standards. That is but the first step . . .

are positioned, the analyst can call up previously described data element specifications from the dictionary to be applied to the field or variable. Such specifications would define attributes like *size*, *format*, and *value ranges*. The analyst can alter these attributes and even add additional design attributes such as *display attributes* (for example, reverse video and blinking), *additional editing rules, help messages,* and *error diagnostics*. The screen can be chained to other screens to define a sequence of screens.

Once completed, the screens can be demonstrated to the end-users. They can sit at the terminal or microcomputer and use the screens. The screens use the underlying dictionary to simulate normal operation. Thus, the end-user can receive and interpret help messages, react to error diagnostics, enter data, and even save that data, which can then be passed to prototype reports created via rapid prototyping. The end-users can tell the analyst what they don't like and what needs to be added, deleted, or changed. Once the design is approved, the analyst can usually generate code for several alternative technical environments.

This entire process can be completed in a relatively small number of hours! Other aspects of systems design would be handled by traditional physical design or systems prototyping methods.

Systems prototyping requires the use of a true fourth-generation language/applications generator. The physical design tasks that were described earlier in the chapter are still valid. The only difference is that they are completed via prototyping. The process goes something like this:

1. A prototype database would be designed, using whatever constraints are imposed by the 4GL/AG's underlying database. The prototype database would be loaded with a sufficient collection of test data.

2. The following tasks could occur in parallel or in any sequence:
   a. Prototype outputs can be created using the report generator of the 4GL/AG. Report generators allow new reports to be quickly defined. Report fields will be filled with test data from the prototype database.
   b. Prototype inputs can be created and generated using the screen or report generator of the 4GL/AG. Screens can be chained to form a dialogue. Normally, the input screens would be designed with minimum data editing and no security features.

3. Once the inputs and outputs are completed, they would be integrated around some sort of user-friendly shell. The most common shell consists of menus and submenus.

Once the completed prototype system has been accepted, the analyst can add data editing and security features to the system, unless the prototype will be discarded in favor of an implementation using a more traditional language like COBOL. The analyst can also experiment with the database structure to improve systems efficiency.

It is difficult to demonstrate prototyping in a book. Prototyping is a "live" computing technique. You need the prototyping technology to fully appreciate it. Still, as you progress through the design chapters, we will at least show you some sample prototyping screens that will help you contrast prototyping with traditional approaches.

Summary

Systems design is the process whereby the end-users' requirements are transformed into a software package and/or a specification for a computer-based information system. Systems design consists of two phases that can be successfully completed through a series of well-defined tasks that are common to all projects.

The purpose of the acquisition phase is to evaluate and select specific software packages and/or computer equipment that fulfills requirements. The decision to select a software package instead of writing the programs in-house was made during the selection phase of the life cycle (Chapter 6). The most important document of the phase is the request for proposals or request for quotations. These documents communicate our needs to prospective vendors, who will respond with formal proposals.

The most detailed phase of systems development is the physical design phase. The purpose of this phase is to generate detailed specifications for the computer elements of the new information system (or for modifications and enhancements to a software package). These design specifications will be passed on to the computer programmers for implementation. Obviously, the degree to which computer programmers will be able to construct the system without further assistance is dependent on the completeness and clarity of the design specifications. Although the ultimate goal of systems design is to communicate specifications to programmers for implementation, the importance of end-user participation cannot be overstressed. Systems design is the phase in which the outputs, inputs, and on-line dialogues take form. An understanding of the importance of human engineering and end-user acceptance is crucial to overall project success.

Prototyping has emerged as a preferred strategy for physical design. Prototypes are working models of a system. Analysts can quickly build prototypes using modern fourth-generation languages and applications generators. The prototyping strategy is not a substitute for the life cycle. Each phase of the life cycle is still essential to successful systems development. Prototyping does, however, consolidate portions of the definition, physical design, and construction phases of the traditional life cycle. Consequently, prototyping accelerates productivity. There are two types of prototyping, rapid and systems. Rapid prototyping builds models of distinct systems components. Systems prototyping builds models of an entire working system.

In the following chapters, you will be introduced to tools and techniques for accomplishing the systems design tasks presented in this chapter.

Problems and Exercises

1. How can a successful and thorough systems analysis be ruined by a poor systems design? Answer the question relative to
   a. The impact on the subsequent implementation (in other words, the construction and delivery phases, which you studied in Chapter 5)
   b. The lifetime of the system after it is placed into operation
   c. The impact on future projects

2. What skills are important during systems design? Create an itemized list of these skills. Identify other computer, business, and general education courses that would help you develop or improve your skills. Prepare a plan and schedule for taking the courses. (If you are not in school, prepare a plan for using available corporate training resources, reading appropriate books, enrolling in seminars or continuing education courses, and so on.) Review your plan with your counselor, adviser, or instructor.

3. How does your information systems pyramid model aid in systems design?

4. What by-products of the systems analysis phases are used in the systems design phases? Why are they important? How are they used? What would happen if they were incomplete or inaccurate?

5. What are the end products of the acquisition and physical design phases? What is the content of each end product?

6. United Films Cinemas has asked you to help them select microcomputer systems for their theaters and main office. Write a letter that proposes a disciplined approach to selecting an appropriate system. Assume that your end-user is inclined to ignore a disciplined approach and would prefer to go to the local computer store and just buy something. In other words, defend your approach.

7. Distinguish between *validation* and *evaluation* as the terms apply to the selection of computer equipment and software.

8. What would you do if a vendor refused to respond to an RFP or RFQ using the following argument?
   "This thing is not useful to you or me. It rarely tells me what you really want or need. I can do a better job by visiting your business and configuring a system to meet your needs. Also, it takes too long for me to answer all the questions in this RFP. And even if I do, you may not fully understand or appreciate the answers and their implications."

9. A programming assignment in the classroom is a subset of a systems design. Obtain a copy of a programming assignment from a current course. Evaluate the design from the perspective of the systems design phase tasks and the completeness of the design specification.

10. Obtain a copy of a computer programming assignment. Assume that the assignment is to be implemented on a microcomputer that has not been acquired. Estimate the costs necessary to complete the project (hardware, programming, and so forth). State your assumptions about salaries, supplies, and whatever else seems relevant.

Projects and Minicases

1. Make an appointment with or write to a hardware and software vendor. Tell them you would like to see and discuss a typical request for proposals. Ask the vendor how they feel about RFPs. If they don't like them, find out why. How could RFPs be improved from the vendor's point of view? Do the vendor's attitudes about RFPs help the vendor, the end-user, or both?

2. Make an appointment to discuss physical design standards of a local information systems operation. Does it have standards? Does it follow them? Why or why not? Does the company use fourth-generation languages or applications generators to prototype systems? Why or why not? If it does prototype systems, has the approach proved successful?

3. The city of Granada's art museum recently purchased an IBM PS/2 Model 50 microcomputer. They read an article about an art collection inventory system software package that they want to put on that computer. You, having experienced end-users who too hastily purchased software that didn't fulfill promises and expectations, are concerned that they are jumping the gun and should approach the software selection decision with great care. Write a letter to the museum's board of trustees that expresses your concerns and proposes a better approach.

4. Write a letter to your last (or favorite) programming instructor. Suggest a disciplined approach to developing a systems specification to guide the programming assignments for the next term. Your goal should be system (of programming) specifications that will eliminate or drastically reduce the need for students to request clarification from the systems analyst, played by the instructor. Defend your approach.

Annotated References and Suggested Readings

Boar, Benard. *Application Prototyping: A Requirements Definition Strategy for the 80s*. New York: Wiley, 1984. This is one of the first books to appear on the subject of systems prototyping. It provides a good discussion of when and how to do prototyping, as well as thorough coverage of the benefits that may be realized through this approach.

Connor, Denis. *Information System Specification and Design Road Map.* Englewood Cliffs, N.J.: Prentice-Hall, 1985. This book compares prototyping with other popular analysis and design methodologies. It makes a good case for not prototyping without a specification.

Isshiki, Koichiro R. *Small Business Computers: A Guide to Evaluation and Selection.* Englewood Cliffs, N.J.: Prentice-Hall, 1982. Although it is oriented toward small computers, this book surveys most of the better-known strategies for evaluating vendor proposals. It also surveys most of the steps of the selection process, although they are not put in the perspective of the entire systems development life cycle.

Joslin, Edward O. *Computer Selection.* Rev. ed. Fairfax Station, Va.: Technology Press, 1977. Although somewhat dated, the concepts and selection methodology originally suggested in this classic book are still applicable. The book provides keen insights into vendor, customer, and end-user relations.

Lantz, Kenneth E. *The Prototyping Methodology.* Englewood Cliffs, N.J.: Prentice-Hall. Provides excellent coverage of the prototyping methodology.

Martin, James. *Fourth Generation Languages.* 2 vols. Englewood Cliffs, N.J.: Prentice-Hall. Volume 1 covers principles that underlie mainframe fourth-generation languages and applications generators. It is highly recommended for anyone planning a career in systems. Volume 2 surveys the 4GL/AG marketplace. It can help you better appreciate the capabilities of 4GLs and AGs. You may find that your school or business owns one of these packages.

SOUNDSTAGE

File, Database, and Output Design for the New System

When we left Sandra and Bob, they had documented the detailed requirements for the new system. Now they must design the files and/or databases in the new system and the outputs to be produced by the system.

EPISODE 6 Three weeks have passed since the feasibility report was completed. The steering committee has approved the recommendation to proceed to the design phase. Sandy and Bob have been working on the design of the new system. Because the requirements statement thoroughly defined all the data flows and data stores, Sandra and Bob have been working independently—Bob on the design of computer files and databases and Sandra on the design of computer outputs. They will cross-check their final specifications in order to maintain consistency.

Bob has finished the first draft design specifications for the files and databases required by the new system.

File/Database Specifications for the New System

Bob has been designing the files and databases for the new system.

He did not directly involve the end-users because of the highly technical nature of file and database design. We join Bob as he is finalizing the design documentation.

Sandy enters Bob's office. "Hi, Bob. Is that the IDMS database schema you are working on?" Sandy was referring to the diagram on Bob's desk (see Figure E-6.1).

"Yes it is. I am just checking the final design. Are you finished with outputs?"

Sandra sighed and said, "No, I still have to conduct a walkthrough with the end-users in order to verify my report designs."

"Be sure and let me know if you identify any new data storage requirements in your walkthrough," Bob responded. "Maybe you could help me review this schema. I am having trouble deciding how to relate products to ordered and backordered products."

"But I thought we were going to use the existing VSAM file with supplier, product, and title information," interrupted Sandy.

"Oops," Bob said. "I knew something wasn't right. Thanks for helping me find it. I'll buy lunch and we can look over the rest of these specifications."

Output Specifications for the New System

Meanwhile, Sandy has been designing the outputs to be produced by the new system. Sandra, Joe Bosley, Ann Snyder, and Sally Hoover are reviewing the proposed reports in Sally's office. "You look tired, Joe. What's the matter?" asked Sandy.

"I stayed up late last night watching the baseball game. Four extra innings and then my team loses. I'm still upset."

Episode 6, continued▶

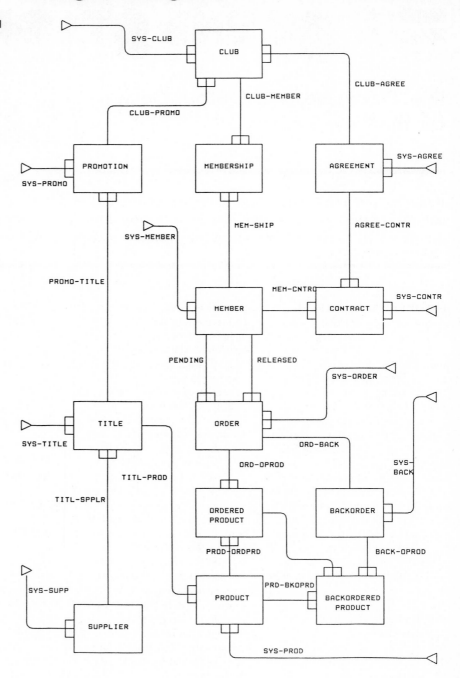

FIGURE E6.1
Typical IDMS schema

SOUNDSTAGE

"Well, I will try to keep this meeting as short as possible. I'd like to review some proposed reports with all of you."

Sally was obviously enthusiastic. "I can hardly wait to hear about the new system and exactly what it will do."

Sandy smiled. "Let's begin with the Order To Be Filled form. Based on the requirements statement we created earlier, I've drawn a sketch of what the form might look like (see Figure E-6.2).

If it's ok, I'd like to confirm a few facts before we review this form."

Since there weren't any objections, Sandra continued. "The DP department will print these forms each day, but I need to know how many could be printed in a single day."

"Currently, we never process more than 4,000 a day," Sally answered.

"But what about growth?" asked Ann. "We expect to continue adding 100 new members a

month. Maybe we had better plan on 5,000 orders a day."

Joe watched as Sandy made some notations on a form (see Figure E-6.3). "Sandra, what's the form you're writing on?"

"This is a list of output specifications I'm making for the information systems people. It helps them to anticipate the impact your outputs will have on their facilities. It will also help us to choose the best printer to meet your needs,"

Episode 6, continued▶

FIGURE E6.2   Sketch of the new ORDER TO BE FILLED

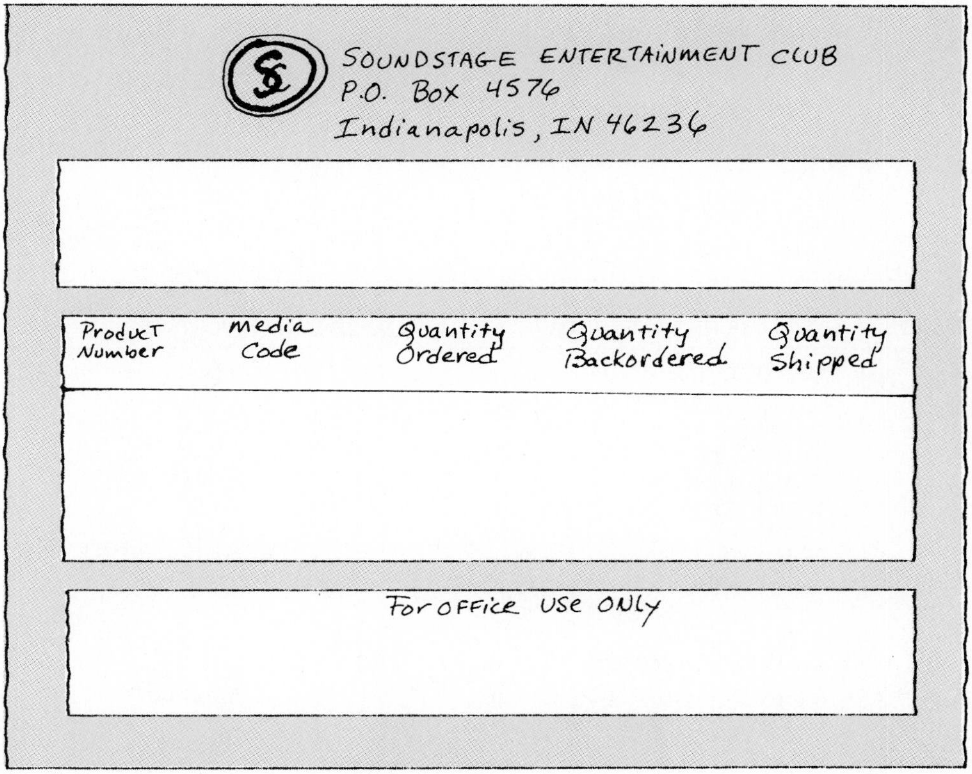

SOUNDSTAGE

OUTPUT DATA DICTIONARY FORM

NAME OF OUTPUT: *Order to be filled*
PREPARED BY *Sandy Shepard*   DATE:
DESCRIPTION: *Four part document describing a sales order*

MEDIUM: *Paper*   OUTPUT TYPE: *External/Form*
OUTPUT CHARACTERISTICS: *Preprinted, 8.5" x 8.5", designed for mailing*

FREQUENCY PREPARED: *Twice daily at 8 AM and 1 PM*
VOLUME: *5,000/day*   NUMBER OF COPIES: *3*
COPYING METHOD: *Chemical carbon*
OUTPUT RECIPIENT(S): *Order entry (all copies)*

SPECIAL INSTRUCTIONS: *Quantity shipped is hand entered. Quantity backordered may be modified by hand.*
OUTPUT COMPOSITION:

> MEMBER NUMBER
> ORDER NUMBER
> MEMBER NAME
> ORDER DATE
> MEMBER ADDRESS

> 1 to 15 OCCURENCES OF:

MEMBER ADDRESS is composed of:

>> PRODUCT NUMBER
>> MEDIA CODE
>> QUANTITY ORDERED
>> QUANTITY SHIPPED
>> QUANTITY BACKORDERED

> P.O. BOX
> STREET
> CITY
> STATE
> ZIP CODE

REF. NO.	DATA ELEMENT NAME	DATA TYPE	EDIT SIZE	EDIT MASK	SOURCE
1	MEMBER NUMBER	A/N	6	X(6)	
2	ORDER NUMBER	A/N	6	X(6)	
3	MEMBER NAME	A/N	30	X(30)	
4	ORDER DATE	A/N	8	MM/DD/YY	
5	P.O. BOX	A/N	10	X(10)	
6	STREET	A/N	15	X(15)	
7	CITY	A/N	15	X(15)	
8	STATE	A	2	AA	
9	ZIP CODE	A/N	9	X(9)	
10	PRODUCT NUMBER	A/N	7	X(7)	
11	MEDIA CODE	A/N	2	XX	
12	QUANTITY ORDERED	N	4.0	Z,ZZ9	
13	QUANTITY SHIPPED	N	4.0	hand entered	
14	QUANTITY BACKORDERED	N	4.0	Z,ZZ9	

FIGURE E6.3   Typical output data dictionary specifications

Sandra responded. "Which brings us to my next question. Do you really need four copies of each form?"

"Absolutely!" Sally answered. "But separating each of the copies from the carbon paper is a real nuisance."

Sandra smiled and said, "There are lots of different ways to produce copies. Maybe chemical carbon paper that is treated with a chemical that darkens under pressure would work best. Since no actual carbon paper is used, you only have to separate the four copies. Okay, let's look at the sketch. What do you think about this design?"

"It looks real good," Ann replied. "But why is FORM 33 in the upper right-hand corner?"

"That is the number of this new form we designed. It's used for information systems control purposes and uniquely identifies this form from all other forms."

"Do we have to fill in each product order line by hand?" Sally asked.

"No," answered Sandy. "The system will do it. The only hand-entered field on the form is the quantity shipped completed by the shipping clerks."

"The QUANTITY BACK-ORDERED and the QUANTITY SHIPPED are backwards. The form

we use now is the other way around. That could create some confusion," observed Joe.

"Oops! I accidentally reversed them. Now you see how important it is for you to verify these sketches." Sandra made corrections to a different form (see Figure E-6.4).

Ann noticed the form Sandy was writing on. "That looks awfully complicated—what is it?"

"This is a printer spacing chart," Sandra responded. "It is a technical tool I use to communicate the final design of your outputs to the computer programmers. Getting back to the form—

Episode 6, continued▶

FIGURE E6.4   Typical printer spacing chart

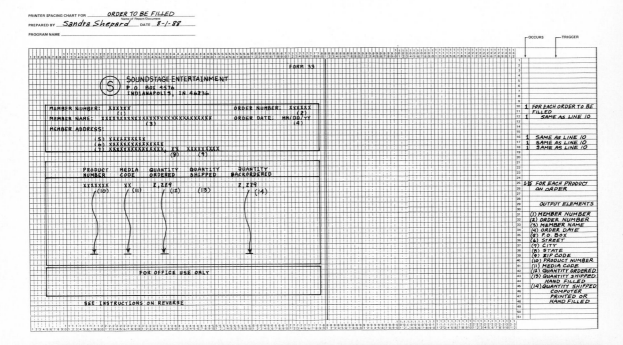

are there any other corrections? Let's look at the product order lines..."

The walkthrough continued. Eventually, Sandy completed the specifications for all the outputs. After reviewing the first set of sketches with the users, Sandy asked them to draw other outputs. This created user interest and increased the users' sense of involvement and ownership in the system. Sandy checked their designs carefully and reconciled any differences with the original requirements statement. As the sketches were finalized, Sandra prepared the more technical design specifications for the programmers.
requirements statement. As the sketches were finalized, Sandra prepared the more technical design specifications for the programmers.

Where Do We Go from Here?

This episode introduced two tasks that must be accomplished during systems design—the design of computer files/databases and the design of computer outputs. In the next few chapters, you will learn about the design issues and tools introduced in this episode. In Chapters 13 and 14, you will learn how to design computer files and databases. You will learn how to design computer outputs in Chapter 15.

This episode also reemphasized the importance of good communication skills. Sandra spent considerable time conducting walkthroughs with users to get their approval of the design of outputs. To learn more about making presentations and conducting walkthroughs, we suggest you read Part Four, Module C, "Communications Skills for the Systems Analyst."

CHAPTER THIRTEEN

Designing Conventional Computer Files and Controls

McKillip and Westin Insurance Brokerage

Ron Kurtz, a programmer/analyst for McKillip and Westin Insurance Brokerage, was sitting at his desk reviewing design specifications for the new computer-based requisition processing system. Ron had been responsible for this project from the very beginning. Most of the programs for the system had been implemented, but the final two programs had been returned by the programmer with a note explaining that they were impossible to code according to the specifications. Because Ron had been in a hurry to implement this new system, design specifications had been passed along to the programmer as soon as they were completed. Now, Ron was frustrated. His boss, John Floren, entered the office.

"Hi, Ron. What's the matter? You look like you've just lost your best friend."

Ron paused and then responded, "You're not going to be happy. I may have really messed up this time. I've worked so hard implementing that darn requisition system. But I've encountered a problem with the last two programs."

John tried not to show anger that the system might not meet its schedule. "What type of problem? What do the programs involve?"

Ron answered, "One of them produces a management report and the other an on-line inquiry response."

John responded, "That hardly seems like an overwhelming problem. Why can't you just generate the programs to produce the reports? I don't understand, we do that all the time. You know that. The data's there in the files, right?"

"Well, it's not all that simple," Ron replied. "The inquiry program requires quick response. The file I designed is organized as a sequential file. I didn't anticipate this query requirement. It wasn't in the end-user's initial request."

Hearing no response from John, Ron continued, "The other report requires data to be retrieved and printed by product number. That presents a problem because the file is organized as an indexed file with requisition number as the primary key. I didn't set this file up with any secondary indices because none of the previous programs needed to access the file by anything other than the requisition number. You see . . ."

John now seemed visibly uneasy. "You mean you have already designed and implemented the files?"

Ron knew he blew it! "Yes, we were behind schedule. I figured that the highest volume of transactions against the files would be executed by the programs that maintained the files. I organized the files to make those programs run efficiently. Then I passed the specifications for the files and those particular programs on to the programmers so we could begin writing those programs."

John responded, "Not smart, Ron. Now you find that the file organization is not best for the way the new programs need to access the file. First, I'm a bit surprised that you took such a shortcut. It's not like you. Second, if you were to have approved such a shortcut, my staff would have strongly encouraged that you implement both files as indexed sequential structures with secondary keys to minimize the impact of just such problems."

Ron was silent, so John continued. "There are no shortcuts. You're just going to have to choose between the lesser of two evils. First, we could write programs to extract data from the existing files and place it in a temporary file whose organization suits the new programs. I know, I'm not crazy about that either. It looks to me like you're going to have to temporarily halt implementation so that you can redesign the files. I need this system on time, Ron. Looks like you'll be working overtime. Oh yes, and tell the programmers that they will likely be working overtime also—once you have the files redesigned. I'm sorry, but you brought this on yourself and your team. Let's just try to pick up the pieces and not let it happen again. Don't be too discouraged. I can accept mistakes. Just learn from them, okay?"

Ron answered, "Yes sir. I'm sorry. It won't happen again. The project will come in on time, even if I have to work on some of the code myself."

Discussion

1. What should Ron do?

2. Ron made a big mistake. Where did he go wrong?

3. Can you describe how the two files should have been implemented?

What Will You Learn in This Chapter?

This is the first of two chapters on the design of physical data stores. This chapter teaches you how to design conventional computer-based files. The next chapter will look at a modern alternative to computer-based files, the *database*. You will know that you have mastered the tools and techniques of computer-based file design when you can

1. Design fixed- and variable-length logical records for files in a computer-based information system.
2. Determine the optimal storage format for fields, given the constraints imposed by common computer systems.
3. Explain why records are blocked, and determine the blocking factor for a given file.
4. Identify and differentiate between several types of files typically encountered in information systems.
5. Determine the best file organization for a given file by studying that file's required access methods and usage.
6. Explain how a data model can aid in designing conventional computer-based files.
7. Design internal controls into computer files.
8. Define file design requirements, and record those requirements in the project dictionary.

Conventional files are collections of data that are designed to support specific applications and their computer programs. And despite the trend toward modern databases (covered in the next chapter), conventional computer files remain a viable way to implement data stores.

Conventional files are the lifeblood of many information systems. Almost every computer program will require access to files. The design of computer files can be difficult because the storage and organization of data on computer media require the analyst to consider complex and often conflicting issues, such as storage capacity and performance. In this chapter, you will learn how to design computer files.

Conventional files will remain popular so long as their advantages exist. First, they are relatively easy to design and build, especially when you are using file-based languages like COBOL. Second, they still offer the advantage of faster processing than their database counterparts. For these reasons, file design remains an important skill for analysts.

Technical Concepts for File Design

Many of the technical issues important to file design are taught in different courses. We will try to focus on those technical issues that are pertinent to the systems analyst's responsibilities. Since we don't know whether those concepts are still fresh in your mind, we'll briefly review some of them.

Fields

Field is another name for a data element (introduced in Chapter 8). **Fields** are the smallest unit of data to be stored. As you define fields for storage, you will need to consider *field types* and how data will be stored in fields. There are five types of fields that can be stored: primary keys, secondary keys, descriptors, audit fields, and security fields.

Primary keys, introduced in Chapter 8, uniquely identify an entity in the business. For instance, CUSTOMER NUMBER uniquely identifies a customer, and ORDER NUMBER uniquely identifies an order. The important concept is *uniqueness.*

Secondary keys are alternate indexes into a file. A secondary key's values may or may not be unique to records. If unique, the secondary key can be thought of as an alternate primary key. If not unique, the value of the secondary key may divide entity occurrences into subsets. For example, GENDER defines two subsets of the EMPLOYEE entity, male employees and female employees. A file may have several secondary keys.

Finally, most fields, including keys, are **descriptors** of business entities such as EMPLOYEE, WORK TICKET, and MACHINE. For example, given the business entity EMPLOYEE, some descriptor fields include EMPLOYEE NAME, DATE HIRED, PAY RATE, and YEAR-TO-DATE WAGES.

Keys and descriptors were defined when you performed data modeling in systems analysis (Chapter 8). Another type of field is not added until systems design. **Audit fields** are added to records to track the usage of records. Examples include DATE CREATED, WHO CREATED THE RECORD, DATE LAST MODIFIED, WHO LAST MODIFIED THE RECORD, NUMBER OF TIMES ACCESSED YEAR-TO-DATE, RECORD STATUS (for example, *active* or *inactive*), and NUMBER OF TIMES UPDATED YEAR-TO-DATE. These fields are rarely, if ever, seen by end-users. They establish **internal controls** for monitoring who accesses and updates a given record.

Similarly, **security fields** limit access to or use of records in a file. For instance, we could add a SECURITY LEVEL field to a file. The value range of SECURITY LEVEL might limit access to different end-users as follows: *read and write access, read-only access but no write, no read or write access,* or other similar possibilities.

Because space may be a limited resource in any computer system, the analyst is responsible for economizing space when storing any field. This requires a knowledge of the way in which data is stored.

Data Storage Formats   There are three basic field storage formats: binary codes, fixed-point numbers, and floating-point numbers. A **binary code** is a unique combination of ones and zeros that represent a character, number, or symbol. There are three common binary codes: EBCDIC, ASCII, and packed decimal. During file design, your choice of storage formats for fields will determine how much disk or tape capacity will be needed for your files.

EBCDIC (Extended Binary Coded Decimal Interchange Code) is a popular code used on IBM mainframe computers and compatibles. Each letter, number, and special symbol is represented by an eight-bit code. Each code consists of a four-bit zone and a four-bit digit. Thus, a field value "Bill" would be stored as "11100010" (B), "10001001" (i), "10010011" (l), and "10010011" (l). This field requires four bytes of storage (a byte is eight bits long on most business-oriented machines).

ASCII (American Standard Code for Information Interchange) is another popular code used by many computer manufacturers. ASCII represents each letter, number, or symbol with a unique seven-bit code. An eighth bit is added, but the last bit is called a *parity bit* and is used only to check a stream of ASCII characters for correct transmission from one place to another, such as from computer to printer. Standard tables for EBCDIC and ASCII codes are widely available.

Both EBCDIC and ASCII are also commonly referred to as *character codes* since each code represents one character, number, or symbol.

Although character codes can be used for numeric fields, most computers do not use such codes in arithmetic operations. Some computers perform arithmetic only on fixed- and floating-point binary numbers. These are pure binary formats that significantly economize on storage space. On the other hand, IBM mainframe computers and others do arithmetic on packed-decimal numbers. **Packed decimal** is equivalent to EBCDIC, except that the zone bits are not used. By storing numeric fields in packed-decimal format, the field requires only half as much storage space.

Implications of Data Storage Formats for the Systems Analyst How fields are stored will determine the size of files. Therefore, it is important for you to understand the storage formats available on your computer system. Files are managed through a file management system, usually integral to the programming language—such as COBOL—used. This software may limit or even predetermine the data storage formats.

To economize on storage requirements, you will often make use of elaborate coding schemes. Codes allow you to store large amounts of nonarithmetic data in a relatively small amount of space. Because codes should be meaningful to business, coding schemes were discussed in Chapter 10, a systems analysis chapter. On the other hand, coding schemes can be easily implemented and disguised from end-users who don't like codes. The programs need only convert the codes into meaningful interpretations that appear on screens and reports.

Field storage decisions must be recorded in the project dictionary (Figure 13.1). Notice that field design decisions are recorded in the data element

FIGURE 13.1 Sample Data Element Dictionary Description As fields are defined, their associated data element descriptions in the project dictionary must be updated.

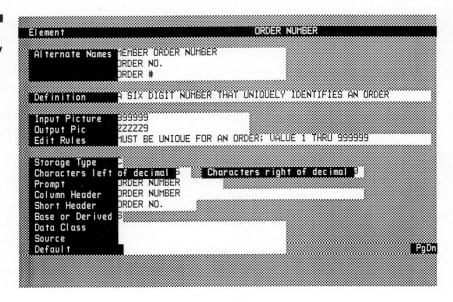

screens (*Storage Type* is equivalent to *storage format*). The CASE product used to create the data element dictionary in Figure 13.1 permits the following storage types:

C   Character, either ASCII or EBCDIC
B   Binary
P   Packed
F   Floating Point
D   Date

Most information systems exclusively use character and packed formats for storage.

Records

CUSTOMER NUMBER
CUSTOMER NAME
CUSTOMER ADDRESS
DATE INITIATED
CUSTOMER CREDIT
   LIMIT
CUSTOMER BALANCE
BALANCE PAST DUE

A **record** is a collection of fields arranged in a predefined format. It is also the smallest unit of data storage that is operated on by a computer program (most computers can read records, but not individual fields from a record). A record for CUSTOMER PAYMENT may be described by the fields listed in the margin. The primary key is underlined.

The concept of a record was introduced in Chapter 8 when you learned about entities and data models. At that time you defined logical records that describe a business entity's attributes. In this chapter, the logical record(s) is (are) transformed into a physical record for storage in computer files.

FIGURE 13.2 Fixed-Length Versus Variable-Length Record All occurrences of a fixed-length record will be the same size and contain the same fields. Variable-length records are usually composed of a fixed-size root segment followed by some number of occurrences of the other fields.

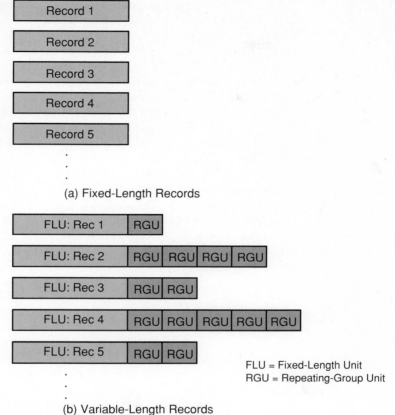

(a) Fixed-Length Records

(b) Variable-Length Records

FLU = Fixed-Length Unit
RGU = Repeating-Group Unit

Record Storage Formats During systems design, records will be classified as either fixed-length or variable-length records. The distinction will become important during design. Occurrences of **fixed-length records** will be equal in length. Normally, every record occurrence will also be defined by the same fields. The fixed-length record will contain no repeating elements or groups.

If a record contains a repeating group of elements, the record is designed as a **variable-length record**. For example, an ORDER record will contain some fields that OCCur once for any given order. Examples include ORDER NUMBER, ORDER DATE, and CUSTOMER NUMBER. On the other hand, a single order may contain multiple occurrences of the following fields: ORDERED PART NUMBER, UNIT PRICE, and QUANTITY ORDERED. Note that variable-length records normally contain a fixed-length group of fields and a repeating group of fields. Figure 13.2 demonstrates the difference between occurrences of fixed- and variable-length records. The length of any given occurrence of the record will depend on how many products are on the order.

For the analyst, the specification of fixed- or variable-length records will affect file size and performance in ways that we'll study later in this chapter. Variable-length records also have an adverse effect on flexibility of data, which is not surprising since a variable-length record is not, by definition, in first normal form.

As records are designed, decisions must be recorded in the project dictionary. Figure 13.3(a) demonstrates a typical record layout. Notice that the *OCCurs* column suggests that the fields composing the subrecord ORDERED PRODUCT repeat. Thus, this dictionary entry describes a variable-length record. If all the *Occ* entries were 1, the record would be fixed length. Each field on the record explodes to a data element description as previously illustrated (Figure 13.1). The subrecord ORDERED PRODUCT explodes to a separate record layout description [Figure 13.3(b)]. Note that the TYPE entry does not refer to *storage type*. Its values are K = key, E = element, and R = subrecord.

An alternative record layout is shown in Figure 13.4. This is a record layout chart. These charts, though not as common as they once were, may still be encountered in the documentation of older systems.

Record Blocking   Consider the following question: When you execute a read or write instruction to a file from your computer program, what are you actually reading or writing? One record? Think about that! Wouldn't it be more efficient to read **blocks** of records at a time? That is what most systems do! The operating system reads and writes blocks of records to and from a workspace in your program. When the workspace has been processed, a revised block of records can be written to the file or a new block of records can be read from the file. The system handles this feature for you.

The number of records in a block is called the **blocking factor** for the file. As a systems analyst, you may have to specify the blocking factor during file design. On some computer systems, the blocking factor is determined or optimized by the computer's operating system—thus, blocking factors may be beyond your control.

You can't make the blocking factor equal to the number of records in the file. Why? The computer's memory (also called RAM) is nearly always only a fraction of the size of most files. Also, the storage device itself—for example, disk—may impose limitations on the maximum size for a block. For instance, many systems cannot read or write more than one disk track at a time. This would limit the blocking factor to the number of records that can be placed on a single track.

Blocking factors are normally a compromise between the above factors. Block size may also be restricted to be some fixed portion of a track, such as 1/4 track or 1/3 track. The blocking factor should try to utilize as much of that portion as possible because any leftover space is wasted (not available to

FIGURE 13.3 Sample Record Dictionary for the ORDERS FILE
Fields (equivalent to data elements) must be organized into physical records for a computer file.

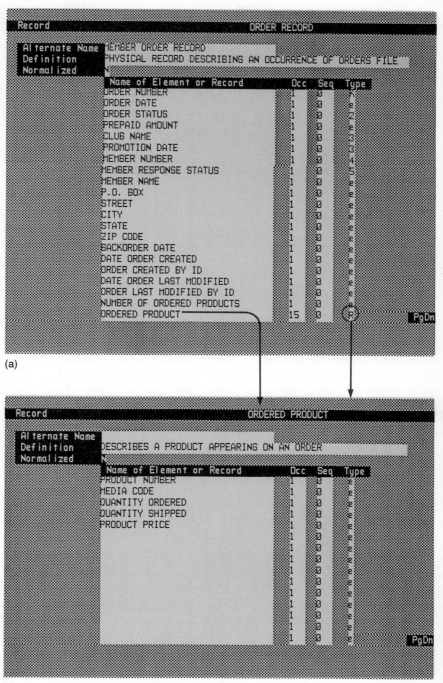

(a)

(b)

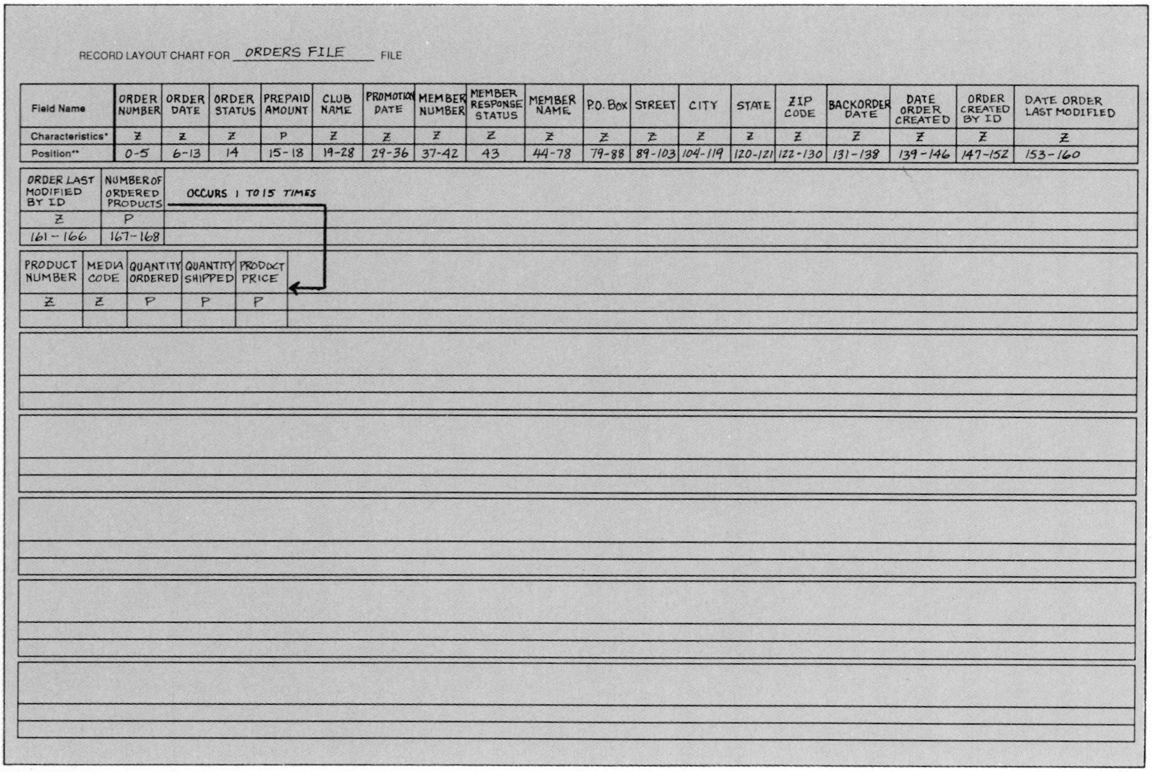

RECORD LAYOUT CHART FOR _ORDERS FILE_ FILE

Field Name	ORDER NUMBER	ORDER DATE	ORDER STATUS	PREPAID AMOUNT	CLUB NAME	PROMOTION DATE	MEMBER NUMBER	MEMBER RESPONSE STATUS	MEMBER NAME	P.O. Box	STREET	CITY	STATE	ZIP CODE	BACKORDER DATE	DATE ORDER CREATED	ORDER CREATED BY ID	DATE ORDER LAST MODIFIED
Characteristics*	Z	Z	Z	P	Z	Z	Z	Z	Z	Z	Z	Z	Z	Z	Z	Z	Z	Z
Position**	0-5	6-13	14	15-18	19-28	29-36	37-42	43	44-78	79-88	89-103	104-119	120-121	122-130	131-138	139-146	147-152	153-160

ORDER LAST MODIFIED BY ID	NUMBER OF ORDERED PRODUCTS	OCCURS 1 TO 15 TIMES
Z	P	
161-166	167-168	

PRODUCT NUMBER	MEDIA CODE	QUANTITY ORDERED	QUANTITY SHIPPED	PRODUCT PRICE
Z	Z	P	P	P

FIGURE 13.4   Record Layout Chart for the ORDERS FILE   Record layout charts are sometimes used as an alternative to Figure 13.3 and to communicate record designs to the programmer.

other records and files). You'll learn how to calculate the blocking factor for a file later in this chapter.

Files

A **file** is the set of all occurrences of a designed record. Typically, several types of files are encountered in information systems.

- **Master files** contain records that are relatively permanent. Once a record has been added to a master file, it remains in the system indefinitely. The values of fields for the record will change over its lifetime, but the record occurrence normally remains active for a long period of time. Examples of master file records include CUSTOMER, PRODUCT, and BUILDING. Most master files contain fixed-length records.

- **Transaction files** contain records that describe business events. The data describing these events normally has a limited useful lifetime. For instance, an INVOICE record is ordinarily useful until the invoice has been paid or written off as uncollectible. In information systems, transaction records are frequently retained on-line for some period of time. The inactive, off-line records are retained as an archive file (see below). Examples of transaction files include ORDER, INVOICE, and MATERIAL REQUISITION. Many, if not most, transaction files contain variable-length records.

- **Archive files** contain the off-line records of master and transaction files. Records are rarely deleted; they are archived when they are not accessed as frequently. Archives would be used when data must be recalled for subsequent audit or analysis.

- **Table files** are used to store tabular data that changes relatively infrequently. Examples include payroll system tax tables and insurance actuary tables. Tables can also be used to convert coded elements into meaningful interpretations. Table files are typically loaded directly into the programs that use them. Those programs don't usually update such tables.

- **Scratch files** (also called *work files* or *temporary files*) are special files that contain temporary duplicates or subsets or alternate sequencing of a master or transaction file. A scratch file is created, used by the appropriate computer program, and then disposed of. In other words, it is created for a single task and must be recreated each time that task is performed.

- **Log files** are special records of updates to other files, especially master and transactions. They are used in conjunction with archive files to recover "lost" data.

One of the major responsibilities of the systems analyst is to determine file access and organization. Files are a shared data resource. Because several programs will likely use and maintain the same file, the records should be organized so each program can easily access them. File organization is a function of file access. Therefore, we will begin with access.

File Access The method by which a computer program will read records from a file is called **file access**. Every computer program will access a file in one of two ways: sequentially or directly. The **sequential access method** starts reading or writing with a record in the file (normally the first) and proceeds—one record after another—until the entire file has been processed. Sequential access is used when a program needs to look at a relatively high percentage of all records in a file. For instance, a monthly invoice program may need to access virtually every record in the ACCOUNT file. Sequential access is normally required for batch transaction processing as well as for management reporting.

FIGURE 13.5
Alternative Sequential
File Organization of
Records There are two
ways to store records in
a sequential file. First,
you can store the
sequential records one
after the other, as in
Figure 13.2(a). Second,
you can implement the
sequence using a linked
list—called a logically
sequential file—of
records, as in Figure
13.2(b).

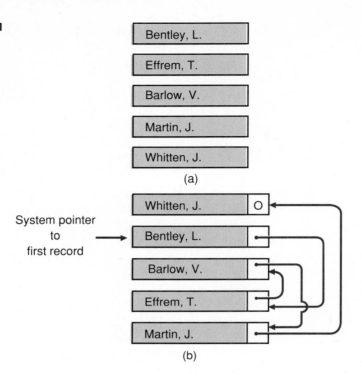

The **direct access method**—also called *random access*—permits access to any record in a file without reading all previous records in that file. Direct access is appropriate for programs that need to access only one or a few records in a file (queries, for example). This approach is normally needed for on-line transaction processing and decision support.

File Organization A **file organization** defines how records in a file are related to one another. Different file organizations are designed to optimize performance for one or both file access methods. There are several common file organizations:

- **Sequential** file records are usually arranged in a sequence determined by the value of a key field. For instance, customer records could be organized sequentially according to CUSTOMER NUMBER (a primary key) or CUSTOMER NAME (a secondary key). Sequential records can be stored adjacent to one another, as in Figure 13.5(a), or they can be arranged sequentially with a linked list; see Figure 13.5(b). The figure illustrates the possible use of multiple linked lists that support different ways to sequentially process the file. Sequential files are optimized for situations where high percentages of all the records in a file must be read by all programs that use the file.

- **Relative or direct** file record occurrences are physically located at an address that is calculated from its primary key field. The calculation of the address from the key field is called **hashing**. Thus record occurrences are scattered on the storage device instead of being arranged next to one another or linked via pointer fields. Records can be retrieved rapidly by applying the hashing formula to the primary key field of the desired record. Unfortunately, records cannot be processed sequentially. Although records are scattered, they are generally stored in preallocated areas that, when full, must be reallocated.

  Relative files are optimized for situations where rapid access to a small percentage of the records in a file is needed by all programs. Thus, relative files work well for on-line programs, whereas sequential files work better for high-volume batch programs.

- An **indexed** file organization is illustrated in Figure 13.6. Records are pointed to by indexes. The advantage of an index is that it is relatively small and easy to update. At a minimum, the file will have one index per primary key. However, Figure 13.6 also shows the use of indexes to implement secondary keys, each value pointing to multiple records. A file that has several indexes—for primary and secondary keys—is said to have an **inverted file** organization. For all indexed files, there is an inherent cost of maintaining the indexes. If not properly maintained (and sometimes even when maintained), the indexes can become corrupt and require rebuilding.

What if some programs need sequential access while others need direct access? The **indexed sequential** file organization is a special variation on the indexed file concept. Records are physically arranged in sequence to allow sequential access, in contiguous locations or using a linked list or both. However, the file also contains an index of record keys and their physical addresses that can be used to provide semi-direct access to records within the sequential file. Each program determines how it wants to access the records, sequentially or directly (via the index).

Indexed sequential file organization is supported by the operating system software or program compilers so that when records are added to or deleted from the sequential file, the index is also maintained automatically. Two of the more common environments are VSAM (Virtual Storage Access Method) and ISAM (Indexed Sequential Access Method). VSAM and ISAM differ primarily in the manner in which the physical files are organized. This has a greater impact on the programmer than on the analyst.

The selection of the best type of organization for a file is a function of the file access methods required by programs that will use that file. Today, it is

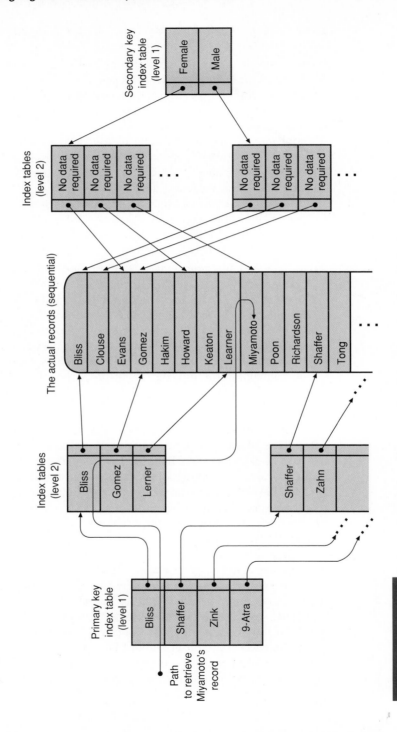

FIGURE 13.6 Indexed File Organization This employee file is organized as an indexed file. In indexed files, the actual records—shown in the middle of the figure—are usually stored sequentially, although some approaches, such as VSAM, offer other options. As records are entered into the sequential file, index tables are built that allow the sequential files to be retrieved without sequentially reading through the whole file. The path for retrieving Miyamoto's record is indicated.

frequently mandated that all master and transaction files must be organized as indexed sequential files. Why? Because future requirements may dictate a combination of sequential and direct access even if that combination is not needed today. ISAM and VSAM files offer such flexibility.

As was the case with fields and records, file design decisions must also be recorded in the dictionary. A file is the physical implementation of a data store; therefore, design specifications are recorded in a data store entry. Notice that design decisions such as blocking factors and file organization, to name only two, are recorded in the dictionary. We'll demonstrate how to complete this dictionary entry later in the chapter.

File Design Considerations

We'd like to impress on you two important and related issues that will affect file design: redundancy and internal controls. These issues influence the value and usefulness of the files you will design.

Minimizing Redundancy in Files   The minimization of redundant data in files has long been a suggested philosophy of file design. Ideally, no single field would be included in more than one record. Take a moment and ask yourself why. Yes, redundantly stored fields will use considerably more storage space. But the problem is even more significant. If the same field is stored in more than one record, the data must be updated in more than one place when it changes. Otherwise the *same* field could take on different values in different records.

So why do systems frequently store redundant data? The reasoning is not unsound. First, there are natural relationships between files. If we need to use the data in one file (for example, CUSTOMER) to cross-reference the data in another file (say, INVOICE), we redundantly store some customer fields in both files. Second, to improve performance of some programs, we can reduce the number of files needed by a program—thus reducing the number of files opened, read from, and written to—by redundantly storing some common data in more than one file.

Redundancy, as a term, has an undeserved reputation. Redundancy is not necessarily wrong. *Uncontrolled redundancy* is the more common problem! We suggest you restrict duplication of fields in different records to primary key fields (needed to cross-reference files) and relatively stable descriptor fields (those that don't change values often). For instance, CUSTOMER ADDRESS is relatively stable, but CUSTOMER BALANCE is not. The more volatile the field, the greater the danger in storing that field redundantly in different files.

The problems of data integrity and redundancy can be addressed by basing your file design on a third normal form data model as presented in Chapter 8. Conventional files, as you will soon learn, tend to compromise that ideal data model; however, the compromises can be more carefully thought out using the data model as a foundation. There is one other way: the use of modern database technology. That alternative will be presented in the next chapter.

Internal Controls for Files   **Internal controls** are a requirement in all computer-based systems. They ensure that the computer-based system is protected against accidental and intentional errors and use, including fraud. In this book, we will cover controls separately for the design components in their respective chapters.

Internal controls are designed into files to ensure the integrity and security of the data in those files. Internal control issues for files include the following:

- **Access controls** should be specified for all files. Access controls determine who will be able to read and write the file. Some end-users will have read and write access. Other end-users will have only read access. If corrections must be made to file data, *who* will authorize those changes? The data belongs to the end-user; therefore, authorization should not come from the analyst.

- **Length of retention** of files is another control issue. Master file records are retained indefinitely. However, transaction records should be retained in the file only so long as the data is immediately useful. Otherwise, valuable storage capacity will be wasted. When should such records be archived, or moved off-line to tape or diskette? How long should archived records be retained? In some cases, government regulations may dictate a long retention cycle.

- **Backup methods and procedures** should be defined for all data files. Backups are used to recover data when files have been lost or destroyed. All master and transaction files should be periodically copied to tape or diskette.

- **Audit trails** should be established for all files. An audit trail makes a permanent record of any update—added record, deleted record, or changed record—that is made by a program. If such changes are processed only in batches, the program needs to print a report of the updates. Today, on-line processing of updates is more common. On-line programs need to post updates to a special audit file that can be printed at convenient times. It should be acknowledged that many file management systems automatically create audit trails, relieving the analyst and programmer of that responsibility. Be sure to check!

How to Design and Document Conventional Files

In this section, we discuss how files are designed. And we'll design a file to demonstrate this important systems design task.

Computer files that need to be designed are easily identified by studying either the entity relationship diagrams or logical data flow diagrams prepared during systems analysis. You learned to define data storage requirements with these tools in Chapters 8 and 9.

An entity relationship diagram (or data model) depicts the data to be stored in files. The actual files will usually be constructed from combinations of entities. The ERD also describes mandatory relationships between different types of data to be stored.

On the logical data flow diagram, files appear as *data stores*. In reality, physical data stores will likely be constructed from combinations of logical data stores. For example, the logical stores ORDER and ORDERED PRODUCT will normally be implemented as a variable-length ORDER record and file. DFDs also demonstrate what data is used by various systems processes.

Let's study a file design example to see how this procedure is applied. We'll identify several files to be designed; however, we'll demonstrate the complete design for only one file. We will continue to use our CASE product, Excelerator, to demonstrate one way of recording design decisions into the dictionary.

Step 1: Identify Files to Be Designed

The best way to define conventional files to be designed is to study the data model. The entity relationship data model for the Entertainment Club is reproduced in Figure 13.7. You could partition the data model into files in many different ways. One *possible* partition of the model into files has been indicated on the data model. We call your attention to the following annotated items in the figure:

(A) **The PROMOTIONS FILE (Fixed-Length Records)** Data describing the PROMOTION entity and its relationships with the CLUB and TITLE entities will become the PROMOTIONS FILE (see margin). Notice that the boundary of the file also includes the relationships. Why? Because we want the file's records to not only describe promotions, but also to describe the club that sponsors the promotion and the title to be promoted. This will be implemented through planned redundancy, namely the inclusion of some CLUB and TITLE fields in the PROMOTION record. As described earlier in the chapter, the redundancy will be restricted to duplicated key fields—from CLUB and TITLE—and a few relatively stable descriptor fields.

Promotion
data

Data relating
to CLUB for a
PROMOTION

Data relating
to TITLE for a
PROMOTION

PROMOTIONS FILE
(Fixed-Length Record)

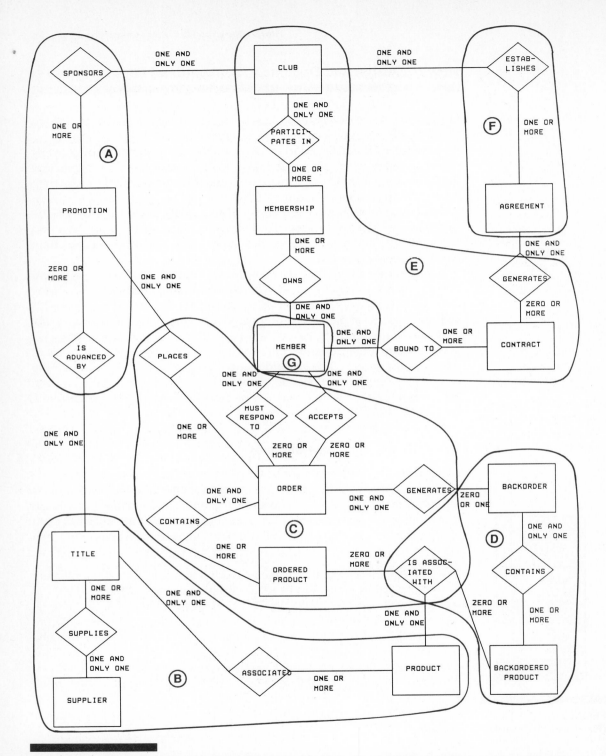

FIGURE 13.7 Partitioned Entity Relationship Data Model for the Entertainment Club The partitioned areas represent files that will show the scope of implementation for separate files.

TITLE
data

SUPPLIER
data

PRODUCT 1
data

PRODUCT 2
data

.
.
.

PRODUCT N
data

**INVENTORY FILE
(Variable-Length
Record)**

Ⓑ **The INVENTORY FILE (Variable-Length Records)** The entities TITLE, PRODUCT, and SUPPLIER and their corresponding relationships will be consolidated into a single INVENTORY file. Our decision to implement entities in a common file is based on a number of factors. We realized that most end-user reports and inquiries that use TITLE data also require PRODUCT data. In addition, since there is only one SUPPLIER for a TITLE, the SUPPLIER data could be merged into the same record, too.

Do you see why the record will be variable length? For each TITLE there will always be one or more occurrences of PRODUCT, one per medium—for example, cassette tape, compact disc, and so on. The PRODUCT fields will repeat as many times as there are media for the title (see margin at left). As a rule of thumb, the need for variable-length records is easily identified by one-to-many relationships between two entities on the data model.

Have we lost anything by not implementing PRODUCT and TITLE as separate files? To be truthful, yes. Those PRODUCT fields that will be stored in the repeating portion of variable-length records cannot serve as keys for retrieving records from the file. Thus, we can't retrieve PRODUCT data without first retrieving TITLE data. For this file, that is not a serious drawback.

Ⓒ **The ORDERS FILE (Variable-Length Records)** The ORDER and ORDERED PRODUCT entities and various relationships with others will be consolidated into an ORDERS FILE. The ORDERS FILE will consist of variable-length records (see margin, page 452). Data describing the ORDER will appear in the fixed portion and data describing the ORDERED PRODUCT entity will reside in the repeating portion.

Choosing to store these entities in a single, variable-length file requires accepting a trade-off. We lose the ability to directly retrieve all ORDERS for a particular PRODUCT. This is because most file management systems do not permit secondary keys on fields in the repeating portion of the record. The need could, however, be fulfilled indirectly by building a temporary, fixed-length scratch file from the ORDERS FILE.

The multiple relationships between the ORDER and MEMBER entities can be implemented in two ways. First, we could implement two separate order files, one for AUTOMATIC ORDERS (orders that the members MUST RESPOND TO) and one for ACCEPTED ORDERS (orders that a member has accepted by response or by failure to respond by the required date—the ACCEPTS/GENERATED FOR relationship).

We chose a second, simpler solution that just requires storing an additional element, MEMBER RESPONSE STATUS, to differentiate between the two types of ORDERs represented by the two relationships.

The GENERATES/GENERATED FOR relationship between ORDER and BACKORDER will be implemented by merging the unique portion of the

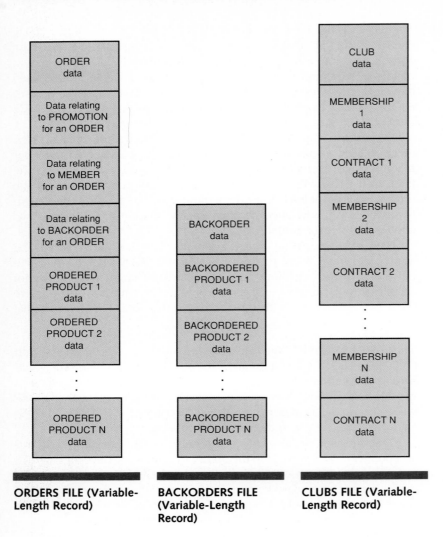

ORDERS FILE (Variable-Length Record)

BACKORDERS FILE (Variable-Length Record)

CLUBS FILE (Variable-Length Record)

BACKORDER entity's key (BACKORDER DATE) into our new ORDER record. The trade-off is that, for most ORDERs, the field will be blank—but take up storage space—since most ORDERs are never backordered.

Finally, note that the relationship IS ASSOCIATED WITH/IS FOR is partitioned into two files. This simply indicates that both files will implement that common relationship. The ORDERS FILE will implement the relationship between its entity, ORDERED PRODUCT, and the entity PRODUCT.

Implementing the IS ASSOCIATED WITH/IS FOR relationship between ORDERED PRODUCT and PRODUCT requires us to introduce intentional redundancy. We will simply include some of PRODUCT's keys and static descriptor fields into the ORDERED PRODUCT portion of our ORDERS record.

Ⓓ **The BACKORDERS FILE (Variable-Length Records)** The BACKORDER and BACKORDERED PRODUCT entities and two relationships will be consolidated into a BACKORDERS FILE. The BACKORDERS FILE will contain variable-length records, with the fixed portion containing BACKORDER data and the repeating portion containing BACKORDERED PRODUCT data (see diagram, middle left). The IS ASSOCIATED WITH/ IS FOR relationship will be implemented in the same manner as for the ORDERS FILE.

Ⓔ **The CLUBS FILE (Variable-Length Records)** The CLUBS FILE will contain data describing the CLUB, MEMBERSHIP, and CONTRACT entities and numerous relationships. The CLUBS FILE will contain variable-length records with CLUB data as the fixed portion and MEMBERSHIP and CONTRACT data as the repeating portion (see diagram, left).

We chose to store MEMBERSHIP and CONTRACT data in the same file because the relationships between MEMBER and MEMBERSHIP and between MEMBER and CONTRACT are mutually inclusive. That is, for every occurrence of the entity MEMBERSHIP, there is also an occurrence of the entity CONTRACT, and vice versa. The amount of duplicate data seemed minimal.

Ⓕ **The AGREEMENTS FILE (Fixed-Length Records)** The entity AGREEMENT and its relationship with the CLUB entity will compose the AGREEMENTS FILE. The AGREEMENTS FILE will consist of fixed-length records (see margin).

Ⓖ **The MEMBERS FILE (Fixed-Length Records)** The MEMBERS FILE will consist solely of data describing the MEMBER ENTITY. The MEMBERS FILE will consist of fixed-length records (see margin, below). It will contain no fields from any of the entities to which it is related.

You've just learned how a logical data model can be used in planning conventional file design. There are many other ways to partition the data model into files. In fact, you may have been wondering why we chose not to implement each entity appearing on the entity relationship diagram as an individual file. This would be a valid yet undesirable alternative design solution for conventional files. The large number of files would present efficiency and maintenance trade-offs that most analysts and organizations are not willing to accept. Besides, this solution would be better accomplished through a relational database system that will be described in Chapter 14.

Finally, you may be asking why our file designs reflect some duplication of data and repeating elements. We went to great lengths in the coverage of

AGREEMENT data

Data relating to CLUB for an AGREEMENT

AGREEMENTS FILE (Fixed-Length Record)

MEMBER data

MEMBERS FILE (Fixed-Length Record)

data modeling in Chapter 8 to ensure that our logical model did not exhibit these characteristics. So why did we reintroduce those seemingly undesirable characteristics into our file designs?

In designing conventional files, duplication of data is usually necessary. It is a judgment call. We acknowledge that different experts, instructors, and perhaps yourself would have partitioned the data model differently. For instance, duplication of primary key elements is necessary to implement relationships between entities. Also, some data elements are relatively stable, and for the sake of efficiency, it may be advantageous to strategically duplicate elements across files to avoid the necessity of opening and accessing multiple files.

Step 2: Determine the Data Content of the File

The next step in file design is to determine the actual data to be stored in the files. For demonstration purposes, the remaining steps will focus on steps toward designing the ORDERS FILE. Let's examine the substeps required to define the contents of the ORDER record.

Step (2a): Identify Data Elements to Be Stored in the File   According to our partitioned entity relationship diagram, the ORDERS FILE is to contain data describing the ORDER and ORDERED PRODUCT entities, as well as a number of relationships. There's no need to start from scratch and brainstorm the data element contents of the file.

Think back. Dictionaries were initiated in systems analysis to define the logical, or implementation-independent, contents of each entity (see Chapters 8 and 10). The contents were described in a logical record description in the project dictionary. For your convenience, the data elements for each entity are listed in Figure 13.8(a). Keys are underlined.

Step 2(b): Consolidate the Data Elements into a Single Record   A single occurrence of the final ORDERS record should contain data elements describing one occurrence of the entity ORDER plus one or more occurrences of the entity ORDERED PRODUCT. Figure 13.8(b) shows the integrated data elements from the two entities. Notice that the data elements from the ORDERED PRODUCT entity appear in the repeating portion of the record. The data element NUMBER OF ORDERED PRODUCTS was added to the list to indicate the actual number of occurrences of the repetition of data elements for any given order. Also, recall that data elements appearing in the repeating portion of a variable-length record cannot be keys in most conventional files. Therefore, ORDER NUMBER, PRODUCT NUMBER, and MEDIA CODE are no longer underlined.

We also eliminate any redundancy in the resulting record. Thus, ORDER NUMBER is deleted from the group of repeating elements since it is the key of the fixed-length portion of the record. Some data redundancy is less evi-

FIGURE 13.8 Preliminary Data Element Content of ORDERS FILE (a) identifies the data element content for the two entities contained in the ORDERS FILE. These 3NF data element lists were compiled during data modeling; (b) represents the consolidated (from ORDER and ORDERED PRODUCT) data element content for a record in the ORDERS FILE.

ORDER is composed of the following data elements:

> ORDER NUMBER
> ORDER DATE
> ORDER STATUS (added by end-users)
> TOTAL ORDER PRICE (sum of extended price)
> PREPAID AMOUNT

ORDERED PRODUCT is composed of the following data elements:

> ORDER NUMBER
> PRODUCT NUMBER
> MEDIA CODE
> QUANTITY ORDERED
> QUANTITY SHIPPED
> PRODUCT PRICE

(a)

An ORDERS FILE record contains the following data elements:

> ORDER NUMBER
> ORDER DATE
> ORDER STATUS (added by end-users)
> ~~TOTAL ORDER PRICE (sum of extended price)~~
> PREPAID AMOUNT
> ~~ORDER NUMBER~~
> PRODUCT NUMBER
> MEDIA CODE
> QUANTITY ORDERED
> QUANTITY SHIPPED
> PRODUCT PRICE

(b)

dent. Examine the data element TOTAL ORDER PRICE. TOTAL ORDER PRICE can be calculated as the sum of QUANTITY ORDERED times PRODUCT PRICE for all products on the order. Therefore, we can save storage space by simply deleting TOTAL ORDER PRICE.

Step 2(c): Add Data Elements to Implement Relationships to Other Entities Let's refer back to our partitioned entity relationship diagram. The ORDERS FILE is to store additional data elements used to implement relationships (six relationships, to be exact). Figure 13.9 revises our ORDERS FILE record to describe the required relationships. The six relationships are to be implemented as follows:

- The relationship between ORDER and PROMOTION is implemented by adding the key of PROMOTION (CLUB NAME and PROMOTION DATE) to the record.

- The relationship between ORDER and MEMBER is implemented by adding two elements. The data element MEMBER NUMBER is used to associate an ORDER to the MEMBER that placed the order. We also brought in the elements MEMBER NAME, P.O. BOX, STREET, CITY, STATE, and ZIP CODE. These are relatively stable data elements; therefore, the redundancy is not very risky. The benefit is that we may not have to open the MEMBERS FILE to print most reports about orders, since the data we need about members is also in the ORDERS FILE. Planned redundancy! Planned redundancy, of course, must be documented if it is to be controlled.

  The new data element MEMBER RESPONSE STATUS (first described in step 1) will distinguish between the two relationships between ORDER and MEMBER. In other words, MEMBER RESPONSE STATUS describes whether the order is awaiting the member's response or released to be filled after receiving the member's response.

- The data element BACKORDER DATE will implement the relationship between ORDER and BACKORDER. If it takes on a value, then we know that a backorder record exists in the BACKORDERS FILE.

- The data element PRODUCT NUMBER will be used to establish the relationship between ORDERED PRODUCT (in the ORDERS FILE) and TITLE. Specific occurrences of the entity PRODUCT will then be obtained by using the MEDIA CODE data element common to ORDERED PRODUCT and PRODUCT. If you wish, you could also duplicate some relatively static elements from TITLE into this record.

Step 2(d): Add Audit and Control Fields   Finally, we may choose to add audit and control fields to the content of our record. This is not always necessary since some operating systems and file management software automatically provide audit and control mechanisms. Figure 13.10 represents the contents for the ORDERS record.

Now we can describe our record to the project dictionary as a record and data element. Typical forms or screens were shown earlier in the chapter. The ORDER record must specify the maximum number of occurrences for each field for one occurrence of the record. Each field should be described as a data element, including specifications for storage format and length (which we learned is dependent on storage format). After entering the record and describing each data element in the record, we can print out or prepare a detailed record layout (Figure 13.11).

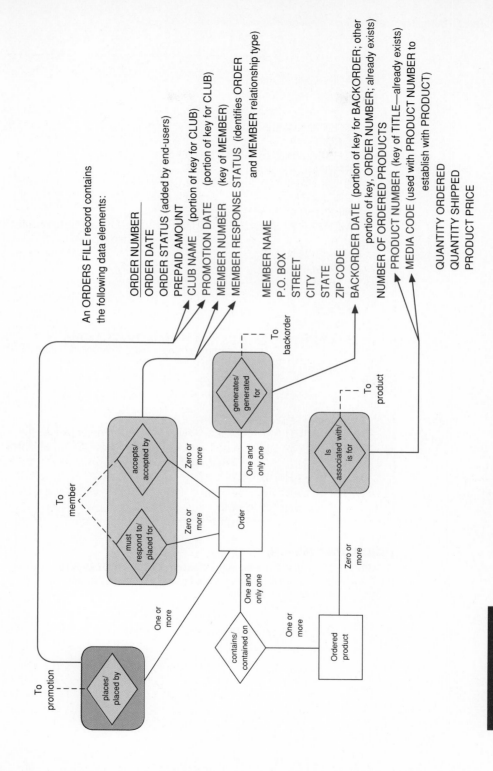

An ORDERS FILE record contains
the following data elements:

ORDER NUMBER
ORDER DATE
ORDER STATUS (added by end-users)
PREPAID AMOUNT
CLUB NAME   (portion of key for CLUB)
PROMOTION DATE   (portion of key for CLUB)
MEMBER NUMBER   (key of MEMBER)
MEMBER RESPONSE STATUS (identifies ORDER
                          and MEMBER relationship type)

MEMBER NAME
P.O. BOX
STREET
CITY
STATE
ZIP CODE
BACKORDER DATE (portion of key for BACKORDER; other
                 portion of key, ORDER NUMBER; already exists)
NUMBER OF ORDERED PRODUCTS
PRODUCT NUMBER (key of TITLE—already exists)
MEDIA CODE (used with PRODUCT NUMBER to
            establish with PRODUCT)

QUANTITY ORDERED
QUANTITY SHIPPED
PRODUCT PRICE

FIGURE 13.9 Implementing Relationships Through Data Elements Relationships between the ORDERS FILE
and entities appearing in separate files are implemented by redundantly storing the keys of those external
entities.

457

An ORDERS FILE record contains
the following data elements:

    <u>ORDER NUMBER</u>
    ORDER DATE
    ORDER STATUS (added by end-users)
    PREPAID AMOUNT
    CLUB NAME (portion of key for CLUB)
    PROMOTION DATE (portion of key for CLUB)
    MEMBER NUMBER (key of MEMBER)
    MEMBER RESPONSE STATUS (identifies ORDER and MEMBER
                                      relationship type)
    MEMBER NAME
    P.O. BOX
    STREET
    CITY
    STATE
    ZIP CODE
    BACKORDER DATE (portion of key for BACKORDER; other
                  portion of key, ORDER NUMBER; already exists)

    NUMBER OF ORDERED PRODUCTS
        PRODUCT NUMBER (key of TITLE—already exists)
        MEDIA CODE (used with PRODUCT NUMBER to establish with
                PRODUCT)
        QUANTITY ORDERED
        QUANTITY SHIPPED
        PRODUCT PRICE

FIGURE 13.10   Final Data Element Content of the ORDERS FILE   The
final content of the ORDERS FILE contains data elements describing
entities, relationships, and audits.

```
DATE: 28-OCT-88                          RECORD - EXPLOSION
TIME: 10:53                              NAME: ORDER RECORD

NAME:                        ORDER RECORD
ALIAS:                       MEMBER ORDER RECORD
```

ELEMENT/RECORD	OFF	OCC	TYPE	LEN
ORDER NUMBER	000	001	K	006
ORDER DATE	006	001	E	008
ORDER STATUS	014	001	2	001
PREPAID AMOUNT	015	001	E	004
CLUB NAME	019	001	3	010
PROMOTION DATE	029	001	3	008
MEMBER NUMBER	037	001	4	007
MEMBER RESPONSE STATUS	044	001	5	001
MEMBER NAME	045	001	E	035
P.O. BOX	080	001	E	010
STREET	090	001	E	015
CITY	105	001	E	015
STATE	120	001	E	002
ZIPCODE	122	001	E	009
BACKORDER DATE	131	001	E	008
DATE ORDER CREATED	139	001	E	008
ORDER CREATED BY ID	147	001	E	006
DATE ORDER LAST MODIFIED	153	001	E	008
ORDER LAST MODIFIED BY ID	161	001	E	006
NUMBER OF ORDERED PRODUCTS	167	001	E	002
ORDERED PRODUCT	169	015	R	
PRODUCT NUMBER	169	001	E	007
MEDIA CODE	176	001	E	001
QUANTITY ORDERED	177	001	E	003
QUANTITY SHIPPED	180	001	E	003
PRODUCT PRICE	183	001	E	004

Record length is 439.

FIGURE 13.11 Typical CASE Dictionary Report This report describes the complete record structure (including sub-records—indented) and key field attributes.

Step 3: Update the Logical Requirements for the File

Are we now ready to design a file? Not just yet! There are a number of imple-mentation-independent file attributes that must first be reviewed and, if nec-essary, updated. These attributes will have an impact on any design solution. Figure 13.12 shows the logical requirements for our ORDERS FILE. Pay par-ticular attention to the following:

- The **Explodes to:** entry points to the record description we just finished.

- The **Volume** entries describe the number of occurrences of ORDER that will be active at any given time. Volume data should have been recorded when the entity ORDER was defined in the systems analysis phase of the project. We'll soon need this data to calculate disk space requirements for our ORDERS FILE.

Step 4: Design the File

We can now complete the design of our ORDERS FILE. Figure 13.12 also pre-sents the remaining physical specifications for the ORDERS FILE. Skim through these dictionary entries and then direct your attention to the following:

(A) This file is to be implemented as a variable-length VSAM record. VSAM—an indexed file organization—was chosen to maximize file access flex-ibility since it provides both sequential and direct access opportunities.

(B) The record size is specified as a minimum and maximum number of bytes. Minimum record size is easy to calculate. Because every ORDER must have at least one ORDERED PRODUCT, we simply sum the sizes—specified in bytes—of the individual data elements appearing in the ORDER record.

Maximum field size is a little more difficult to determine than min-imum. Maximum record size is based on the maximum number of times that the repeating group can occur (in our case, 15). The maximum record size equals the sum of the field sizes in the fixed-length portion of the record plus the sum of 15 times the sum of the repeating elements.

$$169 \frac{\text{bytes}}{\text{record}} + (18 \frac{\text{bytes}}{\text{record}} \times 15 \text{ groups}) = 439 \frac{\text{bytes}}{\text{record}}$$

(C) There is usually a limit to the size of a block—for example, half of a track on the disk. And some information systems departments establish a standard for block size. For any given disk drive, there are tables that provide blocking reference data for that disk drive. Different tables are used for files to be stored with and without keys. The blocking factor table tells you how big the block can be. For example, for files that require keys, the following block sizes might be found (page 462):

```
                    DATA STORE — OUTPUT
                    NAME: ORDERS FILE


TYPE Data Store                         NAME ORDERS FILE

                        EXPLODES TO ONE OF:
    Label ORDERS FILE       Record              ORDER RECORD
                        Data Model Diagram
                        ERA Diagram


                Location          ORDER PROCESSING
                Manual or Computer C
                Number of records  17300

                Index Elements    ORDER NUMBER
                                  ORDER STATUS
                                  CLUB NAME + PROMOTION DATE
                                  MEMBER NUMBER
                                  MEMBER RESPONSE STATUS


  Description
  AN AUTOMATIC, DATED ORDER GENERATED IN RESPONSE TO A PROMOTION. IT MAY
  BE REVISED, APPROVED, OR CANCELED VIA MEMBER RESPONSE.

  LOGICAL, IMPLEMENTATION—INDEPENDENT ATTRIBUTES OF THIS DATA STORE

    VOLUME
      AVERAGE: 17,300      GROWTH RATE (PER TIME PERIOD): 5% PER YEAR
      PEAK:    20,034      WHEN: NOVEMBER THROUGH CHRISTMAS (DECEMBER)

    USER REFERENCES:  SALLY HOOVER, DIRECTOR OF ORDER PROCESSING
                      ORDER—PROCESSING STAFF

    CONSTRAINTS: NONE
  PHYSICAL, IMPLEMENTATION—DEPENDENT ATTRIBUTES OF THIS DATA STORE
 (A) IMPLEMENTED AS:  VARIABLE—LENGTH, VSAM      LOCATION: IBM 3081
    COMPUTER NAMES OR IDs:  ORDERS.DAT
    MAXIMUM PHYSICAL RECORD SIZE:  439 (B)
    BLOCKING CONSTRAINTS: 1/3 TRACK BLOCK
    BLOCKING FACTOR: 5 (C)      BLOCK SIZE: 2603
    FILE SIZE (IN BYTES): 10,430,221 (D)
    FILE SIZE (IN TRACKS OR INCHES): 1336 (E)
    FILE SIZE (IN CYLINDERS):  149 (F)
    AVERAGE RECORD LIFETIME UNTIL ARCHIVE: 2 MONTHS
    BACKUP AND RECOVERY
       BACKUP TIMING:  DAILY 8:00 PM      BACKUP MEDIA: TAPE
       BACKUP RETENTION: 1 YEAR
       AUDIT TRAIL METHOD: UPDATES LOGGED TO DISK, PRINTED AT END OF DAY
    SECURITY: ONLY ORDER—PROCESSING STAFF MAY UPDATE
```

FIGURE 13.12   Detailed Logical Requirements and File Design Specifications for the ORDERS FILE

Track Size	Minimum Block Size (bytes)	Maximum Block Size (bytes)
1	4,026	8,293
1/2	2,604	4,025
1/3	1,892	2,603
1/4	1,466	1,891

These numbers apply to one particular disk drive and the use of primary keys. For demonstration, we used the above table. We assumed that Entertainment Club's Information Systems Department has adopted a 1/3-track blocking standard. Therefore a block size consists of 2,603 bytes. Blocking factor (BF) is calculated as follows:

$$\text{blocking factor} = 2,603 \frac{\text{bytes}}{\text{block}}/439 \frac{\text{bytes}}{\text{record}}$$

$$= 5.93 \frac{\text{records}}{\text{block}}$$

$$= 5 \frac{\text{records}}{\text{block}}$$

This is read "5 records *per* block." Blocking factors are always rounded down because a fraction of a record cannot be stored in a block.

(D) File size is an important calculation because we can't store data for which we don't have capacity. To calculate file size, we first determined, by consulting the end-users, that there are currently 17,300 ORDERS and that the base is growing 5 percent per year. To calculate the file size in bytes, tracks, and cylinders, the following calculations were used:

$$\text{bytes} = 17,300 \text{ records} \times 1.158/5 \frac{\text{records}}{\text{block}} \times 2,603 \frac{\text{bytes}}{\text{block}}$$

$$= 10,429,388 \text{ bytes of storage}$$

Note that the blocking factor had to be considered because the wasted storage in any block cannot be reclaimed, since block size is fixed at 2,603 bytes per physical record. Also note that we included anticipated growth (5 percent = 1.05) over three years (1.05^3 = 1.158) in our formula. Finally, after dividing by the blocking factor, that result was rounded up. Why? Once again, a file and a block must both contain complete records.

(E) It is also useful to express file size in terms of tracks and cylinders, because those tracks and cylinders might have to be dedicated to the file.

$$\text{tracks} = 17,300 \frac{\text{records}}{\text{volume}} \times 1.158/5 \frac{\text{record}}{\text{blocks}}/3 \frac{\text{blocks}}{\text{track}}$$

$$= 1,336 \text{ tracks} \qquad (\text{we rounded up})$$

(F) To determine the number of cylinders required to store the file, you need to understand another characteristic of the disk packs used by Entertainment Club. The disk packs currently in use have 9 tracks per cylinder (this may vary for other disk packs).

$$\text{cylinders} = 1,336 \frac{\text{tracks}}{\text{file}}/9 \frac{\text{tracks}}{\text{cylinder}}$$

$$= 149 \text{ cylinders/file}$$

You've seen how to design a conventional computer-based file containing variable-length records. Design of a file containing fixed-length records is accomplished in the same manner. The blocking factors and file-size calculations would be simplified since all records are fixed length.

Summary

Computer files consist of physical records, records, and fields. A field or data element is the smallest unit of data to be stored. There are three types of fields: primary keys, secondary keys, and descriptors. Most business-oriented systems store fields as binary codes. There are three common binary codes: EBCDIC, ASCII, and packed decimal. An understanding of storage formats is important during file design because storage format affects file size.

A record is a collection of fields arranged in a predefined format to describe a single entity. A record is also the unit of data storage that is operated on by a computer program. Records for a system are specified by the systems analyst. Depending on the data structure to be stored, the record will have fixed or variable length. The record for a file is normally documented using a record layout chart.

Occurrences of records are blocked to improve input and output efficiency. A block is the smallest number of occurrences of records that can be read or written at one time. The analyst often determines the number of records in the block because this can affect performance.

Finally, occurrences of records make up files. There are several types of files, but the two most important classes are master files, whose records— usually fixed length—describe basic business entities (such as parts, cus-

THE NEXT GENERATION

Looking Beyond Files and Databases: Can We Store Expertise?

First there were files and then databases. Today we are seeing data banks, large databases owned by a company that sells access to the data banks (sort of a computerized library—a good example is The Source, which can be accessed via long-distance telephone calls from most popular microcomputers). Both databases and data banks store data and information.

But consider the possibility of storing knowledge and expertise. That's the next generation.

Knowledge is the ability to apply information. When you generate a report for a manager, that report is information. Putting that report to use requires knowledge and expertise. Wouldn't it be nice to be able to capture and store that expertise? For example, wouldn't it be nice

if the computer could periodically scan the business environment the way that a manager does, examine various databases, recognize the need for a decision or immediate attention, and make the decision—or at least inform a manager of the need for the decision and provide options and analyses of those options?

This is exactly what researchers and vendors are looking at right now! The current buzzword for this concept is *expert systems*. In an **expert system**, the expertise and knowledge associated with decision making are stored in a knowledge base. Programs are then written using an expert systems shell to define rules that access the knowledge bases. These rules define conditions,

combinations of conditions, and rules for dealing with the conditions and combinations.

Conventional files and databases provide the data. Knowledge bases provide the rules. Programs apply the rules to the data to inform managers of decision needs and alternative actions (computerized chess programs have always been based on this idea). The programs attempt to define solutions to problems, evaluate the impact of those solutions, possibly predict how the environment may react to those solutions, and select a course of action. The expert systems applications are based on the development of artificial intelligence technology, both hardware and software. These systems will require extremely

tomers, and employees), and transaction files, whose records—which are frequently variable length—describe business events (such as orders and invoices).

One of the most important performance decisions made by the systems analyst is file organization. File organization defines how records in a file are related to one another. Three common file organizations are sequential, direct, and indexed. The best way to determine an optimal file organization is by studying the file access methods required by the programs that will use and update the file.

To ensure the integrity and security of the data, internal controls should be designed for files. Internal controls for files include access, retention, backup, and recovery controls.

File designs should be based on the data model defined during systems analysis. This helps the analyst ensure that the resulting files implement most

THE NEXT GENERATION

Looking Beyond Files and Databases: Can We Store Expertise?, Continued

rapid data and knowledge access methods and processing speeds.

Knowledge bases must be combined with databases to make expert systems. Some such systems have already been built. Medical expert systems allow doctors and nurses to input symptoms and conditions. The expert system simulates the expertise of many more doctors than a hospital or clinic can afford to keep on staff. It gives the medical staff access to diagnoses and alternative treatments that may be unfamiliar to some doctors. And in all cases, the system provides a second opinion. Similar expert systems are being used to provide advanced expertise to pharmacists—for example, which drugs don't mix with other medications—and to geol-

ogists, perhaps indicating where to drill for oil.

On the business front, future expert systems may try to simulate the expertise of managers and staff specialists. For instance, production schedules, quality control, accounting, marketing, and other business fields may benefit from expert systems. This is especially true if the systems could be designed not only to respond to data stored in databases but also to sense business conditions via cameras and analog sensing devices (this has already been done in applications such as environmental control, manufacturing process controls, and robotics).

As systems analysts, we may become increasingly involved in applying artificial intelligence technology to develop expert

systems. We may be called on to interface these systems with current information systems. And we will have some tough decisions to make. For instance, expert systems could potentially put a lot of managers and end-users out of work. In other words, there are some social implications to this trend. And lest you take this too lightly, don't forget—systems analysts (and all information systems professionals) are themselves experts whose knowledge can be incorporated into expert systems.

Indeed, we are already seeing some CASE products that not only automate modeling and dictionaries, but provide expert analysis of your work as you do it. Knowledgeware's Information Engineering Workbench is an example of such a product.

of the business data entities and relationships. Initially, the data model should be partitioned to identify the files that must be designed.

For each file to be designed, the analyst must define the record's contents. The essential requirements are already known since the entities were defined during systems analysis. However, the contents of a physical file often require consolidation of entities. Furthermore, the analyst must usually add some new fields from other entities to implement relationships and install internal controls.

Once a file's record has been established, the analyst can review logical requirements (such as number of occurrences that must be stored) and complete the physical design. The physical design includes specification of file organization, blocking factors, and disk space requirements.

Problems and Exercises

1. Define the terms *field*, *record*, and *file*.
2. What are the advantages of record blocking? How is the blocking factor for a file derived?
3. Identify three types of fields and give several examples of each.
4. Identify five types of files and give several examples of each.
5. Define file access and identify and describe two file access methods.
6. The selection of the best type of organization for a file is a function of the file access methods that will be used for that file. Identify three types of file organization and how these files organizations are accessed.
7. Differentiate between fixed- and variable-length records. What impact does a record storage format have on a file design?
8. Explain the implications of data modeling for file design. How might file designs be affected by a failure to do data modeling?
9. Describe how a data model is used to plan file design.
10. Given the field LIBRARY BOOK CODE, a six-digit number that uniquely identifies the books in a company library, how would you store this field? (*Hint:* Is the field numeric or alphanumeric?)
11. Calculate the blocking factor for a fixed-length record file, given the following factors:

    1/4-track blocking (1/4 track = 1,891 bytes)
    record size = 192 bytes

    How much storage space is wasted in each physical record? How much storage space is required to store 2,500 records? Specify your answer in bytes.
12. Calculate the blocking factor for a variable-length file record, given the following factors:

    1/3-track blocking (1/3 track = 2,603 bytes)
    minimum record size = 90 bytes
    maximum record size = 275 bytes

    How much storage space is wasted in each physical record? Specify minimum and maximum.
13. Describe the relationship between file access and file organization.

14. Design a file for the following data store structure. State all assumptions, and prepare complete documentation in a data dictionary.

EMPLOYEE consists of the following elements:	FIELD SIZE (in characters)
<u>EMPLOYEE NUMBER</u>	5
<u>SOCIAL SECURITY NUMBER</u>	9
EMPLOYEE NAME	20
EMPLOYEE HOME ADDRESS, which consists of	
STREET ADDRESS	20
CITY	10
STATE	2
ZIP CODE	5
EMPLOYEE HOME PHONE	10
EMPLOYEE BUSINESS PHONE	10
DEPARTMENT CODE	2
DATE EMPLOYED	6
DATE OF BIRTH	6
MARITAL STATUS	1
SEX CODE	1
RACE CODE	1
One of the following:	
MONTHLY SALARY	5
HOURLY RATE	2.2
VACATION DAYS DUE	2
SICK DAYS DUE	2
TAX EXEMPTIONS	2
GROSS PAY YEAR-TO-DATE	6.2
FEDERAL TAX YEAR-TO-DATE	6.2
STATE TAX YEAR-TO-DATE	5.2
FICA TAX YEAR-TO-DATE	6.2

15. Design a file for the following data store structure. State all assumptions and prepare complete documentation in a data dictionary.

	FIELD SIZE (in characters)
VENDOR PART, which consists of	
<u>VENDOR NUMBER</u>	6
VENDOR NAME	15
VENDOR ADDRESS, which consists of	
P.O. BOX NUMBER (optionally)	2
STREET ADDRESS	20
CITY	10
STATE	2
ZIP CODE	9
VENDOR PHONE	10
VENDOR TERMS, which consists of	
DISCOUNT RATE FOR EARLY PAYMENT	1.1
EARLY PAYMENT PERIOD	2
NET PAYMENT PERIOD	2
1 to 15 occurrences of the following:	
MATERIAL NUMBER	5
MATERIAL DESCRIPTION	10
UNIT OF MEASURE	2
UNIT PRICE	4.2
QUANTITY REQUIRED FOR DISCOUNT	4
QUANTITY DISCOUNT RATE	2.1
LEAD TIME	2

Projects and Minicases

1. Robert Williams, a systems analyst for Future Ventures, is facing a perplexing file design problem. He is attempting to design a vendors file. This file is to contain variable-length records. The fixed portion of the record contains data describing a vendor. The repeating portion contains data describing a part supplied by the vendor. Future Ventures uses a standard of 1/3-track blocking for all file designs. However, when Robert calculated the record size, he found that its 3,077 bytes exceeded the 2,603-byte limit of 1/3-track blocking. In other words, one logical record would not fit within standard block size. What can he do? Explain at least two alternatives Robert might choose from. Explain the implications of each alternative.

2. The partitioning of the SoundStage data model into the files presented in this chapter represents only one possible set of files. Repartition the data model into a different set of files and justify your decisions. Redesign the physical file that will contain the ORDER entity. This time, assume 1/2-track blocking.

Annotated References and Suggested Readings

Fitzgerald, Jerry. *Internal Controls for Computerized Information Systems.* Redwood City, Calif.: Fitzgerald, 1978. This is our reference standard on the subject of designing internal controls into systems. Fitzgerald advocates a unique and powerful matrix tool for designing controls. This book goes far beyond any introductory systems textbook—must reading.

Johnson, Leroy F., and Rodney H. Cooper. *File Techniques for Data Base Organization in COBOL.* Englewood Cliffs, N.J.: Prentice-Hall, 1981. We haven't found too many books that exclusively cover file structures and processing (there are several books that cover the underlying data structures). This book is both readable and comprehensive, even if you've never had a COBOL course or experience with COBOL. It explains the various file and database structures in relatively simple terms.

CHAPTER FOURTEEN

Designing Modern

Computer

Databases

MicroWorld Software, Inc.

Steven Friedmont, President of MicroWorld Software, Inc., was outraged at the transparency being displayed on the projector screen. It was a copy of a cover story in a major publication—"Users Outraged with MicroWorld Upgrade Service."

MicroWorld is a reasonably successful producer of microcomputer software such as word processors, spreadsheets, graphics presentations, project managers, database managers, and the like. They had established a loyal following for their products and a reputation for outstanding performance- and functionality-to-price ratios.

Three months ago, major new releases were announced for all products. These releases provided a common, windows-based user interface for all their products. The new interface conformed to industry specifications intended to standardize user interfaces across products and vendors.

But the bottom fell out on their service reputation. One year ago, MicroWorld had consolidated the separate data files from the registration, sales and upgrade, and distribution systems into an integrated database that would provide superior customer service and support. When the upgrades were announced, marketing notified all legally registered customers about the upgrade and the limited time of special prices for the upgrade. Customers were told how many registered copies were eligible for the upgrade.

Shortly thereafter, many customers returned their upgrade response cards. Upgrade orders were sent to distribution, which promptly notified many of the same customers that MicroWorld's records showed that the customers' copies were not registered and implied that the customers might have pirated (illegal) copies of the software. The customer response was vocal and livid. On the one hand, they were told that because they properly registered their software, they were now eligible for this special upgrade. Then MicroWorld told them that they were not only not registered, but possibly criminals.

Discussion

1. How could such a mistake happen? Could such a mistake have been avoided if the data files were not integrated into a common database?
2. What are the implications of this "honest" data mistake? Who do you think might be at fault? Why?

What Will You Learn in This Chapter?

This is the second of two chapters on the design of physical data stores. This chapter teaches you how to design data stores as modern computer databases, the alternative to conventional computer-based files. You cannot possibly master database design in one chapter. Instead, you should strive to familiarize yourself with this popular approach and then seek additional education concerning databases and database methods. You will know that you have mastered this chapter when you can

1. Compare and contrast conventional files and modern databases.
2. Differentiate between applications, subject, and end-user databases.
3. Describe the architecture of a database management system. Also, describe how this architecture supports the prototyping approach to systems analysis and design.
4. Describe three alternative database implementation models: relational, network or linked, and hierarchical.
5. Transform a simple, logical data model into simple versions of each of the following database schemas:
   • A relational or tabular schema
   • A network or linked schema
   • A hierarchical schema

Conventional files may be the lifeblood of many information systems; however, they are slowly being replaced with databases. A **database** may loosely be thought of as a set of related files. By related we mean that records in one file may be associated with the records in a different file. For example, a CUSTOMER record may be physically linked to all of that customer's ORDER records. In turn, each of those ORDER records may be linked to relevant PRODUCT records. This linking allows us to eliminate most of the need to redundantly store fields in the various files. Indeed, in a very real sense, conventional files are consolidated into a single file, the database.

The idea of relationships between different collections of data was introduced in Chapter 8. In that chapter, you learned to analyze and model data as entities and relationships. Database is the modern implementation of those entities and relationships.

Database is a relatively new subject for introductory systems analysis and design courses. But today, so many applications are being built around database technology that database design has become an important skill for the analyst. Indeed, database technology, once considered important only to the largest corporations with the largest computers, is now common for applications developed on microcomputers. In this chapter, you will learn the basics of how to design and document databases (from a systems analyst's perspective).

Database Concepts for the Systems Analyst

The history of information systems has led to one inescapable conclusion: data is a resource that must be managed. Very few experienced information systems staffs have avoided the frustration of uncontrolled growth and duplication of data stored in their computer files and systems. As systems were developed, implemented, and maintained, the common data needed by the different systems was duplicated in multiple, conventional files. This duplication carried with it a number of costs: extra storage space required, duplicated input to maintain redundantly stored data and files, and data integrity problems (for example, the ADDRESS for a customer not matching in the various files that contain customer ADDRESSes).

Out of necessity, database technology was created so an organization could maintain and use its data as an integrated whole instead of as separate data files. We can now develop a shared data resource that can be used by several information systems.

Let's survey the relative advantages and disadvantages of database versus conventional files. We'll also study the terminology and current state of practice in database.

Conventional Files Versus the Database

Before you conclude that the conventional file is undesirable, we should compare the file and database alternatives. Look at Figure 14.1. It illustrates the fundamental difference between the file and database environments. In the file environment, data storage is built around the applications that will use the files. In the database environment, applications will be built around the integrated database. Ideally, the database is not dependent on the applications that will use it. Each environment has advantages and disadvantages. Let's examine these differences.

Pros and Cons of Conventional Files Conventional files are relatively easy to design and implement because they are normally based on a single appli-

**FIGURE 14.1
Conventional Files
Versus the Database
Approach** File-based
environments emphasize
the system or
application. As
applications are
developed, custom files
are built for them. These
files may not be suited
to future needs or even
future systems.
Database environments
emphasize the data
independently of the
applications that use the
data. The applications
are allowed to evolve
around a database
designed such that it
can adapt to changing
needs

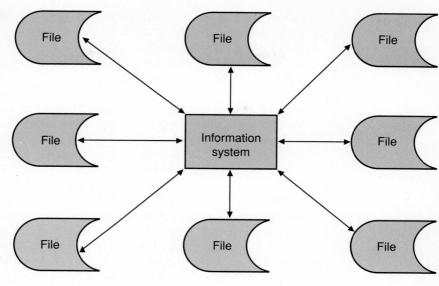

Conventional file environment

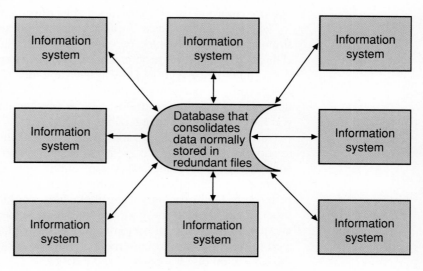

Database environment

cation or information system, such as accounts receivable or payroll. If you understand the end-user's output needs for that system, you can determine the data that will have to be captured and stored to fulfill those needs and define the best file organization for those requirements. That is the approach you learned in Chapter 13.

Another advantage of conventional files is processing speed. Database technology uses complex indices, linked lists, trees, and other data structures. Such data structures are sometimes too slow to handle large volumes of transactions with an adequate throughput. It should be noted that this limitation of database technology is rapidly disappearing, though.

Still, even if database technology matches or exceeds the performance of file management systems, numerous file systems are already in place. Given the shortage of information systems personnel and growth in demand for new information systems, many firms cannot afford the time or money required to redesign all of their current file-based systems as database systems. To accelerate the change, makers of CASE tools are beginning to provide technology to assist with the database design and conversion processes.

It is likely that some day database technology will replace file technology. Conventional files have numerous disadvantages. As noted in Chapter 13, duplication of data between files is normally cited as the principal disadvantage of file-based systems. Files tend to be built around single applications without regard to other applications, thus causing the redundancy.

Duplicate data results in duplicate input, duplicate maintenance, duplicate storage, and possibly integrity errors. And what happens if the data format changes? Consider the problem faced by many firms if all systems must support a nine-digit zip code. Do you have any idea how many redundant files and synonyms would have to be located and changed in a typical organization? Add to this the enormous volume of programs that use the field, and you have a nightmare maintenance project.

Another disadvantage of files is inflexibility. Files are typically designed to support the end-user's current requirements and programs. New needs—such as reports and queries—often require files to be redesigned because existing files cannot easily support the new needs. If those files were to be restructured, all programs using those files would have to be rewritten. In other words, the current programs have become dependent on the files, and vice versa. Since reorganization is impractical, redundant files are often created to meet the new requirements. Thus, we escalate the redundancy and the associated disadvantages of redundancy.

Most of these problems are the result of file design inadequacies. In the last chapter you learned how to base conventional file designs on sensible data models developed during analysis. That approach can overcome many of the above problems. In fact, the approach you learned for file design is a database-influenced approach. But in the final analysis, true database makes fewer compromises.

Pros and Cons of Database Neither is database a solution to every problem. We've already stated the principal advantage of database—the sharing of data. A common misconception about database is that you can build a single database that contains all data items of interest to an organization. This notion, however desirable and theoretical, is not currently practical. The reality of such a solution is that it would take forever to build such a complex database. Realistically, most organizations build several databases, each one sharing data with many information systems. Yes, there will be some redundancy between databases. However, this redundancy is both reduced and controlled—even more so than in well-designed conventional files.

Database technology also offers the advantage of storing data in more flexible formats. This is made possible because databases are defined externally from the programs that will use them. Theoretically, this allows us to use the data in ways not originally specified by the end-users. Care must be taken to achieve this data independence. Different combinations of the same data can be easily accessed to fulfill new report and query needs. When fully realized, data independence permits data formats and structure to change (recall our zip code example) without having to change any of the computer programs that currently use that data. Thus, new fields and record types can be added to the database without affecting current programs.

On the other hand, database technology is more complex than file technology. Special software, called a **database management system (DBMS)**, is required. The flexibility provided by a DBMS usually makes it slower (again, the speed differentials between conventional files and databases are getting smaller). Thus many organizations have to buy bigger computers.

The cost of developing databases is higher because analysts and programmers must learn how to use the DBMS. Also, in order to achieve data independence, analysts and database specialists must adhere to rigorous design principles. The DBMS technology itself can be expensive to acquire and maintain.

Another problem with database is the increased vulnerability inherent in the use of databases. Because you are using a shared data resource, you are literally placing all your eggs in one basket. Therefore, backup and recovery procedures and security issues become more complex and costly.

Despite the problems discussed, database usage is growing by leaps and bounds. The technology will get better, and performance limitations will disappear. Design methods and tools will also improve. But for the time being, we'll have to adjust to a world that uses both files and database systems.

Database Environments

Data becomes the central resource in a database environment. Information systems are built around this central resource to give both computer programmers and end-users flexible access to data. Let's examine this environment more closely by studying two different types of database situations.

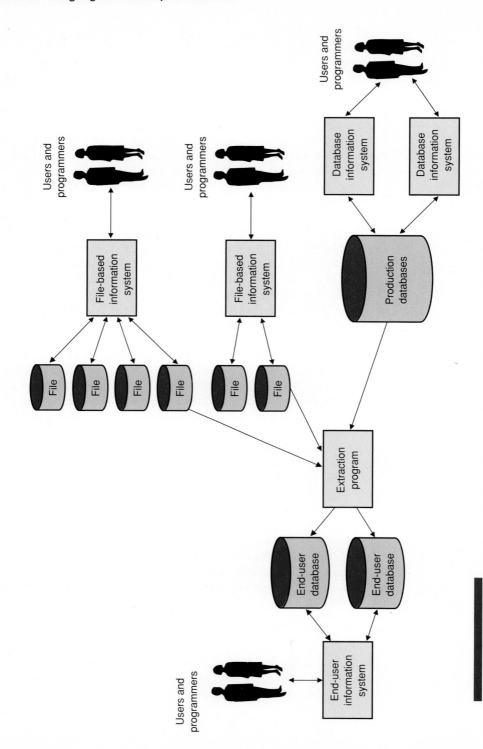

FIGURE 14.2 A Typical Database Environment By today's standards, a typical database environment still includes numerous conventional files. Additionally, it may contain both production databases and end-user databases. Databases are frequently loaded and updated from the conventional files.

Figure 14.2 illustrates the way many companies have evolved into a database environment. Note that most companies have numerous conventional file-based information systems, most of which were developed prior to the popularity of database. In many cases, the processing efficiency of files or the projected cost to redesign files prevents conversion of these systems to database. Other information systems have been built around databases. Access to these databases is limited to computer programs that use the DBMS to process transactions, maintain the data, and generate regularly scheduled management reports. Some query access may also be provided. These databases are often referred to as the **production databases** because they are heavily used to support the transactions for major information systems.

Many information systems shops hesitate to give end-users access to production databases for queries and reports. The volume of unscheduled reports and queries could overload those databases. Instead, **end-user databases** are developed, possibly on separate computers (including micros). Data, extracted from the production databases and conventional files, is stored in these end-user databases. Both conventional and fourth-generation programming languages are used to generate reports and queries off these databases. Admittedly, this scenario is advanced, but many firms are currently using variations of it.

You might be disturbed by what appears in the scenario to be a lack of commitment to the database concept. But just remember that economics, not theory, dictates how most information systems evolve. Existing file-based applications cannot always be quickly and economically converted to database. In many cases, the decision to convert files to databases comes when a system is due to be totally redesigned anyway.

Database Architecture

Let's briefly examine how the database environment is implemented. The discussion is simplified but nonetheless adequate for an introductory-level treatment.

In a database environment, the notion of conventional files disappears. In its place, we have a structure where records of one type, such as ORDERs, can be related to records of different types, such as CUSTOMERs. If this all sounds similar to the data modeling concepts you learned in Chapter 8, that is precisely correct! Database physically implements logical entities and relationships.

How does database work? The database is created, accessed, and controlled through specialized computer software called *database management systems*, which are available from numerous computer vendors.

Figure 14.3 depicts the database technical environment. A systems analyst, or database analyst, designs the structure of the data in terms of record types, fields contained in those record types, and relationships that exist between record types (once again, this should sound like data modeling). The analyst then uses the DBMS's **data definition language (DDL)** to physically establish those record types, fields, and relationships. Additionally, the DDL defines *views* of the database. Views restrict the portion of a database that may be used or accessed by different users and programs. DDLs record the definitions in a permanent data dictionary. Some data dictionaries include formal, elaborate software that helps database specialists keep track of everything stored in the database—including generation of data dictionary reports, analyses, and inquiry responses (very similar to our CASE concept).

Computer programs are then written to load, maintain, and use actual data. These programs may be written in a host programming language—such as COBOL, PL/1, or BASIC—that is supported by the DBMS. Using the host language, the programs call subroutines in the DBMS's **data manipulation language (DML)** to retrieve, create, delete, and modify records, and navigate between record types—for example, from CUSTOMER to ORDERs for that customer. The programmers don't have to understand how the data is physically stored (file organization) or accessed. The DBMS takes care of such details. The DML refers to the data dictionary (DDL) during execution.

Alternatively, many DBMSs don't require a host programming language. They provide their own self-contained programming language that includes a DDL and a DML. Generally speaking, these self-contained languages greatly simplify applications prototyping and development. These languages and features are typically designed to be simple to learn and use, so much so that experienced programmers can be replaced by analysts and end-users. This alternative is also depicted in Figure 14.3.

Many mainframe DBMSs greatly simplify internal controls by automatically logging all updates and enforcing security as defined by the database analysts. Eventually, this should even be true of microcomputer DBMSs.

Multiple-user DBMSs frequently include a **teleprocessing** or **TP monitor**. This is specialized software that supervises and controls access to the database via terminals in on-line environments. Most such systems can also interface with TP monitors other than their own, such as IBM's CICS.

A DBMS allows us to physically implement a logical data model. The physical data model is usually called a **schema**. Any given DBMS supports two schemas, a logical schema (which, unfortunately, is semiphysical as constrained by the DBMS itself) and a physical schema (which truly describes the physical storage of the data).

The **physical schema** defines records, files, access methods, file organizations, indices, blocking, pointers, and other physical attributes. That's right, database management systems don't replace file structures—they just hide them from the programmers and end-users. This aspect of the database is not of concern to most systems analysts.

The **logical** (or semiphysical) **schema** defines the database in simpler terms as seen by end-users and programmers. This schema is very similar,

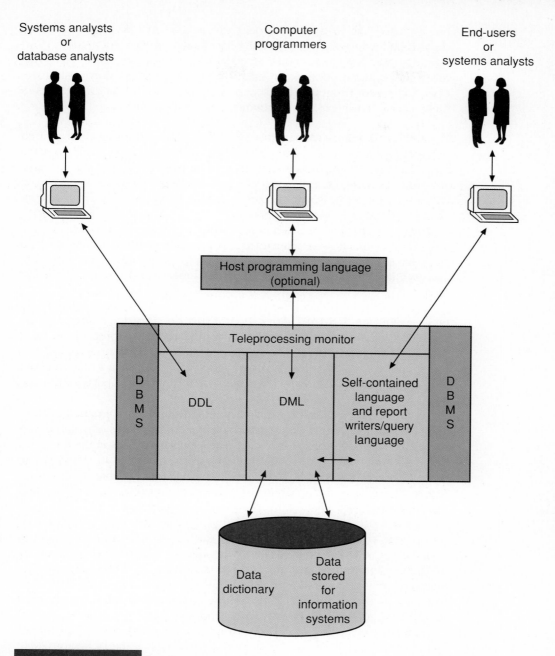

FIGURE 14.3 Architecture of an Ideal DBMS Not all database management systems contain all of the components shown in this diagram; however, the diagram is representative of a typical DBMS. The key components are the data definition language and the data manipulation language, the common components of all DBMSs.

but not identical, to the data models introduced in Chapter 8. It defines records and associations, just like the entity relationship data models. However, the logical schema is constrained by one of three popular DBMS structures: hierarchical, network or linked, and relational. As previously mentioned, different programs or users may be restricted to different views of the database. The views are sometimes called **subschemas**.

Hierarchical Database Management Systems   The earliest DBMSs required different record types to be interrelated only via hierarchical or treelike structures. Each tree is implemented as its own variable-length file that can take advantage of high-performance file organizations like ISAM and VSAM. The DBMS, however, will isolate the programmers from the need to write conventional ISAM or VSAM programs.

In Figure 14.4, we see a hierarchical schema. Hierarchical record types—also called segments—are referred to as parents and children. A parent record can have several children, but a child can have only one parent. A record, such as ORDER, can be a parent and child as long as it doesn't violate the last rule.

The ORDERED PART segment implements the many-to-many relationship between ORDER and PART. Data about a part *on* an order is stored in the sequent. Data about the part, independent of orders, is stored in the PART segment. Using the customer hierarchy, we can retrieve all orders for a customer, all ordered parts for an order, and specific part data for an ordered part. Using the PART hierarchy, we could retrieve all ordered parts for a single part.

As records are retrieved, we have access to their fields. In another hierarchy, we see that for a part, we can retrieve all occurrences of ordered part records as well as historical sales data. The identical records in the two trees only look redundant. Only one of the two trees (files) will in fact store the record. The other tree (file) will actually store a logical pointer to the other record. However, to the programmers, both files appear to physically contain the record type.

Hierarchical databases are normally implemented with complex, but transparent, conventional file structures. Each hierarchy can be considered as one file, albeit more complex than the conventional files you learned about in Chapter 13. For instance, the records (segments) in the customer order hierarchy might be stored as a file as shown in Figure 14.5. Each "bar" represents a record in the file. Thus the file contains dissimilar record types (CUSTOMER, ORDER, ORDERED PART) and qualifies as a database.

Examples of commercial DBMSs that use the hierarchical data model include IBM's Information Management System (IMS), Intel's System 2000, and Information Builder's FOCUS (which is not exclusively hierarchical, though).

Network or Linked Database Management Systems   Network database management systems provide a somewhat more flexible data model. The

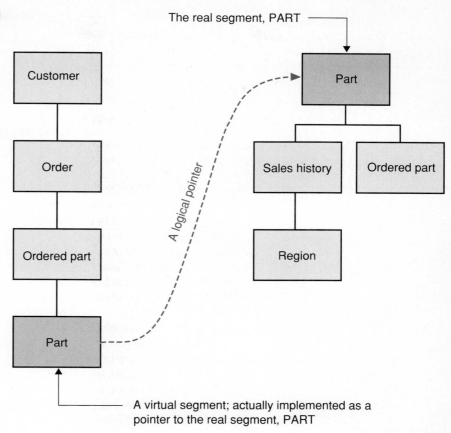

FIGURE 14.4
Hierarchical Data Structure Hierarchical DBMSs require that related record types be linked in treelike structures. No child record can have more than one physical parent record.

The real segment, PART

A logical pointer

A virtual segment; actually implemented as a pointer to the real segment, PART

schema for a typical network DBMS is illustrated in Figure 14.6. Record types are physically linked to other record types by the DBMS through linked lists. The linked lists are sometimes called sets. For example, the set CUSTORD allows us to either find the CUSTOMER linked to an ORDER or all ORDERs linked to a CUSTOMER. The storage of multiple linked lists (pointer chains) is demonstrated in Figure 14.7. The complex pointer chains generally make a network DBMS somewhat slower than its file-based, hierarchical DBMS counterpart. Additionally, the pointers can consume considerable disk storage capacity.

A key difference between the network database and the hierarchical database is that a child record type in a network database can have more than one parent record. Thus, in our example, we didn't need two separate structures to depict the records, in contrast to the hierarchical example.

EOR = end of record or segment

EOF = end of file

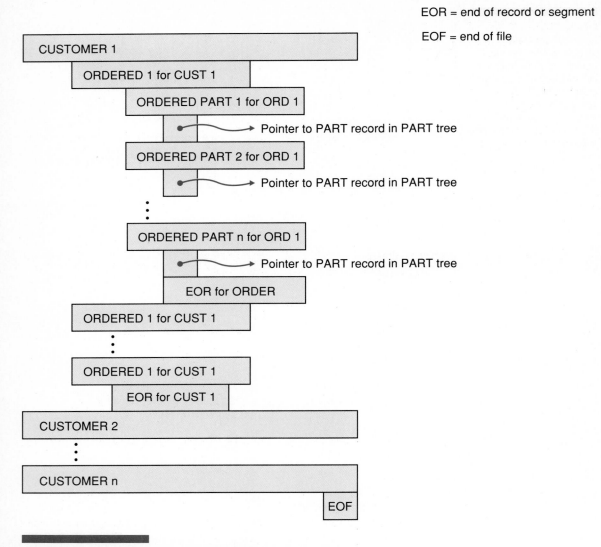

FIGURE 14.5 How Records in One Hierarchical Tree Are Stored
Each tree in a hierarchy is stored as a single, complex file.

Examples of network DBMSs include Cullinet's IDMS and Cincom's TOTAL. IDMS conforms to a network database standard called CODASYL. You can learn more about network structures in any good database textbook or course.

Relational Database Management Systems Relational databases use a model intended to greatly simplify the end-users' and programmers' views of a data-

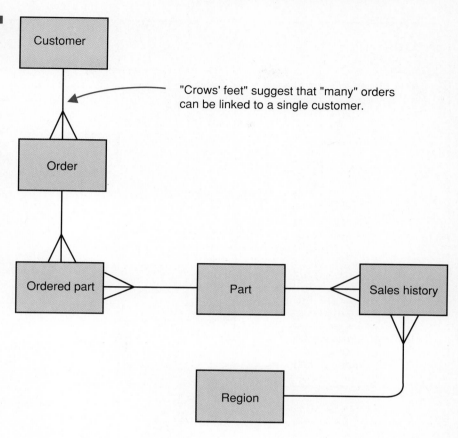

FIGURE 14.6 Network Database Structure A network DBMS allows record types to have any number of relationships to other record types. Each connection logically represents a linked list that connects records of one type to appropriate records of another type. The "crow's-feet" indicate that many occurrences of the record type at the end of the feet may be linked to a single occurrence of the other record type.

"Crows' feet" suggest that "many" orders can be linked to a single customer.

base. As shown in Figure 14.8, files are seen as simple tables, also known as *relations*. The rows are record occurrences—also called *tuples*—and the columns are fields, also called *domains*. Relationships are handled quite differently from the way they are in hierarchical and network models. Instead of physically denoting relationships by using pointers, location, or indexes, true relational databases do not store relationships. Instead, relationships are inferred when needed. How? Relationships are determined from intentionally redundant fields, usually keys, common to the different tables. This concept requires further explanation.

The DML for a relational DBMS doesn't navigate paths and pointers, which is what happens with hierarchical and network databases. Instead, to write reports and answer inquiries, the DML lets the programmer or end-user perform simple table operations to create temporary tables. These operations include

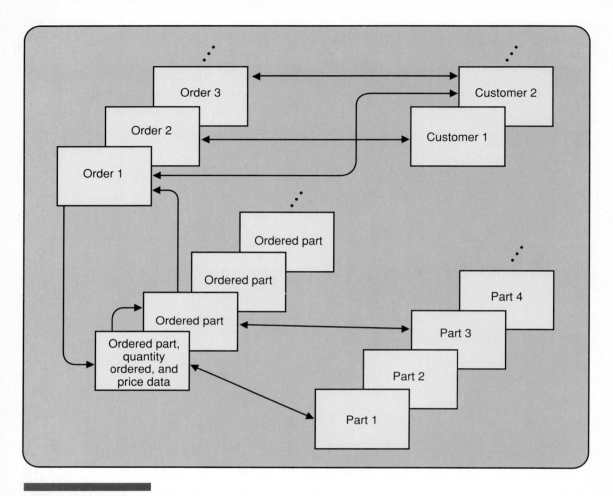

FIGURE 14.7 Record Occurrences in a Network or Linked Database Network or linked database records are stored as linked lists with complex pointer chains. All this, of course, is isolated from the programmer. The DBMS manages the structure.

- **SELECT**ing specific records from a table and creating a new, but temporary, table that contains only those occurrences. Criteria can be set to determine which records to select from the initial table.

- **PROJECT**ing out specific fields from a table, creating a temporary table that has fewer fields.

- **JOIN**ing two or more tables across a common field (this is the same as navigating relationship paths in a hierarchical or network database). Again, a temporary table is created.

Customer Table			
Cust. No.	Cust. Name	Cust. Address	...
10112	Lucky Star	Ann Arbor	
10113	Pemrose	Grand Rapids	
	⋮		

Order Table		
Order No.	Order Date	...
A6334	6 - 3 - 89	
A6335	6 - 4 - 89	
	⋮	

Part Table		
Part No.	Part Desc.	Quant. on Hand
77B12	Widget	8000
77B13	Widget	0
	⋮	

Part records describe data about parts in general. Ordered Part describes data about specific parts on specific orders.

Ordered Part Table			
Order No.	Part No.	Quantity Ordered	...
A6334	77B12	50	
A6334	77B13	100	
A6335	77B13	25	
A6336	77B12	4	
	⋮		

FIGURE 14.8   Relational Data Structures   A relational DBMS depicts record types as simple tables, much like those seen in spreadsheets that exhibit relational-like qualities. Tables are related to one another via intentionally redundant fields, usually keys. Relational DBMSs provide operators that build temporary working tables from the tables illustrated that are physically stored.

In all cases, the above relational operators—which, unfortunately, have different names in the various relational DBMSs on the market—create temporary, working tables that will go away when you exit the program. If changes are made to the data in those tables, those changes will be updated into the permanently stored tables before the working tables are discarded (assuming the end-user has update authority).

Also, note that the relational commands can be combined. For instance, let's say we need a table of orders for certain customers, perhaps to produce a report. First, we can use SELECT on the CUSTOMER table to create a working table of those customers needed. Second, we can PROJECT out only those fields needed in both the CUSTOMER and ORDER working tables to reduce their size. Finally, we JOIN the two working tables to give us the final working table needed for the report.

Relational databases are definitely the trend. They present a simpler viewpoint to both programmers and end-users. Query languages and report writers that are easy to learn and use have been built around relational databases. Most relational databases are currently converging to supersets of a de facto standard language called SQL that is utilized to create, update, and use tables. SQL includes the SELECT, PROJECT, and JOIN operators discussed earlier.

Examples of relational DBMSs include IBM's DB2 and SQL/DL, ADR's DATACOM, Relational Technology's Ingress, Oracle Corporation's Oracle, and Information Builder's FOCUS (which is also hierarchical). Additionally, most microcomputer DBMSs are relational. Examples include Ashton-Tate's dBASE IV (which recently embraced true SQL relational standards) and Microrim's R:BASE for DOS. Many relational DBMSs are offered in both mainframe and micro versions. FOCUS is one such DBMS.

That's about all we want to say about the three common data models. Database textbooks and courses will offer entire chapters and units on each of the three models. We do, however, want to demonstrate a simplified translation of a logical data model into each of the three physical data structures we have introduced. But first we need to describe the roles of the database design participants.

The Database Administrator

Systems analysts and database analysts are becoming increasingly involved in database design. However, in most cases, they rarely have the final say on database designs. Who manages the database environment? That depends on the size of the organization. In many shops, a **database administrator (DBA)** oversees a staff of database specialists. These specialists have final say over all database designs because they have the global, application-independent view that most systems analysts lack. DBAs also load and maintain databases, establish security controls, perform backup and recovery, and maintain the DBMS software. In addition, they plan and control database definition to minimize redundancy and keep track of where all data is stored and how various systems use that data. In smaller shops, a systems analyst may perform some or most of these duties.

THE NEXT GENERATION

Reverse Database Engineering

There is a growing crisis in the commercial database applications arena. For many years, most databases were implemented with hierarchical DBMSs, particularly IBM's IMS. Thousands of databases were built. These hierarchical databases have survived mostly because they provided much better transaction performance than other databases.

Network database management systems like Cincom's TOTAL and Cullinet's IDMS offered non-IMS shops the opportunity to take the database plunge without the complex overhead of hierarchical systems. Hierarchical systems still provided better transaction throughput; however, thousands of network databases have been built.

All the while, researchers were touting a better way, the relational data model. But the relational database seemed more of a promise for the future since performance could not approach that of either hierarchical or network systems.

But the technology of relational database kept improving. It still is. Today, IBM's relational DBMS, DB2, is approaching the throughput of IBM's kingpin, the hierarchical IMS. Other relational DBMSs have also

improved. The writing is on the wall. Relational databases, as once predicted, should replace both hierarchical and network databases.

But there is one final roadblock—the thousands of nonrelational databases that have been written and the programs that were written to update and use those databases! It would take years, perhaps decades, to convert those hierarchical and network databases to relational equivalents.

Enter the next generation of CASE, reverse database engineering. Reverse database engineering, in its simplest form, will read the data definition language (DDL) for one database and automatically convert it to the DDL for another database— for example, IMS hierarchical to DB2 relational.

That sounds impressive, but to work, the reverse engineering must also automatically rewrite the original computer programs to replace all data manipulation language (DML) instructions with the DML instructions for the new database.

We are very close to seeing CASE tools with these capabilities. But it will get even better. With the improvement of artificial intelligence, CASE products can eventually incorporate

expert systems capabilities that will help systems analysts and database administrators correct analysis and design errors in the original databases. For example, many of those databases were conceived before the practice of normalization was common. We might be able to undo those errors.

Thus, CASE reverse engineering may soon provide a migration path from one DBMS's first generation to its second, as well as providing the opportunity for correcting old design errors. In fact, reverse engineering may be the only hope for converting those old hierarchical and network databases. It probably can't be done with manual methods!

Some organizations have split the DBA position into two positions. The *database management systems administrator* manages the technical environment of database. The *data administrator* manages the data itself, keeping track of where all data is stored and what programs and end-users require access to what data.

How to Design and Document Databases

In this section, we discuss how databases are designed. We'll introduce some important issues and guidelines. And we'll design a few databases to demonstrate this important systems design task.

Computer-Assisted Systems Engineering (CASE) has been a continued theme throughout this book. There exist specific CASE products that address database analysis and design. Examples include Chen & Associates' ER Modeler and Bachman's Data Designer. Database design facilities are also finding their way into the more popular CASE tools for systems analysis and design. For instance, we have extended our use of Index Technology's Excelerator to the database designs presented in this section. The schemas presented use a customized version of that product. For a look at future directions of CASE and database design, also see the Next Generation box feature for this chapter.

The design of any database will usually involve the DBA and database staff. They will handle the technical details and cross-application issues. Still, it is useful for the systems analyst to understand the basic design principles for relational, network or linked, and hierarchical databases. Keep in mind that the designs presented here are simplified. It doesn't serve our purpose to get into the technical idiosyncrasies of any specific database management system. Database courses and textbooks usually cover the technical side of the subject in greater detail.

Also, note that each of the designs is representative of typical relational, network or linked, and hierarchical designs. Different products in each class will entail different terminology and present alternative technical constraints. However, all databases of the same class will be based on the concepts described earlier in the chapter.

Step 1: Review Database Requirements

As was the case for conventional file design, the system's data model—in our case, an entity relationship diagram or ERD—serves as a starting point. It identifies the entities and associations that must be designed. For your convenience, we have reproduced the systems pictorial data storage requirements in Figure 14.9. These requirements were determined during the definition phase of systems analysis in our life cycle. Recall that each entity is further described in the project dictionary by a record that defines the data

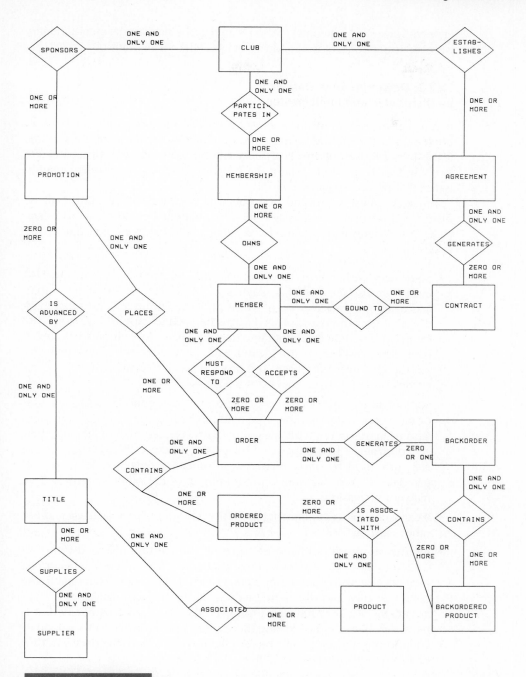

FIGURE 14.9 SoundStage Data Model Developed During Systems Analysis This entity relationship data model was developed during the systems analysis phases of our project. It is the foundation for database design.

elements describing the entity. Also recall that the data model and dictionary were subjected to a rigorous data analysis called normalization.

Step 2: Determine How Data Will Be Distributed and Implemented

Recall that the selection phase of our life cycle identifies different ways to implement a system and assesses the operational, technical, and economic feasibility of those options. Feasibility analysis methods are covered extensively in Part Four, Module D of the book. Let's survey the implementation issues that precede database design.

First, we must determine how the data described in the ERD will be distributed. We don't necessarily want to implement the entire ERD as a database. Distribution is determined according to the following constraints and issues:

- Some data, perhaps a great deal of data, may have to be stored manually—even if we also use conventional files and databases. For example, a radioactive isotope tracking system works under federal regulations that mandate that government-approved forms be used to capture and store data for periods of time extending into decades. We cannot eliminate the manual forms; however, we can glean frequently used data from those forms, put the data into a database, and provide faster and more flexible access to the commonly used data via the database.

- Some data may already be implemented as conventional computer files or databases. A decision must be made regarding the existing implementation. Do we redesign the files or database? Remember, the decision may be constrained by politics or budget. For instance, if existing files are currently used by other programs in other systems, a redesign would mandate the redesign and reprogramming of the other system(s). That may not be feasible.

- We can choose to implement part of the data model as a database and use other existing files as is.

- Alternatively, we could choose to retain the existing files and transport data from those files to a more flexible database. This solution is becoming very popular with microcomputers. Data is downloaded from mainframe computer files and distributed to a small, local micro-database. Yes, we are talking about redundancy; however, the redundancy is planned. It is necessary because the existing files cannot be eliminated. Also, the other applications may eventually be redesigned for use with the database, thus phasing out the conventional files.

- Databases are sometimes distributed to multiple processors. For instance, a major portion of the ERD might become the main database on a mainframe computer. Periodically, portions of that database might be downloaded to a microcomputer database (which might also implement other portions of the ERD) for local processing.

Other implementation decisions are generally constrained by the database management systems currently in use. Most organizations standardize on one or two mainframe database management systems and insist that they be used. Additionally, many organizations limit microcomputer DBMS choices to products approved by the Information Center or Microcomputer Standards Committees. Systems analysts rarely make unilateral decisions to purchase a DBMS.

Again, the distribution and implementation decisions are a prerequisite to database design. The distribution decisions that were made for the Entertainment Club data model are shown in Figure 14.10. Because the existing system is only partially computerized (a simple dBASE IV system), we have decided to implement the entire data model as a single database.

Notice, however, that we discovered that the current inventory system has already implemented a conventional inventory file; therefore, that portion of our database will overlap their file. We could not force that system to rewrite its programs for our database. And we chose not to use their existing file since its structure would not be as flexible for our reporting and query needs.

To maintain consistency between our database and their file, we will have to design a bridge from their file to our database. Each day, we will download the inventory portion of our database from their inventory file. Perhaps a future systems project will convert their inventory information system to a database, at which time they can build on our existing database design. Then, the conventional inventory file can be discarded. This whole approach is consistent with the way most companies evolve from file-based systems to database systems.

Step 3: Design the Logical Schema for the Database

The analyst's role in physical database design is usually restricted to design of what DBAs call the logical schema. Recall that the *logical schema* is a picture or a map of the records and relationships to be implemented by the database. Although it is similar to the ERD, the logical schema reflects the choice of DBMS: relational, network, or hierarchical. The DBA or DB staff will evaluate that schema, make appropriate modifications, generate or modify the DDL, and generate the test database for subsequent prototyping and development. The exception to this procedure is microcomputer databases, which are frequently designed and implemented by the systems analyst.

The schema design and associated dictionary specifications differ for different classes of DBMS technology. We must, therefore, demonstrate step 3 separately for relational, network or linked, and hierarchical databases. In practice, you would only use that database technology which you have, or

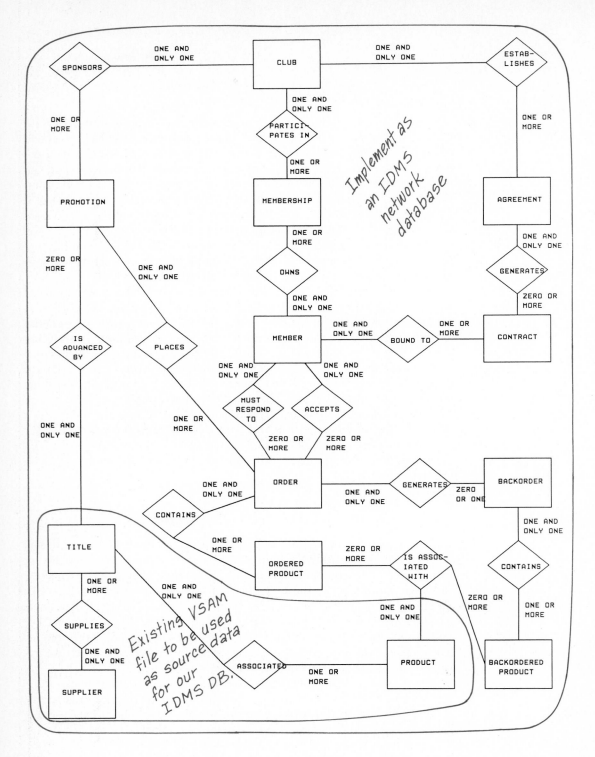

SPONSORS

ONE AND ONLY ONE

CLUB

ONE AND ONLY ONE

ESTAB-LISHES

ONE OR MORE

ONE AND ONLY ONE

PARTICI-PATES IN

ONE OR MORE

ONE OR MORE

AGREEMENT

PROMOTION

MEMBERSHIP

ONE AND ONLY ONE

ONE OR MORE

ONE AND ONLY ONE

GENERATES

ZERO OR MORE

OWNS

ZERO OR MORE

ONE AND ONLY ONE

IS ADVANCED BY

PLACES

MEMBER

ONE AND ONLY ONE

BOUND TO

ONE OR MORE

CONTRACT

ONE AND ONLY ONE

ONE AND ONLY ONE

ONE AND ONLY ONE

MUST RESPOND TO

ACCEPTS

ONE OR MORE

ZERO OR MORE

ZERO OR MORE

ONE AND ONLY ONE

CONTAINS

ORDER

ONE AND ONLY ONE

GENERATES

ZERO OR ONE

BACKORDER

ONE OR MORE

ONE AND ONLY ONE

TITLE

ORDERED PRODUCT

ZERO OR MORE

IS ASSOC-IATED WITH

CONTAINS

ZERO OR MORE

ONE OR MORE

ONE OR MORE

ONE AND ONLY ONE

ONE AND ONLY ONE

SUPPLIES

ONE AND ONLY ONE

ASSOCIATED

ONE OR MORE

PRODUCT

BACKORDERED PRODUCT

SUPPLIER

Implement as an IDMS network database

Existing VSAM file to be used as source data for our IDMS DB.

◄ FIGURE 14.10
Distributed Data Model
for SoundStage This
version of the original
data model reflects the
approved design
decisions. Note that
some of the database
will be loaded and
maintained via
extraction from
conventional files
(similar to Figure 14.2).

plan to have. Analysts rarely select a DBMS since the selection of a DBMS involves factors that extend beyond any one application on which the analyst is currently working.

Step 3 for a Relational Database We'll begin with relational databases, since they represent the current trend and because they are generally easier to design.

Initially, we record our decision to implement the logical model as a relational database. Recall that a database is a physical implementation of a data store. Therefore, we record our implementation decision in a data store description in our project dictionary (Figure 14.11).

Schema design is relatively easy. Ideally, each entity on the Distributed ERD (Figure 14.10) is implemented as a relational table. This is demonstrated in Figure 14.12. Some entities might be split into master tables and secondary tables to separate more frequently accessed fields from other fields in order to improve performance. Why follow the ERD's entities so closely? Recall that the ERD was carefully normalized to minimize redundancy and maximize flexibility. The practice of normalization was derived from early research into

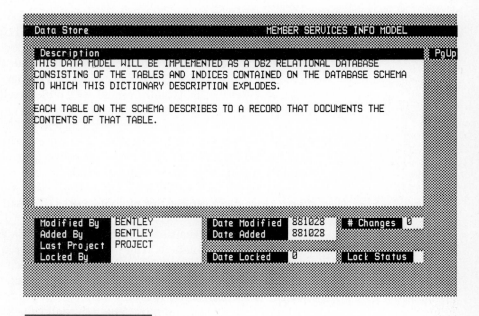

FIGURE 14.11 Project Dictionary Entry for a Global Data Store A database is a global data store. A dictionary entry should be initiated to reflect the chosen database technology.

relational database design. Ideally, we want to retain the advantages of normalization in our database design!

Might you ever want to compromise the third normal form entities when designing the database? For example, might we ever want to combine two third normal form entities into a single table (that would, by default, no longer be in third normal form)? Usually not! Although a DBA may create such a compromise to improve database performance, he or she should carefully weigh the pros and cons. The pros usually manifest themselves as convenience through fewer tables or better overall performance. The major con is the possible loss of data independence—should future, new fields necessitate resplitting the table into two tables, programs will have to be rewritten. As a general rule, combining entities into tables is not recommended.

How are relationships in the ERD implemented by a relational database? Recall that relational databases do not, as a rule, store physical pointers between tables. Instead, associations are established at run time by using combinations of SELECT, PROJECT, and JOIN commands—the command names vary from one relational DBMS to another—to build working tables. Specifically, relationships are implemented by JOINing tables using one or more key fields that the tables have in common. We have represented these virtual relationships on our relational database schema as dashed lines. The lines are labeled to name the fields that would be needed to JOIN the tables.

In some cases, a common element does not exist. For example, there were no data elements common to the CLUB and AGREEMENT tables. In order to join these tables (the need for which was documented on the ERD), we must duplicate the primary key in one table into the other table, where it is called a **foreign key**. Unlike a primary key, a foreign key does not uniquely identify any record in its table. Instead, it identifies a related record in another table.

In our case, we have a choice. We could add AGREEMENT NUMBER—the primary key of the table AGREEMENT—to the CLUB table. This, however, would take the CLUB table out of first normal form since for a CLUB, there could exist multiple agreements. Instead, we will add CLUB NAME as the foreign key in the AGREEMENT table. This won't compromise first normal form, since there can be one and only one CLUB NAME for one AGREEMENT record.

Most relational databases improve performance by allowing (or requiring) that indexes be established on any table from which you might want to SELECT specific records. At a minimum, such an index would be built on primary and secondary keys. We have indicated indexes by diamonds, labeled with the names of the key fields. A solid diamond represents a permanently stored index. A dashed diamond represents a dynamic index, one that must be built each time you want to run a program that will use that index. The "crow's feet" at the end of the line from the index to the table is a classic technique for saying that the index points to many (in fact, most or all) of the records

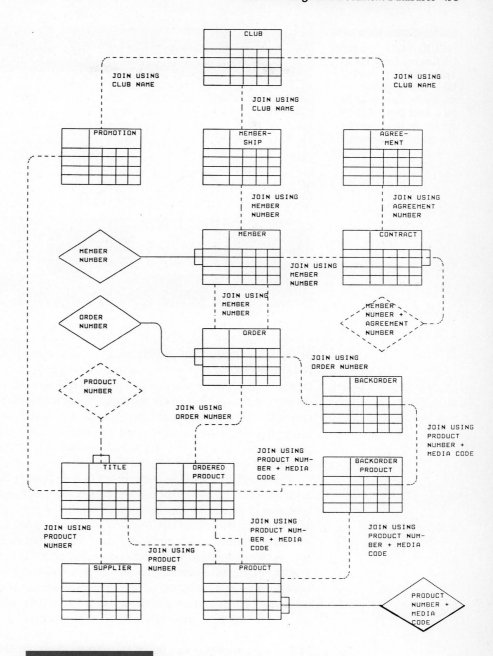

FIGURE 14.12 A Relational Database Schema for SoundStage This database schema illustrates how the SoundStage data model (Figure 14.10) would be implemented as a relational database.

FIGURE 14.13 Project Dictionary Entry for a Table (Part 1) Each table in a relational database represents its own data store that should be described in the project dictionary.

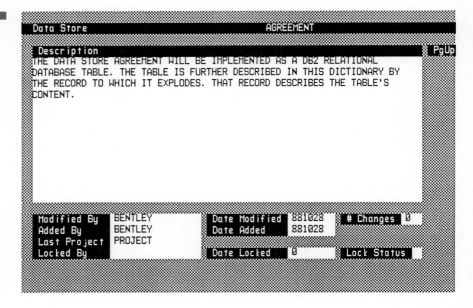

in the table. Purists will rightfully argue that true relational systems do not require indexes. We loosely based our design on IBM's relational database DB2, which does not require indexes. We established permanent indexes for those tables most frequently used and dynamic indexes for medium-volume tables. We did not place indexes on tables with small numbers of records (for example, CLUB), or on tables best accessed through JOINs with other tables (for example, ORDERED PART would normally be accessed by JOINing it with either ORDER or PRODUCT, possibly both).

Each table in the schema explodes to a dictionary description. Each table is the physical implementation of a data store. A typical data store description is shown in Figure 14.13. The contents of the table (data store) are described as records to the dictionary. A printout of the AGREEMENT table's record is shown in Figure 14.14. A sample record is shown in Figure 14.13. This record is a modified version of the entity contents defined during systems analysis. You can eliminate elements that you have decided not to store—for whatever reason—and add elements as needed (for example, the audit fields added at the end of the example). Also note that we added the aforementioned foreign key, CLUB NAME, to this AGREEMENT table record.

Step 3 for a Network or Linked Database Next, we'll design the same database using network database technology. As we mentioned earlier, one example of a network DBMS is Cullinet's IDMS, which conforms to a network DBMS standard called CODASYL. As it turns out, IDMS is the DBMS in use at Sound-

FIGURE 14.14 Project Dictionary Entry for a Table (Part 2) The contents of a typical record in a relational table should be described in the project dictionary.

```
                                                    RECORD - EXPLOSION
                                                    NAME: AGREEMENT

        NAME:                       AGREEMENT
        ALIAS:                      PROMOTIONAL AGREEMENT

        ELEMENT/RECORD                              OFF  OCC  TYPE


        AGREEMENT NUMBER                            000  001   K

        CLUB NAME                                   000  001   E

        MAXIMUM PERIOD OF OBLIGATION                010  001   E

        NO OF PURCHASES REQUIRED IN PERD            010  001   E

        NUMBER OF BONUS CREDITS/PURCH               010  001   E

        NUMBER OF MEMBERS ENROLLED                  010  001   E

        NO OF MEMBERS WHO HAVE FULFILLED            010  001   E

        Record length is 10.
```

Stage Entertainment Club. As was the case with relational DBMSs, different network systems use different terminology and enforce different technical restrictions.

Once again, we would record our IDMS implementation decision in the project dictionary. And once again, we derive our schema from the ERD data model. The resulting schema is shown in Figure 14.15. Each entity on the distributed ERD is implemented as a **record type**. As was the case during relational database design, some entities might be split into two or more record types to separate commonly accessed fields from others, thus improving performance. This decision should be made by the DBA. As much as possible, the analyst should resist the temptation to compromise third normal form by combining multiple entities into single record types. This would compromise data independence and database adaptability to future needs.

Unlike relational databases, you should recall that network associations between different record types are stored as sets of linked lists or pointer chains. We have implemented each relationship from the ERD as a named set (connection between record types) in Figure 14.15. We adopted a naming convention based on abbreviations for the participating record types. The connection uses "crow's-feet" to indicate that there can be many occurrences of one record type for a single occurrence of the connected record type. The connections contain no arrows since the schema does not show flow. All sets can be navigated in both directions.

In addition to record types and sets, a network database requires either entry points or indexes into the database. For example, if we want to find all

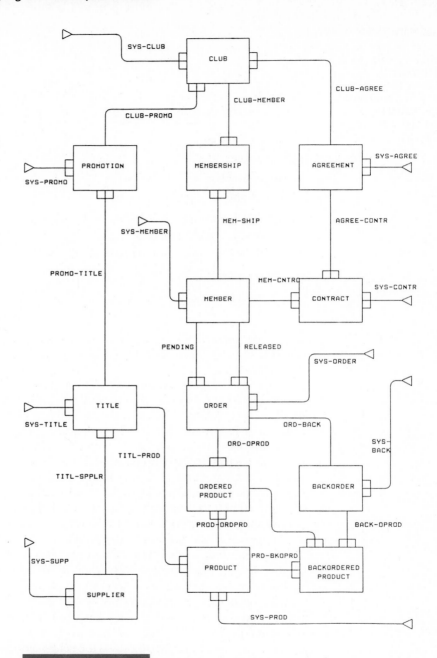

FIGURE 14.15   A Network or Linked Database Schema for SoundStage   This database schema illustrates how the SoundStage data model (Figure 14.10) would be implemented as a network or linked database.

ORDERs for a MEMBER(s), we must first retrieve a MEMBER. Depending on the DBMS, an entry point might be implemented as an index or hashed address (for direct access to records) or another set, called a *system or index set* (for sequential access). For our example, system sets are documented by the small arrowheads connected to various record types. They are all named "SYS-<abbreviated record type name>" to distinguish them from other set types used to navigate the database.

As with relational database tables, our network database record types should be exploded to a record description in the project dictionary. Elements could be added or deleted as necessary, so long as you don't compromise third normal form.

Step 3 for a Hierarchical Database   The design of new hierarchical databases is becoming less common; however, hierarchical databases are still numerous. This is mostly due to the large number of IMS (an IBM product) and FOCUS (an Information Builders product) installations that have built up over the years.

Hierarchical databases are not nearly as simple to design, due in part to the complex technology and terminology as well as the numerous file organization options that are permitted. We are going to greatly oversimplify the material to make it at least somewhat comprehensible.

The overriding constraint in hierarchical databases is that trees can only implement one-to-many and one-to-one relationships. More complex relationships must be established by creating multiple trees and linking them using pointers.

After recording the IMS implementation decision in the dictionary, we offer one possible hierarchical design of our database, which is illustrated in Figure 14.16. You should immediately notice that the schema actually consists of several hierarchies. As explained earlier in the chapter, each hierarchy is stored as conventional but complex indexed files.

How did we determine the number of hierarchies? Most hierarchical databases present numerous DBMS constraints. Hierarchies must be navigated from the root segment down. Therefore, every segment for which we need direct access must be a root segment in its own tree. (Recall that a segment is equivalent to a relational table or a network record type.) We chose four such segments: MEMBER, TITLE, CLUB, and ORDER. The segments must be roots in their respective trees. Furthermore, root segments must be real, not virtual, segments.

Solid rectangles represent real segments, meaning that the segments are actually stored in the file that implements that hierarchy. Dashed rectangles represent virtual segments. They are not really stored in the file. Instead, virtual segments are the pointers to other hierarchies where the segments are *really* stored. This is how a hierarchical database implements many-to-many—or nonhierarchical—associations that exist in the business. There cannot be adjacent virtual segments in any tree. Thus, one pointer segment may

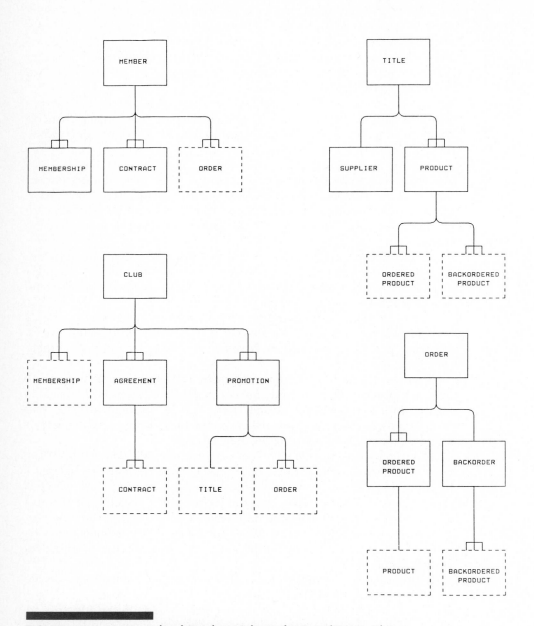

FIGURE 14.16 A Hierarchical Database Schema for SoundStage This database schema illustrates how the SoundStage data model (Figure 14.10) would be implemented as a hierarchical database.

not be connected to another pointer segment; it can only be connected to a real segment.

The combination of virtual segments and logical pointers in a hierarchical schema will be totally transparent to end-users and programmers. All segments will look real to them. They may be virtual segments in the database, but the programmers and end-users don't have to know that—the DBMS takes care of these details.

Segments can be described to the dictionary in a manner similar to record types (in a network DB) and tables (in a relational DB). Most segments are described in CASE dictionaries as records containing elements.

Step 4: Prototyping the Database (If Desired or Mandated)

Prototyping is not an alternative to carefully thought out database schemas. On the other hand, once the schema is completed, a prototype database can usually be generated very quickly. Most modern DBMSs (mainframe and microcomputer alike) include powerful, menu-driven database generators that automatically create a DDL and generate a prototype database from that DDL. A database can then be loaded with test data that will prove useful for prototyping and testing outputs, inputs, screens, and other systems components.

Even in the absence of a mainframe database management system, a prototype database built with a relatively inexpensive micro-DBMS like dBASE IV can prove indispensable as the design proceeds, even if you have no intention of implementing that database with the micro-DBMS. Throwaway prototypes have long been used in engineering circles.

Summary

A database is an alternative to multiple conventional files. It can be thought of as a single file of dissimilar records. These dissimilar records (such as CUSTOMERs, ORDERs, and PARTs) can be linked or associated with one another such that programmers and end-users can navigate the records (for example, find all ORDERs for a CUSTOMER or all PARTs on an ORDER). Navigation is bidirectional, so that for instance we can also find all ORDERs for a PART. Both databases and conventional files have their relative advantages and disadvantages; however, there is a consistent trend toward the use of databases in information systems environments.

Databases are made possible through the use of special software called database management systems (DBMSs). A DBMS uses a data definition language (DDL) to define a database and gives programmers and/or end-users a data manipulation language (DML) to navigate, use, and maintain the database. Some DMLs must be used in conjunction with a host language like

COBOL. The result is to provide programmers and end-users with tools to structure, maintain, and use their data with greater ease and flexibility.

There are three classes of DBMSs. Hierarchical databases organize dissimilar records into easy-to-understand hierarchies or trees that depict the relationships between the records. Network or linked databases organize dissimilar records using linked lists to implement relationships between the records. The current trend is toward relational databases, which store each record as a simple table. Dissimilar records are not linked. Instead, associations are formed when they are needed by joining dissimilar records using fields that they have in common.

Database design, as performed by analysts, is usually restricted to recommending the logical schema or structure of the database needed. A database administrator or DBA will fine-tune that design to consider performance and technical idiosyncrasies. Database design is greatly simplified when it is based on data models developed in the systems analysis phase. Such a data model, such as an entity relationship diagram, is transformed into a database schema that considers the constraints of the DBMS that will be used to implement the database.

Problems and Exercises

1. Dan, an Accounts Receivable manager, is considering a database management system for his microcomputer. He's not certain that he really understands what a database is. In college, he took an introductory computer course and learned about files. He assumed database is the current buzzword for a collection of files. Write him a memo explaining the difference between a file and database environment. What are the advantages and disadvantages of each environment?

2. Explain the pros and cons of conventional files versus databases.

3. What is the difference between a production database and an end-user database?

4. Briefly explain the differences between a *data definition language*, a *data dictionary*, a *host programming language*, and a *data manipulation language*.

5. What is the difference between a DBMS's physical structure and its logical structure? What are three popular logical views supported by DBMSs? Prepare a simple schema for each type of logical view and explain how they differ.

6. Visit your local computer store and ask for a demonstration of a microcomputer database management system. Get some literature on the product and prepare a summary description (one or two pages) for someone unfamiliar with computer technology and terminology.

7. Visit a local information systems shop that uses a database management system. Describe the existing database environment—do they have production-oriented databases or end-user databases? What host programming language(s) is utilized to load, maintain, and use the data? Ask the systems analyst or database administrator to give you a brief orientation on the physical and logical structures supported by the DBMS. To what extent are conventional files used?

8. Discuss the implications that a database would have on each phase of the systems development life cycle.

9. Visit a local information systems shop that operates in a strictly conventional file environment. Ask the systems analyst for information describing several of the master and transaction files. Do some of the files contain duplicate data? Is this data input several times? What impact has the duplicated data had on maintenance? Do they experience problems with data integrity? Have the analyst explain the impact of changing the format of one of the files.

10. If databases were created with the ability to solve many of the problems characteristic of conventional file-based systems, why aren't all information systems shops operating in a database environment?

11. For the database environment at a local business (Exercise 7), discuss the implications of the database on systems development. Do the analysts do anything differently than when using conventional files? How do the analysts interact with the database administrator?

12. Choose a simple system (such as a course registration, order-processing, or payroll system) and identify the basic business entities and relationships to be stored in a database. Try to identify at least four entities for each. Draw a data model diagram to depict the relationships.

13. The data manipulation language for hierarchical and network database management systems navigates paths and pointers to access data for producing reports and answering inquiries. How does the DML for a relational DBMS allow programmers or end-users to access data necessary for reports and inquiries?

14. What is the relationship between the systems analyst and the database administrator?

15. Explain how a logical data model aids in the design of databases.

16. What constraints and issues should be addressed in determining how data depicted on an entity relationship diagram—or other logical data modeling tool—should be distributed?

17. How are relationships appearing on an entity relationship diagram implemented by a relational database management system?

18. Define a foreign key. Explain the difference between a primary key and a foreign key. Give an example of each.

Projects and Minicases

1. Sunset Valley Distributors recently completed a major conversion project. Several months ago Sunset made the decision to move into the database era. Many of its computer-based files had become unreliable, difficult to maintain, and too inflexible to be used to fulfill many end-user reporting and inquiry requests. A database management system seemed to be the obvious solution. Two systems analysts were primarily responsible for the conversion project, which took several months to complete. The systems analysts had decided to simply implement each of the computer-based files as a separate table in their relational database. Once the conversion was completed, the same problems that existed with the file-based system reappeared in the database system. Reports contained inaccurate data, report and inquiry requests could not easily be obtained, and data maintenance was still difficult. A consultant was hired to investigate their problems. The consultant acknowledged that many of the problems resulted because the analysts failed to do data modeling. Explain the importance of doing data modeling ahead of time when designing databases.

2. Design the logical schema for a relational, network or linked, and hierarchical database using the entity relationship diagram on the next page. The primary keys of the entities are as follows:

   CUSTOMER (CUSTOMER NUMBER)

   RENTAL (RENTAL NUMBER)

   RENTAL ITEM (RENTAL NUMBER AND EITHER TITLE, TAPE NUMBER, OR RECORDER SERIAL NUMBER)

   MOVIE TITLE (TITLE)

   VIDEOTAPE (TAPE NUMBER)

   VCR (RECORDER SERIAL NUMBER)

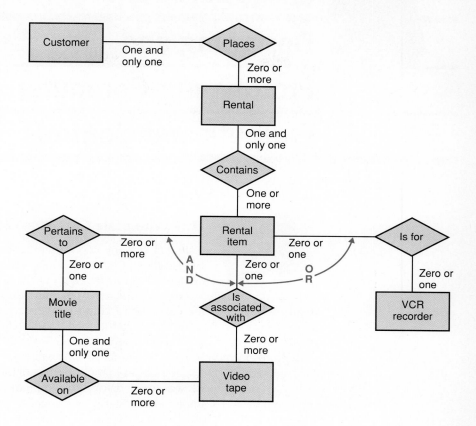

Annotated References and Suggested Readings

There are numerous good textbooks on the subject of database. We suggest you consult your own school's database instructor to add a copy of the required database book.

Hawryszkiewycz, I. T. *Database Analysis and Design*. Chicago: Science Research Associates, 1984. This book deals exclusively with the design of databases.

Kroenke, David. *Database Processing*. 2d ed. Chicago: Science Research Associates, 1983. This is a popular introductory database textbook that surveys database concepts, technology, design, and implementation.

Martin, James. *Managing the Database Environment*. Englewood Cliffs, N.J.: Prentice-Hall, 1983. Martin is one of the most noted authorities, writers, and lecturers in the database field. No database list would be complete without one of his many titles. We chose this title because of its management orientation and readability.

CHAPTER FIFTEEN

Designing and

Prototyping Computer

Outputs and Controls

The Wholesale Cost-Plus Club

Kathleen Li Wang, a programmer/analyst for the Wholesale Cost-Plus Club, had an early afternoon appointment
with Linda Pratney, the Accounts Payable Assistant Manager.
Wholesale Cost-Plus Club is a large, citywide warehouse outlet that sells virtually any type of merchandise to club members for a cost very close to
wholesale price (significantly below the retail prices charged by grocery stores,
drug stores, department stores, and other retail stores). Kathleen had been
largely responsible for the Accounts Payable system implemented last fall.
Accounts Payable pays off invoices to the suppliers of the club's merchandise.
The meeting purpose was a mandatory post-implementation review of the
new system. (Post-implementation reviews occur one month after a new
system replaces an old system.) It was no company secret that Linda was
displeased with some aspects of the new system.

Linda spoke first. "Kathleen, I guess you heard that I'm having a little
trouble with this new payables system. You told me that these computer
reports would make my job much easier, and that just hasn't happened."

Kathleen replied, "I'm really sorry it hasn't worked for you. But let me
make it right. I'm willing to work overtime to make this system work the way
you need it to work. What isn't working for you? I thought we included all
the fields you requested." She asked as an afterthought, "Is it that you're not
getting the reports on time?"

Linda answered, "That's not the problem. And the fields are there. It's just
that I find the report difficult to use. For example, this report doesn't tell me
what I need to know. I spend most of my time trying to interpret it."

"Why don't you tell me how you use the report?" Kathleen suggested.

Linda responded, "Well, first I read down the report, line by line, and
attempt to count and classify the number of invoices that are less than 10 days
old, between 10 and 30 days old, and between 31 and 60 days old. You see,
I have this graph here that shows the totals for each category over the past

12 months. I compare the new totals from this report with the totals on the graph to identify any significant trends in payment activities."

"Why?" asked Kathleen.

"Why the classification? Because we pay some invoices off early to gain a 2 percent discount. Others we defer to the final due date of the supplier, which is usually 30 or 60 days after receipt of the invoice."

"I see. Go on."

Linda continued, "I also use the report to identify the costs of discounts that we did not take, choosing instead to defer payment until close to the final due date. To do this, I locate the invoice on the report and look up that same invoice on the previous copy of the report. I then log the amount as a lost discount to a particular supplier. At the end of the week, I sum the lost discounts by supplier and send the report to my superiors. It really helps. I can think of several occasions when my superiors authorized a change in payment policy for a particular supplier in order to take better advantage of discounts."

Kathleen paused and then replied, "When I wrote down specifications for this report, I assumed I understood what information you were requesting. Sounds as if there is some additional data that should have been included on the report. You're wasting a lot of time looking for information that could be automatically generated by the system."

Linda seemed reassured that Kathleen was sympathetic to her problem. "I hope you can do something about it. I'm pretty frustrated with that report."

Kathleen said, "I'm sure I can have it fixed in no time. I'm really sorry. I guess I just didn't understand what information you wanted and how you would be using it. Let's see, how should we proceed? I don't want to spend a lot of time designing a new report that isn't precisely what you need. I've got an idea. I just got a new package on my office microcomputer. It's called Lotus 1-2-3. I'm pretty sure that I can quickly mock up and, if necessary, change a sample of the report you want. I can even simulate this pie chart graph you want. Once we get the reports and graphs looking the way you want, we'll use them as models for the programmers."

"That's a good idea," Linda acknowledged.

"Okay, what about this other report?" Kathleen was looking at another report that Linda had brought to the meeting.

Linda answered, "I'm getting some flack about this report I asked you to produce for my clerks. It was supposed to list those supplier accounts that we owe payment on. Anyhow, my clerks claim the report is too cluttered and is difficult to use. All they really need to know is the supplier's account number, current balance due, discount date, and the final due date. As you can see, there are a number of unnecessary fields, and the report lists two supplier accounts on the same line. Could you clean this report up too?"

Kathleen responded, "No problem. Again, I'm sorry I was so off target on those reports. I can probably use Lotus to mock up these reports as well."

Discussion

1. What type of reports was Linda receiving?
2. What type of reports does Linda really need?
3. What did Kathleen do wrong? What erroneous assumption did Kathleen make regarding data requirements for the report?
4. What do you think of Kathleen's new strategy of using Lotus 1-2-3 for mocking up new reports? Does the approach sound similar to a strategy that has been frequently described in this book?
5. Why do you suppose the report for Linda's subordinates was inappropriate? Who's to blame, Linda or Kathleen?

What Will You Learn in This Chapter?

In this chapter, you will learn how to design and document computer outputs. You will know that you have mastered the tools and techniques of output design when you can

1. Define the appropriate format and media for a computer output.
2. Differentiate between internal, external, and turnaround outputs.
3. Apply human factors to the design of computer outputs.
4. Design internal controls into computer outputs.
5. Identify data flows on the DFD that must be designed as computer outputs.
6. Define output design requirements and record those requirements in a project dictionary.
7. Describe various techniques for prototyping outputs.
8. Prepare layouts to communicate both paper and screen outputs to programmers who must implement those outputs.

Outputs present information to end-users. Outputs, the most visible component of a working information system, are the justification for the system. During systems analysis, you defined output needs and requirements, but you didn't design those outputs. In this chapter, you will learn how to design effective outputs for end-users.

 **EXTERNAL OUTPUTS**

Invoices
Paychecks
Course Schedules
Airline Tickets
Boarding Passes
Travel Itineraries
Telephone Bills

There are two basic types of computer outputs. **External outputs** leave the system to trigger actions on the part of their recipients or confirm actions to their recipients. Examples of external outputs are listed in the margin. Most external outputs are created as preprinted forms that are designed and duplicated by forms manufacturers for use on computer printers. Some are designed as **turnaround documents** so that all or part of the form eventually reenters the system as an input. The revolving charge account invoice depicted in Figure 15.1 is a typical external turnaround document.

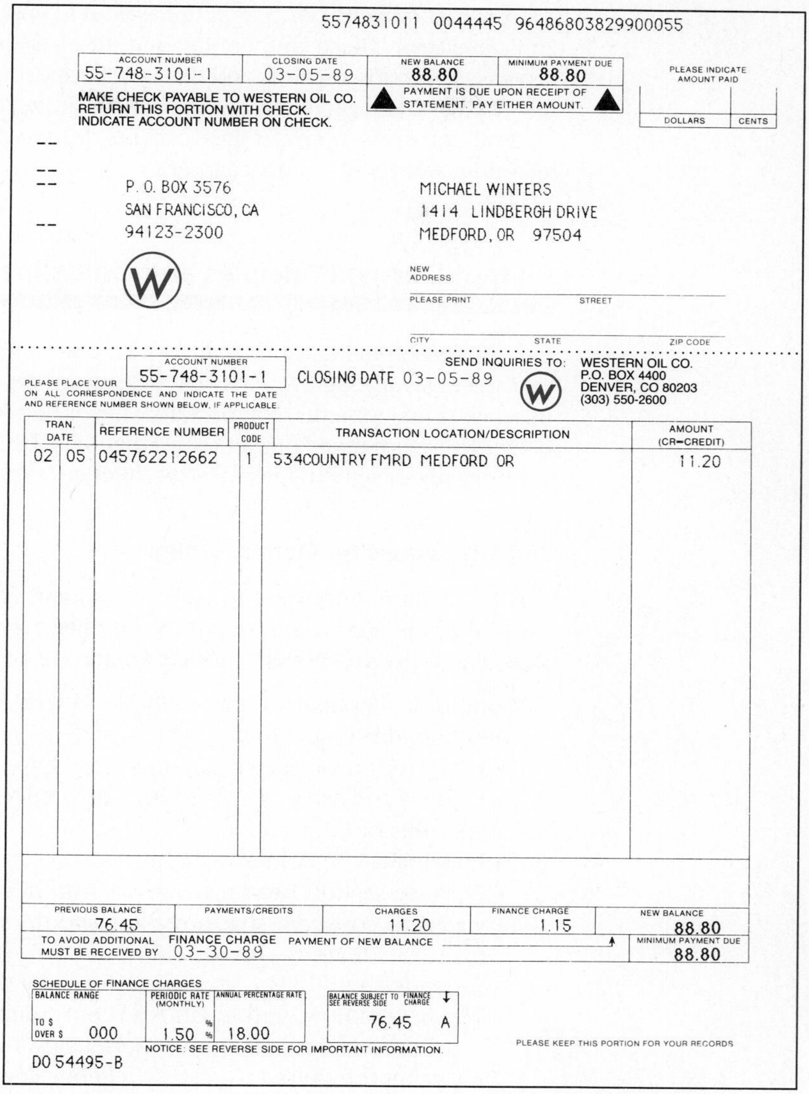

5574831011 0044445 96486803829900055

ACCOUNT NUMBER	CLOSING DATE	NEW BALANCE	MINIMUM PAYMENT DUE
55-748-3101-1	03-05-89	88.80	88.80

MAKE CHECK PAYABLE TO WESTERN OIL CO.
RETURN THIS PORTION WITH CHECK.
INDICATE ACCOUNT NUMBER ON CHECK.

PAYMENT IS DUE UPON RECEIPT OF STATEMENT. PAY EITHER AMOUNT.

PLEASE INDICATE AMOUNT PAID

DOLLARS | CENTS

P. O. BOX 3576
SAN FRANCISCO, CA
94123-2300

MICHAEL WINTERS
1414 LINDBERGH DRIVE
MEDFORD, OR 97504

NEW ADDRESS
PLEASE PRINT STREET
CITY STATE ZIP CODE

PLEASE PLACE YOUR ACCOUNT NUMBER 55-748-3101-1 ON ALL CORRESPONDENCE AND INDICATE THE DATE AND REFERENCE NUMBER SHOWN BELOW, IF APPLICABLE.

CLOSING DATE 03-05-89

SEND INQUIRIES TO: WESTERN OIL CO.
P.O. BOX 4400
DENVER, CO 80203
(303) 550-2600

TRAN. DATE		REFERENCE NUMBER	PRODUCT CODE	TRANSACTION LOCATION/DESCRIPTION	AMOUNT (CR=CREDIT)
02	05	045762212662	1	534COUNTRY FMRD MEDFORD OR	11.20

PREVIOUS BALANCE	PAYMENTS/CREDITS	CHARGES	FINANCE CHARGE	NEW BALANCE
76.45		11.20	1.15	88.80

TO AVOID ADDITIONAL FINANCE CHARGE PAYMENT OF NEW BALANCE MUST BE RECEIVED BY 03-30-89

MINIMUM PAYMENT DUE 88.80

SCHEDULE OF FINANCE CHARGES

BALANCE RANGE	PERIODIC RATE (MONTHLY)	ANNUAL PERCENTAGE RATE
TO $	%	
OVER $ 000	1.50 %	18.00

BALANCE SUBJECT TO FINANCE CHARGE SEE REVERSE SIDE: 76.45 A

NOTICE: SEE REVERSE SIDE FOR IMPORTANT INFORMATION.

PLEASE KEEP THIS PORTION FOR YOUR RECORDS

DO 54495-B

FIGURE 15.1 Sample External Turnaround Document This output is *external* because it initiates a transaction (payment). It is also a turnaround output since a portion of the output is returned, with payment, as input.

Internal outputs stay inside the system to support the system's end-users and managers. These outputs include all of the detailed reports, summary reports, exception reports, and decision support inquiries you learned about in Chapter 3. Sample internal outputs are shown in Figure 15.2.

Both of these output types can be designed using the principles and techniques covered in this chapter.

Output Design Principles and Guidelines

Designing outputs is more than working up a few layout charts for reports and screens. Although you may have designed reports for computer programming class assignments, you probably never had to justify your design to end-users. There are many output issues that must be addressed *before* the output is physically designed. And most of them are people-oriented.

End-User Issues for Output Design

It is particularly important to consider human factors when designing computer outputs. End-users must find outputs easy to use and useful to their jobs. The following general principles are important:

1. Computer outputs should be simple to read and interpret. These guidelines enhance readability:
   - Every report or output screen should have a title.
   - Reports and screens should include section headings to segment large amounts of information.
   - Information in columns should have column headings.
   - Because section headings and column headings are frequently abbreviated to conserve space, reports should include legends to interpret those headings.
   - Legends should also be used to formally define all fields on a report. You never know whose hands a report might end up in! (*Note:* Legends can be built into on-line outputs using function keys to temporarily interrupt the output to display legends and help.)
   - Computer jargon and error messages should be omitted from all outputs, or at the very least, relegated to the end of the output.

   On many computer outputs, the above guidelines are ignored or overlooked; consequently, the outputs appear cluttered and disorganized.

2. The timing of computer outputs is important. Outputs must be received by their recipients while the information is pertinent to transactions or decisions. This can affect how the output is designed and implemented.

3. The distribution of computer outputs must be sufficient to assist all relevant end-users.

SUMMARY OF MONTHLY
BANK MACHINE TRANSACTIONS
FOR THE PERIOD 03/01/89 TO 03/31/89

MACHINE NUMBER	BRANCH NUMBER	NUMBER OF DEPOSITS	AMOUNT OF DEPOSITS	NUMBER OF WITHDRAWALS	AMOUNT OF WITHDRAWALS	NUMBER OF TRANSFERS	AMOUNT OF TRANSFERS	NUMBER OF LOAN PAYMENTS	AMOUNT OF LOAN PAYMENTS
01	01	192	57,600.32	672	31,213.50	140	14,025.33	23	1,725.86
02	03	134	43,756.45	478	23,144.75	63	6,192.88	11	1,545.38
03	05	112	47,650.44	462	24,897.26	43	5,023.61	13	1,195.76
05	07	155	49,864.04	567	27,875.00	97	11,729.58	15	2,304.42
06	08	234	61,768.34	748	37,563.73	153	17,688.93	26	2,112.45

TOTAL DOLLARS DEPOSITED $260,639.59
TOTAL DOLLARS WITHDRAWN $144,694.24
TOTAL DOLLARS PAID ON LOANS $ 8,883.87
TOTAL DOLLARS TRANSFERRED $ 54,660.33

10/02/89   INTERNATIONAL MANUFACTURING COMPANY   PAGE 1
                                    BONUS REPORT

CLOCK NUMBER	SHIFT	CLOCK HOURS	INCENTIVE HOURS	DOWN TIME	BONUS PERCENT	BONUS HOURS
1000	1	35.5	30.0	05.5	150	15.0
1010	1	40.0	32.0	08.0	110	03.2
1020	2	40.0	31.5	08.5	142	13.2
1030	2	36.5	20.3	16.2	113	02.6
1040	3	09.4	08.2	01.2	144	03.6
1050	3	10.2	02.8	07.4	107	00.2
2000	1	55.0	45.3	09.7	134	15.4
2010	1	50.0	33.2	16.8	139	12.9
2020	3	12.1	03.4	08.7	132	01.1
2030	2	20.4	17.9	02.5	125	04.5
3000	1	16.8	12.6	04.2	127	03.4
3010	2	40.5	30.1	10.4	104	01.2
3020	3	40.0	29.0	11.0	143	12.5
3040	3	32.0	29.5	02.5	147	13.9
3050	1	07.0	03.8	03.2	141	01.6
4000	2	60.2	47.8	12.4	150	23.9
4010	1	61.4	50.3	11.1	117	08.6
4020	3	14.7	08.5	06.2	121	01.8
5000	1	50.0	44.1	05.9	100	00.0
5010	1	52.5	40.0	12.5	133	13.2
TOTALS:		684.2	520.3	163.9		151.8

FIGURE 15.2 Sample Internal Outputs Internal outputs include detailed, summary, and exception reports. They are intended primarily for use by managers.

4. The computer outputs must be acceptable to the end-users who will receive them. An output design may contain the required information and still not be acceptable to the end-user. To avoid this, the systems analyst must understand how the recipient plans to use the output. This appeared to be the problem Linda experienced with her reports in the chapter minicase.

Choices for Media and Formats of Computer-Generated Outputs

We assume you are familiar with different output devices, such as printers, plotters, computer output on microfilm (COM), and cathode-ray tube (CRT) display terminals. These are standard topics in most introductory information systems courses. In this chapter, we are more concerned with the actual output than with the device. A good systems analyst will consider all available options for implementing an output, especially medium and format.

A **medium** is what the output information is recorded on, such as paper. The **format** is the way the information is displayed on that medium—for instance, columns of numbers. To select an appropriate medium and format for an output, consider the human factors discussed earlier, paying particular attention to how the output will be used and when it is needed.

Alternative Media for Presenting Information   The most common medium for computer outputs is paper; such outputs are called printed output. Currently, paper is the cheapest of the three media we will survey. Although the paperless office (and business) has been predicted for several years, it has not become a reality. Perhaps there is an irreversible, psychological dependence on paper as a medium. In any case, paper output will be with us for a long time.

Paper is bulky and requires considerable storage space. To overcome the paper storage problem, many businesses have turned to the use of **film** as an output medium. There are two film formats: microfilm and microfiche. **Microfilm** is a roll of photographic film that is used to record information in a reduced size. **Microfiche** is a single sheet of film that is capable of storing many pages of reduced output. The use of film does present its own problems—microfiche and microfilm can only be produced and read by special equipment.

The fastest growing medium for computer outputs is **video**, the *on-line* display of information on a visual display device, such as a CRT terminal or microcomputer display. Although this medium provides convenient end-user access to information, the information is only temporary. When the image leaves the screen, that information is lost unless it is redisplayed. If a permanent copy of the information is required, paper and film are superior media.

Alternative Formats for Presenting Information   There are several formats you can consider for communicating information on a medium. **Tabular** columns of text and numbers are the oldest and most common format for computer outputs. This format presents information as columns or in designated areas. Most of the computer programs you've written probably generated tabular reports.

Zoned output is similar to tabular output. It places text and numbers into designated areas of a form or screen. Zoned output is often used in conjunction with tabular output. For example, an order output contains zones for general customer and order data in addition to tables (or rows of columns) for ordered items.

Graphic output is becoming an increasingly popular alternative format for information. To the end-user a picture can be more valuable than words. Bar charts, pie charts, line charts, step charts, histograms, and other graphs can help end-users grasp trends and data relationships that cannot be easily seen in tabular numbers. The popularity of graphics output has been stimulated by the availability of low-cost, easy-to-use graphics printers and software, especially in the microcomputer industry.

Another current trend is the increased use of the **narrative** format. In this type of format sentences and paragraphs replace or supplement standard text, numbers, and pictures. Word processing technology has exploited the narrative format for reports, business letters, and personalized form letters. For example, an accounts receivable system might interface with a word processor to provide names, addresses, and past due data for personalized credit reminder letters.

Internal Controls for Outputs

Internal controls, a continuing theme throughout the design chapters of this book, are a requirement in all computer-based systems. They ensure that the system is protected against accidental and intentional errors and use, including fraud. Output controls ensure the reliability and distribution of the outputs generated by the computer. The following guidelines are offered for output controls:

- *The timing and volume of each output must be precisely specified.* You cannot simply state that a report is needed daily. When daily? 8:00 A.M.? 10:30 A.M.? 2:00 P.M.? Computer facilities offer limited resources, and the systems analyst must frequently negotiate an appropriate schedule with the computer operations staff.

- *The distribution of all outputs must be specified.* For each output, the recipients of all copies must be determined. A distribution log, which provides an audit trail for the outputs, is frequently required.

- *Access controls are used to control accessibility of video (on-line) outputs.* For example, a password may be required to display a certain output on a CRT terminal.

- *Control totals should be incorporated into all reports.* These controls can be compared with the input controls that will be discussed in Chapter 16. The number of records input should equal the number of records output. These control totals are compared before the outputs are distributed. If a discrepancy is found, the outputs are retained until the cause has been determined and corrected.

We'll apply these controls in the next section.

How to Prototype and Design Computer Outputs

In this section, we'll discuss and demonstrate the process of output design. We'll introduce some tools for documenting output design, and we'll also apply the concepts you learned in the last section.

We will continue to use our Computer-Assisted Systems Engineering (CASE) product, Excelerator, to record our design decisions into our project dictionary. And we'll also demonstrate how CASE and other computer tools can be used to prototype outputs and layouts to end-users and programmers. As usual, each step of the output design technique will be demonstrated using examples drawn from our SoundStage case study.

Step 1: Review Output Requirements

Output requirements should have been defined during systems analysis. A good starting point for output design is the data flow diagrams for the new system. Look at the DFD in Figure 15.3. How is this DFD different from those you have already encountered? The difference is the **person-machine boundary** that has been added to the DFD. It represents an alternative implementation for the requirements shown on the DFD.

Everything inside the boundary will be computerized. Every data flow that travels across the boundary of the DFD, from the machine side to the person side, is a net output that must be designed. Thus, the DFD identifies our output requirements. We have noted all outputs with shaded bullets on the boundary in Figure 15.3.

Each of these output data flows should also have a corresponding project dictionary entry that was recorded during systems analysis. The entry describes the essential contents of the output data flows, independently of their implementation. Now we are ready to address those physical, implementation details that tell the end-users and programmers *how* the output will or should be implemented.

FIGURE 15.3   A Bounded Data Flow Diagram for SoundStage   When evaluating alternative solutions, the systems analyst will add the person-machine boundary to a logical data flow diagram. The systems analysts may also annotate the DFD with specific implementation requirements. Outputs to be designed are indicated by the bullets on the boundary. ▶

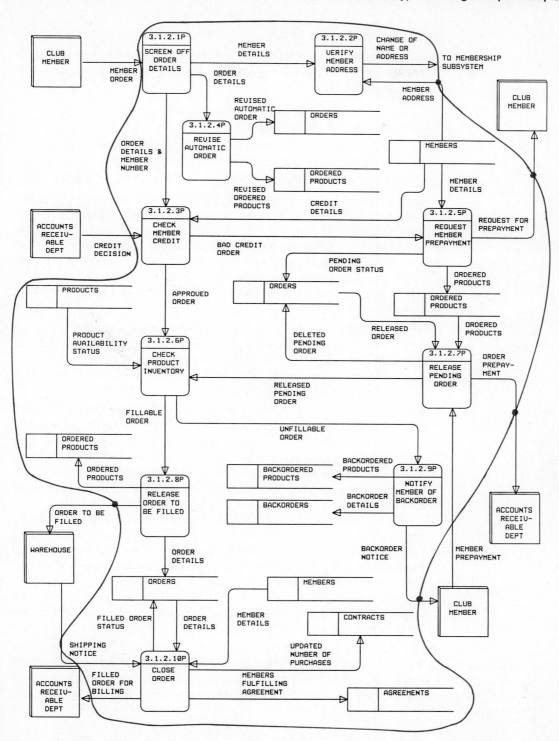

Step 2: Decide How the Output Data Flow Will Be Implemented

Recall that the selection phase of the life cycle will determine how the output data flows will eventually be implemented. Relative to outputs, the decisions will be based on two sets of criteria:

- Based on type and purpose of the output, what are the best medium and format for the design and implementation?
- Based on technical and economic feasibility, what are the best medium and format for the design and implementation?

Since feasibility is important to more than just outputs, the techniques for evaluating feasibility are covered separately (in Part Four, Module D). The first set of criteria, however, is described in the following paragraphs.

First, you must understand the type and purpose of the output. Is the output an internal or external report? If it's an internal report, is it a historical, detailed, summary, or exception report? If it's an external report, is the form a turnaround document? After assuring yourself that you understand what type of report the output is and how it will be used, you need to address several design issues.

1. *What medium would best serve the output?* Various media were discussed earlier in the chapter. You will have to understand the purpose or use of the output to determine the proper medium. You can select more than one medium—for instance, video with optional paper. All of these decisions are best addressed with the end-users.

2. *What would be the best format for the report?* Tabular? Zoned? Graphic? Narrative? Some combination of these? After establishing the format, you can determine what type of form or paper will be used. Computer paper comes in three standard sizes: 8½ by 11, 11 by 14, and 8 by 14 inches. Many printers can now easily compress 132 columns of print into an 8-inch width. You need to determine the capabilities and limitations of the intended printer.

   If a preprinted form is to be used, requirements for that form must be specified. Should the form be designed for mailing? What will be the form's size? Will the form be perforated for bursting into several sections? What legends and instructions need to be printed on the form (both front and back)? What colors will be used and for which copies?

   Incidentally, form images can be stored and printed with modern laser printers, thereby eliminating the need for dealing with forms manufacturers in some businesses.

3. *How frequently is the output generated?* On demand? Hourly? Daily? Monthly? For scheduled outputs, when do end-users need the report? Scheduled reports have to be worked into the information systems operations schedule. For instance, a report the end-user needs by 9:00 A.M. on Thursday

may have to be scheduled for 5:30 A.M. Thursday. No other time may be available.

4. *How many pages or sheets of output will be generated for a single copy of a report?* This data is necessary to accurately plan paper and form consumption.

5. *Does the output require multiple copies?* If so, how? Photocopy (doesn't tie up printer)? Carbons (most printers can make no more than six legible carbons)? Duplicates (requires the most printer time, although laser printers are changing this situation)?

   For external documents, there are also several alternatives. Carbon and chemical carbon are the most common duplicating techniques. *Selective* carbons are a variation where certain fields on the master copy *will not be printed* on one or more of the remaining copies. The fields to be omitted must be communicated to the forms manufacturer. *Two-up printing* is a technique whereby two sets of forms—possibly including carbons—are printed side by side on the printer.

6. *For printed outputs, distribution controls should be finalized.* For on-line outputs, access controls should be determined.

7. *For elements contained on the output, what format should be followed?*

After a design is determined feasible and approved by end-users and management, the preceding design decisions are recorded in the project dictionary. As noted earlier, an output is a data flow. During systems analysis the logical data flow was recorded. Now, the physical, implementation decisions are recorded. Let's study an example.

ORDER TO BE FILLED is an output data flow to be designed for the SoundStage Member Services system. Discussions with all end-users reveal that the output needs to be designed as a form with three additional copies. The original will be called the "master," and the three copies will be called "picking," "packing," and "shipping." To fulfill this requirement, we will design a preprinted form with carbon copies. Today, most carbons are produced using chemically coated paper rather than the messy ink carbons.

The design specifications for ORDER TO BE FILLED are presented in Figure 15.4. Note that the physical, implementation specifications have simply been appended to the logical, implementation-independent details documented during the analysis phases. This figure is a printout (not a screen dump) from our CASE product, Excelerator. Take a moment to scan through the entries on this form.

The entries are relatively self-explanatory. They address the issues presented earlier in this chapter and section. All of the criteria were developed with and approved by the end-users.

There are also a number of design considerations for data elements that are included in an output. Figure 15.5 is a printout of the data flow's contents

FIGURE 15.4 Design Specifications for ORDER TO BE FILLED After the design is determined feasible and approved by end-users, the output design decisions are recorded in the project dictionary.

```
                          DATA FLOW - OUTPUT                      PAGE  1
                          NAME: ORDER TO BE FILLED


    TYPE Data Flow                          NAME ORDER TO BE FILLED

       Label  ORDER TO            EXPLODES TO ONE OF:
              BE FILLED           Record                 ORDER TO BE FILLED
                                  Data Model Diagram
                                  ERA Diagram



       Description

       DESCRIPTION:  A FOUR-PART DOCUMENT DESCRIBING A SALES ORDER
       MEDIUM:  PAPER                    OUTPUT TYPE:  EXTERNAL/FORM
       VOLUME: 5,000 PER DAY             NUMBER OF COPIES: 3
       COPYING METHOD:  SELECTIVE CARBON
       FREQUENCY PREPARED:  DAILY AT 8:00 A.M. AND 1:00 P.M.
       OUTPUT CHARACTERISTICS:  PREPRINTED 8 1/2" X 8 1/2 ", DESIGNED
                                FOR MAILING. THE 3 COPIES SHOULD HAVE
                                DIFFERENT COLORS:
                           ORIGINAL/MASTER = WHITE    PICKING COPY  = YELLOW
                           PACKING COPY = PINK        SHIPPING COPY = GREEN
       OUTPUT RECIPIENT(S): ALL COPIES WILL BE PICKED UP BY ORDER ENTRY
       SPECIAL INSTRUCTION: QUANTITY ORDERED IS HAND ENTERED
                            QUANTITY BACKORDERED MAY BE MODIFIED BY HAND
```

(the record to which the data flow in Figure 15.4 explodes). Each of these elements is further described by the printout in Figure 15.6. Note the following items on Figure 15.6:

(A) For each data element, we specified the data TYPE as follows:
  • Packed (P)—a *numeric* data element. Why is packed designated as a type for numeric fields? Most numeric elements in business information systems are stored and manipulated in packed decimal format (which was described in Chapter 13).
  • Character (C)—also known as *alphanumeric,* a data element that contains any combination of alphabetic letters, nonarithmetic numbers, and special characters.
  • Binary (B) and Floating Point (F)—alternative formats for numeric data; these are usually encountered in scientific and engineering applications and are rarely used in information systems.
  These codes were dictated by the CASE product being used for our project dictionary. Other CASE products and manual techniques may adopt different codes—for instance, AN = alphanumeric, N = numeric, and so forth.
  We recommend that nonarithmetic data elements that consist only of numbers be specified as character, alphanumeric, instead of packed,

FIGURE 15.5 ORDER TO BE FILLED Record
This report identifies the data element content of the record. This report was produced using the CASE product Excelerator. The report truncated the definitions of the record and each data element.

```
                                        RECORD — EXPLOSION
                                        NAME: ORDER TO BE FILLED

NAME:                        ORDER TO BE FILLED
ALIAS:

ELEMENT/RECORD                          OFF  OCC  TYPE  LEN


ORDER NUMBER                            000  001   1    006

MEMBER NUMBER                           006  001   7    007

MEMBER NAME                             013  001   8    030

ORDER DATE                              043  001   8    008

ORDER STATUS                            051  001   8    007

PREPAID AMOUNT                          058  001   9    005

ADDRESS                                 063  001   R
  P.O. BOX                              063  001   E    010
  STREET                                073  001   E    015
  CITY                                  088  001   E    015
  STATE                                 103  001   E    002
  ZIPCODE                               105  001   E    009

ORDERED ITEM                            114  015   R
  PRODUCT NUMBER                        114  001   E    007
  MEDIA                                 121  001   E    008
  CREDIT                                129  001   E    001
  QUANTITY ORDERED                      130  001   E    004
  QUANTITY SHIPPED                      134  001   E    004
  PRODUCT PRICE                         138  001   E    005
  EXTENDED ORDERED PRODUCT PRICE        143  001   E    006

TOTAL ORDER CREDITS                     639  001   E    002

TOTAL ORDER PRICE                       641  001   E    007

BONUS CREDITS NOT USED                  648  001   E    002

Record length is 650.
```

numeric (for example, PRODUCT NUMBER, CUSTOMER NUMBER, ZIP CODE). Why? By specifying the element as alphanumeric, we make it clear to the programmer that arithmetic should not be performed on the element.

(B) For each data element, we specified size in terms of CHARACTERS LEFT and CHARACTERS RIGHT of the decimal point. We do not include special editing symbols, such as hyphens, commas, decimal points, and slashes, when determining this size. For character or alphanumeric elements, only the CHARACTERS LEFT will be used.

(C) You may recognize the entries under PICTURE as being standard COBOL picture entries. COBOL picture clauses are a relatively common way to

```
                    REPORT OF DATA ELEMENTS CONTAINED                    PAGE    1
                    IN ORDER TO BE FILLED RECORD
                              (A)                    (B)              (B)       (C)
  ELEMENT NAME                TYPE CHARACTERS LEFT   CHARACTERS RIGHT OUTPUT PICTURE

  MEMBER NUMBER                 C         7                   0       9999999
  ORDER NUMBER                  C         6                   0       ZZZZZ9
  MEMBER NAME                   C        30                   0       X(30)
  ORDER DATE                    C         8                   0       MM/DD/YY
  ORDER STATUS                  C         7                   0       XXXXXXX
  PREPAID AMOUNT                C         3                   2       ZZZ.99
  P.O. BOX                      C        10                   0       X(10)
  STREET                        C        15                   0       X(15)
  CITY                          C        15                   0       X(15)
  STATE                         C         2                   0       AA
  ZIPCODE                       C         9                   0       X(9)
  PRODUCT NUMBER                C         7                   0       9999999
  MEDIA                         C         8                   0       XXXXXXXX
  CREDIT                        P         1                   0       9
  QUANTITY ORDERED              P         4                   0       ZZZ9
  QUANTITY SHIPPED              P         4                   0       ZZZ9
  PRODUCT PRICE                 C         3                   2       ZZ9V99
  EXTENDED ORDERED PRODUCT PRICE C        4                   2       ZZZ9V99
  TOTAL ORDER CREDITS           P         2                   0       Z9
  TOTAL ORDER PRICE             C         5                   2       ZZZZ9.99
  BONUS CREDITS NOT USED        P         2                   0       Z9
```

FIGURE 15.6   Data Element Details of ORDER TO BE FILLED Output   This report details the data elements to appear on the output. The report was produced using the CASE product Excelerator.

communicate element print formats to programmers. Figure 15.7 presents these editing symbols and samples of resulting output for your review.

Where did the details about the elements come from? During systems analysis—when the analysts were learning and documenting business requirements—they recorded most of these details in the project dictionary using data element screens like the one depicted in Figure 15.8. Any new elements added to outputs should be similarly recorded in the data element dictionary.

Step 3: Prototype the Layout for End-Users

After design decisions and details have been recorded in the project dictionary, we must create the format of the report. The format or layout of an output directly affects the end-user's ability to read and interpret it. The best way to lay out outputs is to sketch or, better still, generate a sample of the report or document. We need to show that sketch or prototype to the end-user, get feedback, and make modifications to the sample. It's important to use realistic or reasonable data and demonstrate all control breaks.

SYMBOL	EDITING REQUIREMENT	EDIT MASK	SAMPLE DATA	HOW DATA WILL APPEAR TO USER
9	Print any numeric digit (0–9) in this position. Used for numeric fields.	999	000 005 346	000 005 346
X	Print any character in this position. Used for alpha-numeric fields.	XXXXX –or– X(15)	HORSE NO. 3	HORSE NO. 3
A	Print any letter of the alphabet (or blank space) in this location. Used for alphabetic fields.	AAAAA –or– A(5)	CAT CAT	CAT CAT
Z	Suppress leading zeros. Use only for numeric fields.	Z9 ZZ	07 00 10 00	7 0 10
B	Print a blank character at this position.	99B99 XXBXX AABAA	1234 FRT6 SSNO	12 34 FR T6 SS NO
$	Print a dollar sign at this position. Used only for numeric fields.	$999 $ZZZ $$$$	015 015 015	$015 $ 15 $15
.	Print a decimal point at this location. Use only for numeric fields.	99.99 ZZ.ZZ $$.99	05.75 05.75 05.75	05.75 5.75 $5.75
/	Print a slash at this location in the field.	XX/XX 99/99 AA/AA	AB61 5533 SSSN	AB/61 55/33 SS/SN
–	Print a hyphen at this location in the field.	XX–XX 99–99 AA–AA	FRT6 1266 SSNO	FR–T6 12–66 SS–NO
,	Print a comma at this position if the digit to the left is 1–9. Use for numeric fields only.	9,999 Z,ZZ9	1356 2241 0225	1,356 2,241 225

FIGURE 15.7   COBOL Syntax for Defining Editing Requirements

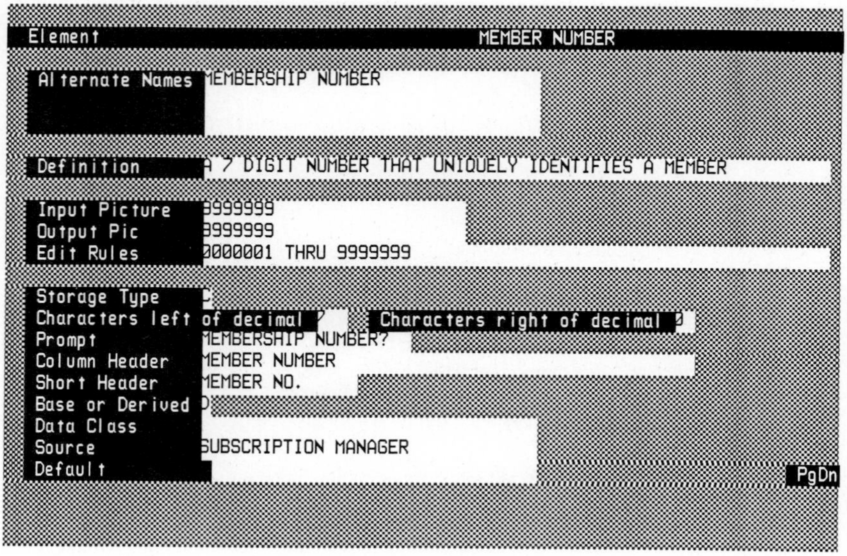

FIGURE 15.8 Sample Data Element Screen
This is a sample Excelerator screen for defining the details of a data element.

Prior to the availability of prototyping tools, analysts could only sketch rough drafts of outputs to get a feel for how end-users wanted outputs to look. With modern tools, we can develop more realistic prototypes of these outputs. Let's study some tools and techniques for prototyping output layouts. We'll demonstrate each technique with examples of outputs from the SoundStage project.

Simple Prototyping Perhaps the least expensive and most overlooked prototyping tool is the common **spreadsheet**. Examples include Lotus 1-2-3, Microsoft's Excel, Borland's Quattro, and Microsoft's Multiplan. A spreadsheet's tabular format is ideally suited to the creation of rapid prototypes. And since arithmetic and logical formulas and functions can be placed in cells (*note:* a cell is the intersection of a row and column), spreadsheets can automatically calculate and recalculate some cells to make the information accurate. Finally, most spreadsheets now include facilities to quickly convert tabular data into a variety of popular graphic formats. Consequently, spreadsheets provide an unprecedented way to prototype graphs for end-users.

Figure 15.9(a) is a Microsoft Excel–generated prototype for a SoundStage summary report. Figure 15.9(b) is a graphic bar chart version of the same information. The prototype chart was created in less than one minute.

Prototyping with CASE Tools Many CASE products support or include facilities for report and screen design and prototyping via a project dictionary created during systems analysis. For example, Pansophic's Telon helps analysts and end-users create rapid prototypes and, once the prototypes are approved, converts those prototypes into COBOL or PL/1 code, including interfacing code for CICS (a teleprocessing monitor), IMS (a hierarchical database management system), or DB2 (a relational database management system).

Products like Telon assume the end-users have defined their requirements and data needs. Throughout this book, we've demonstrated the use of Excelerator as a CASE product to define requirements and data needs. Excelerator requirements can be directly interfaced to Telon. On the other hand, Excelerator also provides its own internal facility (called *screens & reports*) for prototyping outputs and screens.

A screen design-in-process is shown in Figure 15.10(top). Normally, the prototype would use the entire screen. In this case, however, the analyst has told Excelerator that a new field is to be added to the screen. The bottom third of the screen has been replaced with a field definition form. The analyst knows this field has been previously defined in the project dictionary as a data element. By typing that element's name in the *Related ELE* space, Excelerator will look up the element and automatically copy pertinent specifications into the field definition form. The analyst can change these specifications for *this* implementation of the element as well as adding new attributes in the field definition—for example, blinking, reverse video, and so on. The final prototype screen (with sample information) is depicted in Figure 15.10(bottom).

After all fields have been placed on the screen, the analyst simply completes the prototype by entering values. Excelerator conveniently checks the values against the allowable value ranges defined for each field to ensure the data is reasonable. After showing the data to end-users and gaining their approval, the screen can, if desired, be automatically converted to its equivalent COBOL picture.

Prototyping Output Layouts with Database Management Systems or Fourth-Generation Languages Recall that most database management systems or fourth-generation languages include powerful applications generators for quickly prototyping fully functional systems. If a prototype file or database was created during file or database design (Chapters 13 and 14), the test data stored in the file or database can be used to prototype outputs and screens. Most systems prototyping tools include report writers and query languages that allow analysts (or end-users) to quickly design and generate samples of outputs. This capability can be found in expensive mainframe databases such as Cullinet's IDMS, in relatively inexpensive, microcomputer databases such as Ashton Tate's dBASE IV, and in 4GLs such as Information Builder's FOCUS and PC/FOCUS.

Even though the SoundStage system will eventually be implemented in COBOL with the IDMS database management system, the analysts quickly

Member Response to Selection of Month by Club Category September, 1989

Category	Potential Orders	Selection of Month	Alternate Selection	Selection of Month + Alternates	No Order
Pop/Rock	6342	2410	824	241	2867
Country	3577	1538	644	154	1241
Easy Listening	954	181	38	18	716
Classical	1486	877	45	88	477
Jazz	540	389	54	39	58
Show/Comedy	104	9	54	1	40

LEGEND

Category: Club to which members belong
Potential Orders: Number of members in club who received selection of the Month promotion
Selection of Month: Number of members who selected ONLY the Selection of the Month
Alternate Selections: Number of members who selected titles other than the Selection of the Month
Selection of Month + Alternates: Number of Members who selected both Selection of Month plus alternates
No Order: Number of members who rejected Selection of Month and ordered no alternates

(a)

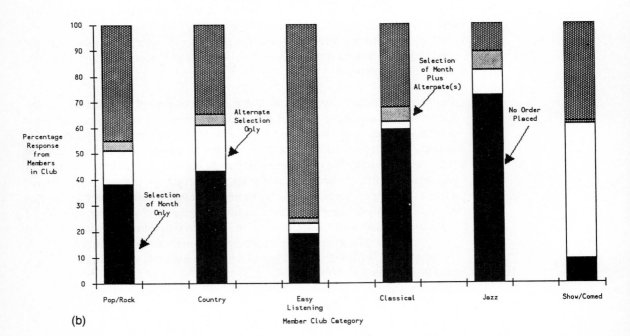

(b)

◄ **FIGURE 15.9 Sample Prototypes Generated from a Spreadsheet** The summary report prototypes were generated using Microsoft's Excel. A facility was used to quickly convert the tabular report into the equivalent bar chart.

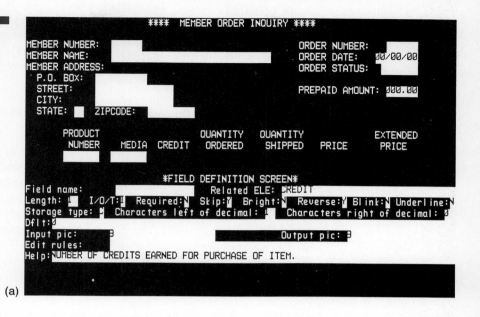

(a)

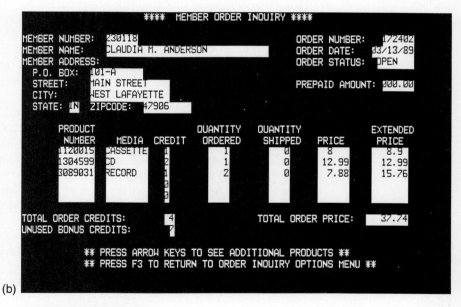

(b)

FIGURE 15.10 Prototyping Screens from a CASE Tool

built prototypes using an inexpensive microcomputer database management system, dBASE III (which was handy since SoundStage had already implemented a crude MEMBER database using that DBMS). Thus, the prototypes were developed both quickly and inexpensively, two properties of a good prototyping environment.

Step 4: If Necessary, Convert User-Oriented Layouts into Programmer-Oriented Layouts

In some cases, the prototype layout is sufficient for programmers to implement the design. You would simply say, "Make it look like this." But some prototypes cannot be used as final specifications. The prototype may contain compromises that were imposed by the tools used to create the prototypes. Also, the sample information on the prototypes may not be complete—in other words, the prototype doesn't accurately communicate how to handle all data formats.

In these situations, or if standard procedures dictate it, the analyst may have to produce a more technical layout for the programmer. This requires the use of classic layout charts.

Using Printer Spacing Charts to Document the Format of Printed Outputs For printed forms and reports, the **printer spacing chart** (Figure 15.11) remains a popular tool to describe the format of printed outputs. The form consists of a **printing grid** that indicates the print positions by row and column. These charts can accommodate almost any size output you might want to design (however, the grid is not to scale—in other words, one inch on the grid does not necessarily equal one inch on the final output).

Carriage control, the column to the left of the printing grid, identifies the first and last print lines on a form. For most printers, carriage controls for standard paper sizes are stored in a small memory in the printer. Programmers specify print locations and spacing by using the carriage controls within their programs. For special forms, the analyst must specify the carriage controls to be loaded into the printer whenever that form is mounted on the printer. Depending on the printer used, up to 12 controls can be defined for a form. The number 1 is normally reserved for the first line *to be printed* and 12 for the last line *to be printed*. The remaining numbers can be defined by the analyst to skip—sequentially only—to specific lines on the form.

The **occurs** column specifies the number of times that the corresponding line may be printed *on one page* of the report or document. The **trigger** column describes the conditions for printing the corresponding lines on the printer spacing chart. For example, some of the lines recorded on the chart might only be printed at the top or bottom of each page or whenever a control break occurs.

The printing grid is used to formally specify the layout of both constant and variable information for the report or document. We'll discuss this later

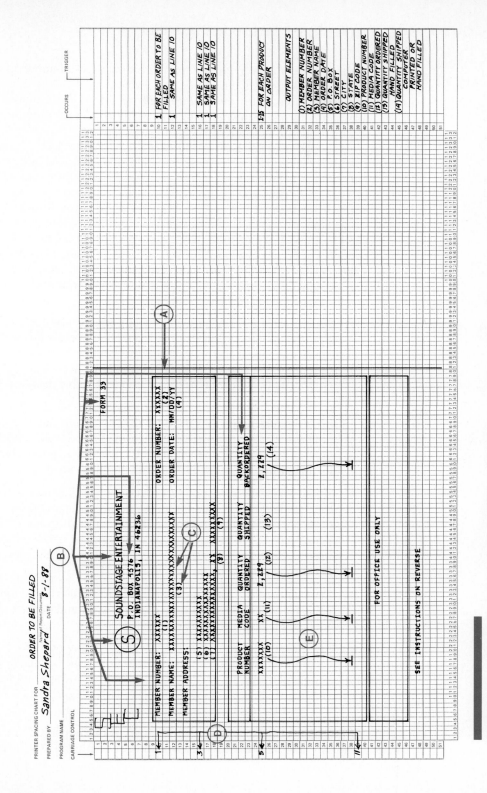

FIGURE 15.11 **Printer Spacing Chart for ORDER TO BE FILLED** A printer spacing chart is used to design the layout of printed outputs.

in the chapter. Caution! Printer spacing charts are not intended for end-user verification. End-users tend to be somewhat put off by the editing symbols and conventions. After the end-user has approved the format, finalize that format as a formal printer spacing chart with which to communicate the final design to the programmer!

Let's take a closer look at the example in Figure 15.11. Printer spacing charts can be used for virtually any type of printed internal or external output. Our sample is an external output, specifically, a preprinted ORDER TO BE FILLED form. Thus, this design will be passed along to a forms manufacturer who will create the form and duplicate it for use in the SoundStage printers.

If different copies of the output were to look different, we would have used multiple spacing charts. Notice that the format includes title, headings, data elements, spacing, page breaks, and other details of importance to the computer programmer who must implement the report. This information may be difficult to derive from a simple prototype. Additionally, the following details are called to your attention:

(A) The size of the form, in terms of number of lines and columns, was indicated on the spacing chart with bold lines. Perforations can be drawn with dashed lines. The boundary of the form was drawn in by the analyst. Remember, the printer spacing chart is *not* a scale drawing. The actual size should be recorded in the dictionary for this data flow.

(B) We drew and printed the constant information to be preprinted by the forms manufacturer. This includes title, company name, address, phone numbers, logos, form name and form number, unique identification number (if that is to be preprinted), lines and blocks that divide the document into sections, and column and field headings.

(C) We used our edit masks from the data element dictionary to record fields to be printed by the computer. The number in parentheses refers back to the corresponding field in the data element dictionary ("REF"). Don't record pictures in fields that will be entered by hand—that would confuse the programmer. In other words, only record picture clauses for fields to be printed by the computer.

(D) Carriage controls were established because this is a nonstandard form size. The printer needs to have its top-of-page and bottom-of-page redefined when printing this form.

(E) There may be from one to twenty parts on any one order. Notice how we used vertical lines with arrowheads to indicate that the line describing a single part may be printed repeatedly. The line spacing will be single.

What if single spacing is inappropriate? Figure 15.12 demonstrates how alternative spacing might be shown.

We again remind you that the documentation prepared in this example is primarily intended for the analysts and programmers. The picture clauses

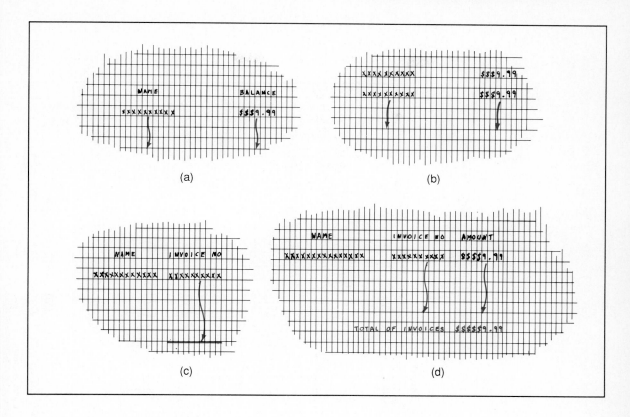

FIGURE 15.12   Spacing Requirements for Detail Lines

recorded on our printer spacing chart will likely be too difficult for end-users to interpret and evaluate.

Using Display Layout Charts to Document the Format of Visual Outputs

Terminal screen **display layout charts** (see Figure 15.13) are used to document typical screens (80-column limit). For screens wider than 80 columns, printer spacing charts can be used.

The **display grid** is used in exactly the same way as the printing grid in the printer spacing charts. You may have noticed that the number of lines (43) is greater than the number of lines that can be displayed on most screens (24). Why? The additional space allows you to indicate additional information that can be scrolled up as the end-user *pages* through the output. To format the screen output you use the same COBOL picture clauses and conventions that you use in printer spacing charts.

TERMINAL SCREEN DISPLAY LAYOUT FORM

APPLICATION _CUSTOMER SERVICES_

SCREEN NO. _4_ SEQUENCE _4_

☐ INPUT
☑ OUTPUT

COLUMN

```
                                            *** MEMBER ORDER INQUIRY **

MEMBER NUMBER:  XXXXXX              ORDER NUMBER:  XXXXX
MEMBER NAME:  XXXXXX               ORDER DATE:  MM/DD/YY
CLUB MEMBERSHIP: XXXXXXXXXXX       X  ORDER STATUS: XXXXXXX
MEMBER ADDRESS:
 P.O. BOX  XXXXXXXXXX               PREPAID AMOUNT: ZZ9.99
 STREET:  XXXXXXXXXXXXXXX
 CITY:  XXXXXXXXXXXXXXX
 STATE: XX  ZIPCODE: XXXXXXXXX

PRODUCT              QUANTITY  QUANTITY          EXTENDED
NUMBER  MEDIA CREDIT ORDERED  SHIPPED  PRICE     PRICE
XXXXXXX XXXXXXXX  9  Z,ZZ9  Z,ZZ9  ZZZ.99  Z,ZZZ.99


TOTAL ORDER CREDITS:  ZZ9     TOTAL ORDER PRICE: ZZ,ZZZ.99
UNUSED BONUS CREDITS:  ZZ9

        *** PRESS ARROW KEYS TO SEE ADDITIONAL PRODUCTS ***
        *** PRESS F3 TO RETURN TO ORDER INQUIRY OPTIONS MENU ***
```

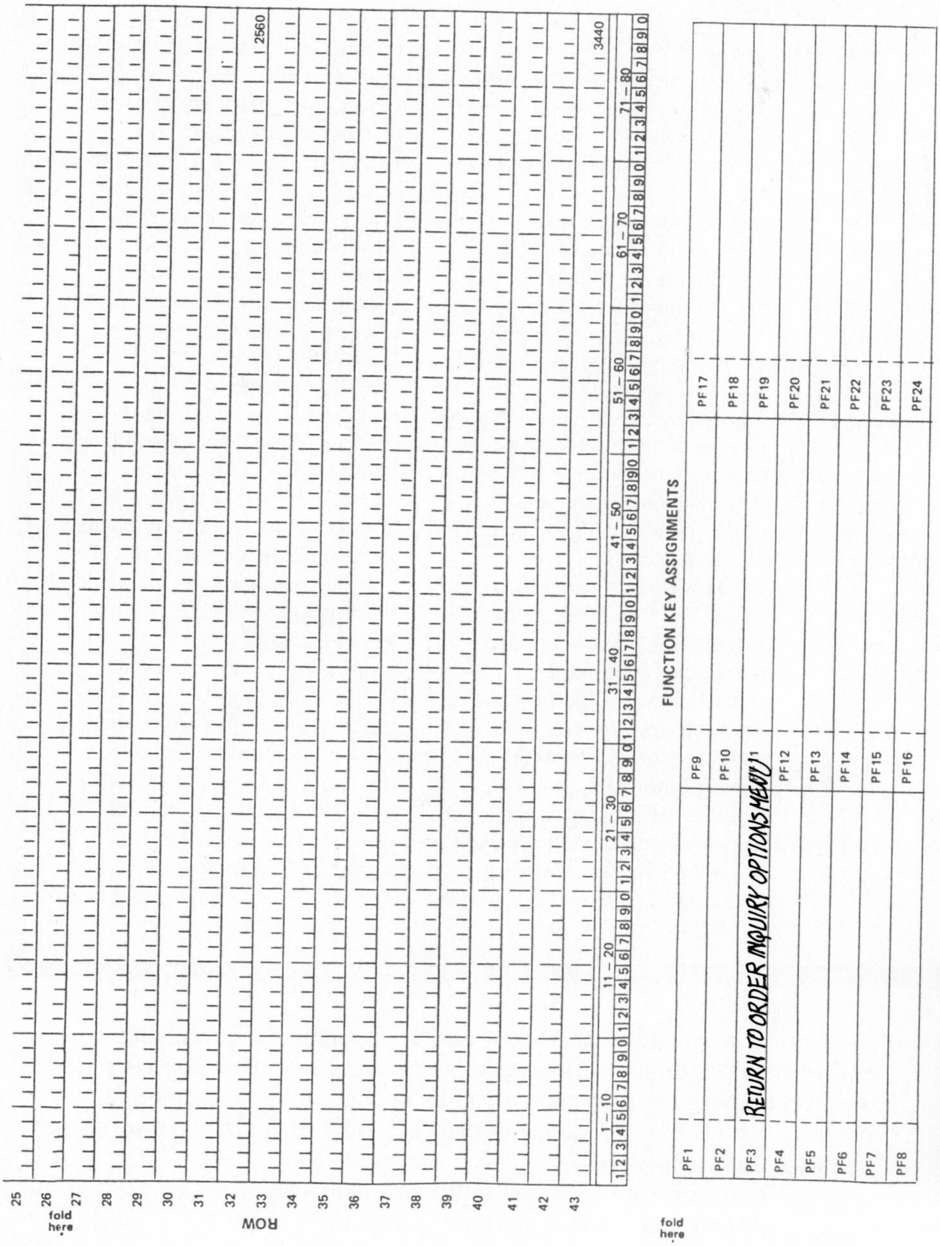

ROW

25
26 fold here
27
28
29
30
31
32
33 2560
34
35
36
37
38
39
40
41
42
43 3440

1—10 11—20 21—30 31—40 41—50 51—60 61—70 71—80

FUNCTION KEY ASSIGNMENTS

PF1		PF9		PF17	
PF2		PF10		PF18	
PF3	RETURN TO ORDER INQUIRY OPTIONS MENU¹	PF11		PF19	
PF4		PF12		PF20	
PF5		PF13		PF21	
PF6		PF14		PF22	
PF7		PF15		PF23	
PF8		PF16		PF24	

fold here

FIGURE 15.13 Display Layout Chart A display layout chart is used to design the layout of display outputs.

THE NEXT GENERATION

Whose Job Is It, Anyway?

Whose job is output design? The analyst's? Looking to the future, perhaps not.

A continuing theme in this book is the availability of powerful computer tools for systems development. Of these tools, perhaps the most commonly used tools—by analysts—are fourth-generation (4GL) or database languages. To date, we've focused mostly on the prototyping advantages of these tools; however, their full potential extends beyond prototyping. These tools make it possible to radically shift responsibility for output design to the end-users.

Most 4GLs contain powerful report writers and query languages. These facilities are generally easy-to-learn and easy-to-use—so easy, in fact, that end-users are making use of them to fulfill their own output requirements against existing databases. Examples of 4GLs that include such facilities are FOCUS, RAMIS, SAS, and dBASE.

Just how much of the output design responsibility can be shifted to end-users? At this time, those external outputs that depend on high-volume use of preprinted forms will likely remain the responsibility of analysts. They involve complex carriage controls and spacing concerns and require efficient processing.

On the other hand, the vast majority of summary reports, exception reports, and inquiries could be designed and implemented by the end-users. Simply give those end-users read-only access to the necessary files and databases, and give them 4GL tools to implement the reports and inquiries.

This is most certainly a trend. The development of information centers in most companies is ultimately intended to provide consultation to end-users who will fulfill their own needs through, among other tools, 4GLs.

The technology is also improving. Most 4GLs are built around database technology. And most databases are evolving toward a de facto or unwritten standard called SQL (Structured Query Language). Together with its cousin, QBE (Query by Example), SQL is a formal, simplified tool that is being incorporated into numerous report writers and query facilities contained in 4GLs. They can be used in lieu of traditional programming languages that are hard to learn and use.

The ultimate impact is one that has been a recurring theme in these "Next Generation" box features—the shift in responsibilities between analysts and end-users. End-users are playing a greater role in design and implementation. Analysts are not playing a smaller role, however. Instead, analysts are becoming more concerned with the data itself—capturing the data, editing the data, securing the data, maintaining the data, and ensuring that the data will be available and accurate when end-users "dream up" some new report or inquiry to address a new problem!

At the bottom of the chart **function key assignments**, available in most CRT terminals and microcomputers, can be described. Function keys can be assigned to perform a number of tasks at the stroke of a single key. For instance, the analyst can specify a *help* key or a *save data* key to be programmed into the system.

Before we bring this chapter to a close, let's take a closer look at the example in Figure 15.13. Keep in mind that in this chapter we are only interested in the net output. Chapter 17 will address the menus or terminal dialogues that result in this output.

Note that we divided the display area into zones, called *windows*. For your convenience, the zone borders are highlighted in a second color. Zones help the end-user know where to look for certain things. One zone is set aside to display the title, date, and page number. Another window is used to display the body of the report. The report will page or scroll through this window. A third window offers instructions to the reader for paging forward and backward, returning to the beginning of the document, and terminating the output program.

Summary

Outputs present information to end-users. There are two basic types of computer outputs. External outputs leave the system to trigger actions on the part of their recipients or confirm actions to their recipients. Internal outputs stay inside the system to support the system's end-users and managers. When designing outputs, both external and internal, the systems analyst must consider a variety of human factors.

Human factors or end-user issues for design outputs include the readability, timeliness, relevance, and acceptability. Choices of media, including paper, film, and video display, as well as the format for presenting information such as tabular, zoned, graphic, and narrative are also important considerations for designing outputs. Finally, good output design involves addressing internal controls to ensure the reliability and distribution of outputs generated by the computer.

Output design requires four basic steps. First, review output requirements. Output requirements should have been defined during systems analysis. Outputs were depicted as data flows crossing from the machine side to the person side of a bounded data flow diagram of the new system. During systems analysis, outputs should also have been described as entries in the project dictionary. Second, decide how the output data flow will be implemented. The implementation decisions will consider the appropriate media and format, as well as a number of design criteria. Third, prototype the layout for the end-user(s). Rather than simply sketch the sample output, the systems analyst may draw on a wide number of prototyping tools including spreadsheets, CASE products, and database management systems or fourth generation languages. Finally, if necessary, convert user-oriented layouts into programmer-oriented layouts. Printer spacing charts and display layout charts are still effective tools for communicating the format of printed and visual outputs.

Problems and Exercises

1. Differentiate between internal and external outputs. Give several examples of each. Why is it important that a systems analyst recognize an output as either internal or external?

2. Identify four end-user issues that an analyst should address when designing outputs.

3. To what extent should end-users be involved in output design? How would you get your end-users involved? What would you ask them to do for themselves?

4. What are the three most commonly used media for outputs? What are the advantages and disadvantages of each?

5. Obtain sample outputs of each of the following format types: zoned, tabular, graphic, and narrative. Was the format type of each output the most effective of the four alternatives? If not, why? What format type, or combination of format types, would you have chosen to implement the output? Sketch the layout you would have chosen.

6. Prepare an expanded data dictionary to describe the following outputs:
   a. Your driver's license
   b. Your course schedule
   c. Your bank statement
   d. Your phone bill
   e. Your W-2 statement (for taxes)
   f. A bank or credit card account statement and invoice
   g. An external document printed on a computer
   You may invent numbers for timing and volume. Don't forget internal controls.

7. Using a sample output from Exercise 5, document the layout of the output using a printer spacing chart. Be sure to make appropriate entries in the carriage control, occurs, and trigger columns for all computer-printed lines. List any improvements that could be made to improve the readability, interpretation, and acceptability of the output.

8. How would the expanded data dictionary in Exercise 6 differ if the external output were designed as an internal visual output? Use a display layout chart to describe the format of a visual version of the sample output.

9. Prototype the output from Exercise 6 or 7. If you don't have access to a spreadsheet, CASE product, or fourth-generation language, you may prototype the output using a word processor or simply sketch the output.

10. Prepare an expanded data dictionary for one or more internal outputs from any computer programming assignment. If a printer spacing chart was not supplied by your instructor, prepare one. If you had to change your output to make it more readable and usable, how would you do it?

11. Explain why a systems analyst might choose to document the format of a printed report using a printer spacing chart after he or she completed a prototype of the report and verified it with the end-user.

12. Identify three types of tools that can be used to prototype output layouts for end-users.

13. What data flows crossing the person-machine boundary on a bounded data flow diagram should be designed as computer outputs?

14. What are some of the effects that may be caused by the lack of well-defined internal controls during output design?

Projects and Minicases

1. The sales manager for SoundStage Record Company has requested a daily report. This report should describe the nearly 1,000 customer order responses received for a given day. A response is a member decision on whether to accept the record-of-the-month selection, request an alternate selection, request both, or request that no selection be sent that month. The report is to be sequenced by MEMBERSHIP NUMBER and CATALOG NUMBER. The data dictionary for the report follows:

>The ORDER RESPONSE REPORT consists of the following elements:
>DATE *of the report
>PAGE NUMBER
>1 to 1,000 of the following:
>>MEMBERSHIP NUMBER *5 digits
>>MEMBER NAME *which consists of the following:
>>>MEMBER LAST NAME *15 characters
>>>MEMBER FIRST NAME *15 characters
>>>MEMBER MIDDLE INITIAL *1 character
>>MUSICAL PREFERENCE *possible values are
>>>"EASY LISTENING" "TEEN HITS"
>>>"CLASSICAL" "COUNTRY" "JAZZ"
>>SELECTION OF MONTH DECISION * possible values are
>>>"YES" "NO" "NONE"
>1 to 15 of the following:
>>CATALOG NUMBER *5 digits
>>MEDIA * possible values are
>>>"RECORD" "CASSETTE"
>>>"AUDIOPHILE" "8 TRACK" "REEL"
>>NUMBER OF PURCHASE CREDITS NEEDED *2 digits
>>PERIOD AGREEMENT EXPIRES * date membership expires

What type of output was being requested by the sales manager? Prepare an expanded data dictionary for the output. Prototype the requested output. Verify the output with your instructor (serving as the sales manager or end-user). Once you've obtained the instructor's approval, use a printer spacing chart to lay out the format for the printed report. Be sure to include appropriate report headings, edit masks with reference numbers to the expanded data dictionary, occurs, and trigger entries.

2. The sales manager has also requested that the sales staff be able to obtain information concerning a *particular* customer's order response at any time during normal working hours. Prepare an expanded data dictionary for the output. Prototype the requested output. Verify the output with your instructor (serving as the sales manager or end-user). Once you've obtained the instructor's approval, prepare a display layout chart for the visual output CUSTOMER ORDER RESPONSE.

Annotated References and Suggested Readings

Fitzgerald, Jerry. *Internal Controls for Computerized Information Systems.* Redwood City, Calif.: Jerry Fitzgerald & Associates, 1978. This is our reference standard on the subject of designing internal controls into systems. Fitzgerald advocates a unique and powerful matrix tool for designing controls. This book goes far beyond any introductory systems textbook—it is *must* reading.

SOUNDSTAGE

Input and User Interface Design for the New System

When we left Sandy and Bob, they had finished the design of outputs and files/databases for the new member services system. Now they must design the inputs and terminal dialog for the new system.

EPISODE 7 We begin this episode in the conference room where Sandy and Bob have scheduled a Saturday morning meeting to review input design specifications with the order-processing staff. Sally Hoover suggested a Saturday morning overtime meeting because she wanted her entire staff to become familiar with the new input and on-line methods that would be used in the new system.

Input Specifications for the New System

Sandy called the meeting to order. "I think we should go ahead and get started. I know all of you are anxious to enjoy your weekend, so we'll make this as brief as possible. First, Bob and I would like to thank you for your cooperation so far. We realize that we have spent a great deal of time defining what you need in order to do your jobs. We have been deliberately avoiding specific detail of how to do things because we wanted to build

a system that would make your jobs easier. It is your system, and it is shaping up very well."

"Today, we want to review system inputs. Let's begin with a member order." Sandy placed a transparency on the overhead projector and continued. (See Figure E7.1.) "This is the proposed Member Order Form 40. It is similar to the old Form 40. You won't have to enter the order date anymore, the system will do that for you using the current date. Also notice . . ."

Sally Hoover interrupted, "What if the order date is different from the date when the order is entered?"

"Why would they be different?" Sandy responded.

"Sometimes orders are sent to the wrong department, or they don't get entered on the day they are received due to a high volume of orders. Since we generate automatic orders to members based on the date, it's very important that the date on a member's order be correct."

Bob interrupted. "That's my fault. Sally and I discussed this but I forgot to make the change to the specifications. If we have the system fill the date field with the current date but allow the order-processing staff to override that value we can solve that problem and save them a lot of typing at the same time."

"I have another question, Sandy. Do we have to enter all of those fields in order to begin processing an order?" Sally asked.

"The order number is automatically generated by the system. Also, when you enter the member number, the member's name will appear for you to verify. In addition to the order date we just talked about, the selection of the month accepted field will be sent to 'YES' and the product number inserted to save you some typing."

"But 70% of our members reject the selection of the month," objected Joe Bosley. "That means they will have to change those fields most of the time."

Episode 7, continued ▶

SOUNDSTAGE

SOUNDSTAGE ENTERTAINMENT CLUB
P.O. BOX 4513
INDIANAPOLIS, INDIANA 43279

FORM 40

MEMBER NUMBER [   ]   ORDER NUMBER [   ]

MEMBER NAME: ORDER DATE:

CLUB: PREPAID AMOUNT:

ADDRESS

SELECTION OF THE MONTH ACCEPTED?	PRODUCT NUMBER	MEDIA CODE	QUANTITY ORDERED	UNIT PRICE	EXTENDED PRICE

For office use only
DATE RECEIVED
DATE ENTERED
CLERK

Signature of customer

SUBTOTAL	
TAX	
TOTAL	

FIGURE E7.1   Sketch of a Source Document

"That's interesting," Sandra admitted. "We will change the default value to 'NO' and leave the product number blank." Noting that there were not any more questions, Sandy continued. "What we need to do now is to verify the size of each field and get an idea of the range of values each field can assume. This will help us write programs to check the accuracy of the data before it's processed. Each of the fields you'll enter is recorded on this transparency (see Figure E7.2). Don't be intimidated

by this form. Most of this information is intended for the computer programmers. All we have to do is verify the size and editing requirements for each field. Each of you will have a computer terminal to enter member orders directly into the computer."

"Oh no!" groaned some unidentified voice.

"Don't worry!" Sandy said. "I remember my first experience with a computer terminal. I thought if I did something wrong, I'd break the machine or some-

thing, but that can't happen. First, we are going to develop a pleasant conversational dialog between you and the computer. As much as possible, we will try to make you forget that you are talking with a machine. We will be walking you through that dialog next week. Right now, I'd like to discuss the final input from that dialog, the member order. May I continue?"

Noting that they were no more questions, Sandra continued. "In order to make your job easier, we made the input very similar to the

SOUNDSTAGE

Data Element Dictionary: Input

Name of Input   MEMBER ORDER

Ref.	Field Name	Type	Size	Memo	Edit Mask	Editing/Validation
1	MEMBER NUMBER	A/N	7		9(7)	NUMERIC
2	ORDER NUMBER	A/N	6		9(6)	NUMERIC, UNIQUE
3	MEMBER NAME	A/N	30		X(30)	MUST CONTAIN VALUES
4	ORDER DATE	A/N	8		MM/DD/YY XX/XX/XX	VALID DATE
5	CLUB NAME	A/N	10		X(10)	MUST CONTAIN VALUES
6	PREPAID AMOUNT	N	3.2		999.99	NUMERIC ≥ 0
7	P.O. BOX	A/N	10		X(10)	OPTIONAL
8	STREET	A/N	15		X(15)	MUST CONTAIN VALUES
9	CITY	A/N	15		X(15)	" " "
10	STATE	A	2		AA	STANDARD CODES
11	ZIPCODE	A/N	9		9(9)	MUST BE NUMERIC

— CONTINUED

FIGURE E7.2   Typical Specifications for Input Fields

Member Order Form 40 we discussed earlier. All you'll have to do is fill in the blanks."

Pam Turner, a clerk in the Order-Processing Department, interrupted. "That sounds good, but I don't think we will all know what to do. I have worked on systems like this at another company and we were always getting stuck. The computer people would tell us to press this key or do that, but we were usually confused, especially if we didn't have our reference guides."

"I know what you mean, Pat," Sandy responded. "I've seen some of those poorly designed systems. What would you think if there wasn't any reference guide and everything you needed to know was right on the screen?"

"That would be fantastic," replied Pam. "But you would need an awfully large screen."

"Not really," Sandy countered. "All we have to do is make sure that simple key assignments and instruction always appear on the screen. Look at this transparency

(Figure E7.3). Note that the instructions appear in lines 2 through 4. They will always appear there. As you enter each field, you press the tab or arrow keys to move to the next field. After you enter the last field, press the enter key and the data will be input to the computer for processing. After you enter all the fields, the system will automatically check the data against the value ranges we discussed earlier. If the value is

Episode 7, continued ▶

SOUNDSTAGE

```
                        ** MEMBER ORDER **
        ENTER THE FOLLOWING ITEMS FROM A MEMBER ORDER (FORM 40). USE TAB OR
        ARROW KEYS TO MOVE FROM ITEM TO ITEM. PRESS ENTER KEY WHEN DONE.

MEMBER NUMBER:                                        ORDER NUMBER:
MEMBER NAME:                                          ORDER DATE:
CLUB NAME:                                            PREPAID AMOUNT:
MEMBER ADDRESS:
  P.O. BOX:
  STREET:
  CITY:
  STATE:       ZIPCODE:

            SELECTION OF          PRODUCT    MEDIA     QUANTITY
            MONTH ACCEPTED?        NUMBER     CODE      ORDERED




      ** PRESS F2 FOR HELP. PRESS F3 TO RETURN TO ORDER PROCESSING OPTIONS MENU **
```

FIGURE E7.3 Typical Prototype of On-Line Input

```
                        *MEMBER ORDER*
                        HELP SCREEN


        The prepaid amount is a right-justified, six character,
        positive field in the form of 999.99.  The decimal point
        must be entered.  For example, a prepaid amount of $21.89
        would be entered as 21.89.



        PRESS ANY KEY TO RETURN TO MEMBER ORDER ==>
```

FIGURE E7.4 Typical Prototype of On-Line Help

```
              - MEMBER SERVICES SYSTEM MENU -


        [1] INQUIRE ON PRODUCTS, ORDERS, AND MEMBERS
        [2] PROCESS MEMBER ORDERS
        [3] REPORTS


        SELECT DESIRED OPTION ==>
```

```
        PRESS "F10" TO TERMINATE SESSION
```

FIGURE E7.5 Typical Menu Screen Prototype

rect, the system will give you an appropriate error message and ask you to reenter the field."

"What happens if I enter something several times and the computer won't take it but I don't understand what I'm doing wrong?" asked Sally.

"That's a good question, Sally," Bob answered. "Notice the 'PRESS F2 FOR HELP' message at the bottom of the screen. Anytime you need more information about a particular field, you can press the F2 key. The system remembers what field you were on and displays a help screen for that field. If, for example, you were trying to

enter a prepayment amount and requested help, you'd see something like this. (See Figure E7.4.) That way you can enter the data so that the system will accept it."

"That's great!" Sally exclaimed. "If the system rejects the data, at least we can find out why."

Sandy concluded her discussion of the on-line member order input. Similar discussions were held for each of the on-line inputs.

Terminal Dialog Specifications for the New System

One week has passed since the input design walkthrough meeting.

Sandra and Bob have prepared some sketches of sample screens of real terminal dialog situations. Various end-users are being walked through typical terminal sessions. Sandy conducts these sessions while Bob makes any necessary modifications to the design specifications. Let's listen in on one walkthrough session.

Sandy began, "Sally, this is a sample of what the screen will look like after you correctly enter your password. (See Figure E7.5.) If you wanted to enter a member order, what would you do next?"

Episode 7, continued ▶

SOUNDSTAGE

```
                    - MEMBER SERVICES SYSTEM MENU -


        [1] INQUIRE ON PRODUCTS, ORDERS, AND MEMBERS
        [2] PROCESS MEMBER ORDERS
        [3] REPORTS


        SELECT DESIRED OPTION ==> 4







        INVALID OPTION - CHOOSE 1, 2, OR 3

        PRESS "F10" TO TERMINATE SESSION
```

FIGURE E7.6   Typical Menu Screen Prototype with Error Message

"Well," Sally began, "normally I'd enter 2, but what would happen if I accidentally choose 4 instead?"

"That's a good question," Sandy answered, as Bob showed Sally a new sample terminal dialog screen. (See Figure E7.6.) "As you can see, the system has identified your response as invalid and is waiting for you to enter another response. Notice the error message at the bottom of the screen. Now what are you going to do?" asked Sandy.

"Okay," Sally conceded, acting somewhat surprised that Sandy and Bob had anticipated incorrect responses. "I'll select option 2."

"That will take you to the member order-processing menu, which looks like this," said Bob, as he showed Sally another sample dialog screen.

"Then I'll choose option 1 for Member Order," added Sally.

"That will take you to the Member Order input screen (refer back to Figure E7.3), which you may remember from last week when we looked at input screens," Sandra said. "You can see that we incorporated several of the changes you suggested in that meeting. You already know how to use that input screen. Let's move on to another sample terminal session."

The walkthrough continued in this manner until all dialog sessions had been simulated with the promotions, membership, and order-processing staff.

Where Do We Go from Here?

This episode introduced two very important tasks of systems design: the design of computer inputs and terminal dialog design. Several important issues must be faced when designing inputs and dialog, including editing of inputs to prevent entry of invalid data and the use of terminal dialog that is "friendly" to the user and easy to use. You have been exposed to several of the tools systems analysts use to document the design of input and terminal dialog specifications.

In Chapter 16, you will learn about issues and tools for design-ing computer inputs. Important terminal dialog design issues will be discussed in Chapter 17.

Communication skills are very important aspects of designing computer inputs and terminal dialog because of the extensive end-user interaction involved in their design and approval. In addi-tion to obtaining an understanding of the issues and tools used for input and terminal dialog design, we strongly encourage you to read Part Four, Module C, "Interper-sonal Communication Skills."

CHAPTER SIXTEEN

Designing and

Prototyping Computer

Inputs and Controls

Greater Metropolis TV Cable Company (Part 1)

Don McAllister is manager of the Information Systems Department at Greater Metropolis TV Cable (GMTVC). He has just returned to his office after an hour meeting with the Customer Services Department. The Customer Services Manager was irate about the new services information system, and Don had taken a lot of heat.

Now, Don is faced with the problem of smoothing things out. Unfortunately, Don didn't know much about the recently installed system. He had turned the project over to Lisa Klemme, who had been a programmer at GMTVC for five years before her recent promotion to systems analyst. This had been her first systems project. Lisa entered Mr. McAllister's office.

"Good afternoon, Don," Lisa seemed apprehensive. "What did you want to see me about? I heard there may be some problems with my last project."

Don replied, "That's right. Lisa, I'm going to get straight to the point. The Customer Services Department is completely dissatisfied with the new order and service entry system you installed. Before they officially cut over to the new system, they want some changes."

Lisa answered, "What are they unhappy about? That may be the first on-line system, but it's a darn good one."

"It is," said Don. "I've studied it in detail. If they ever get to those new reports and inquiries you designed, they will be impressed."

"If?" said Lisa, with a genuinely puzzled look on her face.

"That's right. They can't seem to get past the input functions of the system," Don replied. "They claim the system is not easy to use—that it takes them longer than it should to enter a sales or service order because the dialogue is confusing. They also claim that they're experiencing an increase in customer complaints."

Lisa responded, "I don't see how that can be. The program is very easy to use. You should see it. I used a lot of fancy terminal functions—blinking screens, reverse video, and things like that—to make their job of entering sales orders more interesting. And besides, what do their customer complaints have to do with my system?"

Don shuffled through some papers. "This memo states the new system isn't recording the customer sales orders accurately. They . . ."

Lisa seemed defensive as she responded, "Impossible! That can't be. The program asks the data entry clerk to enter all the information that's on the form. If they filled the form out right, there wouldn't be any problems."

Don answered, "Okay. That makes sense. Let's see if we can't get to the bottom of this. How did the clerks feel about that form?"

"I'm not sure. They said they wanted to do orders on the computer. I sampled the form and then re-created it on their terminal screen."

Don again asked, "But how did they feel about the form to begin with?" After a long pause, Don said, "I think I'm getting to the bottom of this. Sit down. I think there are a few things we need to talk about."

Discussion

1. Is it possible that Lisa has made some mistakes?
2. What do you think Lisa should have done differently?

What Will You Learn in This Chapter?

In this chapter, you will learn how to design and prototype computer inputs, both batch and on-line. You will know that you have mastered input design tools and techniques when you can:

1. Explain the difference between *data capture*, *data entry*, and *data input*.
2. Define the appropriate method and medium for a computer input.
3. Apply human factors to the design of computer inputs.
4. Design internal controls into computer inputs.
5. Identify data flows, on a DFD, that must be designed as computer inputs.
6. Define input design requirements and record them in the project dictionary.
7. Describe various techniques for prototyping inputs.
8. Design a source document for data capture.
9. Prepare layouts to communicate both record- and screen-oriented inputs to programmers who must implement those inputs.

"Garbage in, garbage out!" This overworked expression is no less true today than it was 20 years ago. So far, you've studied the design of computer files, databases, and outputs. Outputs are produced from data that is either input or retrieved from files and databases. And data in files must have been input to those files and databases. In this chapter, you are going to learn how to design the inputs. Input design serves an important goal: Capture and get the data into a format suitable for the computer.

First, we'll study fundamental concepts. Then we'll study the tools and techniques of input design.

Methods and Issues for Data Capture and Input

Information = f (data, processing)! Do you remember that formula from Chapter 3? Well, this section is about data. Where does data originate? How is data captured? How is data input to the computer? And how do we know the input is valid? These questions are addressed in this section.

Data Capture, Data Entry, and Data Input

When you think of input, you usually think of input devices, such as card readers and terminals. But input begins long before the data arrives at the device. To actually input business data into a computer, the analyst may have to design an input form, design input screens or records, and design methods and procedures for getting the data into the computer (from *customer* to *form* to *data entry clerk*—if necessary—to *disk* to *tape* to *computer*).

This brings us to our fundamental question. What is the difference between data capture, data entry, and data input? *Data happens!* It accompanies business events called transactions. Examples include orders, time cards, reservations, and the like. What we must do is determine *when* and *how* to capture the data. **Data capture** is the identification of new data to be input. *When* is easy! It's always best to capture the data as soon as possible after it is originated. *How* is another story! Creating a **source document** (a term commonly associated with forms used to record data that will eventually be input to a computer) is not easy. Source documents or their equivalent should be easy for the end-user to complete and should facilitate rapid data entry into a machine-readable format.

Data entry is not the same as data capture. **Data entry** is the process of translating the source document into the machine-readable format. That format may be a punched card, an optical-mark form, a magnetic tape, or a floppy diskette, to name a few. Only after the data has been entered and converted to a machine-readable format do we have **data input** for the computer. Let's examine some of the data capture and data entry alternatives you should consider during systems design.

Input Methods and Media

The analyst usually recommends the method and medium for all inputs. Let's compare the different input methods and the medium alternatives available for modern information systems. Input methods can be broadly classified as either batch or on-line.

Batch Methods and Media Batch input is the oldest and most traditional input method. Source documents are collected and then periodically forwarded to data entry operators who key the data using a data entry device that translates the data into the machine-readable format. Figure 16.1 illustrates numerous procedures required for these different media.

For many years, punched cards were the most common medium for batch input data. For the most part, batches of punched cards have been replaced by *magnetic* media. Key-to-disk (KTD) and key-to-tape (KTT) workstations transcribe data to magnetic disks and magnetic tape, respectively. These workstations are much quieter, making life as a data entry clerk more bearable. As each input record is keyed, it is displayed on a screen. The data can be corrected because it is initially placed into a buffer. The final input file, possibly merged from several KTD or KTT workstations, permits much faster data input rates to the computer than those achieved with punched cards.

Figures 16.1(a) and (b) illustrate the key-to-tape and key-to-disk input procedures. The figure also reinforces the concepts from the last subsection. Data capture activities are shown in white. Data entry activities are shown in light blue. Finally, data input activities are shown in dark blue.

Figure 16.1(c) illustrates another batch input medium, the optical-mark form. You may have encountered this medium in machine-scored tests. Optical-mark forms eliminate most or all of the need for data entry. Essentially, the source document becomes the input medium, or so it seems. As the figure illustrates, the source document is directly read by an optical-mark reader (OMR) or optical-character reader (OCR). The computer records the data to magnetic tape, which is then input to the computer. OCR and OMR input are generally suitable only for high-volume input activities. By having data directly recorded on a machine-readable document, the cost of data entry is eliminated.

One characteristic of all of these batch media is the significant possibility of error when moving from the source document to the input medium. Before data can be processed, it must be edited. Have you written edit programs in your programming courses? If so, you know that edit programs frequently require as much or more effort than the processing of the transaction itself. We'll discuss this issue further when we present internal controls for inputs.

On-Line Methods and Media Today, most (but not all!) systems have been converted or are being converted to on-line input methods. This makes sense—capture data at its point of origin, in the business, and directly input that data to the computer, preferably as soon as possible after the data originates.

The most common on-line medium cannot really be classified as a medium; it is the display terminal, or microcomputer display. The on-line system includes a monitor screen and keyboard that are directly connected to a computer system. The end-user directly enters the data when—or soon after—that data

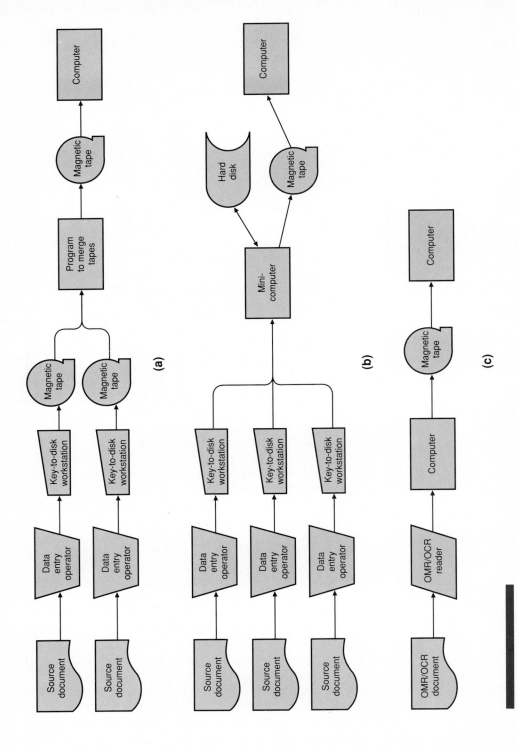

FIGURE 16.1 Alternative Input Procedures for Batch Input Media These illustrations show the similarities and differences between the most typical batch input methods.

Universal Product Code

originates. No data entry clerks are needed! There is no need to record data onto a medium that is later input to the computer; this input is direct! If data is entered incorrectly, the computer's edit program detects the error and immediately requests the CRT operator to make a correction.

On-line data input can become even more sophisticated. With today's technology, we can *completely* eliminate much (and sometimes all) human intervention. Point-of-sale terminals in retail and grocery stores frequently include bar-code and optical-character readers. Everyone has seen the bar codes recorded on today's food store products (see the example printed in the margin). These bar codes eliminate the need for keying data, either by data entry clerks or end-users. Instead, sophisticated laser readers read the bar code and send the data represented by that code directly to the computer for processing.

Batch Versus On-Line   Should all systems be designed for on-line input? The technology is certainly cheaper than it used to be. So why bother with batch input? Not so fast!

No matter how cheap and fast on-line processing gets, an on-line program cannot be nearly as fast as its batch equivalent. Why? Many (but not all) on-line programs require some human interaction, and people are slow, relative to computers. Also, for large-volume transactions, too many CRT terminals and operators may be needed to meet demand. As the number of on-line CRTs grows, the overall performance of the computer declines. Furthermore, many inputs naturally occur in batches. For instance, our mail may include a large batch of customer payments on any given day. Postal delivery is, at least today, a batch operation. Additionally, some input data may *not* require immediate attention. Finally, batch processing may be preferable because internal controls (discussed shortly) are simpler. So you see, batch inputs can still be justified.

There is one compromise solution, the **remote batch**. Remote batch offers on-line advantages for data that is best processed in batches. The data is input on-line with on-line editing. Microcomputers or minicomputer systems can be used to handle this on-line input and editing. The data is not immediately processed. Instead, it is batched, usually to some type of magnetic media. At an appropriate time, the data is uploaded to the main computer, merged, and subsequently processed as a batch. Remote batch is also called *deferred batch* or *deferred processing*.

End-User Considerations for Input Design

Because inputs originate with end-users, human factors play a significant role in input design. Furthermore, if batch methods are used, data entry clerks' needs must also be considered. With this in mind, several human factors should be evaluated.

The volume of data to be input should be minimized. The more data that is input, the greater the potential number of input errors and the longer it takes to input that data. The following general principles should be followed:

1. *Enter only variable data.* Do not enter constant data. For instance, when deciding what elements to include in a SALES ORDER input, we need PART NUMBERs for all parts ordered. But do we need to input PART DESCRIPTIONs for those parts? Think about it! PART DESCRIPTION is probably stored in a computer file. If we input PART NUMBER, we can look up PART DESCRIPTION. Permanent (or semipermanent) data should be stored in files. Of course, inputs must be designed for maintaining those files.

2. *Do not input data that can be calculated or stored in computer programs.* For example, if you input QUANTITY ORDERED and PRICE, you don't need to input EXTENDED PRICE (= QUANTITY ORDERED × PRICE). Another example is incorporating FEDERAL TAX WITHHOLDING data in tables (arrays) instead of keying in that data every time.

3. *Use codes for appropriate data elements.* Codes were introduced in Chapter 10. Codes can be translated in computer programs by using tables as described in the preceding point.

Source documents should be easy for end-users to complete. The following suggestions may help:

- *Include instructions for completing the form.* By the way, did you know that people don't like to have to read instructions printed on the back side of the form?

- *Minimize the amount of handwriting.* Many people suffer from poor penmanship. The data entry clerk or CRT operator may misread the data and input incorrect data. Use check boxes wherever possible so the end-user only needs to check the appropriate values.

Design documents so they can be easily and quickly entered into the system. We suggest the following:

- Data to be entered (keyed) should be sequenced so it can be read like this book, top to bottom and left to right (Figure 16.2a). The data entry clerk should not have to move from right to left on a line or jump around on the form (see Figure 16.2b) to find data items to be entered.

- Ideally, portions of the form that are not to be input are placed in or about the lower right portion of the source document (the last portion encountered when reading top to bottom and left to right). Alternatively, this information can be placed on the back of the form.

Please note that these are only guidelines. End-users should have the final say on source document design!

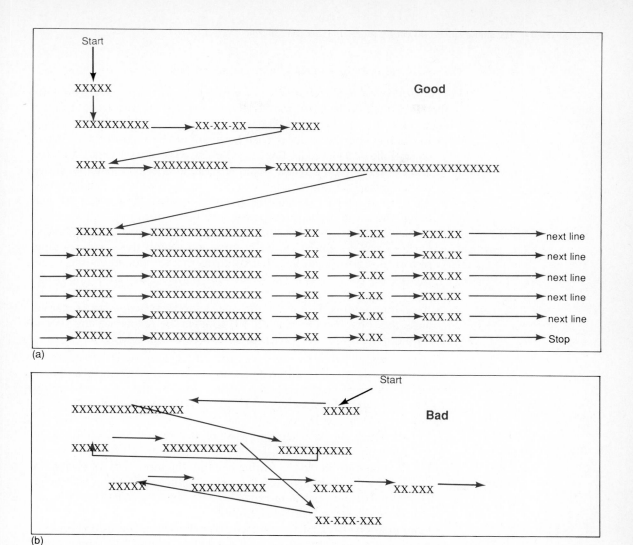

FIGURE 16.2 Keying from Source Documents Source documents should be designed to aid in rapid data entry. The source document (a) was designed to allow the data entry clerk to locate and key data in a more natural top-to-bottom, left-to-right sequence. The source document (b) will more likely negatively effect the data entry clerk's ability to quickly locate and enter data.

Internal Controls for Inputs

Once again, we expand our coverage on the continuing topic of internal controls. Input controls ensure the accuracy of data input to the computer.

The following internal control guidelines are offered:

1. *The number of inputs should be monitored.* This is especially true with the batch method because source documents may be misplaced, lost, or skipped.

    • In batch systems, data about each batch should be recorded on a batch control slip. Data includes BATCH NUMBER, NUMBER OF DOCUMENTS, and CONTROL TOTALS (for example, total number of line items on the documents). Later, these totals can be compared with the output totals on a report after processing has been completed. If the totals are not equal, the cause of the discrepancy must be determined.

    • In batch systems, an alternative control would be one-for-one checks. Each source document would be matched against the corresponding historical report detail line that confirms that the document has been processed. This control check may only be necessary when the batch control totals don't match.

    • In on-line systems, each input transaction should be logged to a separate audit file (described in Chapter 13) so it can be recovered and reprocessed in the event of a processing error or if data is lost.

2. *Care must also be taken to ensure that the data is valid.* Two types of errors can infiltrate the data: invalid data recorded by end-users and data entry errors. Data entry errors include copying errors, transpositions (typing *132* as *123*), and slides (keying *345.36* as *3453.6*). The following techniques are widely used to validate data:

    • **Completeness checks** determine whether all *required* fields on the input have actually been entered.

    • **Limit and range checks** determine whether the input data for each field falls within the legitimate set or range of values defined for that field. For instance, an upper-limit range may be put on PAY RATE to ensure that no employee is paid at a higher rate.

    • **Combination checks** determine whether a known relationship between two fields is valid. For instance, if the VEHICLE MAKE is Pontiac, then the VEHICLE MODEL must be one of a limited set of values that comprise cars manufactured by Pontiac (Firebird, Grand Prix, and Bonneville, to name a few).

    • **Self-checking digits** are a technique for determining data entry errors on primary keys. A *check digit* is a number or character that is appended to a primary key field. The check digit is calculated by applying a formula, such as Modulus 11, to the actual key (see Figure 16.3). How does the check digit verify correct data entry? There are two methods. Some data entry devices can automatically validate data by applying the same formula to the data as it is entered by the data entry clerk. If the check digit entered doesn't match the check digit calculated, an error is

FIGURE 16.3 Modulus 11 Self-Checking Digit Technique Modulus 11 is a very common self-checking digit technique used to verify that the original/source data has been correctly transcribed into machine-processable form. For example, if an end-user read the key field value ''241350'' (the number derived in the Modulus 11 example) and mistakenly keyed in the value ''243150,'' the incorrect data could have been detected by applying the Modulus 11 formula to the key values.

MODULUS 11

The following procedure is used to assign a check digit to a key field:

STEP 1: Determine the size of the key field in digits.

$$2\ 4\ 1\ 3\ 5 = \boxed{5\ \text{digits}}$$

STEP 2: Number each digit location from *right* or *left* beginning with the number "2."

$$2\ 4\ 1\ 3\ 5$$
$$\boxed{6\ 5\ 4\ 3\ 2}$$

STEP 3: Multiply each digit in the key field by its assigned location number.

$$2 \times 6 = \boxed{12}$$
$$4 \times 5 = \boxed{20}$$
$$1 \times 4 = \boxed{4}$$
$$3 \times 3 = \boxed{9}$$
$$5 \times 2 = \boxed{10}$$

STEP 4: Sum the products from step 3.

$$12 + 20 + 4 + 9 + 10 = \boxed{55}$$

STEP 5: Divide the sum from step 4 by 11.

$$55 / 11 = 5\ \ \text{Remainder}\ \boxed{0}$$

STEP 6: If the remainder is less than 10, append the remainder digit to the key field. If the remainder is equal to 10, append the character "X" to the key field.

$$2\ 4\ 1\ 3\ 5\ \boxed{0}$$

---

displayed. Alternatively, computer programs can also validate check digits by using readily available subroutines.

- **Picture checks** compare data entered against the known COBOL picture defined for that data. For instance, the input field may have a picture clause XX999AA (where X can be a letter or number, 9 must be a number, and A must be a letter). The field "A4898DH" would pass the picture check, but the field "A489ID8" would not.

Data validation requires that special edit programs be written to perform checks. However, the input validation requirements should be designed when the inputs themselves are designed.

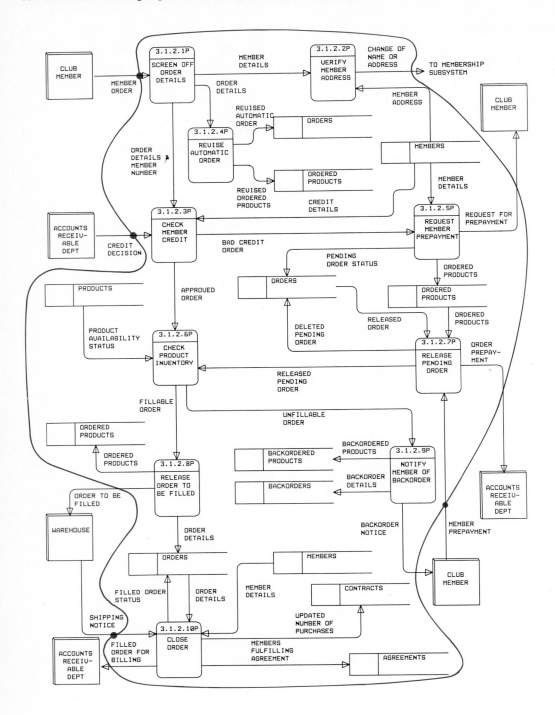

How to Prototype and Design Computer Inputs

◄ **FIGURE 16.4   A Bounded Data Flow Diagram**   Those data flows entering the person-machine boundary of a data flow diagram represent inputs that must be designed. The data flows MEMBER ORDER, CREDIT DECISION, SHIPPING NOTICE, and MEMBER PAYMENT would need to be designed as computer inputs. The analyst must focus on the data capture, data entry, and data input for these data flows.

How do you design on-line and batch inputs? In this section, we'll discuss and demonstrate the process of input design, and we'll apply the concepts you learned earlier in the chapter. We'll continue to use our SoundStage case study to demonstrate the process, and we'll continue to use our CASE product to record our specifications into the project dictionary. We'll also demonstrate how CASE and other tools can be used to prototype inputs.

Step 1: Review Input Requirements

Input requirements, like output requirements, should have been defined during systems analysis. And as was the case for outputs, a good starting point for input design is the data flow diagrams for the new system. The data flow diagram in Figure 16.4 identifies those required inputs to be designed. An overlayed boundary separates the part of the system to be computerized from its manual counterpart. Any data flow that enters the machine side of the boundary is an input to be designed. MEMBER ORDER is an input data flow.

Given an input to be designed, we should review the required data elements. The basic content of these inputs should have been recorded in the project dictionary during systems analysis. If not, we can define input requirements by studying the output and file designs. An output data element that can't be retrieved from files or calculated from elements that are retrieved from files *must be input*! Additionally, inputs must be designed to maintain the files in the system.

Step 2: Decide How the Input Data Flow Will Be Implemented

Recall that the selection phase of the life cycle determines how the system, including inputs, will eventually be implemented. As with outputs, input design decisions will be based on two sets of criteria:

- Based on type and characteristics of the input, what are the best medium and format for the design and implementation?
- Based on technical and economic feasibility, what are the best medium and format for the design and implementation?

Techniques for evaluating the latter (feasibility) are covered in Part Four, Module D. Techniques for the former are described in the following paragraphs. Important considerations for designing inputs include

1. *Identify source documents that need to be designed or modified.*
2. *Determine the input method to be used.* Batch? On-line? What specific medium? Key-to-tape? Key-to-disk? Optical character reader? Remember,

input may be restricted to media available in the current environment, especially if funds are not available for new media or services.

3. *Determine the timing and volume of input.* **Timing** is how often the input will be done (for instance, "Once daily," "Hourly," "On demand, 24 hours," "On demand, 8 A.M.–5 P.M., weekdays"). **Volume** is the number of inputs in a batch or time period. This data is needed to predetermine data entry personnel requirements and schedules or determine how many terminals will be needed.

4. *Specify internal controls and special instructions to be followed for the input* (create an audit trail file for an on-line input, create batch control slips for a batch input, and so forth).

5. *Study the input data elements to determine which data really needs to be input.* Remember the guidelines for data reduction presented earlier in this chapter, and don't forget the value of a good code (refer back to Chapter 10 for more on codes) for reducing data volume.

After a design has been determined feasible and approved by end-users and management, the preceding decisions should be recorded in the project dictionary. Like outputs, inputs are data flows; therefore, our data dictionary entry is for a data flow. Logical, requirements-oriented attributes of the flow were described during systems analysis. The above numbered criteria add physical, implementation-oriented attributes to those same data flows. Let's study an example.

MEMBER ORDER is an input data flow to be designed for the SoundStage Member Services system. The input will initially be designed as a turnaround source document. This means that it will initially be output and mailed to the member, who will respond by completing the order and returning it for input to the order-processing subsystem.

The design specifications for MEMBER ORDER are presented in Figure 16.5. Notice that, as we did for output specifications, the physical, implementation attributes have simply been appended to the logical requirements attributes that were defined during systems analysis. This figure is a printout from our CASE product, Excelerator.

The entries are fairly self-explanatory, addressing the issues presented earlier in this chapter. All the entries were developed with the approval of the end-users.

As with outputs, there are a number of design considerations for data elements to be included in the input. Figure 16.6 is a printout of the input data flow's contents—the record to which MEMBER ORDER explodes. We were careful to delete any elements that could be read from files and databases (except keys, which are needed to read files and databases). Physical attributes of each element are also described to the dictionary. Figure 16.7 is a printout of pertinent attributes of the input. Note the following:

```
     TYPE Data Flow                          NAME MEMBER ORDER

        Label MEMBER              EXPLODES TO ONE OF:
              ORDER                 Record          MEMBER ORDER
                                    Data Model Diagram
                                    ERA Diagram


     Duration Value
     Duration Type


     Description
     SOURCE DOCUMENT:  MEMBER ORDER (FORM 40), A TURNAROUND DOCUMENT
     INPUT METHOD:  ON—LINE
     INPUT MEDIUM:  DISPLAY
     TIMING (frequency prepared):  ON DEMAND WEEKDAYS FROM 8 A.M. TO 5 P.M.
     AVERAGE VOLUME (in records):  500 PER DAY
     PEAK VOLUME AND TIMING:        580 FROM NOV. THROUGH DEC.
     CONTROLS AND SPECIAL INSTRUCTIONS:
             THE INPUT WILL INITIALLY BE DESIGNED AS A TURNAROUND SOURCE
             DOCUMENT. IT WILL INITIALLY BE OUTPUT AND MAILED TO THE MEMBER
             WHO WILL RESPOND BY COMPLETING THE ORDER AND RETURNING IT FOR
             INPUT TO BY ORDER PROCESSING. AN AUDIT TRAIL FILE SHOULD BE
             CREATED AND ALL RETURNED SOURCE DOCUMENTS MUST BE RETAINED.


     Modified By    BENTLEY        Date Modified  881009    # Changes  2
     Added By       BENTLEY        Date Added     881009
     Last Project   PROJECT
     Locked By                     Date Locked    0         Lock Status
```

FIGURE 16.5   Project Dictionary Printout of MEMBER ORDER's Physical Specifications   The input design specifications are recorded in the project dictionary.

```
NAME:                    MEMBER ORDER       DEFINITION:
ALIAS:                   ]                  DESCRIBES THE CONTENT OF AN ORDER TO BE FILLED

ELEMENT/RECORD                         OFF  OCC  TYPE  LEN  DEFINITION
-----------------------------------    ---  ---  ----  ---  ----------------------------------------------------
ORDER NUMBER                           000  001  E     006  A SIX DIGIT NUMBER THAT UNIQUELY IDENTIFIES AN ORDER
ORDER DATE                             006  001  E     002  THE DATE FROM ORDER RESPONSE CARD IN MM/DD/YY FORMAT.
PREPAID AMOUNT                         008  001  E     003  DOLLAR AMOUNT ACCOMPANYING AN ORDER
CLUB NAME                              011  001  E     010  NAME OF THE CLUB IN WHICH A MEMBER IS ENROLLED
MEMBER NUMBER                          021  001  E     007  A 7 DIGIT NUMBER THAT UNIQUELY IDENTIFIES A MEMBER
MEMBER NAME                            028  001  E     030  LAST NAME, FIRST NAME  MIDDLE INITIAL
ADDRESS                                058  001  R
   P.O. BOX                            058  001  E     010  P.O. BOX FOR MAILING TO MEMBER'S RESIDENCE
   STREET                              068  001  E     015  STREET OF MEMBER'S RESIDENCE.
   CITY                                083  001  E     015  CITY OF MEMBER'S RESIDENCE
   STATE                               098  001  E     002  A TWO CHARACTER CODE OF U.S. STATE (SEE STATE CODE TABLE)
   ZIP CODE                            100  001  E     009  A 5 TO 9 CHARACTER IDENTIFYING POSTAL AREA OF RESIDENCE.
MEMBER RESPONSE STATUS                 109  001  E     001  INDICATES WHETHER THE MEMBER DESIRES TO ACCEPT THE PREORDER
ORDERED ITEM                           110  015  R
   PRODUCT NUMBER                      110  001  E     007  A SEVEN DIGIT NUMBER IDENTIFYING A SINGLE TITLE
   MEDIA CODE                          117  001  E     001  A ONE CHARACTER CODE IDENTIFYING THE MEDIUM OF A PRODUCT
   QUANTITY ORDERED                    118  001  E     004  QUANTITY ORDERED FOR A  TITLE APPEARING ON AN ORDER

Record length is 290.
```

FIGURE 16.6   MEMBER ORDER's Contents   The data flow MEMBER ORDER's contents are described in this record description. The elements were initially defined during systems analysis. The contents shown here may not include all those elements, since elements that can be read from files and databases—except for keys— should not be reinput.

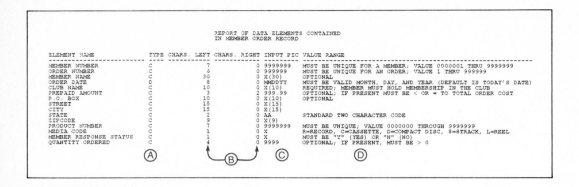

FIGURE 16.7   Data Element Dictionary for the Contents of MEMBER ORDER   This dictionary printout shows details about the elements to be input from a MEMBER ORDER.

(A) For inputs, all fields—numeric, alphabetic, and alphanumeric—are usually input as character strings (ASCII or EBCDIC). Therefore, TYPE is set to C in all but the most unusual of circumstances.

(B) We also specified field size in terms of characters (or positions) to the left and right of the decimal point. For elements with type "character"— meaning "alphanumeric"—the CHARS. RIGHT attribute is not used.

(C) The concept of a PICTURE clause was first introduced in Chapter 15. An input PICTURE should not include such special editing symbols as dollar signs and dashes because, as we have already noted, they are not included in the field. However, the PICTURE should indicate which positions of a field can be numbers and characters. The PICTURE also indicates the decimal point position in some numeric fields.

(D) The last column describes the legitimate values or VALUE RANGEs that the element can assume. It will be helpful to programmers who must write the input edit routines. Additionally, this column describes special validation checks to be performed. For instance, key fields may require check digits. Other fields may require limit checks (record the value ranges).

Step 3: Design or Prototype the Source Document (If One Is Needed)

If a source document will be used to capture data, we prefer to design that document first. The source document is for the end-user. In its simplest form, the prototype may be a simple sketch or an industrial artist's rendition. Guidelines for source document design were outlined earlier in the chapter.

FIGURE 16.8 Source Document Design Zones A source document for input can be designed in zones such as those indicated on this template. The locations are fairly typical with the exception of the instructions zone, which may be located almost anywhere on the form (or separate from the form).

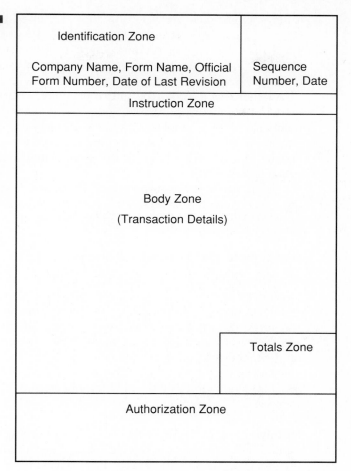

Identification Zone

Company Name, Form Name, Official Form Number, Date of Last Revision

Sequence Number, Date

Instruction Zone

Body Zone

(Transaction Details)

Totals Zone

Authorization Zone

A well-designed source document will be divided into zones. Some zones are used for identification; these include company name, form name, official form number, date of last revision (an important element that is often omitted), and logos. Other zones contain data that identifies a specific occurrence of the form, such as form sequence number (possibly preprinted) and date. The largest portion of the document is used to record transaction data. Data that occurs once and data that repeats should be logically separated. Totals should be relegated to the lower portion of the form because they are usually calculated and, therefore, not input. Many forms include an authorization zone for signatures. Instructions should be placed in a convenient location, preferably not on the back of the form. One possible template for source document design is provided in Figure 16.8.

Prototyping tools have become more advanced in recent years. Spreadsheet programs like Microsoft's Excel can make very realistic models of forms. These tools give you outstanding control over font styles and sizes, graphics for logos, and the like. Laser printers can produce excellent-quality printouts of the prototypes.

Another way to prototype source documents is to develop a very rough model using a word processor, and pass the model to one of the growing number of desktop publishing systems that can transform the rough model into impressive looking forms (so impressive, in fact, that some companies now develop forms this way instead of subcontracting their design to a forms manufacturer).

Finally, forms-processing software packages are starting to become popular. This exciting new prototyping and implementation technology is featured in this chapter's "Next Generation" box feature.

Step 4: Prototype On-Line Input Screens

After design decisions have been recorded in the dictionary and any source documents have been designed, you can now prototype input screens. This is only appropriate, of course, for on-line or remote batch inputs. The best way to design screens is to sketch, or better still, prototype those screens. The sketches or prototypes can be shown to end-users and modified based on their feedback.

Modern prototyping tools have made screen design an infinitely easier task. Let's study some of the tools and techniques of prototyping as they apply to inputs.

Prototyping with CASE Tools   Many CASE products include facilities for rapid prototyping of input screens. They are especially useful since they can use the project dictionary data recorded during systems analysis. Some products, such as Pansophic's Telon, can ultimately transform approved prototypes into COBOL and PL/1 code.

Products like Telon assume you have already defined your input requirements (logical, implementation-independent requirements). Throughout this book, we've used the CASE product Excelerator to capture those requirements. Excelerator can pass the requirements to a product like Telon or provide some simple prototyping capabilities of its own, as you learned in the output design chapter. These capabilities were demonstrated (as best we could without the live product) in the output chapter. Approved Excelerator screen prototypes can be used to generate equivalent PICTUREs in several programming languages.

Prototyping Input Screens with Database Management Systems or Fourth-Generation Languages   A continuing theme in this book has been the use of database management systems and fourth-generation languages to proto-

THE NEXT GENERATION

Forms-Processing Software: Simplified On-Line Inputs for Traditional Paper Forms

The most popular of on-line input processing is *forms filling*. The form appears on the screen and the end-user simply enters data in the appropriate blanks. Implementing forms filling has involved extensive programming, especially to provide adequate editing and instructions. But that is changing.

One of the newest software categories is *forms-processing software*. Forms-processing software eliminates the need for tedious programming and, even better, permits forms prototyping. The resulting forms can be every bit as detailed and complex as their paper equivalents (they can even be printed and duplicated). Think of it as a tool that permits duplicating actual paper forms on a computer screen. Here's how it works.

Forms-processing software packages generally provide numerous, modifiable templates for commonly encountered business forms. The software also includes powerful, easy-to-use facilities (compared to programming) for changing the

forms. Generally, you are given control over text styles, font sizes, shading, borders, and the like. You can also sometimes import scanned logos and graphics. The tools allow you to customize a form (from template or scratch). Virtually any type of form can be created or recreated.

So far, it sounds just like desktop publishing . . . but there's more. Items on any form can be designated as input fields. These fields can be described to the form processor's internal dictionary. In this dictionary, you can specify whether the field is mandatory or optional, any default values, legal value ranges, data and time stamps, data verification rules, and so on. You can also attach help messages, error messages, and sometimes even detailed instructions.

The forms processor becomes the actual input subsystem. End-users can initially study and try out the screens. Data will automatically be edited and verified, which is useful for

creating test data files for subsequent programming. And there's more!

Depending on the forms processor, the edited data will be copied to a file (similar to the concept of remote batch files). This file, of known structure, may be imported to one or more technical programming environments. For example, the file may automatically conform to VSAM, IMS, or dBASE standards—or, at the very least, an ASCII file that can be read by any program you care to write.

You still write the programs to use the data, but you've just eliminated the need to write complex input, editing, and verification routines.

Many of these packages are starting to appear for microcomputers. But their data files (if ASCII) could easily be uploaded to a mainframe system, thus creating a micro-mainframe version of remote batch processing. The prospects for forms processing as an input prototyping and development environment look promising.

type systems. Virtually all such tools include powerful screen design facilities that make it possible to quickly lay out prototype screens, map screen fields to a dictionary that defines edit rules, and define help messages and screens. These screens can be directly tested by end-users and modified to reflect their opinions. Most 4GL-developed prototypes can eventually evolve into finished production systems, although some must be reprogrammed in traditional languages to improve processing efficiency or security. These 4GL prototyping capabilities can be found in both mainframe and microcomputer databases.

For the SoundStage project, a microcomputer database called dBASE IV Plus was used to prototype screens that would eventually be coded in COBOL with the IDMS database management system. dBASE IV is a relatively inexpensive but frequently overlooked tool for prototyping inputs and outputs.

Step 5: If Necessary, Convert User-Oriented Layouts into Programmer-Oriented Layouts

In some cases, sketches and prototypes are sufficient for programmers to implement the design. But in many cases, they are inadequate as final specifications. For example, the prototyping tools may not have been able to implement some of the features required in the final design. Or perhaps the prototype doesn't convey the full range of possibilities to be implemented. In these cases, you must use more formal specification techniques.

Using Display Layout Charts to Document the Format of Input Screens The display layout chart for on-line outputs was introduced in Chapter 15. Figure 16.9 provides the layout for the on-line input of MEMBER ORDER. This form should be familiar to you. Display layout charts are intended for the programmer, not the end-user. Keep in mind that this chapter deals specifically with the final input screens, not the menus or dialogue that gets the end-user to those screens (covered in the next chapter).

There are two generally accepted ways of designing on-line inputs: question-answer dialogues and form filling. In the **question-answer** mode, the system asks several questions and the operator responds with the appropriate answer. Default answers are often provided in brackets, as in <default=0>. Care must be taken to design input dialogues to include responses that must occur when improper data is entered. The windows concept introduced in Chapter 15 is especially useful for this requirement.

Form filling is another on-line input technique. A blank form is *painted* on the screen. The screen cursor moves to the first field to be entered on the form. When the operator enters that field, the return key sends the cursor to the next field to be entered. The automatic movement of the cursor from field to field is often assisted by dialogue software (such as IBM's CICS, a teleprocessing monitor). Dialogue software and teleprocessing monitors provide functions and subroutines, callable from application programs, that simplify screen and dialogue design. Fields are edited as they are entered, and appropriate error messages are displayed. After all fields have been entered, the operator presses a function key to release the data to the computer program. Most of the screen must be used for the form; however, small windows should be set aside for MESSAGES and ESCAPE. Instructions and expanded messages can be placed in a separate file that can be displayed when a function key assigned to HELP is pressed.

As to the layout of the MEMBER ORDER, let's walk through some of the design. We chose the form-filling method for input. Figure 16.9 represents what the end-user will *initially* see on the CRT screen. The screen cursor will initially be positioned at the location marked "C." As the instructions indicate, the end-user will be permitted to move around the screen from one item to the next or back to the previous item. In the escape area, we show the end-user how to discontinue the input of customer orders.

But what about the actual entry of data elements? A programmer would need more specific input requirements than those depicted in Figure 16.9. A more appropriate display layout chart is provided in Figure 16.10. This chart shows the programmer how data is actually input. The headers for input fields can sometimes be confusing to the programmer. Thus the numbers within parentheses near the elements were added to serve as a cross-reference to specific elements in the MEMBER ORDER record in Figure 16.6. Each element in the MEMBER ORDER record would need to be numbered accordingly. We call your attention to the following:

(A) This portion of the screen will scroll upward, thus allowing the end-user to enter the data for a large number of ordered items. Downward arrows were drawn after elements numbered 13 through 15 to indicate that the end-user may enter several values for these elements.

(B) If at any time the end-user enters incorrect data values for an element, a descriptive error message will appear. We explain these error messages in Figure 16.11. Notice that we've used Structured English to document when specific error messages may occur. No, we don't consider this documentation to be "doing the programmer's job." By conveying these editing requirements in Structured English, we are reducing the time we have to spend conveying the requirements to the programmer.

Using Input Record Layout Charts to Document the Structure of Batch and Remote Batch Inputs The classic **input record layout chart** is used to lay out the format of batch input files—the files that result from keying operations. In conjunction with display layouts, input record layouts also lay out the final format of remote batch input files. Figure 16.12 is a sample input record layout chart for a batch version of our MEMBER ORDER input. The form is not hardware-dependent and can be used to document virtually any

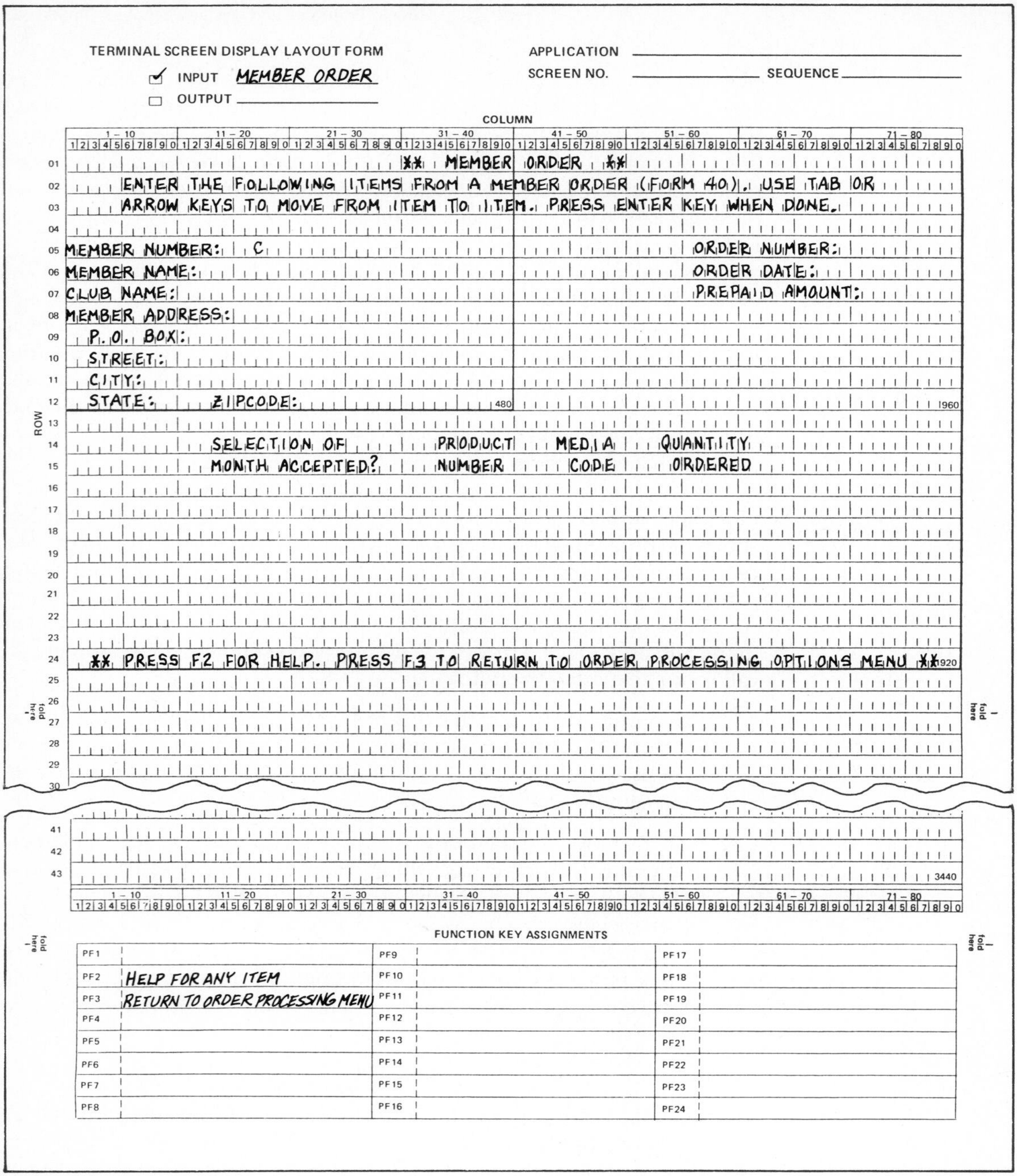

TERMINAL SCREEN DISPLAY LAYOUT FORM

☑ INPUT   *MEMBER ORDER*
☐ OUTPUT _____

APPLICATION _____
SCREEN NO. _____ SEQUENCE _____

COLUMN

```
         1-10        11-20       21-30       31-40       41-50       51-60       61-70       71-80
01                                      ** MEMBER ORDER **
02        ENTER THE FOLLOWING ITEMS FROM A MEMBER ORDER (FORM 40). USE TAB OR
03        ARROW KEYS TO MOVE FROM ITEM TO ITEM. PRESS ENTER KEY WHEN DONE.
04
05   MEMBER NUMBER:   C                                      ORDER NUMBER:
06   MEMBER NAME:                                            ORDER DATE:
07   CLUB NAME:                                              PREPAID AMOUNT:
08   MEMBER ADDRESS:
09     P. O. BOX:
10     STREET:
11     CITY:
12   STATE:      ZIPCODE:                     480                                        1960
13
14          SELECTION OF        PRODUCT   MEDIA   QUANTITY
15          MONTH ACCEPTED?      NUMBER    CODE    ORDERED
16
...
24   ** PRESS F2 FOR HELP. PRESS F3 TO RETURN TO ORDER PROCESSING OPTIONS MENU **  1920
```

ROW (01–29, 41–43)

Row 43: 3440

FUNCTION KEY ASSIGNMENTS

PF1	PF9	PF17
PF2 *HELP FOR ANY ITEM*	PF10	PF18
PF3 *RETURN TO ORDER PROCESSING MENU*	PF11	PF19
PF4	PF12	PF20
PF5	PF13	PF21
PF6	PF14	PF22
PF7	PF15	PF23
PF8	PF16	PF24

FIGURE 16.9   Display Layout Chart for On-Line MEMBER ORDER   The initial screen seen by the end-user is a blank order form to be filled in.

564

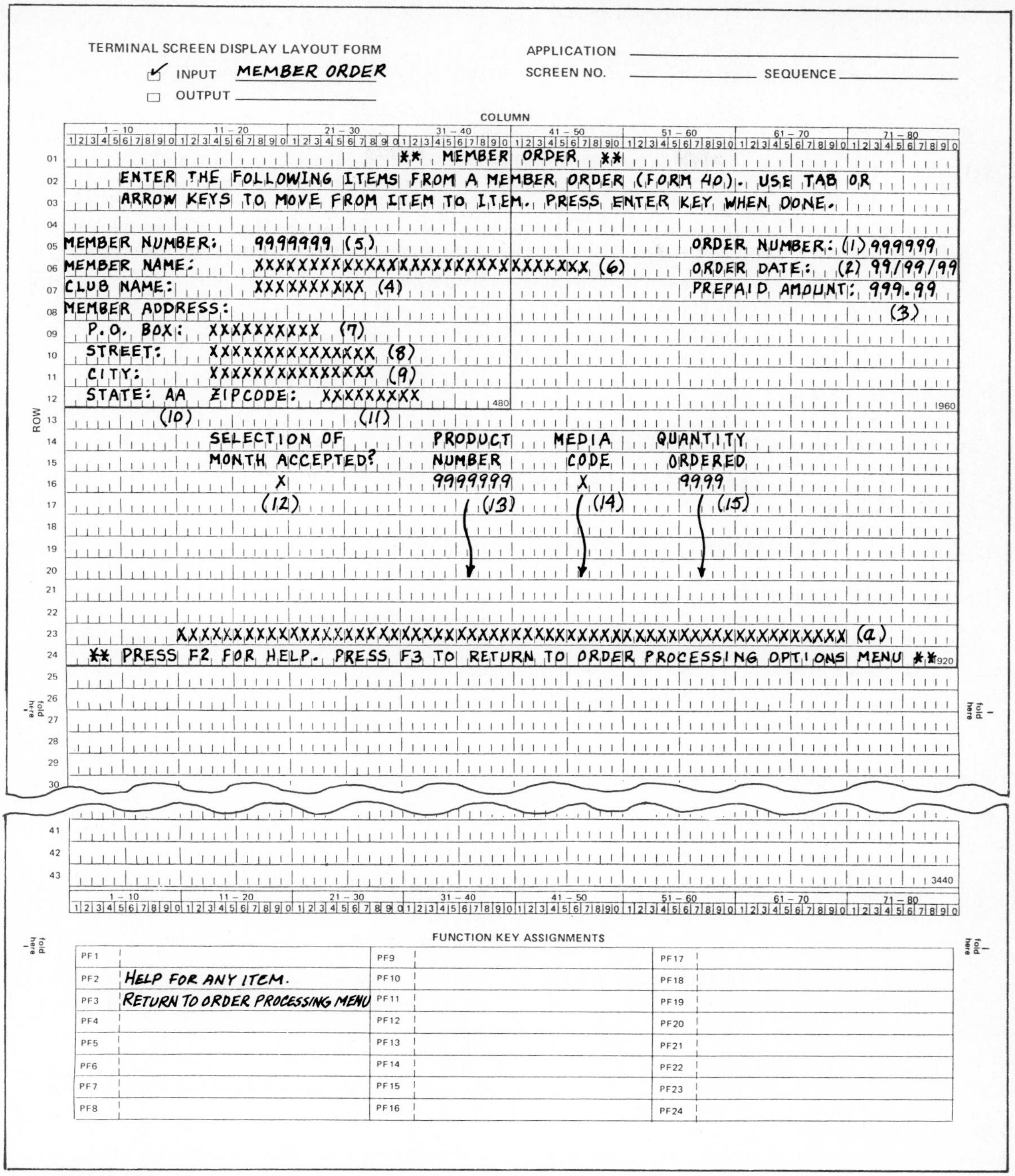

TERMINAL SCREEN DISPLAY LAYOUT FORM

☑ INPUT MEMBER ORDER
☐ OUTPUT _____

APPLICATION _____
SCREEN NO. _____ SEQUENCE _____

COLUMN

Row	Content
01	** MEMBER ORDER **
02	ENTER THE FOLLOWING ITEMS FROM A MEMBER ORDER (FORM 40). USE TAB OR
03	ARROW KEYS TO MOVE FROM ITEM TO ITEM. PRESS ENTER KEY WHEN DONE.
04	
05	MEMBER NUMBER: 9999999 (5) ORDER NUMBER: (1) 999999
06	MEMBER NAME: XXXXXXXXXXXXXXXXXXXXXXXXXXXX (6) ORDER DATE: (2) 99/99/99
07	CLUB NAME: XXXXXXXXXX (4) PREPAID AMOUNT: 999.99
08	MEMBER ADDRESS: (3)
09	P.O. BOX: XXXXXXXXXX (7)
10	STREET: XXXXXXXXXXXXXXX (8)
11	CITY: XXXXXXXXXXXXXXX (9)
12	STATE: AA ZIPCODE: XXXXXXXXX
13	(10) (11)
14	SELECTION OF PRODUCT MEDIA QUANTITY
15	MONTH ACCEPTED? NUMBER CODE ORDERED
16	X 9999999 X 99.99
17	(12) (13) (14) (15)
23	XXXXXXXXXXXXXXXXXXXXXXXXXXXXXXXXXXXXXXXXXXXXXXXXXXXXXXXXXXXXXXXXX (a)
24	** PRESS F2 FOR HELP. PRESS F3 TO RETURN TO ORDER PROCESSING OPTIONS MENU **

ROW

FUNCTION KEY ASSIGNMENTS

PF1	PF9	PF17
PF2 HELP FOR ANY ITEM.	PF10	PF18
PF3 RETURN TO ORDER PROCESSING MENU	PF11	PF19
PF4	PF12	PF20
PF5	PF13	PF21
PF6	PF14	PF22
PF7	PF15	PF23
PF8	PF16	PF24

FIGURE 16.10 Display Layout Chart for On-Line MEMBER ORDER Fields to Be Input This screen shows the proper edit masks for the fields to be entered by the end-user.

MEMO FOR MEMBER ORDER INPUT SCREEN

Ref. No.   Message

a   This field is used to provide the end-user with a descriptive error message when invalid commands or data have been entered. Otherwise the field is not printed. When the message is displayed, it should "blink" to grab the end-user's attention. As a reminder, the editing criteria were explained in the report titled "Report of Data Elements Contained in Member Order Record." The following specifies the conditions and types of messages to be displayed to the end-user.

Select the appropriate case:

Case 1:   MEMBER NUMBER is invalid

If the MEMBER NUMBER is not equivalent to a MEMBER NUMBER of an existing MEMBER then:

     error message = "MEMBER does not exist, please reenter."

Case 2:   ORDER NUMBER is invalid

If the ORDER NUMBER is equivalent to the ORDER NUMBER of any previous MEMBER ORDER then:

     error message = "ORDER NUMBER was assigned to previously entered MEMBER ORDER, please reenter."

Case 3:   ORDER DATE is invalid

Select appropriate case:

     Case 2.1   ORDER DATE contains no values, then:

         error message = "The order date must be provided on all orders, please enter."

     Case 2.2   ORDER DATE contains invalid values for month, day, year, then:

         error message = "The order date is not valid, please reenter."

FIGURE 16.11   Specifications for Display Attributes and Error Messages for MEMBER ORDER Input Screen   These notes are used to describe display attributes—such as blinking fields—and error messages.

MEMO FOR MEMBER ORDER INPUT SCREEN, continued

Case 4:   **PREPAID AMOUNT entered invalid, then:**

If PREPAID AMOUNT is greater than the TOTAL MEMBER ORDER COST (sum of EXTENDED PRICE, plus SALES TAX), then:

error message = "PREPAID AMOUNT exceeds the total costs of the member order, please reenter."

Case 5:   **PRODUCT NUMBER is not valid product number, then:**

error message = "Entered an incorrect part number (does not exist), please reenter."

Case 6:   **MEDIA CODE is invalid, then:**

error message = "media code is invalid, please reenter or press F2 key for list of valid codes and meanings."

Case 7:   **QUANTITY ORDERED is not greater than 0, then:**

error message = "Quantity ordered must be greater than 0, please reenter."

Case 8:   **MEMBER RESPONSE STATUS is not Y (yes) or N (no), then:**

error message = "Enter Y (yes) or N (no) in regard to member's acceptance of selection of the month offering."

FIGURE 16.11   (continued)

type of medium, including punched card images and key-to-tape—or disk—images. Four 80-column punched card images (a common punched card size) or one 320-character record (for magnetic media) can be accommodated. The form is read and prepared in the same manner as a record layout chart. The input record layout chart is primarily intended for the computer programmer, not the end-user. Note the following in Figure 16.12:

(A) We added the data element NUMBER OF ITEMS ORDERED. Why? Because, as you should recall from the previous chapter, variable-length records require a repeating factor field.

It's also quite common for a systems analyst to design a batch input file that contains more than one type of input record. For instance, a system may

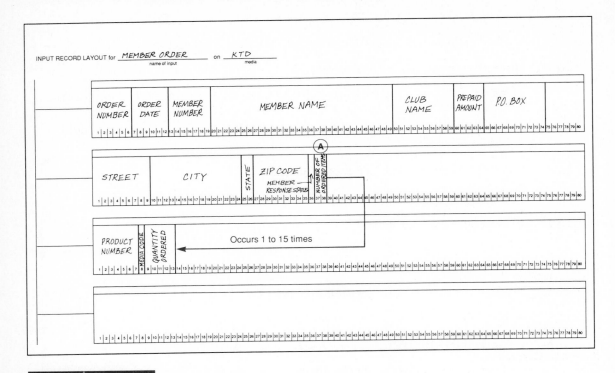

FIGURE 16.12 Input Record Layout Chart for Batch MEMBER ORDER Input

include a program that allows the marketing department to update our TITLE/PRODUCT file. The program might (1) place new products into the file, (2) delete from the file products that are to be discontinued, (3) change the contents of a product record to reflect a new price or unit of measurement. The input data for these three tasks would vary. How would this be handled on an input record layout chart?

In Figure 16.13, you can see how the design of the input layout record could be completed. Because the content and format of the input record are dependent on the particular type of update, we simply showed the three possible design layouts. Let's examine the record layout closely.

(A) This portion represents the contents and layout of the input record that is required when a new product is to be added to the product file. Notice that we included in the record a special field called TRANSACTION CODE. This field will be included in each of the three input record types. The value of this field will identify the particular type of input task to be performed.

INPUT RECORD LAYOUT for _PRODUCT UPDATE_ on _KTT_
name of input · media

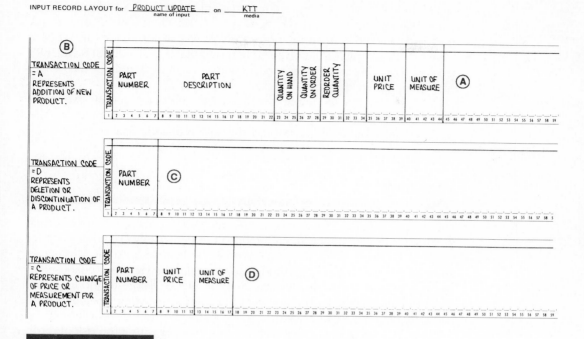

FIGURE 16.13 **Sample Input Record Layout Chart for Multiple Input Records**

Ⓑ This label is provided to let the programmer know which record layout corresponds to which input record type.

Ⓒ This portion of the input record layout chart specifies the data elements and layout for an input record that represents the deletion of an existing product record in the product file.

Ⓓ This portion of the input record layout chart specifies the data elements and layout for an input record that represents updates to the contents of an existing product record in the product file.

The Implications of Input Design for Output Design   In Chapter 15, you learned how to design computer-generated outputs. Recall that the data flow diagram helped you identify the outputs that must be designed. Be careful, though. Because DFDs show only net data flows (input or output), not all outputs that must be designed will show up on the DFD!

For instance, if we select a batch input implementation of MEMBER ORDER, it necessitates some type of control output, in this case a DAILY MEMBER

ORDER ERRORS REPORT. This edit report would identify all MEMBER ORDERS that contain incorrect data. The Member Services clerk will be able to use this report to follow up on the orders to ensure that corrections are made.

Another example would be an audit report that lists all MEMBER ORDERS processed on a given day. This report establishes an audit trail for the orders that were processed.

Because you already know how to design computer outputs, we won't spend time discussing the design of these reports. Just recognize that the design of any input could necessitate the design of additional outputs.

Summary

The goal of input design is to capture data and get that data into a format suitable for the computer. Input methods can be broadly classified as either batch or on-line. The systems analyst must be familiar with the advantages and disadvantages of each method as well as with the various media used to implement both methods. Because input is highly visible to the end-user, analysts should consider a number of *human factors* when designing computer inputs. The volume of data to be input by the end-user should be minimized because every data element input carries with it the risk of error. Source documents for capturing data should be designed for easy completion by end-users and for rapid data entry by data entry clerks and CRT operators.

Internal controls are also essential for inputs. Internal controls should be established for monitoring the number of inputs and for ensuring that the data is valid. Internal control techniques for ensuring the validity of data include completeness checks, limit and range checks, combination checks, self-checking digits, and PICTURE clauses.

To design computer inputs, the systems analyst should begin by identifying the input requirements of the new system. Data flow diagrams identify data to be captured and input to the system. The project dictionary defines the basic content of the inputs to be designed. Next, the analyst must specify the design parameters—issues concerning data capture, data entry, and data input—for the input. Then the analyst designs or prototypes the source document. For on-line inputs, the analyst would next prototype the input screens. Finally, the analyst may have to lay out the format of screens or batch input files using either display layout charts or input record layout charts, respectively.

Now that we've designed inputs and outputs, we must design any terminal dialogue that leads the end-user to these screens (assuming, of course, that an on-line system is being designed). This subject is covered in the next chapter.

Problems and Exercises

1. Explain the difference between *data capture, data entry,* and *data input.* Relate the three concepts to the processing of your school's course request or course registration.

2. To what extent should the end-user be involved during input design? What would you ask the end-user to do? When? What would you do for the end-user? When?

3. Define an appropriate input method and medium for each of the following inputs:
   a. Customer magazine subscriptions
   b. Hotel reservations
   c. Bank account transactions
   d. Customer order cancelations
   e. Customer order modifications
   f. Employee weekly time cards

4. What data flows crossing the person-machine boundary of a bounded data flow diagram should be designed as computer inputs?

5. What effects can be caused by lack of internal controls for inputs?

6. Obtain a copy of an application form—such as a loan, housing, or school form—or any other document used to capture data (such as a course scheduling form, credit card purchase slip, or time card). Do not be concerned whether the application is currently input to a computer system. How do the people who initiate or process the form feel about it? Comment on the human engineering. How well is it divided into zones? Comment on the suitability of the application for data entry. Are elements that wouldn't be keyed properly located? What changes would you make to the form?

7. Design a source document for the MEMBER ORDER input that was referred to throughout this chapter. Remember, the MEMBER ORDER input uses a turnaround source document. Also, the source document is initially output and mailed to the member (thus, you may wish to review Chapter 15).

8. What implications does input design have on output design?

Projects and Minicases

1. The order-filling operation for a local pharmacy is to be automated. The pharmacy processes 50 to 200 prescriptions per day. Customer prescriptions are to be entered on-line by pharmacists. Prepare an expanded data dictionary and display layout chart for documenting the design of the on-

line input PRESCRIPTION. A working data dictionary for PRESCRIPTION follows:

A PRESCRIPTION contains the following elements:

> CUSTOMER NAME
> DOCTOR NAME
> 1 to 10 occurrences of the following:
>> DRUG NAME
>> QUANTITY PRESCRIBED
>> MEDICAL INSTRUCTIONS
> RX NUMBER * a federal licensing number—6 digits

1 to 10 occurrences of the following * added by pharmacist

> DRUG NUMBER * a number that uniquely identifies a prescription drug—6 characters

> LOT NUMBER * a number that uniquely identifies the lot from which a chemical was produced—6 characters

> DOSAGE FORM * the form of the medication issued, such as "pill."
> P = PILL C = CAPSULE L = LIQUID I = INJECTION R = LOTION

> UNIT OF MEASURE * G = GRAMS   O = OUNCES   M = MILLILITERS

> QUANTITY DISPENSED

> NUMBER OF REFILLS

and optionally:

> EXPIRATION DATE

2. A moving company maintains data concerning fuel tax liability for its fleet of trucks. When truck drivers return from a trip, they submit a journal describing mileage, fuel purchases, and fuel consumption for each state traveled through. This data is to be batch input daily to maintain records on trucks and fuel stations. The TRIP JOURNAL data dictionary follows:

A TRIP JOURNAL consists of the following elements:

    TRUCK NUMBER
    DRIVER
    CODRIVER NUMBER
    TRIP NUMBER
    DATE DEPARTED
    DATE RETURNED
    1 to 20 of the following:
        STATE CODE
        MILES DRIVEN
        FUEL RECEIPT NUMBER
        GALLONS PURCHASED
        TAXES PAID
        STATION NAME
        STATION LOCATION

Design the batch input TRIP JOURNAL. Be sure to design an appropriate source document. Fuel receipts are to be stapled to the source document.

Annotated References and Suggested Readings

Fitzgerald, Jerry. *Internal Controls for Computerized Information Systems.* Redwood City, Calif.: Jerry Fitzgerald & Associates, 1978. This is our reference standard on the subject of designing internal controls into systems. Fitzgerald advocates a unique and powerful matrix tool for designing controls. The discussion of input controls is especially thorough. This book goes far beyond any introductory systems textbook—must reading.

CHAPTER SEVENTEEN

Designing and Prototyping
the User Interface and
On-Line Terminal Dialogues

Greater Metropolis TV Cable Company (Part 2)

In Chapter 16 we listened in on a conversation between Don McAllister, manager of the Information Systems Department, and Lisa Klemme, a systems analyst. Don was telling Lisa about complaints the Customer Services Department had voiced about the new services information system Lisa had implemented. In the subsequent weeks, Lisa worked on improving the on-line portion of the system. Most of her efforts were spent adding data editing features to screens used to input member orders.

Customer complaints decreased in the weeks that followed. Lisa had done a better job of ensuring the accuracy of the data input by the customer services clerks. But now she was constantly receiving phone calls concerning other problems the clerks were encountering.

"What does it mean when the terminal says HIT FUNCTION KEY 5?"

"Lisa, I really goofed! I pressed the wrong key and a new screen appeared on the CRT. How do I get back to the previous screen?"

"I keep typing in a date and that dumb computer won't accept it. Can you come fix it?"

"Lisa, I'm using the terminal to do something I've never done before. I have a screen that appeared, and I don't know how to go about typing in the information. Can you give me directions?"

"I keep trying to see one of our customer orders, but the screen runs it by me so fast I don't have a chance to see all the information I need to see."

"Lisa, isn't it about time you make the terminal screen stop displaying that joke of yours? I get tired of seeing that thing appear across the screen."

"Lisa, what is a fatal error? Am I in trouble?"

These types of phone calls come at all hours of the day. Needless to say, Lisa still has improvements to make to the terminal dialogue of the services information system.

Discussion

1. Can you tell from the phone calls what mistakes Lisa has made in designing the on-line user interface?
2. What are some of the other mistakes a systems analyst might make when designing on-line user interface?

What Will You Learn in This Chapter?

In this chapter you will learn how to design the on-line user interface that results in the on-line outputs and inputs that were designed in Chapters 15 and 16, respectively. Today, there are two commonly encountered on-line interfaces:

- terminals (or microcomputers behaving as terminals) used in conjunction with mainframes and minicomputers
- display monitors connected to microcomputers

In any case, you will know that you've mastered on-line user interface design when you can

1. Determine which features on available terminal and microcomputer displays can be used for effective on-line user interface design.
2. Identify the backgrounds and problems encountered by different types of terminal and microcomputer users.
3. Design or evaluate the human engineering in an on-line user interface for a typical information system.
4. Apply appropriate on-line user interface strategies to an information system.
5. Use a dialogue chart to plan and coordinate an on-line user interface for an on-line information system.
6. Describe how prototyping can be used to design an on-line user interface.
7. Use display layout charts (for net inputs and outputs from Chapters 15 and 16) to format the user interface screens in an on-line system.

On-line systems are most definitely the trend. Consequently, the design of an on-line user interface has taken on great importance. You have already learned how to design on-line inputs and outputs. The design of a user interface plays a pivotal role in user acceptance of on-line and microcomputer systems. Consequently, the amount of time spent designing an integrated, easy-to-use interface often exceeds the combined time to design the input and output screens themselves. In this chapter, you will learn how to bring these net inputs and outputs together into a dialogue controlled by the end-user. First, let's learn some of the underlying principles that affect on-line user interfaces.

When People Talk to Computers

On-line user interface design is the specification of a conversation between the end-user and the computer. This conversation generally results in either input or output—possibly both. What makes an on-line user interface good? Does the available technology limit or enhance dialogue possibilities? How can, or should, an on-line user interface be organized?

We know you recognize the difference between a terminal and a microcomputer. But we suspect that many of you are not familiar with the various features offered by their displays (also called monitors). Because display features affect on-line user interface design, we'll begin with an overview of features. Then we can examine some of the fundamental human factors and design strategies that underlie on-line user interface design.

Display Terminals and Monitors: Features That Affect User Interface Design

The design of an on-line user interface can be enhanced or restricted by the available features of your terminal display or monitor/keyboard. Let's examine some of these features.

Display Area The size of the display area is critical to on-line user interface design. The two most common display areas are 25 (lines) by 80 (columns) and 25 by 132. Some displays can be easily shifted between these two sizes. Some newer displays are designed to show more lines, for example, 65 lines on one page. Some terminals can show four complete 25 by 80 character screens simultaneously.

Character Sets and Graphics Every display uses a predefined character set. Most displays use the common ASCII character set that was introduced in Chapter 13. Some displays allow the programmer to supplement or replace the predefined character set. Additionally, many displays offer graphics capabilities. Graphics capabilities must be supported by graphics controllers and software that allow the programmer to take advantage of the graphics capabilities.

Paging and Scrolling   The manner in which the display area is shown to the end-user is controlled by both the technical capabilities of the display and the software capabilities of the computer system. Paging and scrolling are the two most common approaches to showing the display area to the end-user.

Paging displays a complete screen of characters at a time. The complete display area is known as a page (also called a screen or frame). The page is replaced on demand by the next or previous page; this is much like turning the pages of a book. **Scrolling** moves the display up or down, one line at a time. This is similar to the way movie and television credits scroll up the screen at the end of a movie. We'll discuss the choice of paging or scrolling for a dialogue when we discuss human factors.

Color Displays and Display Attributes   Greater numbers of displays are using color capabilities. Color can be used to highlight specific messages, data, or areas of the screen. Most displays also permit a variety of display and audio attributes. Some of these attributes include

- Double brightness on selected fields or messages.
- Blinking for selected fields or messages.
- Nondisplay for selected fields (for example, passwords).
- Reverse video for selected fields, messages, or display areas. Reverse video permits the color of the background—such as black—and the color of selected fields and messages—such as green—to be reversed.

Each of these features, when available, is activated by predefined codes that the programmer must learn and apply.

Split-Screen and Window Capabilities   Split-screen capability is a variation on the windows concept. In either case, the screen, under software control, can be divided into areas (called windows). Each area can act independently of the other windows, using features such as paging, scrolling, display attributes, and color. Each window can be defined to serve a specific purpose for the user.

Windows are rapidly becoming a standard interface. Many microcomputer products use standard windowing interfaces such as Microsoft WINDOWS and Quarterdeck's DESQVIEW. The future for standard interfaces looks promising. Apple popularized the idea with its MacIntosh microcomputers. All Mac applications, no matter who develops them, use the standard user interface and menu system. IBM is leading a movement that could establish a standard interface for all IBM and IBM-compatible applications—mainframe, minicomputer, and microcomputer! This interface is called Systems Application Architecture (SAA), which is pioneered in IBM's OS/2 operating system using a windowing interface called Presentation Manager. It is very similar to Microsoft WINDOWS. Other interface standards include X-Windows (UNIX) and NEXT.

Keyboards and Function Keys   Although not a display feature, most modern terminals and monitors are combined with keyboards. Keyboards contain special keys called function keys (usually labeled F1, F2, and so on). Others are labeled ALTernate or ESCape. These keys can be used to implement certain common, repetitive operations in an on-line user interface (for example, START, HELP, PAGE UP, PAGE DOWN, EXIT). You program the keys to perform the common functions. We'll discuss some of the more common uses of function keys when we discuss human factors later in this chapter. Part of the SAA standard, mentioned in the last subsection, suggests that function keys be used consistently (for example, SAA requires F1 to be the Help key). In any case, a system's programs should consistently use the same function keys for the same purposes.

Pointer Options   We are no longer restricted to the keyboard as the input technology for displays and terminals. Today, we are encountering other options, such as touch-sensitive screens, voice recognition, and mice. As you probably know, a mouse is a small hand-sized device that sits on a flat surface near the terminal. It has a small roller ball on the underside. As you move the mouse on the flat surface, it causes the cursor to move on the screen. Buttons on the mouse allow you to select objects or commands on which the cursor has been moved.

Human Factors for On-line User Interface Design

Nowhere are human factors as important as they are in on-line user interface design. Just ask the typical systems analyst who spends half the day answering phone calls like the ones Lisa got in the minicase that opened this chapter. That's why we want to discuss the subject of human engineering.

Display users can be broadly classified as either dedicated or casual. A **dedicated end-user** is one who will spend considerable time using specific on-line programs. This user is likely to become comfortable and familiar with the terminal's operation. The **casual end-user**, on the other hand, may only use a specific on-line program on an occasional basis. This user may never become truly comfortable with the terminal or the program. The end-user who hasn't used a terminal or a microcomputer is becoming less common in this computer-literate age. It is difficult to imagine today's youth as being ill at ease with the computer or display terminal. Still, most of today's systems are being designed for the casual end-user, with an emphasis on user friendliness.

General Human Engineering Guidelines   Given the type of end-user for an on-line system, there are a number of important human engineering factors that should be incorporated into the design:

- *The end-user should always be aware of what to do next.* The system should always provide instructions on how to proceed, back up, exit, and the like. There are several situations that require some type of feedback (adapted from Kendall and Kendall, 1988). They are
    - *Telling the end-user what the system expects right now.* This can take the form of a simple message such as READY, ENTER COMMAND, ENTER CHOICE, or ENTER DATA.
    - *Telling the end-user that data has been entered correctly.* This can be as simple as moving the cursor to the next field in a form or displaying a message such as INPUT OK.
    - *Telling the end-user that data has not been entered correctly.* Short, simple messages about the correct format are preferred. Help functions can supplement these messages with more extensive instructions.
    - *Explaining to the end-user the reason for a delay in processing.* Some actions require several seconds or minutes to complete. Examples include sorting, indexing, printing, and updating. Simple messages such as SORTING—PLEASE STAND BY or INDEXING—THIS MAY TAKE A FEW MINUTES. PLEASE WAIT tell the user that the system has not failed.
    - *Telling the end-user that a task was completed or not completed.* This is especially important in the case of delayed processing, but it is also important in other situations. A message such as PRINTING COMPLETE or PRINTER NOT READY—PLEASE CHECK AND TRY AGAIN will suffice.

- *The screen is always formatted so that the various types of information, instructions, and messages always appear in the same general display area.* This way, the end-user always knows approximately where to look for specific information. To achieve this goal, we suggest that zones or areas be defined as indicated in Figure 17.1. Zones don't necessarily have to have displayed borders, although they can help. A sample screen template for planned zones is illustrated in Figure 17.2. This is only one possible layout. Another might group key assignments, messages, help, and the like at the bottom of the screen. The zones concept can easily be implemented with screen formatting software that is generally available for most computers. Even without such software, zones can be defined and followed by using conventional programming techniques.

- *Within the body zone, the dialogue should be limited to one idea per frame, whether paging or scrolling through the zone.* For instance, the zone should display one menu, one input, one report, or one query response.

    The choice between paging and scrolling for the body zone depends on the information content to be displayed in that zone. If the information to be displayed is continuous in nature, such as most reports, scrolling can be used. The cursor can be moved up and down such a listing line by line. If the information to be displayed is to be viewed one record at a time or depicted as a form, paging is preferred.

Suggested Window Definitions

Title window — The title window identifies the screen from the end-user's point of view.

Flag window — The flag window is used to *point* to some specific line in one of the other windows to highlight the location of an error or problem. For example if the end-user has made a mistake, the symbol " " might appear in the flag window on the line in which the mistake occurred. To discover the specific nature of the problem, the end-user should look at the message window (the window described next).

Display attributes such as blinking fields and reverse video can accomplish the same purpose and eliminate the need for a flag area.

Message window — The message window is used to display system messages to the end-user. For instance, error messages and/or suggestions may be recorded in this window. Most of the time, this window would be blank.

Escape window — The escape window is used to suggest how the end-user can exit the current system or subsystem. For example, the escape window may display a message on how to get back to the main system menu or to the previous menu of options.

Body window — The body window is the largest display area. It is in the body that the end-user inputs new data or views output information. The body is also used to display *help* messages that are too long to display in the message or instruction windows. This window is also used to display menu options or direct question and answer type dialogues (more about these options later).

FIGURE 17.1 Zone Areas on the Terminal Screen Terminal screens are easier to read if the screen is partitioned into areas, called zones, into which similar data and specific types of messages are always recorded.

- *Messages, instructions, or information should remain in the zone long enough to allow the end-user to read them.* For instance, data should not be allowed to scroll out of a zone before it can be read. One way to accomplish this is to print only as much information as the zone can display at one time and then freeze the screen. A message to press either any key or some specific key to continue can be displayed in the instruc-

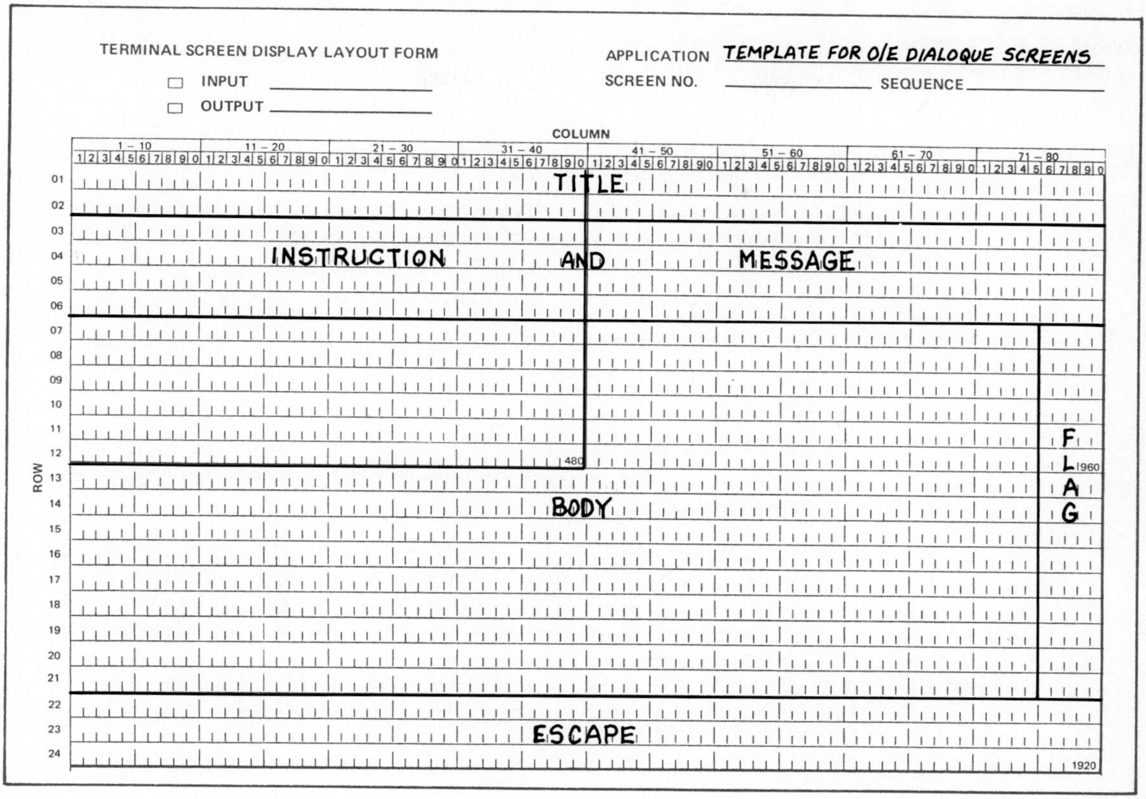

FIGURE 17.2 Sample Zones for a Terminal Screen This is one alternative for zone design on a terminal screen. Different partitions are possible; however, once a partition is created, it should be used consistently.

tion zone. The system can then page or scroll through the next set of information.

Alternatively, to economize on available space, some zones are being eliminated from the basic screen and replaced with "pop-up" windows that temporarily overlay the main screen to provide instructions, help, or messages.

FIGURE 17.3 Common Function Key Uses

TYPICAL FUNCTION KEY ASSIGNMENTS

• START a program or function.

• HELP: display help text.

• *Cursor Movement.* Many systems have predefined keys for moving the cursor forward and backward one character, word, or field at a time as well as for moving the cursor up and down one line or page at a time. If these functions don't exist, they can be developed and used by all systems.

• EXIT *or* TERMINATE *the session.* If data can be lost, appropriate messages and instructions should be used to make sure the end-user hasn't made a mistake.

• ESCAPE from an operation (input or output) that is currently being done. This might be used to "start over" if the operator feels a serious mistake has been made.

• *Keystroke combinations.* Microcomputers and some intelligent terminals can take advantage of special software called keyboard enhancers, such as RoseSoft's Prokey. Keyboard enhancers allow single keys (not restricted to function keys) to automatically execute long sequences of common and predefined keystrokes.

• *Use display attributes sparingly.* Display attributes, such as blinking, highlighting, and reverse video, can be distracting if overused. Judicious use allows you to call attention to something important—for example, the next field to be entered, a message, or an instruction.

• *Simplify complex functions and reduce typing by providing the end-user with function keys.* Some of the functions most commonly defined on function keys are described in Figure 17.3.

• *Default values for fields and answers to be entered by the end-user should be specified.* A common practice is to place the default value in brackets (for example, ORDER DATE? <Today's Date>). The end-user can press the enter key to get the default date.

• *Anticipate the errors end-users might make.* End-users will make errors, even when given the most obvious instructions. If the end-user can potentially execute a dangerous action, let it be known (a message ARE YOU SURE? is nice). An ounce of prevention goes a long way!

With respect to errors, a symbol in the flag zone should point to the error. Also, an appropriate error message should appear in the message zone. The end-user should not be allowed to proceed without correcting the error. Instructions to correct the error can be displayed in the instruction zone. A HELP key can be defined to display additional instructions or clarification in the body zone. In any event, the end-user should never get an operating system message or fatal error. If the end-user does something that could be catastrophic, the keyboard should be locked to prevent any further input. An appropriate instruction to call the analyst or computer operator should be displayed in this situation.

Dialogue Tone and Terminology   The overall tone and terminology of an on-line dialogue are another important human engineering consideration. The session should be user friendly (a frequently unachieved goal) and involve "noncomputerese" (an underemphasized factor). With respect to the tone of the dialogue, the following guidelines are offered:

- *Use simple, grammatically correct sentences.* It is best to use conversational English rather than written English. However, slang and profanity are taboo!

- *Don't be funny or cute!* When someone has to use the system 50 times a day, intended humor is as funny as any joke you've heard for the 50th time.

- *Don't be condescending; that is, don't insult the intelligence of the end-user.* For instance, don't offer rewards or punishment (THAT'S CORRECT or YOU SHOULD KNOW BETTER).

With respect to terminology used during the dialogue, the following suggestions may, however nit-picky, prove helpful:

1. *Don't use computer jargon.*
2. *Avoid most abbreviations.* Abbreviations assume that the end-user understands how to translate the abbreviation. Check first!
3. *Avoid symbolism that may be foreign to the end-user* (such as mathematical notation).
4. *Use simple terms.* Use NOT CORRECT instead of INCORRECT. There is less chance of misreading or misinterpretation.
5. *Be consistent in your use of terminology.* For instance, don't use EDIT and MODIFY to mean the same instruction.
6. *Instructions should be carefully phrased, and appropriate action verbs should be used.* The following recommendations should prove helpful:
    - Try SELECT instead of PICK when referring to a list of options. Be sure to indicate whether the end-user can select more than one option from the list of available options.
    - Use TYPE, not ENTER, to request the end-user to input specific data or instructions.

THE NEXT GENERATION

Natural English as the Perfect User Interface?

The computer technology industry has invested considerable time, effort, and money to try to reach the elusive goal of the perfect user interface. Windowing schemes, menuing strategies, question-answer strategies, and psychological research studies have proliferated. Maybe the answer is right in front of our eyes . . . natural English as the interface.

Some companies are marketing natural language interfaces to their systems. Consider, for example, the *natural language query processors* for database management systems.

Given: a database that is loaded with substantial data about numerous entities of interest to the business. The tool: an English language parser

to generate reports and inquiry responses from the database. Sample inquiry: WHAT CUSTOMERS HAVE A BALANCE OF MORE THAN $500? Interpretation: Find data about customers whose ACCOUNT BALANCE > $500.

The ability of a natural language query processor is dependent on the intelligence of that processor. Most software only allows end-users to enter explicit commands and keywords. This is the nature of classic programming languages like BASIC, FORTRAN, COBOL, Pascal, ADA, and C. Natural language, on the other hand, is hardly explicit. But there are some good interpreters out there. A typical dialogue might occur something like this:

END-USER: WHERE CAN WE GET 3/4 INCH LUG NUTS?

COMPUTER: I'M NOT FAMILIAR WITH THE PHRASE "3/4 INCH LUG NUTS." HOW MANY TERMS IS IT?

END-USER: TWO.

COMPUTER: TYPE THE TWO TERMS, SEPARATED BY A COMMA.

END-USER: 3/4 INCH, LUG NUTS.

COMPUTER: IS "3/4 INCH" THE VALUE OF A FIELD STORED IN THE DATABASE?

END-USER: PROBABLY.

COMPUTER: IN WHAT FIELD WOULD THE VALUE "3/4 INCH" BE STORED? A PARTIAL NAME WOULD HELP.

- Use PRESS, not HIT or DEPRESS, to refer to keyboard actions. Where possible, refer to keys by the symbols or identifiers that are actually printed on the keys. For instance, the <- is used on some terminals to designate the RETURN or ENTER key.
- When referring to the cursor, use the term POSITION THE CURSOR, not POINT THE CURSOR.

On-Line User Interface Strategies

Are there any specific strategies you can employ to design better on-line user interface? Indeed, there are a number of such strategies, and the choice of strategy depends on the nature of the function to be performed and the characteristics of the end-user who will use the terminal. Let's briefly survey these strategies.

THE NEXT GENERATION

Natural English as the Perfect User Interface? Continued

END-USER: SIZE.

COMPUTER: SEARCH DATA DICTIONARY . . . THE STRING "SIZE" WAS FOUND IN 2 RECORD TYPES IN THE DATABASE. ARE YOU REFERRING TO "PRODUCT SIZE" OR "VENDOR PART SIZE"?

END-USER: VENDOR PART.

COMPUTER: IS THE "LUG NUT" OR "LUG NUTS" THE VALUE OF A FIELD STORED IN THE VENDOR PART RECORD?

END-USER: YES.

COMPUTER: IN WHAT FIELD WOULD THE VALUE "LUG NUT" OR "LUG NUTS" BE STORED?

END-USER: DESCRIPTION.

COMPUTER: THE FIELD "VEN-

DOR PART DESCRIPTION" WAS FOUND. IS THIS CORRECT?

END-USER: YES.

COMPUTER: ARE YOU REQUESTING ALL VENDORS FOR WHICH VENDOR PART SIZE = 3/4 INCH AND PART DESCRIPTION = LUG NUT?

END-USER: YES.

COMPUTER: THE FOLLOWING FIELDS CAN BE DISPLAYED OR PRINTED. PLEASE IDENTIFY THOSE FIELDS YOU WISH TO SEE BY NUMBER:
1. VENDOR PART NUMBER
2. VENDOR PART DESCRIPTION
3. VENDOR PART SIZE
4. VENDOR PART QUANTITY ON HAND
5. VENDOR . . .

Interesting, isn't it? While there are no perfect natural language processors currently available, we suggest you research a Cullinet product called On-Line English. This natural language processor can be used with Cullinet's IDMS database system. But this technology is not restricted to mainframes. Try researching a microcomputer database product called R:BASE for DOS (or for OS/2). This popular database offers an optional natural language interface called Clout. Maybe someday we'll stop talking to computers through today's fancy windows and command syntax. Instead, we'll just "talk" to computers like we talk to people!

Menu Selection The most popular dialogue strategy is menu selection. A **menu** of alternatives or options is presented to the end-user. The end-user selects the desired alternative or option by keying in the number or letter that is associated with that option. More sophisticated technology allows menu selection by touching the screen, pointing to the desired item with a light pen, selecting menu options with a mouse, or using cursor keys to the desired alternative.

A classic menu is illustrated in Figure 17.4. Alternative approaches include a *pull-down menu* (Figure 17.5a) and a *pop-up menu* (Figure 17.5b). Pull-down menus are submenus that pull down from a main menu option (like a pull-down window blind). Microsoft WINDOWS and WINDOWS applications (like Aldus Page Maker) are examples of applications that use pull-down menus. Pop-up menus are activated by function keys or combinations of keys pressed simultaneously. When the keys are pressed, the menu pops up from

WARRANTY SYSTEM
MAIN MENU

TYPE NUMBER OF DESIRED REPORT AND PRESS RETURN KEY.

1 PROCESS WARRANTY TRANSACTION
2 DISPLAY WARRANTY REPORTS
3 QUERY WARRANTY STATUS

2

WARRANTY SYSTEM
REPORT MENU

TYPE NUMBER OF DESIRED REPORT AND PRESS RETURN KEY.
SYSTEM WILL ASK FOR ANSWERS TO APPROPRIATE QUESTIONS.

1 WARRANTY TRANSACTION REGISTER
2 PART WARRANTY SUMMARY
3 PROBLEM PART EXCEPTION REPORT

2

WARRANTY SYSTEM
PART WARRANTY SUMMARY

ANSWER THE FOLLOWING QUESTIONS:
PRESS F5 FOR HELP.

WHICH PART NUMBER FOR SUMMARY? (SEPARATE LISTED
PARTS WITH COMMAS AND THEN PRESS RETURN KEY.)

23254433,1325553,2211787,6663211,7015676,4544321

DO YOU WANT A PRINTED REPORT? (NO) YES
TYPE YOUR MAIL ROUTE CODE AND PRESS RETURN: 10023
DISPLAY REPORT AT TERMINAL? (NO) YES

WARRANTY SYSTEM
PART WARRANTY REPORT

PRESS ANY KEY TO SEE NEXT PAGE.
PRESS F1 KEY TO SEE PREVIOUS PAGE.
PRESS F3 KEY TO SEE FIRST PAGE AGAIN.

PART NUMBER   23254433   DESCRIPTION   3.5 HP LAWN ENGINE

WARRANTY CLAIMS:

THIS MONTH	LAST MONTH	THIS YEAR	LAST YEAR	% UP/DOWN
43	52	32	47	+69%

PRESS F6 TO RETURN TO REPORT MENU
PRESS F10 TO RETURN TO MAIN MENU

◀ **FIGURE 17.4 Sample Menu Selection for On-Line User Interface**
Menu selection is the most popular dialogue strategy today. Menu selection is particularly effective when dealing with a casual end-user who knows little about computers. This menu demonstrates a multiple hierarchical menu structure.

nowhere, temporarily overlaying whatever resides on the screen. Completion of a menu command or cancelation of the menu—for example, pressing the ESCape key—will return the end-user to the original screen. Borland's Sidekick is an example of an application that uses pop-up menus.

Menu-driven systems are particularly popular with the casual or semicasual end-user who doesn't use a particular program on a regular basis. Menu-driven systems also place production processing under the control of the end-user. It should be mentioned that by placing control in the hands of the end-users, production efficiency may deteriorate. Menu items should be self-explanatory and should contain neither jargon nor vague abbreviations or statements. If there are so many menu alternatives that the menu screen is too small or becomes cluttered, menus can be designed hierarchically. Small lists of related menu options can be grouped together into a single menu.

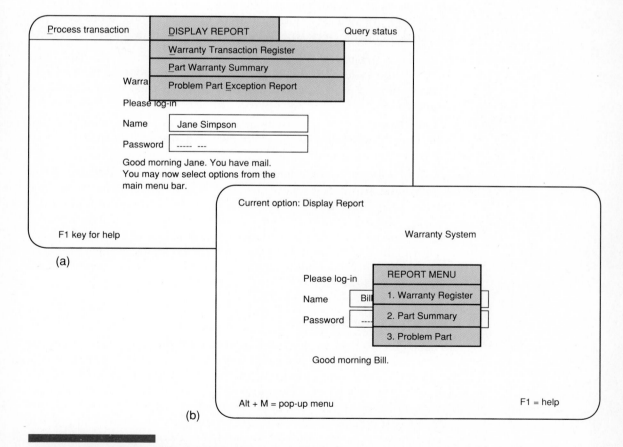

FIGURE 17.5 Pull-Down and Pop-Up Menu Pull-down menus (a) are submenus that pull down from a main menu option. Pop-up menus (b) are activated by function keys or combinations of keys and temporarily overlay whatever resides on the screen.

These menus can then be grouped into a higher-level menu. This approach was applied in Figure 17.4. If the option DISPLAY WARRANTY REPORTS is selected, the submenu WARRANTY SYSTEM REPORT MENU will appear. And if the PART WARRANTY SUMMARY option is then selected, the bottom screens shown in Figure 17.4 will appear. Specific reports can be selected from that screen. There is no technical limit to how deeply hierarchical menus can be nested. However, the deeper the nesting, the more you should consider providing direct paths to deeply rooted menus for the experienced end-user who may find navigating through the multiple levels annoying.

Eventually, the menus will get you to a basic output or input operation or screen, which you learned how to design in Chapters 15 and 16, respectively.

Instruction Sets   Instead of menus—or in addition to menus—you can design an on-line dialogue around an instruction set (also called a *command language interface*). Because the user must learn this syntax, this approach is suitable only for dedicated terminal users. There are three types of syntax that can be defined. Which type is used depends on the available technology.

1. A form of *Structured English* can be defined as a set of commands that control the system. In this type of dialogue, an elaborate HELP system should be created so the end-user who forgets the syntax can get assistance quickly.

2. A *mnemonic syntax* can be defined. A mnemonic syntax is built around meaningful abbreviations for all commands. Once again, a HELP facility is highly recommended.

3. *Natural language syntax* interpreters are now becoming available. When employing natural language syntax, the end-user enters commands using natural English (either conversational or written). The system interprets these commands against a known syntax and requests clarification if it doesn't understand what the end-user wants. As new interpretations become known, the system learns the end-user's vocabulary by saving it for future reference. This emerging technology is featured in this chapter's "Next Generation" box feature.

Question-Answer Dialogues

To supplement either menu driven or syntax-driven dialogues, you can use question-answer strategies where appropriate. The simplest questions involve yes or no answers—for instance, "Do you want to see all records? <NO>." Notice how we offered a default answer! Questions can be more elaborate. For example, the system could ask, "Which part number are you interested in <last part number queried>?" Just make sure you consider all possible correct answers and deal with the actions to be taken if incorrect answers are entered. Question-answer dialogue is difficult because you must try to consider everything that the end-user might do wrong!

Graphics   Graphics capabilities are not only providing new ways to output information (for example, graphs, charts, and so on). It is also allowing friendlier interfaces. For example, many applications now use *icons*, small graphic images that suggest their function. For example, a trash can icon might symbolize a delete command. Selecting the icon with a pointing device like a mouse or light pen executes the function. Also, icons can work in conjunction with one another. For instance, a pointing device can be used to drag the icon of a file folder (representing a named file) to a trash can icon—intuitively instructing the system to delete (or throw away) the file. The Apple MacIntosh interface (called Finder) has popularized the use of icons. IBM's Presentation Manager is the equivalent in the IBM/clone marketplace.

It should be acknowledged that not all computer users like graphic interfaces.

The four strategies we've suggested should be considered together with human factors discussed earlier. If you evaluate your dialogue against these fundamental concepts, you may save yourself from that dreaded 2:00 A.M. phone call, "Betty? Did I wake you? Sorry! But we have a problem with what the system is asking us for . . ."

How to Design and Prototype an On-Line User Interface

The typical approach to designing an on-line user interface is to throw together a few display layout charts. This strategy doesn't fare too well—the final dialogue ends up being designed by the programmer, on the run. It shouldn't surprise you that a typical on-line user interface may involve many possible screens, perhaps hundreds! Each screen can be laid out with a display layout chart. But what about the coordination of these screens?

Screens will occur in a specific order. Perhaps you can move forward and backward through the screens. Additionally, some screens may occur only under certain conditions. And to make matters more difficult, some screens may occur repetitively until some condition is fulfilled. This almost sounds like a programming problem, doesn't it? We need a tool to coordinate the screens that can occur in on-line user interface.

Step 1: Chart the Dialogue

A **dialogue chart** is a screen sequencing variation on program flowcharts and hierarchy charts. A sample dialogue chart is illustrated in Figure 17.6. The arrows indicate that a screen can get to another screen in that sequence. Note that the arrows *can be* bidirectional, meaning that you can get to either screen in either direction. The treelike structure suggests that at some point in the dialogue, the end-user will select an option and execute only those screens

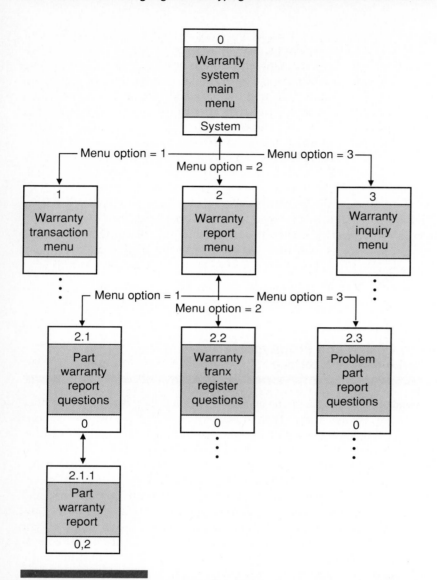

FIGURE 17.6 Sample Dialogue Chart Dialogue charts are used to coordinate on-line user interface screens. The tool can depict the sequence and variations of screens viewed by the end-user.

in that branch of the tree. This if-then structure is perfect for menu-type dialogues. Figure 17.6 shows a simple dialogue chart. Let's study this tool in greater detail.

Dialogue Chart Conventions   The purpose of the dialogue chart is to depict the sequence and variations of screens that can occur when the end-user sits at the terminal. You can think of it as a roadmap. Each screen is analogous to a city. Not all roads go through all cities. Figure 17.7 depicts all of our dialogue chart conventions. The rectangles represent display screens (formatted by display layout charts). The arrows represent the flow of control through the various screens.

(A) Note that the rectangles are divided into three sections. The top section contains a reference number that corresponds to the SCREEN NUMBER that appears on the display layout chart. The letter H after the number indicates that one or more help screens are available for the screen represented by the rectangle. To keep things simple, we'll follow the same numbering convention you learned in the data flow diagram chapters.

(B) The middle section names or describes the screen. Most dialogues permit the end-user to get from one screen to another in either direction—hence the arrowhead at each end of the arrow.

(C) Because it is desirable in many systems to provide shortcuts back to previous screens, the bottom section of the rectangle describes the allowable shortcuts, called escapes.

The rectangles only describe what screens can appear during the dialogue. The arrows describe when these screens occur. Earlier, we stated that the flow of screens in an on-line user interface occurs with almost programming-like precision. Indeed, the three constructs of a well-structured computer program apply equally to the dialogue chart:

(1) Sequence. The screens occur in a natural sequence one after the other.

(2) Selection. Based on the end-user's answer to a question or the selection of a menu alternative or icon, a different path through the screens is selected (similar to programming's if-then-else construct).

(3) Repetition. Until some condition is fulfilled, a sequence of screens is repeated. This is similar to programming's repeat-until or do-while constructs.

To supplement these standard constructs, note the following:

(4) The small triangle that contains a letter indicates the return of control to that circle containing the same letter.

The constructs can be nested just like programming constructs. Unlike program flowcharts, most arrows in a dialogue chart are bidirectional. This indicates that the end-user can also move backward through the screens. The escape numbers in the screen rectangles help us avoid sloppy go-to arrows on the dialogue chart. Using these simple notations, the analyst (and possibly

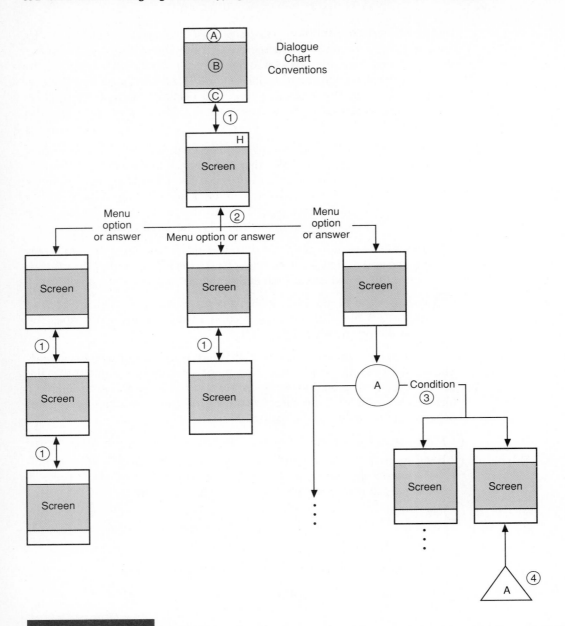

FIGURE 17.7 **Dialogue Chart Conventions** A dialogue chart consists of rectangles that represent display screens and arrows that represent flow of control through various display screens.

end-users) can design a complete dialogue. The dialogue chart serves as a table of contents into the display layout charts for all of the screens we've just discussed.

Let's examine the on-line user interface for some of the SoundStage project. The SoundStage dialogue charts were drawn manually. Our CASE product, Excelerator, doesn't directly include a dialogue graphic option. It could be simulated with other graph types that are supported. Some CASE products include proprietary dialogue graphics. For example, LBMS's Automate-Plus includes a dialogue graphic called Invocation Dialogue Structure Diagrams.

Design of a Dialogue Structure   The on-line portion of the SoundStage Member Services system can be partitioned into several levels of screens with each level aimed toward accomplishing a better-defined function. For example, the on-line portion of the SoundStage system can initially be decomposed into three on-line support capabilities: transaction processing, management reporting, and decision support. Each of these capabilities can be assigned screens that will assist the end-user. In addition, these support capabilities might be decomposed into more specific support functions for which screens are also assigned.

In Figure 17.8, we present a dialogue chart that depicts the sequence and variations of screens that can occur when the end-user sits at the terminal. Notice that some of the screen names appear below a horizontal line rather than inside a rectangle. To keep this chapter down to a reasonable length, we did not design and discuss these segments of the dialogue that appear below the horizontal lines. We call your attention to the following points of interest in Figure 17.8:

(1) The MEMBER SERVICES SYSTEM MENU is the first screen to be viewed by the end-user. Because this is the first screen seen by the end-user, the only escape is to the operating system. The SoundStage system contains several options, conveniently grouped into submenus. The MEMBER SERVICES SYSTEM MENU will provide the end-user with the option of processing a member order, obtaining a report, and performing inquiries. Each of these options, when chosen, will result in the end-user seeing a new screen: the MEMBER ORDER-PROCESSING MENU, the MEMBER SERVICES REPORT MENU, or the MEMBER SERVICES INQUIRY MENU screen. The menu options selected are recorded on the branching lines.

(2) The MEMBER ORDER-PROCESSING MENU screen will offer the end-user a number of transaction-processing options. No escapes were specified for this screen because MEMBER SERVICES SYSTEM is the only previous screen seen by the end-user so far—the back-arrow already handles this escape option.

(3) Option 1 sends the end-user into a repeating screen, MEMBER ORDER (designed in Chapter 16), that can be filled in until there are no more

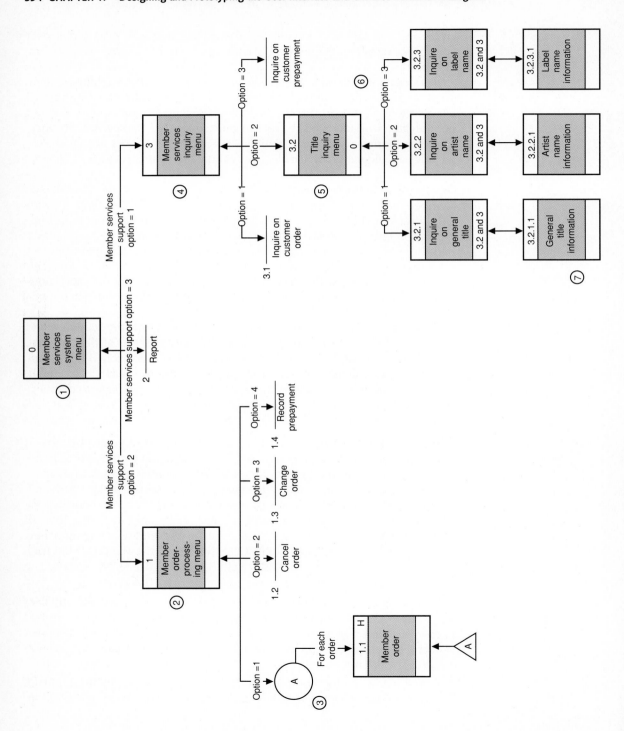

◀ **FIGURE 17.8 Dialogue Chart for SoundStage Member Services Support Functions** This dialogue chart shows the relationship between the many display screens needed to describe the dialogue of the SoundStage Member Services on-line support functions. The labeled horizontal lines represent portions of the Member Services system for which the display screens needed to support a function have not yet been defined.

orders to input—at which time the end-user selects one of the escapes. Notice that help screens are indicated.

(4) The system will allow the end-user to do three types of on-line inquiries. The MEMBER SERVICES INQUIRY MENU screen displays the inquiry options to the end-user.

(5) The end-user will be provided with the TITLE INQUIRY MENU screen if the end-user's choice was to do an inquiry on a SoundStage title or product. The end-user may request three different types of information concerning a title. Screen 3.2, for example, will provide the end-user with the option of retrieving general information on a title.

(6) Either screen 3.2.1, 3.2.2, or 3.2.3 will be displayed to the end-user, depending on the title inquiry option selected from the TITLE INQUIRY MENU screen. The purpose of these screens is to determine which title the end-user wants to receive information on. Notice that the escape options shown on each of these screens allows the end-user to return to the TITLE INQUIRY MENU screen, MEMBER SERVICES INQUIRY MENU screen, or MEMBER SERVICES SYSTEM MENU screen. As we design more levels of screens, the escape options become more beneficial to the end-user.

(7) Screens 3.2.1.1, 3.2.2.1, and 3.2.3.1 represent outputs. These screens were presumably designed in Chapter 15. They always appear immediately after the end-user has supplied appropriate information called for on the previous screen. Notice that there are no escape mechanisms other than the ability to return to the previous screen.

Step 2: If Possible, Prototype or Simulate the Dialogue and Interface

There are several ways to prototype a dialogue and end-user interface. Since a live demonstration in a textbook is impossible, we will briefly describe different tools and techniques and direct you to some examples.

One of the simplest tools for dialogue and interface prototyping is **demonstration software**. One of the best and least expensive tools is Software Garden's Demo. Using Demo, you can prototype screens and provide logic (or a playscript) for moving between screens. This gives the impression that you are watching the actual program execute.

Many **Computer-Assisted Systems Engineering (CASE)** products include dialogue and interface prototyping tools. In previous chapters, we demonstrated how Excelerator permits screen designs to be prototyped. Screen-to-screen interfaces can be exercised, but only sequentially—in other words, conditions (menus) and repetition of screens cannot be exercised in the product.

Other CASE products such as Pansophic's Telon can support both screen design and exercising of all three basic constructs: sequence of screens, conditional branching to screens, and repetition of screens. Telon can also generate COBOL and PL/1 code for approved screens.

Most **database management systems** and **fourth-generation languages** include screen and dialogue generators that can be used to prototype the user interface.

The key advantage of all these prototyping environments is the ability to *test* or **exercise** the user interface. This means that end-users literally experiment and test the design prior to extensive programming to truly implement the working system. Analysts can observe this test to improve on the design.

In the absence of prototyping tools, the analyst should at least simulate the dialogue by walking through the screen sketches with end-users.

Step 3: If Necessary, Produce Programmer-Oriented Layout Charts

Given the dialogue structure, we can design the actual dialogue using display layout charts. These charts, introduced in Chapters 15 and 16, can either be done in lieu of a prototype or completed to more formally specify what you did in a prototype.

Referring back to the dialogue chart shown in Figure 17.8, there are several paths that the dialogue will follow. For the sake of brevity, we will design screens only for the order-processing portion of the on-line system. These screens were chosen because they serve as the dialogue that would be required for the end-user to arrive at the input screens that were designed in Chapter 16. Similar screens would normally be developed for the entire dialogue chart.

If one hasn't already been constructed, we suggest that a template for the zones or areas be defined. We used this template shown in Figure 17.9 for our design.

The initial screen seen by the end-user is presented in Figure 17.10. It is the main systems menu. The screen number on the display layout chart should match the screen number in the dialogue chart. Line 22 is a message field for error messages. The accompanying letter in parentheses is used to direct us to a memo that describes the messages that can be printed. Finally, notice that we used the function key assignments of the display layout chart to define an operation, TERMINATE SESSION.

Let's move on. If our end-user wishes to input some MEMBER ORDERs, then option 2 on the MEMBER SERVICES SYSTEM MENU would be selected. According to our dialogue chart, the end-user would now see screen 1, the MEMBER ORDER-PROCESSING MENU, which is shown in Figure 17.11. This screen is similar to the previous screen, because both screens were designed using our template and because both screens simply offer a menu of options to the end-user. Notice that, in accordance with our dialogue chart, this screen offers the end-user an escape option to return to the previous menu. The end-user who wishes to input MEMBER ORDERs would select option 1.

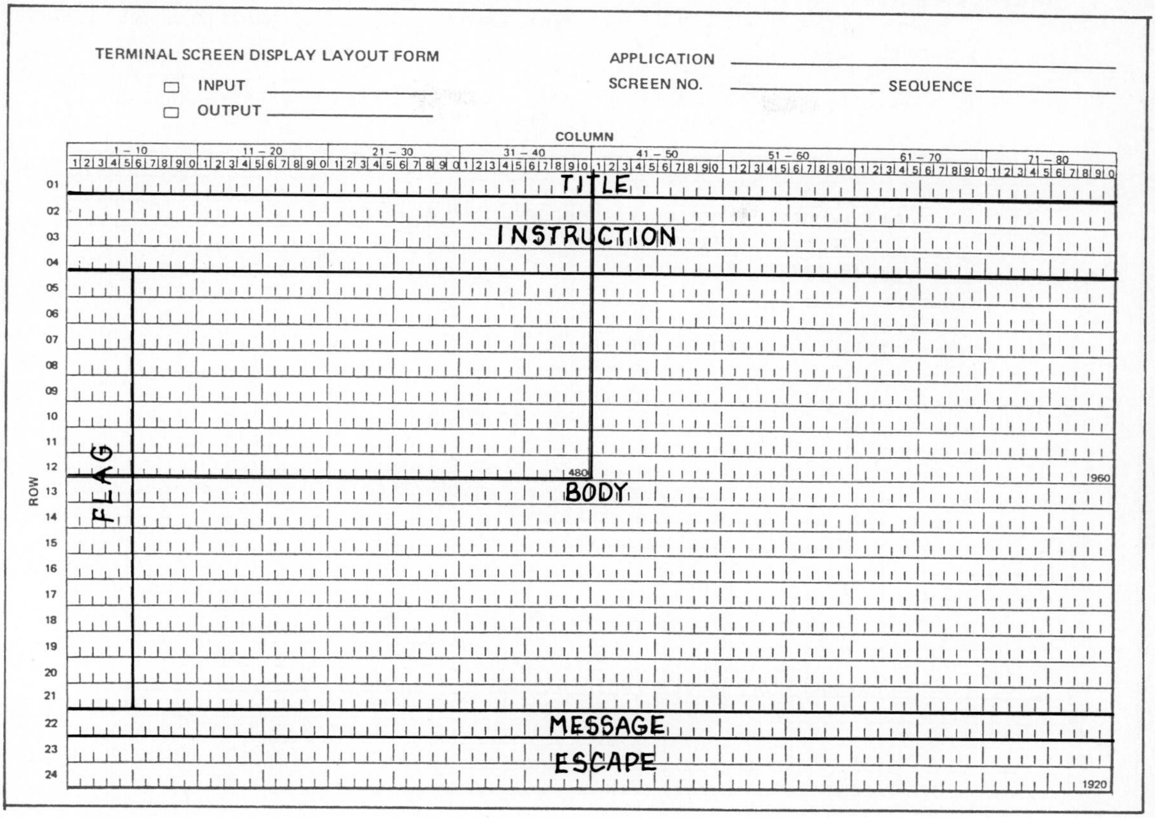

FIGURE 17.9   Template for Member Services System's Screen Zones   We recommend that a template for your zone areas be defined. In fact, create an overlay template using a blank transparency for an overhead projector. The overlay can be placed on each screen you design to check for consistency with the zones.

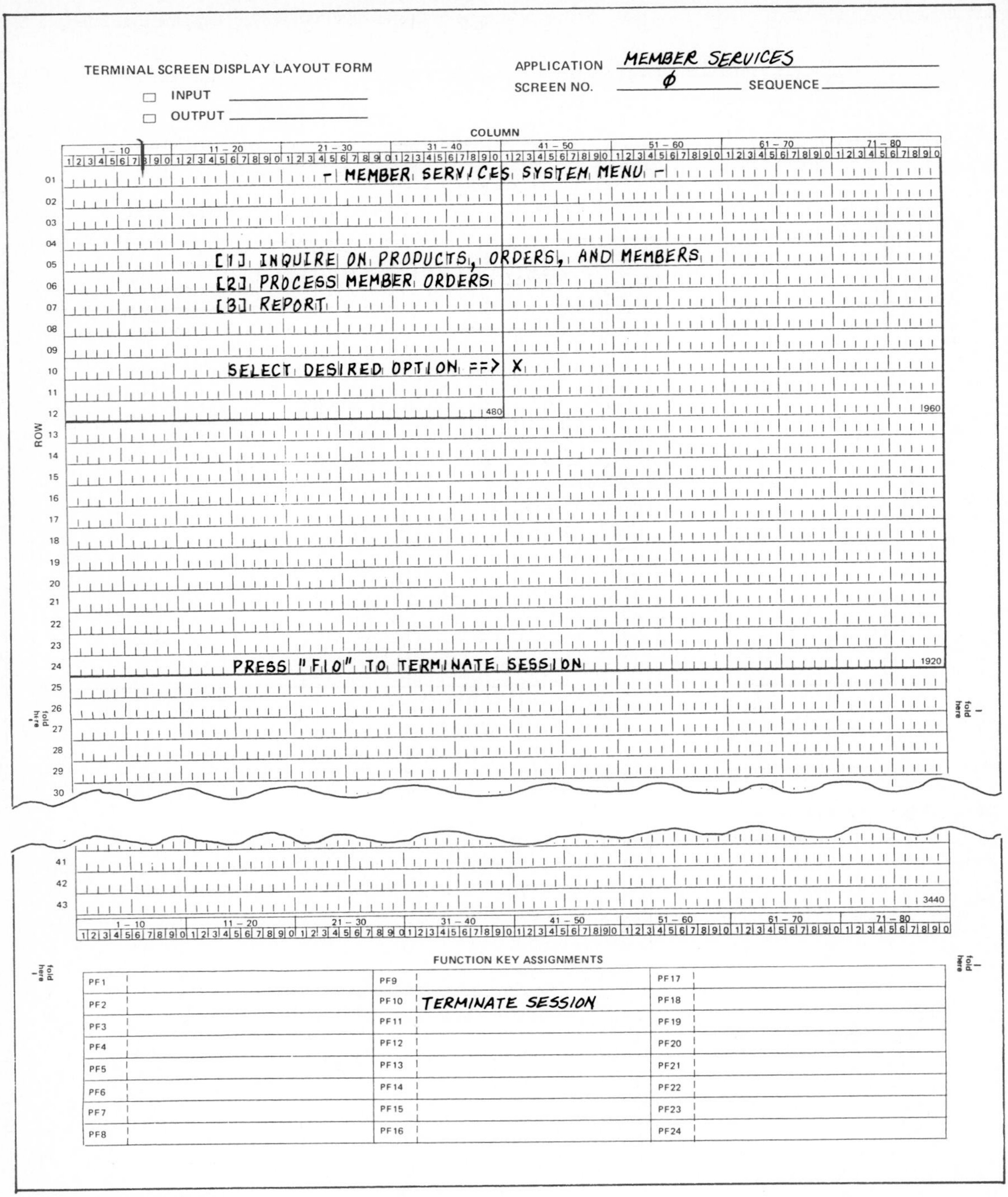

FIGURE 17.10 Display Layout Chart of the MEMBER SERVICES SYSTEM MENU Screen Format This is the first display screen that would be viewed by the end-user. The editing symbols have been added for the programmer's benefit. Notice that the three options appearing on the dialogue chart are displayed within the body of the screen. Selection of either one of the options will result in the display of a new screen from the dialogue.

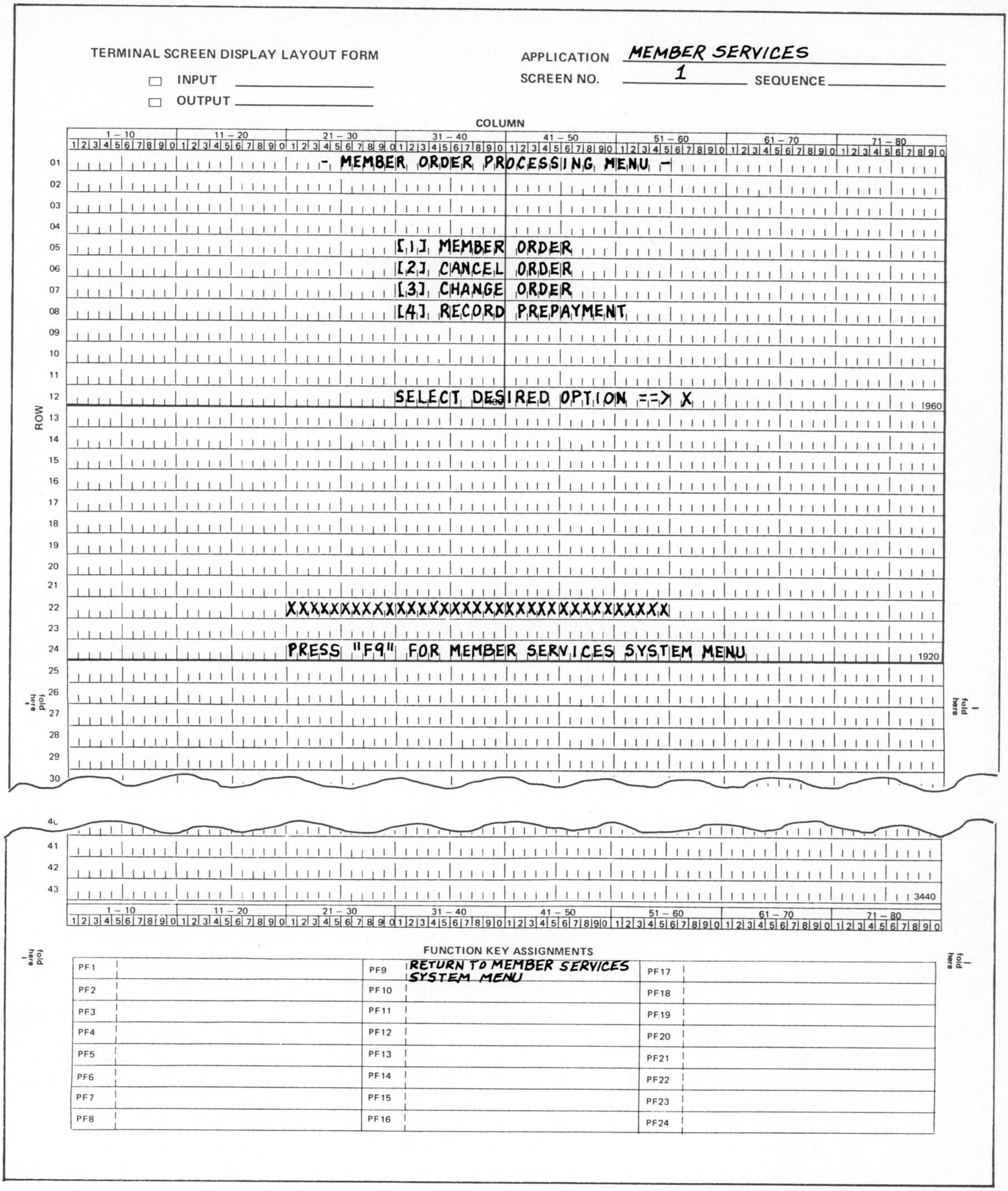

TERMINAL SCREEN DISPLAY LAYOUT FORM

☐ INPUT _____

☐ OUTPUT _____

APPLICATION **MEMBER SERVICES**

SCREEN NO. _____ **1** _____ SEQUENCE _____

COLUMN

```
01              - MEMBER ORDER PROCESSING MENU -
05                     [1] MEMBER ORDER
06                     [2] CANCEL ORDER
07                     [3] CHANGE ORDER
08                     [4] RECORD PREPAYMENT
12              SELECT DESIRED OPTION ==> X                    1960
22         XXXXXXXXXXXXXXXXXXXXXXXXXXXXXXXXXXXX
24         PRESS "F9" FOR MEMBER SERVICES SYSTEM MENU          1920
43                                                             3440
```

FUNCTION KEY ASSIGNMENTS

PF1	PF9 RETURN TO MEMBER SERVICES SYSTEM MENU	PF17
PF2	PF10	PF18
PF3	PF11	PF19
PF4	PF12	PF20
PF5	PF13	PF21
PF6	PF14	PF22
PF7	PF15	PF23
PF8	PF16	PF24

FIGURE 17.11   Display Layout Chart for MEMBER ORDER PROCESSING MENU Screen As depicted on the dialogue chart, this screen will appear if the end-user selects option 2 from the MEMBER SERVICES SYSTEM MENU screen (see Figure 17.10). In accordance with the dialogue chart, the screen will provide the end-user with the four processing options, as well as a fifth option whereby the end-user may return to the MEMBER SERVICES SYSTEM MENU.

TERMINAL SCREEN DISPLAY LAYOUT FORM   APPLICATION _____

☑ INPUT **MEMBER ORDER**   SCREEN NO. _____ SEQUENCE _____

☐ OUTPUT _____

COLUMN

```
                    1 - 10        11 - 20       21 - 30       31 - 40       41 - 50       51 - 60       61 - 70       71 - 80
01                                        ** MEMBER ORDER **
02        ENTER THE FOLLOWING ITEMS FROM A MEMBER ORDER (FORM 40). USE TAB OR
03        ARROW KEYS TO MOVE FROM ITEM TO ITEM. PRESS ENTER KEY WHEN DONE.
04
05  MEMBER NUMBER:   9999999 (5)                           ORDER NUMBER: (1) 999999
06  MEMBER NAME:   XXXXXXXXXXXXXXXXXXXXXXXXXXXX (6)         ORDER DATE: (2) 99/99/99
07  CLUB NAME:                                             PREPAID AMOUNT: 999.99
08  MEMBER ADDRESS:                                                       (3)
09  P.O. BOX:   XXXXXXXXX (7)
10  STREET:   XXXXXXXXXXXXXXX (8)
11  CITY:   XXXXXXXXXXXXXXX (9)
12  STATE: AA   ZIPCODE:  XXXXXXXXX                 480                          1960
13        (10)          (11)
14        SELECTION OF     PRODUCT   MEDIA   QUANTITY
15        MONTH ACCEPTED?  NUMBER    CODE    ORDERED
16            X            9999999    X      9999
17        (12)             (13)      (14)    (15)
18                      Ⓐ
19
20
21
22
23        XXXXXXXXXXXXXXXXXXXXXXXXXXXXXXXXXXXXXXXXXXXXXXXXXXXXXXXX (a)  Ⓑ
24  ** PRESS F2 FOR HELP. PRESS F3 TO RETURN TO ORDER PROCESSING OPTIONS MENU ** 920
25
26
27
28
29
30
```

```
41
42
43                                                                                      3440
     1 - 10        11 - 20       21 - 30       31 - 40       41 - 50       51 - 60       61 - 70       71 - 80
```

FUNCTION KEY ASSIGNMENTS

PF1		PF9		PF17	
PF2	HELP FOR ANY ITEM	PF10		PF18	
PF3	RETURN TO ORDER PROCESSING MENU	PF11		PF19	
PF4		PF12		PF20	
PF5		PF13		PF21	
PF6		PF14		PF22	
PF7		PF15		PF23	
PF8		PF16		PF24	

Option 1 should take us to a new screen. This screen should allow our end-user to start keying in data for MEMBER ORDERs. This is the input screen that we designed in Chapter 16. The display layout chart for MEMBER ORDER has been reproduced for you in Figure 17.12. The dialogue chart indicates that this screen should be displayed to the end-user repeatedly until the end-user has input all the customer orders. The end-user would then press function key F3 to escape back to screen 1.

What happens if the end-user has problems when entering data into the MEMBER ORDER screen? For example, the end-user might not know how the ORDER DATE should be entered or why the screen will not accept the PRODUCT NUMBER that was typed in. On the dialogue chart, we see screen 1.1 should include a special help screen that is associated with the MEMBER ORDER input screen. The user who encounters a problem while keying a MEMBER ORDER order can press the function key, F2, and the HELP screen will be displayed. Figure 17.13 represents the layout of the HELP screen. The possible explanations that may appear on this screen should be documented using a memorandum. Each explanation given to the end-user should correspond to a particular data element the cursor was positioned at when the end-user pressed the F2 (help) key.

Summary

The design of conversational on-line user interfaces has taken on greater importance due to the trend toward on-line information systems. On-line user interface design is the specification of a conversation between the end-user and the computer that results in the input of new data to the information system, the output of information to the end-user, or both. Display terminals and microcomputer monitors have features that affect dialogue design. The systems analyst should be familiar with the current technology of available terminals/monitors and their features, because they can be used to improve dialogue design.

Human factors are also important considerations for good on-line user interface design. Most of today's systems are being designed for the casual end-user, with an emphasis on user friendliness. Human engineering principles can guide the development of user-friendly on-line user interfaces for different types of end-users. There are four strategies commonly used for on-line user interface design: menu selection, instruction sets, question-answer dialogues, and graphics. The choice of strategy depends on the nature of the function to be performed and on the characteristics of the end-users who will use the terminal.

Typical on-line user interfaces may involve many screens. The coordination of these screens is very important; for example, some screens will occur

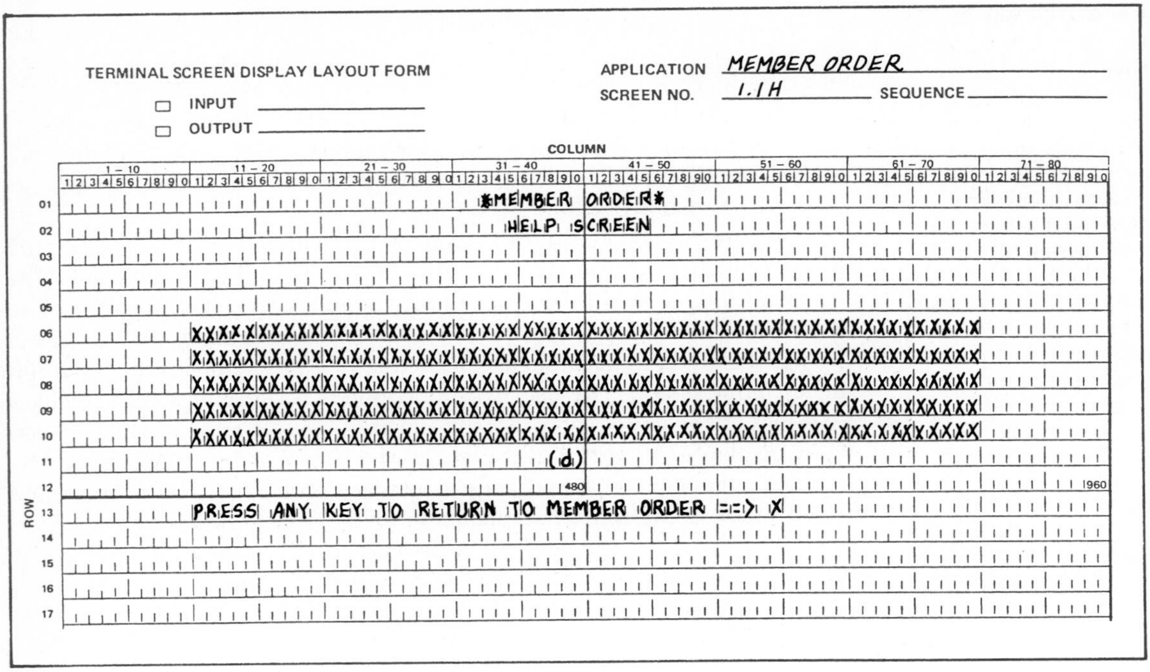

FIGURE 17.13   Display Layout Chart for the MEMBER ORDER HELP Screen   All on-line inputs should have help screens that may be referenced by the end-user. Help screens should be brief, concise, and informative.

in a specific order, whereas others occur under certain conditions. A dialogue chart is a tool used to depict the sequence and variations of the screens that can occur when the end-user sits at a terminal. The dialogue chart can serve as a table of contents for the numerous display layout charts for the screens.

After developing a dialogue, it should be tested on the end-users who will have to use that dialogue. Prototyping tools are the best way to test a dialogue. The end-users should be observed during this test period—you are trying to find flaws in the on-line user interface. A flaw occurs when the end-user does something the analyst didn't consider. In some cases, screen sketches and prototypes must be documented with display layout charts to communicate the formal specifications to the programmer.

Problems and Exercises

1. To what extent should the end-user be involved during on-line user interface design? What would you do for the end-user? What would you ask the end-user to do for you? Detail a strategy that consists of specific steps you and the end-users would follow.

2. Study the features on two visual display terminal(s) or microcomputers. You may need to borrow manuals to complete this assignment. How might these features be used to design effective on-line user interfaces?

3. What documentation prepared during input design and output design is needed during on-line user interface design? How does that input and output design documentation relate to on-line user interface design?

4. Explain the difference between a dedicated and casual terminal user. How would your strategy for designing on-line user interfaces for a dedicated user differ from that for designing on-line user interface for a casual user?

5. Display attributes can be overused and frequently hinder an end-user's performance during a terminal session. Cite some examples where display attributes are appropriate and where display attributes may hinder an end-user's performance.

6. Describe four strategies for designing on-line user interfaces. What criteria would you consider when choosing between the strategies?

7. Arrange to study an on-line or microcomputer application. It may be either a business system (such as an inventory, accounts receivable, or personnel system) or a productivity tool (such as a word processor, spreadsheet, or database system). Analyze the human engineering of the on-line user interface. Analyze the human engineering of the display screens. If possible, discuss the dialogue and screens with end-users. What do they like and dislike about the design?

8. Redesign the application in Exercise 7 to improve or change the on-line user interface and screens. If possible, discuss your improved design with end-users. Do they like your new design better? Did they raise any concerns?

9. Obtain documentation or magazine reviews on an automated screen-design aid. If possible, arrange for a demonstration. How would the product improve your productivity? How would the product decrease your productivity? What features do you dislike or would you prefer to see?

Projects and Minicases

1. An automated record-keeping information system is being designed for an employment agency. Some of the tasks to be automated on-line include

   Transaction Processing
   - A. Processing clients
     - a. Recording new clients
     - b. Matching client with job openings
     - c. Notifying clients of job openings
   - B. Processing jobs
     - a. Recording job openings
     - b. Matching job with clients
     - c. Recording job placements

   Management Reporting
   - A. Reporting of job openings
   - B. Reporting weekly job placements
   - C. Reporting client credentials

   Decision Support
   - A. Query clients
     - a. Query general client information
     - b. Query employee job qualifications
     - c. Query employee job requirements/preferences
   - B. Query job openings
     - a. Query general job opening information
     - b. Query job opening requirements
   - C. Query job placements

   Assume that the terminal input and output screens have already been designed to support these on-line processes. Develop a dialogue chart to depict the sequence and variations of dialogue screens that might occur when an end-user sits at the terminal. Be sure to include help screens for all input screens, escape options for navigating the structure, screen reference numbers, and descriptive screen names.

2. Terminal screens should be designed for consistency. Design a template for the screens in Project 1. The template should clearly indicate zones or areas of the screen used to display common messages. Do you have zoning or similar screen design software on your computer system? If so, describe how you'd implement your zones. If not, how will you ensure that data and messages are displayed in the proper zone?

3. Using the dialogue chart from Project 1 and the screen template from Project 2, design the dialogue screens required for an end-user to arrive at the basic input and output screens. (Assume that the actual input and output screens have already been designed.) Be sure to include explanations of possible error messages that may appear when the end-user makes invalid entries.

4. Test the on-line user interface you prepared in Project 2. Replace all display layout charts with screens that contain actual data and messages instead of editing symbols. Simulate a user terminal session by having someone walk through the dialogue using these sample screens. Challenge them to do something your dialogue wasn't designed to handle.

Annotated References and Suggested Readings

Fitzgerald, Jerry. *Internal Controls for Computerized Information Systems.* Redwood City, Calif.: Jerry Fitzgerald & Associates, 1978. This is our reference standard on the subject of designing internal controls into systems. Fitzgerald advocates a unique and powerful matrix tool for designing controls. This book goes far beyond any introductory systems textbook—must reading.

Kendall, Kenneth, and Julie Kendall. *Systems Analysis and Design.* Englewood Cliffs, N.J.: Prentice-Hall, 1988. Chapter 16 provides another look at user interface design.

Mehlmann, Marilyn. *When People Use Computers: An Approach to Developing an Interface.* Englewood Cliffs, N.J.: Prentice-Hall, 1981. We are indebted to Mehlmann for the concept of zoning a screen into areas. But this book goes far beyond that. Every systems analyst can get something out of this book, which is a modern and comprehensive study of how to analyze and design intelligent on-line user interfaces.

SOUNDSTAGE

Methods, Procedures, and Program Specifications for the New System

We begin this episode in Sandra's office, two weeks later. Sandra and Bob have just arrived and are enjoying their morning coffee. Nancy Willis, Director of Information Systems Services, has just walked into the office.

EPISODE 8 Methods and Procedures Specifications for the New System.

"Hello Nancy!" Sandy said. "How was your vacation?"

"The vacation was great. You just can't beat the Canadian Rockies for spectacular scenery. I'm exhausted, so it must have been a good trip. How are you doing with the new member services system? Is everything still on schedule?" asked Nancy.

"Everything is coming along very well," Sandra responded. "We're just finishing up the design right now. We will start packaging program specifications later today. I hope the programmers are ready, because this project is going to take some time."

"Excellent!" Nancy said. "Just let me know how many programmers you'll need and who you want as chief programmer to work with the two of you. I need to check in with the rest of the teams. Keep up the good work." Nancy left the office.

Bob began to shuffle through some papers on Sandy's desk. "Hey, I didn't know you had already done some of the basic system flowcharts," Bob exclaimed (see Figure E8.1).

"I worked on some of those over the weekend," Sandy explained. "I think we need to get started on backup and recovery procedures."

Bob thought for a moment and then said, "Let's begin with the VSAM supplier, product, and title file and then do the rest of the IDMS database."

Sandy interrupted, "Since the inventory department needs continuous on-line access to that file, why don't we log all transactions to a file?"

"That's a good idea" Bob answered. "Then we can have operations back up both files in the evening when they won't interfere with normal operations. Then, if there is a systems crash, we'll just reload the latest backup tape."

Bob had been drawing a picture of what they had been discussing. He finished after a few moments and handed the drawing to Sandra (see Figure E8.2). "Here, how does this look?"

"That's perfect," Sandy responded. "Now what about the IDMS database? First, . . ."

It's several hours later. Sandra and Bob have finished defining the backup and recovery procedures for the new member services system. All flowcharts have been delivered to operations personnel, who will review them and schedule a meeting to discuss the new procedures. Meanwhile, Sandra and Bob are planning their next task.

Program Design Specifications for the New System

Bob indicated three large notebooks on Sandy's shelves containing analysis and design specifications for the new member services system. "Look at all of that!" he groaned. "Now we have to organize all those specifications so that

Episode 8, continued ▶

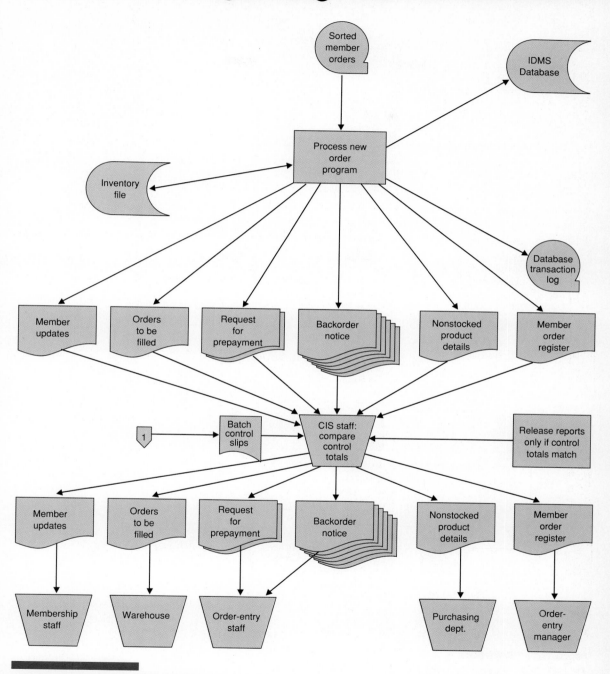

FIGURE E8.1 Typical Systems Flowchart

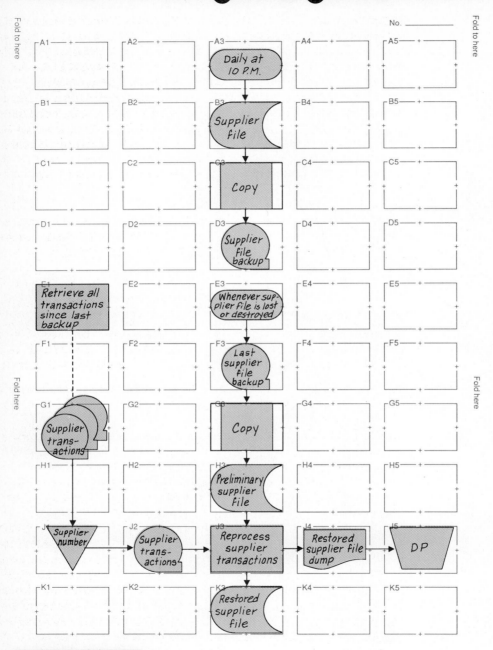

FIGURE E8.2   Sample Systems Flowchart on Flowchart Form

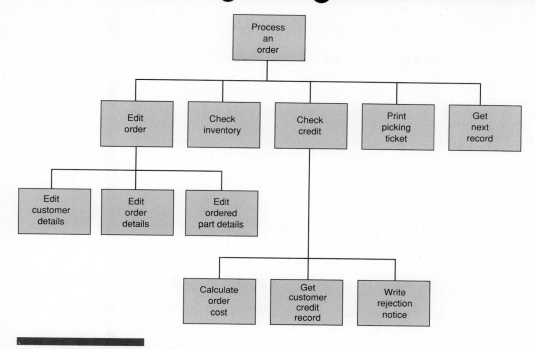

FIGURE E8.3   Typical Program Structure Chart

the programmers can begin writing the programs."

Sandy smiled, "I'll do the implementation planning and start working on the user's manuals. You can package the program specifications; but first, let's finalize the format, okay?"

Bob nodded approval and Sandra continued. "I want Wayne Tomaso to manage the programming team. I've worked with him before and he does an excellent job. Wayne likes all the file specifications to be included in a single packet. That means all the data dictionary forms, record layout charts,

schema diagrams, and other similar documentation. He will assign a programming team to create test files and databases for use by all the members of the programming team."

"How do you organize the specifications for the programmers?" Bob interrupted.

Sandy answered, "Within each program, we normally start with a structure chart (Figure E8.3). The structure chart shows the various modules within a program and their relationships with one another. This is useful to ensure that the programs will be main-

tainable once they have been installed.

"By specifying the modular structure of the programs, some of the actual program code can be reused by other programs. This also allows us to make sure that each module is highly cohesive and loosely coupled; that is, each module performs one and only one task, and a change in one module of a program should have as little effect on the other modules as possible. Then, you need to provide the programmers with a general idea of what is to be

Episode 8, continued ▶

EDIT MEMBER ORDER

Description
FOR ALL INVALID MEMBER ORDER DATA FROM THE USER, DO THE FOLLOWING:
SELECT THE APPROPRIATE CASE:
    CASE 1: MEMBER NUMBER IS INVALID
        IF THE MEMBER NUMBER IS NOT EQUIVALENT TO A MEMBER NUMBER
        OF AN EXISTING MEMBER THEN:
            ERROR MESSAGE = "MEMBER DOES NOT EXIST, PLEASE RE-ENTER"
    CASE 2: ORDER NUMBER IS INVALID
        IF THE ORDER NUMBER = ORDER NUMBER OF A PREVIOUS ORDER THEN:
            ERROR MESSAGE = "ORDER NUMBER WAS ASSIGNED TO PREVIOUSLY
                    ENTERED MEMBER ORDER, PLEASE RE-ENTER"
    CASE 3: ORDER DATE IS INVALID
        SELECT APPROPRIATE CASE:
            CASE 2.1 ORDER DATE CONTAINS NO VALUES, THEN:
                ERROR MESSAGE = "THE ORDER DATE MUST BE PROVIDED"
            CASE 2.2 ORDER DATE CONTAINS INVALID MM/DD/YY VALUES,
                THEN:
                ERROR MESSAGE = "THE ORDER DATE IS INVALID,
                    PLEASE RE-ENTER"
    CASE 4: PREPAID AMOUNT > TOTAL MEMBER ORDER COST, THEN:
        ERROR MESSAGE = "PREPAID AMOUNT EXCEEDS TOTAL COST OF MEMBER
                ORDER, PLEASE RE-ENTER"
    CASE 5: PRODUCT NUMBER NOT VALID, THEN:
        ERROR MESSAGE = "INCORRECT PRODUCT NUMBER (DOES NOT EXIST),
                PLEASE RE-ENTER"
    CASE 6: MEDIA CODE IS INVALID, THEN:
        ERROR MESSAGE = "MEDIA CODE IS INVALID, PLEASE RE-ENTER OR
                PRESS F2 KEY FOR VALID CODES AND THEIR
                MEANINGS"
    CASE 7: QUANTITY ORDERED IS NOT > ZERO, THEN:
        ERROR MESSAGE = "QUANTITY ORDERED MUST BE GREATER THAN ZERO,
                PLEASE RE-ENTER"
    CASE 8: MEMBER RESPONSE STATUS IS NOT Y OR N, THEN:
        ERROR MESSAGE = "ENTER Y (YES) OR N (NO) IN REGARD TO
                MEMBER'S ACCEPTANCE OF SELECTION OF

FIGURE E8.4   Sample Structured English Specification

accomplished in each module. We use something called 'Structured English.' It's a little like the pseudocode you used in school, but not quite as detailed (see Figure E8.4). Any questions?"

"No," answered Bob. "It's not really difficult. All we're doing is reorganizing all this documentation into computer programs."

Where Do We Go from Here?

In this episode, we observed Sandy and Bob completing the design of the new system. They used a new tool, system flowcharts, to document data processing methods and procedures. You will learn how to draw system flowcharts in Chapter 18. Sandy and Bob are beginning to package the final set of design specifications into programs that can be written by computer programmers. This is where all the documentation you have been building comes together. You will learn how to package design specifications in Chapter 19.

Sandy planned an implementation schedule for the new system. Successful implementation depends on your project management skills. We suggest you read Part Four, Module A, "Project Management Tools and Techniques."

CHAPTER EIGHTEEN

Designing Computer-Based

Methods, Procedures,

and Controls

Midwest Imports

Midwest Imports has just completed the automation of its mail order system. Let's take a look at how the operations are performed.

At approximately 8:30 each morning, the Order-Entry Department collects all new sales order forms. These forms include orders received both in the mail and over the phone. Order Entry delivers the batch of forms to Data Entry, where the orders are keyed (typed) and stored on magnetic tape.

Data Entry then delivers the sales order tape to the Computer Information Systems Department, where the tape is read by a program that checks the sales orders for keying errors. This program generates an errors report that identifies incomplete and erroneous sales orders. The report is delivered to order-entry clerks, who correct the incomplete and erroneous sales orders and send them back to Data Entry for rekeying. The same program stores all valid sales orders on a new magnetic tape. When all sales orders have been successfully processed and written to that tape, the tape is read by another program that produces a different tape on which the orders are resequenced according to customer number.

This sorted tape is input to another program that checks an inventory master file to determine the availability and price of products that appear on the sales order. If the sales order cannot be filled, a backorder notice is generated. The next program uses the customer accounts receivable master file to check credit on those orders that can be filled. For customers who have a poor credit rating, a payment overdue notice is printed. A report of these notices is printed for management. The program produces an order confirmation letter for orders that will be filled. A final program produces an invoice with carbons for picking, packing, and shipping the orders that have passed the credit check. That program also produces a sales order transaction file that contains all successfully processed sales orders and a sales order register. Also, all successfully processed sales orders are stored on a new magnetic tape and archived. All reports generated by this program are initially held by the CIS Department. CIS compares the batch totals generated by the last program against a batch control slip that was generated by Data Entry when the orders were keyed. If the totals match, CIS will deliver the reports to Order Entry.

The new system also includes an on-line program that allows the sales manager both to query the inventory file to obtain product prices and to query the customer accounts receivable file to obtain customer credit and invoice information. This program is available to the manager from 8:00 A.M. to 5:00 P.M. each working day.

Customer order cancelations are processed immediately. When the order-entry clerk receives a cancelation request (via the mail or phone), the clerk enters the order cancelation request on the CRT. An on-line program reads the customer accounts receivable file to determine the status of the order. If the order has not yet been filled, the clerk phones the warehouse to have the order terminated. The program also generates order cancelation confirmation letters (printed as a batch at 4:30 P.M. that day).

This narrative description of Midwest Imports' operations is becoming lengthy. We haven't even talked about how the customer payments and billing operations are performed. But let's stop here—we have a challenge for you.

Challenge

Remember, we have previously stated that a picture is worth a thousand words. Draw a picture to describe the information systems methods and procedures being performed in the Midwest Imports' operations, how these operations are being accomplished, and the sequence in which they are occurring.

What Will You Learn in This Chapter?

In this chapter, you will learn how to design and document computer-based methods, procedures, and controls using systems flowcharts. You will know that you have mastered the specification of methods and procedures when you can

1. Differentiate between batch, on-line, remote batch, and distributed processing methods.

2. Describe the general procedures required to implement each of the methods just listed.

3. Define and design methods and procedures for internal controls, including backup and recovery.

4. Explain how systems flowcharts are used for systems design and how they relate to the tools you learned in Chapters 13 through 17.

5. Read, prepare, and present systems flowcharts describing typical processing methods and procedures. (Systems flowcharts should conform to guidelines to enhance their communication value to both technical and nontechnical audiences.)

Information systems require well-defined methods, procedures, and controls to ensure that the systems function properly. Nowhere is the absence of well-defined methods, procedures, and controls more evident than in the recent emergence of microcomputer-based information systems. When end-users design their own systems, they frequently forget about methods and procedures for backup and recovery and other internal controls. In this chapter, we will study the design of methods, procedures, and controls, which are equally applicable to mainframe and microcomputer environments.

Systems flowcharts are a graphic tool used to show the sequence of processing and activities in a computer-based information system. Why not use data flow diagrams for this purpose? Many analysts are making this switch (see the "Next Generation" box feature for this chapter). But systems flowcharts are still typically encountered since many thousands of systems were designed in a pre-DFD era. Also, many older analysts are still more comfortable with systems flowcharts, though this is slowly changing. And whereas systems flowcharts are definitely more complicated than PDFDs, if they are properly drawn and presented, they can be presented to the nontechnical audience.

Information Systems Methods, Procedures, and Controls and Their Implications

What are methods, procedures, and controls? How do we document them? Methods and procedures define the sequence of events that produce outputs from their requisite inputs. Specifically, a **method** is a way of doing something. A **procedure** is a step-by-step plan for implementing the method. Methods and procedures can also be described as answering the questions "*who* does *what* and *when* do they do it?" and "*how* will it be done?" In this section, we will briefly discuss methods, procedures, and controls typically used to implement information systems. Most methods can be broadly classified on each of two scales:

1. Degree of centralized versus distributed processing

2. Degree of batch versus on-line processing

Systems can be partially distributed or partially on-line. Let's examine the two more closely.

Centralized Versus Distributed Processing

At one time, all data processing was centralized, with data recorded, input, and processed at a central computer site. The cost of placing computers closer to the end-user was prohibitive. Today, because computers have become much cheaper, many organizations have decentralized or distributed their processing workload to multiple computer sites. Distributed computing can offer several advantages, including improved responsiveness, better end-user control over data, and reduced costs—especially with microcomputers. As an analyst, you need to be aware of the alternatives and be able to discuss the technology with specialists and end-users.

There are several approaches to distributed processing. Each approach connects the computer systems through one of the network architectures listed in the margin. We'll briefly survey these options. The study of network strategies, technologies, and data communications should become an integral part of your continuing education for a career in systems analysis and design.

✓ **NETWORK STRATEGIES**

Point-to-Point Connection

Bus Networks

Star Networks

Hierarchical Networks

Ring Networks

Point-to-Point and Bus Networks   The simplest networking strategy is to provide a direct link between any two computer systems you wish to connect. This concept is illustrated in Figure 18.1. Notice that the network can contain microcomputers, minicomputers, mainframe computers, and terminals. To completely connect all points between N computers or devices, you would need $N(N-1)/2$ direct paths. Unless each data path is heavily utilized, the cost could be prohibitive. In fact, utilization of any point-to-point data path should be verified before installing that path.

What can you do if the direct paths will not be heavily utilized? You could have specific computers and devices share a single point-to-point data path (see Figure 18.2). The data path in this case is called a **bus**. Only one computer system or device can send data through the bus at any given time. Incidentally, the computer systems and devices are said to be *multidropped off the bus*.

Ethernet, a product developed jointly by Xerox, Intel, and Digital Equipment Corporation, is an example of a bus network strategy. Ethernet also happens to be an example of a special product category called *local area networks*. A **local area network** is a collection of computers, terminals, printers, and other computing devices that are connected through cable over relatively short distances—for instance, in a single building. These computers and devices can transmit data to one another through this network. Ethernet's bus architecture manages point-to-point communication to prevent collisions that can occur when multiple devices try to use the bus at the same time.

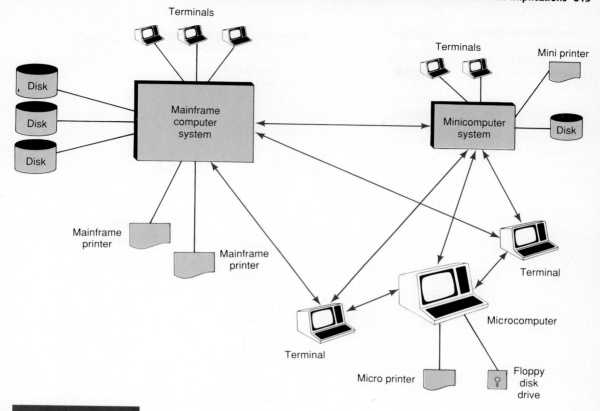

FIGURE 18.1 Point-to-Point Network The simplest distributed processing network architecture is point-to-point, whereby a dedicated data path is placed between two devices. That data path only has to concern itself with understanding the devices on each end.

Star and Hierarchical Networks A **star network** links multiple computer systems—often called *satellite processors*—through a central computer (see Figure 18.3). Some will argue that this is a holdover from the days when centralized computing reigned supreme. However, the truth is that much data in any organization can (and should) be maintained on and shared through the central computer. The central computer is being used as a *traffic cop* to control the transmission of data and information between the distributed processor sites.

A **hierarchical network** can be thought of as a multiple star network. Figure 18.4 illustrates such a network. The top computer system, usually a mainframe computer, controls the entire network. The satellite processors (in this case, minicomputers and microcomputers) have their own satellites

FIGURE 18.2 Bus Network A bus network is similar to point-to-point networks except that multiple devices share a single point-to-point pathway. Only two devices, a sender and a receiver, may use the path at any given time.

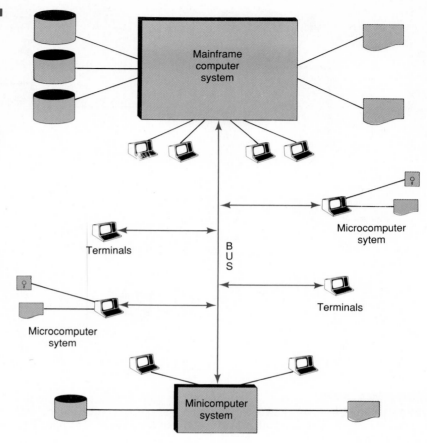

(in this case, microcomputers and terminals). Notice that each satellite may have its own complement of *dedicated* devices, such as disk drives and terminals. IBM's System Network Architecture (SNA) is essentially a hierarchical network.

Ring Networks A **ring network** connects multiple computers, but not other devices, into a ringlike structure (see Figure 18.5). Each computer can transmit data to only one other computer. Every data transmission includes an address, similar to the address you write on an envelope. When a computer in the ring receives a packet, it checks the address. If the address is not for that computer, it passes it on to the next computer in the ring. IBM's Token-Ring Network is an example of a ring network.

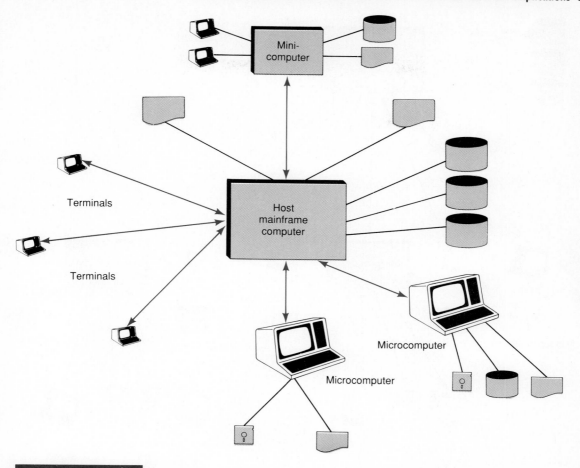

FIGURE 18.3 Star Network In a star network, a central computer plays traffic cop to satellite processors and devices that are trying to communicate with each other and with the central computer.

The Importance of Data Communications All of the network architectures are implemented through data communications technology. Making computers and devices *talk* with one another is one of the most exciting technologies in today's information systems world. Transmitting data through the network is not as easy as it may seem. Data can be directly transmitted through cables (as in local area networks), over telephone via fiber optic wire, via microwave, and even via satellite. Furthermore, the data are subject to complications, such as different codes recognized by different computers and devices. Microcomputer–to–mainframe computer connection is a particularly critical issue today. Over time, standards should evolve to simplify this technology.

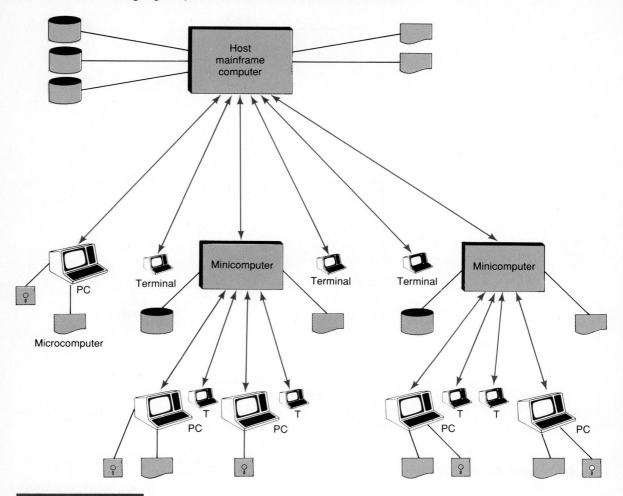

FIGURE 18.4 Hierarchical Networks Hierarchical networks, like IBM's popular SNA, use a host computer to control satellite processors and devices, which in turn may control other satellite processors and devices, and so forth. The host computer supervises the entire network.

The many issues and complexities of data communications are beyond the scope of this book. But as a systems analyst, you often enlist data communications and distributed processing specialists to help answer questions that arise. You'll find it easier to work with such specialists if you begin to plan your continuing education in this rapidly changing technological area. Most curricula offer at least one course on either *data communications* or *distributed processing*. Take the course; you can't go wrong.

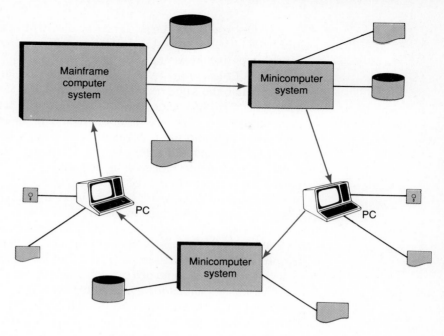

FIGURE 18.5 Ring Networks Ring networks link computers, one to another, but in one direction only. Data transmissions are passed around the ring until they arrive at their intended destination.

Batch Versus On-Line Processing and Their Internal Control Implications

Batch and on-line processing concepts were introduced and surveyed in Chapters 15 through 17, at least with regard to inputs and outputs. Our perspective here is on batch versus on-line *processing*.

Many of the required internal controls for the system were specified when inputs, outputs, and files were designed. We must now incorporate these and other controls into the methods and procedures for the system. The purpose of internal controls for methods and procedures is to ensure the accountability, auditability, and recovery capability for the new information system.

A Review of Batch Processing Methods In the batch processing method, transactions are accumulated into batches for periodic processing. The batch inputs are processed against master files. Transaction files may also be created or updated at that time. Alternatively, a database may be substituted for the conventional master and transaction files. In a batch system, management reports are usually generated on a scheduled basis. That is, either a report is regularly scheduled to be generated on a specific date and time or the report is scheduled *after* it has been requested.

Batch processing procedures are dependent on the organization of the computer master files. Most early files were sequential and stored on tape.

Tape devices could not retrieve records directly because all of the records prior to the desired record had to be read. Therefore, standard procedures required that the input transactions be sorted into the same order as the master files. Consider the PAYROLL system shown in Figure 18.6. TIMECARDS are sorted into the same sequence as the PAYROLL MASTER FILE (keyed on EMPLOYEE NUMBER) before they can be processed. Also note that two copies of master files are used. As a result of processing, an entirely new master file is produced. The old file is normally retained as a backup. Sequential files are not as common today as they were in the past. Unless they are being used as part of systems developed long ago, sequential files have been largely replaced by more flexible direct and indexed files.

When direct and indexed files are used, the processing procedures are greatly simplified (illustrated in Figure 18.7). The input data does not need to be sorted prior to processing because specific master file data can be retrieved directly from disk without the computer having to read the records that precede the desired record (data might still be sorted to improve efficiency). Furthermore, a record can be retrieved, changed, and then rewritten to the same file, eliminating the need for two copies of the master file.

We can also combine the concepts of batch and distributed processing. A batch of records can be created and edited at a distributed site and then transmitted to the central site for processing (see Figure 18.8). This is called *remote batch* (or deferred batch) because the batch originates at a remote site. This processing method is appropriate for situations in which identical transactions are captured at geographically different locations. For instance, each regional sales office may capture, input, and edit its own sales orders. Those orders can be transmitted as remote batches to the central computer site, where they will be merged and processed as a single batch.

Internal Controls for Batch Systems In the past several chapters, you've learned to design numerous internal controls into systems. But we're not through yet. The following additional controls must be designed into batch systems:

1. All inputs and outputs are scheduled and logged. A runbook states when programs should be run, informs the computer operator of any necessary setup—for instance, loading special forms on the printer—and describes which JCL (Job Control Language) file to execute (this concept should be familiar to students of programming). The runbook is also used to record the production runs and list any problems that have occurred.

2. Procedures describe how input errors (edit reports, introduced in Chapter 16) should be distributed to end-users and whether processing can proceed before these errors are corrected. The error correction cycle is included in the procedures.

3. Procedures require batch control totals (see Chapter 16) to be checked

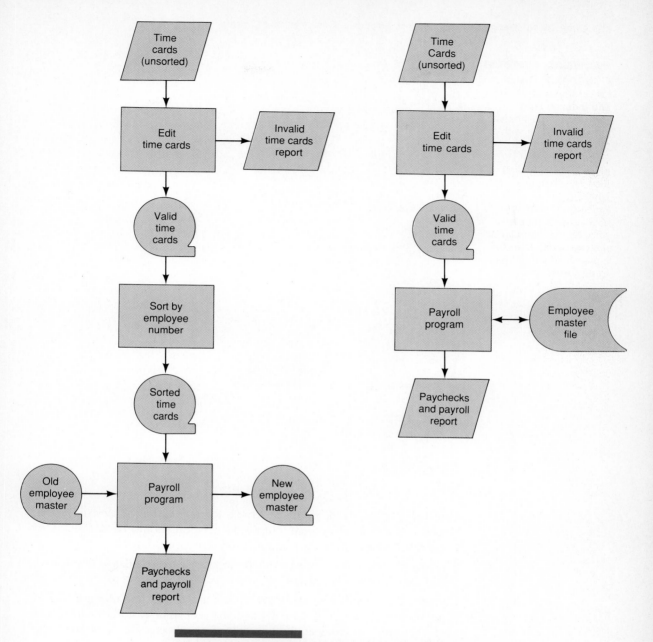

FIGURE 18.6 Batch Processing with Sequential Files   This figure illustrates the batch input method using sequential tape files. Input transactions must be sorted into the same sequence as the master file(s). Furthermore, the old master file cannot be updated. Instead, a new master is produced when transactions are processed.

FIGURE 18.7 Batch Processing with Direct and Indexed Files This picture illustrates batch input processing with direct and indexed (which may be physically sequential) files stored on disk. Unlike tape file processing, disk file processing is simpler since master files can be directly updated and input records don't have to be sorted.

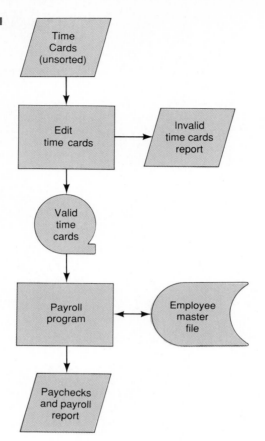

against the historical report generated for transaction processing. If the totals are not in agreement, output distribution should be delayed until the discrepancy has been accounted for.

4. Procedures describe how scheduled outputs are to be collated, duplicated, and distributed to end-users.

5. Procedures specify when and how master and transaction files are backed up. Backup copies of files are needed so they can be restored in the event of a disaster—such as equipment failure or sabotage—that destroys the main copies of those files. For most files, backup copies are written to an off-line medium, such as tape, cassette, or floppy disk.

6. Procedures specify how files will be recovered in the event that they are lost or destroyed. The backup files (just described) will only protect the data up to the point in time when that backup was made. Transactions that occurred after that backup must be *re*processed. There are two methods

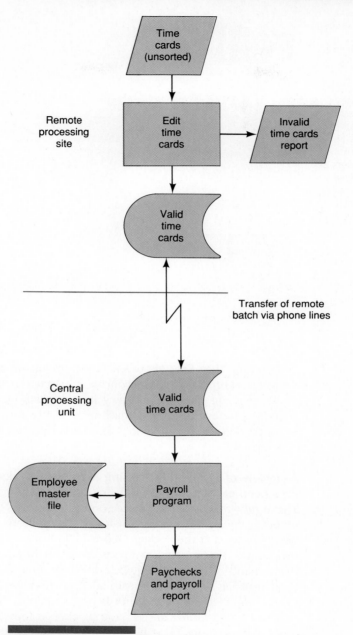

FIGURE 18.8   Remote Batch Processing   In remote batch processing, separate batches are created at remote sites (another batch system or an on-line system) and then transmitted to a central site for processing. The transmission may be made via direct cable for short distances or telephone, microwave, or satellite for long distances.

FIGURE 18.9 On-line Processing With on-line or interactive processing, transactions are processed as they occur. Since transactions can occur during all business hours, on-line programs must be available to end-users at all times. Along those lines, the programs may also be idle—but still available—at some times.

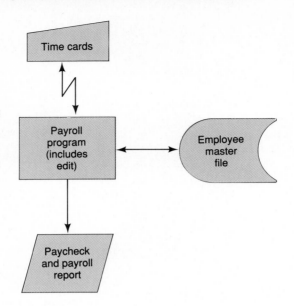

for doing this. One approach is to reinput those transactions from the historical reports produced during transaction processing. This, however, means going through data entry and editing again. Another approach is to record historical data in machine-readable format (during transaction processing) and then simply reprocess the data from this backup transaction file.

A Review of On-Line Processing Methods Modern information systems have been moving away from batch methods and toward on-line methods. That is why this book has placed so much emphasis on the design of on-line inputs, outputs, and terminal dialogue. Transactions and inquiries are often best processed as they occur. Figure 18.9 illustrates the on-line method and procedures. Transactions are processed when they occur (no batches). Notice that on-line processing requires direct or indexed files (or a database). Sequential files simply can't provide adequate response time for on-line systems.

A completely on-line system is not common. The management reports may still be *scheduled* in the same manner as they are in batch processing. Some organizations don't have computing resources that permit end-users to generate large reports as frequently as they desire. Still, more and more reports are produced *on demand* and displayed at a terminal screen. However, printouts are usually scheduled at printing devices when those devices become available. The exception is terminal workstations (especially microcomputers) that include a printer.

Internal Controls for On-Line Systems   The following additional internal controls should be designed into on-line systems:

1. Access to the on-line system is restricted to authorized end-users. Appropriate security measures are defined in the methods and procedures to prevent unauthorized access to data and processing. Frequently, multiple levels of security are built into the system because different end-users are allowed to do different things. For instance, at our local credit union, all tellers are allowed to process your basic deposits, withdrawals, transfers, and loan payments. However, modifying specific transactions, correcting errors, and making large withdrawals is restricted to tellers who have a *super teller* password.

2. Backup and recovery for files are more difficult for the on-line system. Master and transaction files are periodically backed up as was the case with batch processing. However, whereas batch transactions include batch control totals that were verified after a processing run, on-line processing offers no such control. Instead, separate files are maintained for successfully processed transactions. If the main file is lost or damaged, the recovery procedure consists of two steps: (1) load the last backup of the file that was lost and (2) reprocess those transactions that occurred after the last backup of the lost file. These transactions can be processed directly from the tape journal of those transactions.

Additional Internal Controls to Be Designed   As you've gone through the design chapters, you've learned that you must design internal controls for outputs, files/databases, inputs, and on-line terminal dialogue. In this chapter, we've added internal controls for batch and on-line processing. There are some other controls that must be specified for any type of system. They may include

- *Simultaneous Processing Controls.* When two or more end-users try to update the same record in a file or database at exactly the same time, you have a problem. If end-user A retrieves the record, then end-user B retrieves the same record, then end-user A writes the updated record, and then end-user B writes his or her own updated record, the update for end-user A has been lost. The analyst should ensure that this can't happen by making certain that if simultaneous processing is possible, a record can be *locked out* to more than one end-user at a time.

- *Maintenance Controls.* We've discussed this earlier, but as you prepare systems flowcharts, you need to ensure that all updates to master and transaction files were properly performed. Some end-user or manager should get a daily update report to check for any unauthorized updates that might have been recorded. This report is frequently called a *journal*.

- *Physical Security.* In some cases, physical security must be established by the analyst. Who can use the computer equipment and software? This question is becoming more important as systems are installed on micro-computers that are not usually housed in a secure area. The analyst may have to investigate ways to physically restrict both the access to and the movement of computer equipment. And the analyst may also have to investigate software techniques for restricting access to software, files, and databases.

- *Physical Reliability.* The analyst may have to specify controls to prevent accidental or intentional environmental problems. Some systems are susceptible to static electricity. Magnetic fields can erase data recorded on tape. Dust contamination can be a problem in some areas. Humidity and temperature can also be problems. Normally, these issues are addressed by computer operations management, not analysts. However, as micro-computer-based systems multiply, the analyst has become more responsible for such controls.

A complete survey of internal controls is a subject that can only be presented well in its own book (see the suggested reading for this chapter). What we have tried to do throughout Part Three is to briefly discuss the most important controls and to place them into the context of the design tasks—files, database, outputs, inputs, and so forth—during which they should be specified.

Using Systems Flowcharts to Document Methods, Procedures, and Controls

Systems flowcharts were one of the very first tools commonly used by systems analysts and computer programmers. In fact, the American National Standards Institute (ANSI) has established certain symbols that have been widely used in the computer industry to describe the logic of both systems and computer programs. Although the symbols have been standardized, their use has not. Thus, many systems flowcharts look incomprehensible to those who would like to use them. For this reason, systems flowcharts have developed an unfavorable reputation with end-users.

Systems flowcharts are supposed to be the basis for communication between end-user, systems analysts, computer operations personnel, and computer programmers. This is a difficult task, but it's not impossible. You should think of systems flowcharts as a chance to prove (or disprove) that a specific technical solution to the end-user's requirements will work.

In this section, you'll learn how to draw clear systems flowcharts. Although we will follow the ANSI standard for the most part, at some point we will go beyond that standard because it has not progressed to keep pace with advances in information systems methods, such as distributed computing, on-line computing, and newer batch input methods. Furthermore, the ANSI standard does not suggest appropriate guidelines for drawing systems flowcharts. We will

**SYSTEMS
FLOWCHART
✓ SYMBOL
CATEGORIES**

Processing
Batch Input
Batch Output
File/Database
On-line Input and
Output
Miscellaneous

do so. Some of the ideas are our own. Others have been suggested by fellow experts in the systems analysis and design field. We suggest that you adopt and document a standard for your organization, systems group, or individual projects.

Systems Flowchart Symbols and Conventions

Systems flowcharting symbols can be conveniently classified into the six subsets listed in the margin. As we discuss these symbols and patterns, we'll use a cloudlike symbol to represent those symbols that haven't yet been introduced in the narrative. Most symbols and patterns are shown in the margin.

Program name (ID No. is optional)

Utility or library program name

Computer Program Symbols

Manual operation and who does it

Manual Operation Symbol

Auxiliary operation name or required equipment

Auxiliary Operation Symbol

Processing Symbols   There are only three basic symbols for processing: the computer program, the manual operation, and the auxiliary operation. The most important of these symbols, the rectangle, represents a computer program to be written or purchased for the system (top rectangle). If the program already exists (in a library of common utilities and programs), it is depicted as shown in the margin (bottom rectangle).

Sometimes, you need to indicate that a person must perform a manual operation—for example, CORRECT INPUT ERRORS—before a processing sequence can be started or continued. A trapezoid is used to depict such a process. Inside the trapezoid you should record both the operation or task to be performed *and* the person who should perform that operation, as shown in the margin.

The square is a less commonly used symbol that indicates an auxiliary operation performed by auxiliary information systems equipment that is not directly connected to the computer. Examples include punched card sorters, decollators, and magnetic-ink character sorters. The name of the operation or the required device should be recorded in the square.

Batch Input Symbols   Batch symbols should convey how source documents originate and how they are recorded to a computer-readable medium. As you learned in Chapter 16, most data originates on business source documents or forms. Once the data is recorded, a batch of documents are sent to data-entry clerks, who transcribe the data from the forms into a computer-readable format. This operation is called *keying*. The most common media to which the data is keyed are punched cards, magnetic disk (key-to-disk or KTD), and magnetic tape (key-to-tape or KTT). These batch input operations are illustrated for you at the top of the next page.

Punched cards, which are labeled (a) (top of next page), are much less common today than they were several years ago. But when the ANSI standard was adopted, most batch input was done on punched cards and paper punch tape. Today, most batch input is done using key-to-tape and key-to-disk methods. We need to adopt a notation for these methods. We have tried several

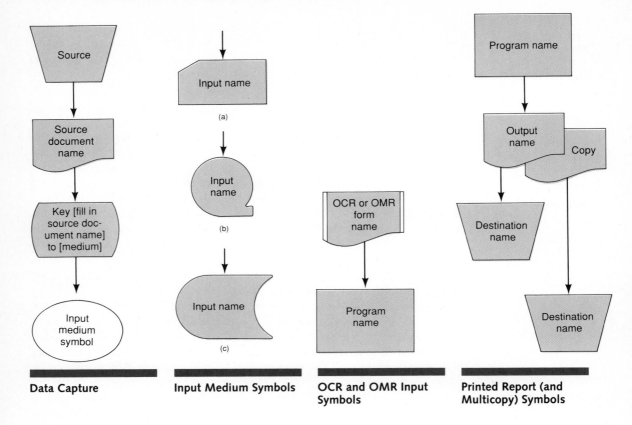

Data Capture — Source, Source document name, Key [fill in source document name] to [medium], Input medium symbol

Input Medium Symbols — (a) Input name, (b) Input name, (c) Input name

OCR and OMR Input Symbols — OCR or OMR form name, Program name

Printed Report (and Multicopy) Symbols — Program name, Output name, Copy, Destination name, Destination name

notations and settled on the simple notations above. The keying symbol—from the preceding paragraph—represents the entire KTD or KTT system. The net output of that computer is the tape (labeled (b) for KTT) or disk (labeled (c) for KTD) file that will be read into the computer program.

There is one more input method that is not clearly defined in the ANSI standard, optical-character recognition (OCR) or optical-mark recognition (OMR). Because the input form is predefined, we suggest using the notation displayed above. Notice that OCR and OMR forms can be directly read by the computer program, assuming you have the appropriate reader. OCR and OMR methods greatly simplify the data capture and data input procedures by eliminating the need for keying.

Batch Output Symbols The most common output is the printed report or form. Both were covered in Chapter 15. In all cases, the output is generated by a computer program (rectangle) and received by a person (trapezoid). Multiple copies of reports or forms are depicted as offsets of the same symbol. Each copy can have a separate destination, or all copies can go to the same destination (see art above).

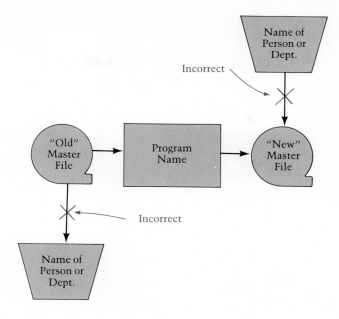

FIGURE 18.10   Tape File Processing on Systems Flowcharts

If the output is to be produced on microfilm, auxiliary equipment will be necessary. Normally, the output is first written to tape. That tape is read by computer output microfilm (COM) equipment to produce the output. Once again, the ANSI standard doesn't offer a notation for this method. We suggest you use the notation depicted on the top left side of the next page.

File/Database Symbols   Systems flowcharts show only those files and databases stored on the computer. Figure 18.10 illustrates tape file processing. Tape files may interface *only* with computer programs and auxiliary equipment; people, through manual operations (trapezoids), may not directly read or write tape files. The name of the file is recorded in the tape file symbol. Recall that tape files are automatically sequential files. The update of a sequential file requires that a new copy of that file be produced. The old copy is usually saved but eventually erased. The updated file is labeled *old* and *new,* as illustrated in Figure 18.10.

Disk devices can be used to store sequential, direct, and the indexed files. Note (see top middle of next page) that the arrow between the file and the program may be single-ended or double-ended. This indicates whether the program reads and uses the file or whether it reads and updates the file. As was the case with tape files, only computer programs may *read from* and

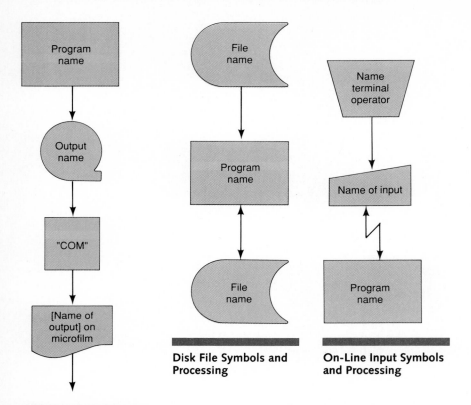

Computer Output on Microfilm Method

Disk File Symbols and Processing

On-Line Input Symbols and Processing

write to disk files. Never use a manual operation (trapezoid) in place of the computer program (rectangle).

A database can be depicted with a single, named disk symbol. However, the schema for a database (see Chapter 14) frequently permits analysts to define subsets called *subschemas* or *views*. The views can overlap. On the systems flowchart, a view can be represented by a disk symbol with a view or subschema name.

On-Line Input and Output Symbols The symbolism for showing on-line inputs and outputs can be somewhat tricky. If we had to show every possible screen (Chapter 17), the systems flowchart would become very, very cluttered. Therefore, we adopt the DFD-like convention of showing only the *net* input and output operations for the on-line system. The ANSI standard has given us two symbols for on-line input and output. You should only use one of the two symbols with any given program. The choice depends on whether the *net* result of the given on-line program is input or output.

A net input program is illustrated above right. The authorized terminal operators are indicated by the trapezoid connected to a keyboard symbol. Note the *ragged* communications line between the program symbol and the keyboard symbol. This double-ended arrow indicates on-line communica-

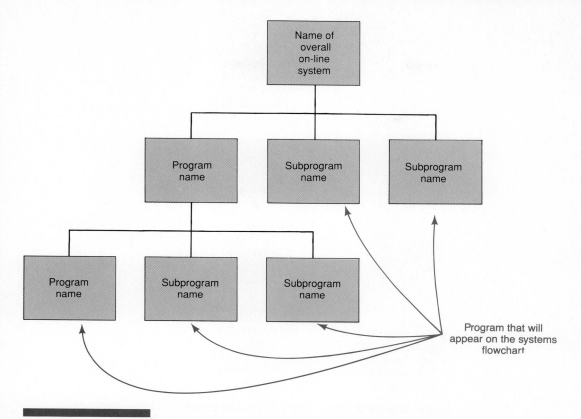

FIGURE 18.11   Hierarchy Chart for an On-line Program   Show each input and output subprogram of an on-line system in the systems flowcharts. Then use a hierarchy chart to show the relationship between these subprograms.

tions. It represents instructions and data from the end-user to the system as well as instructions and messages from the system to the end-user.

A net output subprogram is depicted above (at the top of the next page). Again, the authorized end-users are indicated by the trapezoid. And again, the double-ended-arrow communications line represents instructions, information, messages, and the like.

Now you may argue that you have programs that do both net input and net output, and so you need to show both symbols with a single program. Not typically! What you likely have are two separate on-line programs under the control of a master on-line control program, possibly driven by a menu. We recommend that you show the separate programs on the systems flowchart and use a structure chart (Figure 18.11) to show that the programs are part of a single on-line system. Structure charts are covered more extensively in the next chapter.

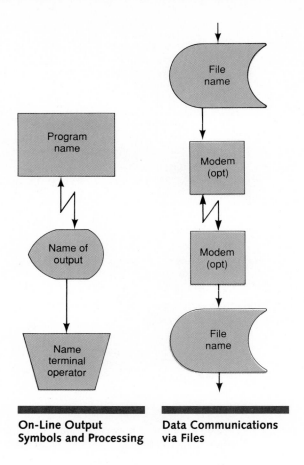

**On-Line Output
Symbols and Processing**

**Data Communications
via Files**

The trend toward distributed computing has also complicated the use of the ANSI standard. For instance, how do you show the uploading or downloading of data between files on different computers? We suggest using the notation shown above left. The ANSI communications arrow is used, but with only one arrow and only between file media. This is the way you would show a remote batch being transmitted to the main site. For distributed systems, a high-level systems flowchart similar to the one shown above right should be drawn to show the interface. Separate detailed systems flowcharts should be drawn for the applications performed on each computer. Modems, for telephone data communications, should be depicted using the auxiliary operation symbol (square).

Miscellaneous Symbols   There are a number of miscellaneous symbols in the ANSI standard. They can be useful for documenting aspects of methods and procedures not covered by the other symbols.

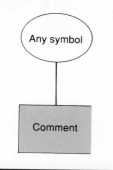

Comment Symbol

Extract symbol (if necessary use comment symbol to describe extraction criteria)

Sort, Merge, and Extract Symbols

Flow-of-Control Symbol

Perhaps the most important miscellaneous symbol is the *comment,* which is shown in the margin. The open-ended box is used to add any needed explanation to any other symbol on the chart. These explanations may describe security features, archiving instructions, timing, or any other important aspect of the procedure. We like to use *comments* to describe the schedule or trigger of a sequence of systems flowchart symbols (for instance, "every 4th Tuesday at 3:00 P.M." or "On demand, weekdays, 8:00 A.M. until 5:00 P.M.").

The *sort, merge,* and *extract* symbols shown in the margin are used to describe file operations. They require files as input and files as output. As a brief reminder, *sorting* is the sequencing of records in a single file. *Merging* is the sequencing of records from more than one file into a single file. These operations were much more common when tape master and transaction files were common. *Extracting,* still commonly used, is the selection of specific records from a larger file. The smaller file is usually used to produce reports and answer inquiries. Also, it is usually deleted after it has been used.

Finally, the single-ended arrow on a systems flowchart indicates the flow of control from one step to another. Flow-of-control arrows, unlike data flow arrows on DFDs, *are not named*!

For your convenience, Figure 18.12 shows all of the systems flowchart symbols and their meanings.

Computer-Assisted Systems Engineering (CASE) for Systems Flowcharts

Computer-Assisted Systems Engineering (CASE) was introduced in Chapter 5 as an emerging, enabling technology for systems analysis and design methods. Most CASE products support a modified subset or superset of the American National Standards Institute (ANSI) standardized systems flowcharting symbol set. For instance, Excelerator uses a subset under the name Presentation Graph. Many CASE vendors have come to the conclusion that ANSI systems flowcharting is not as commonly used today, and perhaps is on its way out (see "Next Generation" box feature, page 635).

Since Excelerator (SoundStage's CASE product) does not support full ANSI systems flowcharting symbolism, we chose not to use the product to develop the remaining figures in this chapter. However, some CASE products do support more productive and qualitative development of systems flowchart. You may wish to explore the opportunities provided by most CASE products to customize the product to include needed systems flowcharting symbols.

Drawing Systems Flowcharts

Systems flowcharts are not difficult to construct, especially if you follow the IPO (Input-Process-Output) concept. In this section, we will examine how

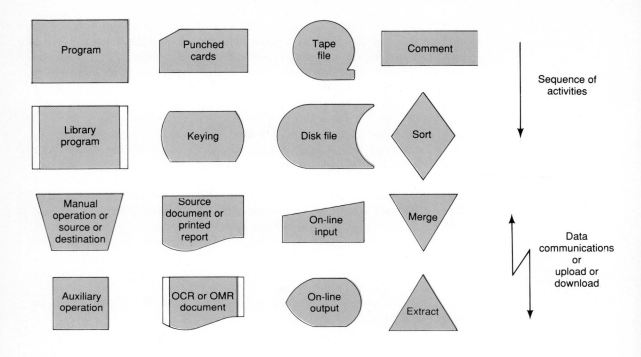

FIGURE 18.12   The Systems Flowchart Symbols and Their Meanings

systems flowcharts are used to document the design of batch and on-line methods and procedures of the SoundStage project's Member Services system.

While previous chapters mentioned that the SoundStage's system will be implemented using the IDMS database management system, for demonstration purposes we also chose to include examples of how systems flowcharts would be developed if the file designs in Chapter 13 were being implemented.

Methods and Procedures for Batch Processing   If you've been using data flow diagrams to define requirements, how do you shift gears to the systems flowchart? The complementary use of these tools is extremely valuable and not very difficult. Your objective is to transform a set of general requirements, expressed by the DFD, into a set of methods and procedures, expressed by systems flowcharts.

A data flow diagram depicting a design solution for a subset of order processing at SoundStage is presented in Figure 18.13. You may recognize it as one of the logical DFDs from Chapter 9. We will now document methods and procedures for the computerized processes in the DFD.

THE NEXT GENERATION

Systems Flowcharts—Going, Going, Gone?

In this box feature, you'll pardon us if we wax philosophical. It's hard for us to suggest the possible departure of an old friend. And systems flowcharts are an old friend. They've gotten us through a lot of information systems development projects. And they've received a lot of undeservedly bad publicity. At one time, systems flowcharts were the only popular tool the analyst had. What does the future hold for the next generation relative to methods and procedures? And what about systems flowcharts? In this chapter, we've suggested that systems flowcharts are not obsolete—they are still useful for designing methods and procedures. And we believe that. But it saddens us to realize that systems flowcharts are finding their way into fewer and fewer books and methodologies. Some of us who have used them successfully will resist. But we may be fighting a losing battle.

You see, if we are totally honest, there is *nothing* that you can't document with systems flowcharts that couldn't be documented with *physical* data flow

diagrams. The DFD is the new kid on the block. And with the current emphasis on structured tools and techniques, the trend is toward using new tools like DFDs and placing less emphasis on systems flowcharts (and, for that matter, program flowcharts—pseudocode and fourth-generation techniques are the new, favored kids on that block). And we can understand the issue at hand. If we use DFDs for systems analysis, why switch tools during design, especially since it isn't really necessary? Using the PDFD conventions discussed in Chapter 7, you could draw any systems flowchart in Chapter 18. You could even use the top-to-bottom, left-to-right sequencing guidelines and the on- and off-page connectors to make them easy to read!

Advocates of a modular program design methodology called *Structured Design* have sometimes suggested that the new system's methods and procedures be specified only with data flow diagrams, not systems flowcharts. You see, Structured Design provides a formal technique for specifying the top-

down modular structure of programs, working from the data flow diagrams for the system. However, Meiler Page-Jones (1980), a leading expert on Structured Design, has clearly demonstrated the use of systems flowcharts within that methodology.

Therefore, for the time being, the tools coexist. There are people out there who have never even heard of data flow diagrams. Others know what they are but are skeptical of their value. Or maybe they just haven't learned or used them. Systems flowcharts have a lot of friends in that crowd. But *you* are the next generation for systems design of methods and procedures. And most of you are being trained in the structured techniques. Consequently, you will eventually be the majority. As for our beloved systems flowcharts, we take comfort in the fact that someday a new tool will come along and probably make the data flow diagram obsolete. But for the time being, you'll forgive those of us who put up a fight . . . R.I.P.?

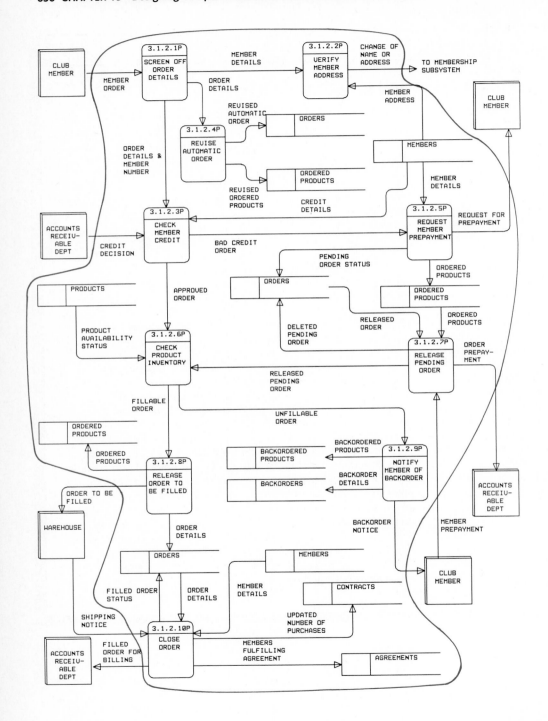

◄ **FIGURE 18.13**
Bounded Data Flow Diagram for SoundStage Entertainment Club's Member Services System A bounded data flow diagram is a useful starting point for drawing detailed systems flowcharts. The DFD conveys design decisions that have already been made. Of particular interest are data flows that cross the boundary (net computer inputs and outputs).

Let's first document the batch transaction, MEMBER ORDER. How do we start flowcharting? In Figures 18.14 and 18.15 we see the systems flowchart for a batch version of PROCESS CUSTOMER ORDER. Let's walk through the systems flowchart. Figure 18.14 illustrates the methods and procedures for batch input of customer orders (from Figure 18.13).

(1) A boundary (on the DFD) is the source of the net input data flow in Figure 18.13. That boundary becomes a manual operation (trapezoid) symbol on the systems flowchart. Because customers can't directly deliver orders to the Computer Information Systems Department, we replaced them with their authorized representative, the ORDER-ENTRY DEPARTMENT.

(2) The data flow MEMBER ORDERS must initially be captured on a source document. For our batch system, each weekday morning the Order-Entry group delivers a batch of member orders to the Data-Entry Department.

(3) For demonstration purposes we'll assume the Data-Entry Department currently uses key-to-tape data entry. Using the key-to-tape system, the Data-Entry Department will record the member orders. Although many KTT workstations may be involved, the net result is a single tape that contains all orders to be processed that day. For internal control, the Data-Entry Department will also prepare a batch control slip. After member orders have been processed, this slip will be compared with the historical report to ensure that all transactions were indeed processed.

(4) The member order tape from the key-to-tape system must be edited to detect keying errors made by the data-entry clerks. The MEMBER ORDER EDIT PROGRAM will check for these errors. Notice that this program was not depicted on the DFD. It was added here because the computerized system required internal controls to ensure valid input data. A MEMBER ORDER ERRORS REPORT will identify erroneous orders. This report will be used by the order-entry group to make corrections and resubmit corrected member orders for rekeying. Valid orders are written to a new tape, VALID MEMBER ORDERS.

(5) To improve processing efficiency, the member orders are sorted by member number. Notice that the field that is to serve as the sort key is recorded in the *sorting* symbol. The resulting SORTED MEMBER ORDERS can now be processed.

Time out! Let's study some of the mechanical guidelines we followed when drawing this systems flowchart. First, notice that the symbols occur in patterns of three: input-process-output. The output frequently becomes an input to the next process, thereby allowing the pattern to repeat itself. Next, notice that the flow proceeds from top to bottom and left to right. You can read the flowchart like a book. By using this simple guideline you could dispel the

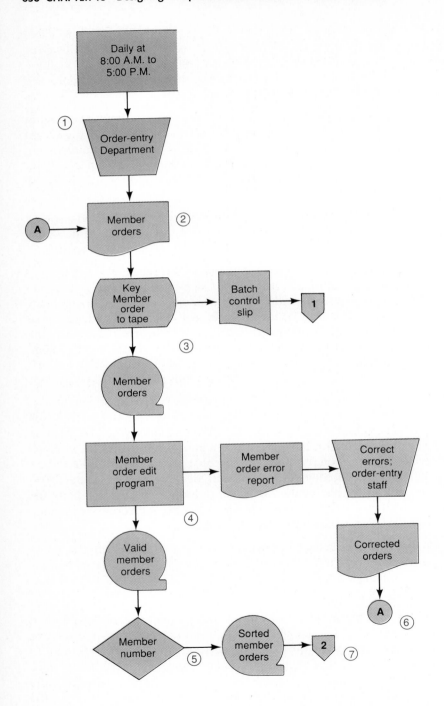

◀ **FIGURE 18.14   Batch Input Methods and Procedures for Member Orders** This systems flowchart depicts the methods and procedures typically employed for batch inputs. Particularly common is the keying/edit/correct loop. Not all batch systems require sorting (especially those that use disk for master and transaction files).

bad reputation systems flowcharts have with many end-users. This brings us to two new symbols that appear on our flowchart.

If we follow the last guideline, how can we indicate return to a previous step? Look again at Figure 18.14. The small circle with the letter *A* in it (labeled ⑥ on the figure) is used to indicate *return of control* to a previous step on the page. Note that we said *on the page*! The circle is called an *on-page connector.* It should only occur *in pairs*—on the same page of a systems flowchart. We used the on-page connector to illustrate the edit and corrections cycle on our systems flowchart.

What about situations in which we can't depict the entire sequence of events on one page? For example, what happens after VALID MEMBER ORDERS have been sorted? How is the BATCH CONTROL SLIP used later in the procedure? Just as we had an on-page connector, so we also have an *off-page connector.* Off-page connectors are demonstrated on Figures 18.14 and 18.15. The off-page connector in Figure 18.14 is labeled ⑦. The number inside the symbol on one page should match with the number in another off-page connector on another page. Make sure you don't violate the IPO pattern when you use on-page or off-page connectors. Neither connector symbol counts as an input, process, or output symbol.

All names and descriptions are written *inside* the symbols. Finally, all arrows are vertical, horizontal, or at a 45° angle. Lines do not cross unless absolutely necessary. All of these conventions enhance the readability of the systems flowchart.

Now we can continue our example. What happens after the member orders are captured, keyed, edited, and sorted? For the answer, we can return to our DFD (Figure 18.13). We have to make a decision as to which processes will be consolidated into specific programs. We have two alternatives. We could treat each DFD process as a separate program. If so, we'd have to introduce temporary files (called *scratch files*) to pass data between the programs. Alternatively, we could group one or more processes into separate programs. This would eliminate the need for scratch files. However, these more complex programs will have to have many more computer files open at one time.

Let's assume we've decided to consolidate all but the CLOSE ORDER and RELEASE PENDING ORDER processes on our DFD into a single program. Figure 18.15 presents this alternative. Note the following:

Ⓐ The off-page connector indicates a continuation of the input sequence depicted in the previous figure. Note that we duplicated the SORTED MEMBER ORDERS file for the sake of clarity. The program produces a MEMBER ORDER REGISTER (a historical report) that reports transactions processed—another control feature we've added to the system.

Ⓑ Remember the BATCH CONTROL SLIP that was created for the input batch in Figure 18.14? It shows up again on this flowchart. The MEMBER ORDER

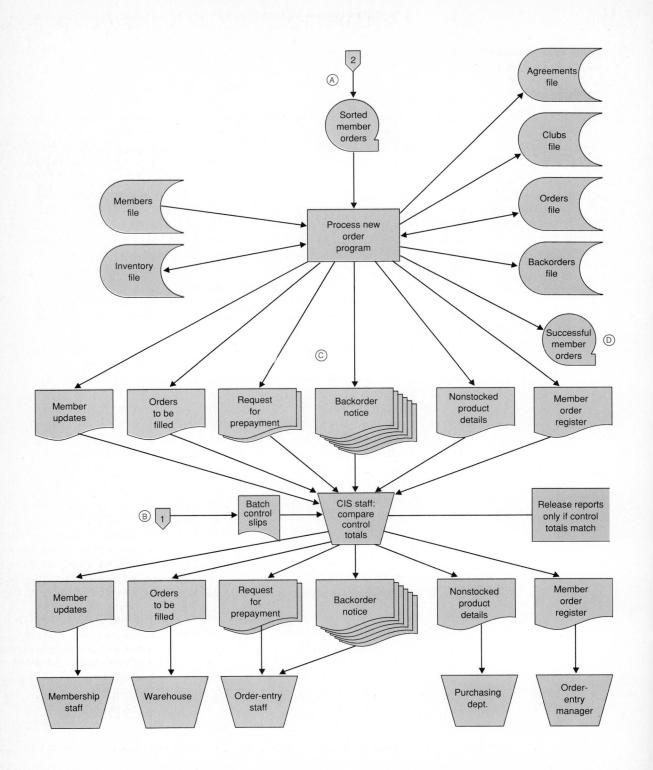

◄ **FIGURE 18.15**
Member Order
Processing with No
Limitations on Number
of Files Open to a
Program   This systems
flowchart is a
continuation of Figure
18.14. The program
illustrated accomplishes
some of the tasks
documented inside the
boundary of the DFD in
Figure 18.13. This
solution requires a
computer system that
allows a large number
of files to be open at a
single time.

REGISTER is compared with the BATCH CONTROL SLIP to see if any transactions have been lost. The reports and documents that were printed are not distributed until the totals match.

Ⓒ How can so many reports be printed at one time? The answer is by spooling. The reports and documents are not printed at the same time. Output data for each report is spooled to a disk file. Reports are printed separately, when the printer becomes available and proper paper or forms is mounted.

Ⓓ All successfully processed member orders were written to a tape file. This tape file is our *audit trail* for transactions. If a master file is lost or destroyed, for any reason, we can reprocess lost transactions from this tape. We'll show this later in this section.

The systems flowchart in Figure 18.15 may not be technically feasible in some environments. Why? Because some computer systems restrict the number of files that can be used (open) by any one program. Suppose, for example, that our computer system allowed us to open only three files per program. We would have to change our procedures. In such a case, our program would be factored into smaller programs—each perhaps corresponding to a single process on our DFD—sharing temporary or scratch files.

How would the systems flowchart be altered for a database environment? For easier comparison, we retained the batch processing mode. Figure 18.14 would not change. However, Figure 18.15 would be replaced by a flowchart similar to Figure 18.16. This solution depicts a combination file and database solution. The inventory file is retained; recall from Chapter 14 that changing that file would be impossible. The other files from Figure 18.15 have been replaced by a single disk symbol representing the MEMBER SERVICES DATABASE (again, recall that Cullinet's IDMS was the chosen database management system).

Methods and Procedures for On-Line Processing

In Figure 18.17 (page 644), we see the systems flowchart for an on-line version of processing orders. We call your attention to the following details:

Ⓐ In the on-line version, member orders are processed immediately. The MEMBER ORDER form has been made optional because the Order-Entry Department will also accept orders via phone. An order-entry clerk will input the member order, via a CRT, to the PROCESS NEW ORDER PROGRAM. This program will edit the input data for errors and immediately notify the CRT operator of errors. Given a valid order, that order will be processed immediately. There is no need for the separate edit and correct cycle that we used for batch processing.

Ⓑ Note the comment indicating that the Order-Entry Department end-users must have a security level of 2. The new system will have three security levels that may be assigned to end-users. Security level 1 will

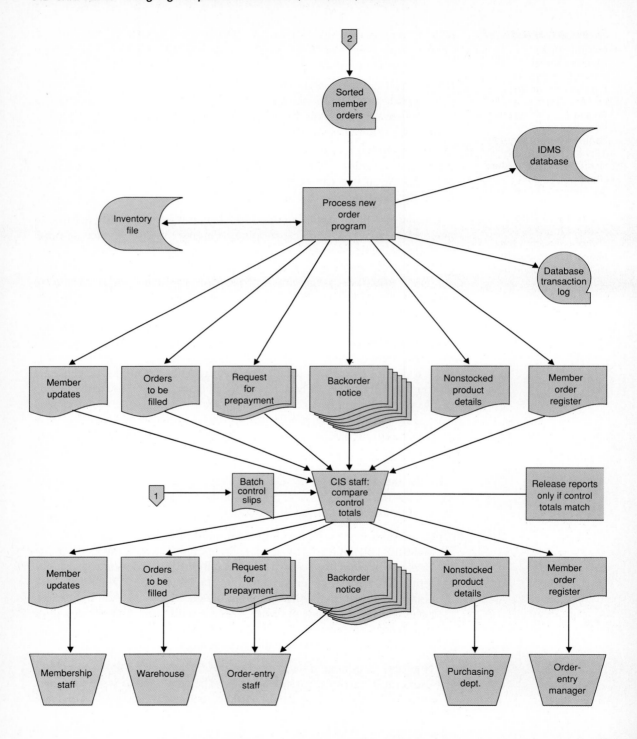

◄ FIGURE 18.16
Member Order
Processing Combining a
File and Database
Solution This solution
depicts a combination
file and database
solution for SoundStage.
The disk symbol is used
to represent both a file
and a database
subschema.

permit the end-user only to display reports and perform select inquiries. End-users who are assigned security level 2 will be allowed to perform transaction processing and select inquiries. Finally, security level 3 permits the end-user to perform transaction processing, management reporting, all inquiries, and file maintenance.

(C) Because the PROCESS NEW ORDER PROGRAM is available from 8:00 A.M. to 5:00 P.M., outputs will be continuously generated during that period. We can't afford to dedicate separate printers to generate each of the reports required during this period. Instead, reports will be dumped to a temporary file from which they will be printed, twice daily.

(D) Internal controls are also very important for on-line systems. As was the case for the batch input system, we produced a historical report and backup file. But because transactions are processed on-line, over a long period of time, we couldn't dedicate a printer or tape drive to record the successful transactions. Instead, we record all successful transactions to an on-line file, SUCCESSFUL MEMBER ORDERS. Once each day, we copy the SUCCESSFUL MEMBER ORDERS file to an off-line medium (tape), and from that we produce our historical report, MEMBER ORDER REGISTER.

What about on-line outputs? The order-entry system allows end-users to inquire against files. One of the inquiries was about member orders. The systems flowchart in Figure 18.18 shows that the Order-Entry staff is allowed access to the MEMBER ORDER INQUIRY PROGRAM. The program prompts the end-user for an ORDER NUMBER (a search key), retrieves order information from one of three files, and displays the information on the CRT. If the end-user wants to get a printout, the dialogue permits him or her to request a printed copy. The output is generated as shown. However, the comment specifies that the printed copy of the output MEMBER ORDER QUERY RESPONSE will not be available until the next morning (overnight printing).

Methods and Procedures for Scheduled Outputs Recall that management reports are typically scheduled in both batch and on-line systems. The systems flowchart for a typical scheduled report is shown in Figure 18.19. The data is retrieved from the files, and the report is produced and distributed. Remember, some systems may limit the number of files that can be open to a program. If so, the program may be split into separate programs. The first program would extract data from one file, creating a scratch file. The second program uses the scratch file plus other files to create the report.

Methods and Procedures for Maintenance, Backup, and Recovery of Files Transaction files are maintained by transaction-processing programs. Similar programs must be designed to maintain the master files in the system

FIGURE 18.17   Member Order Processing in an On-Line System   This is an on-line version of the member order processing subsystem. The program handles all the processing requirements that were depicted on the DFD as well as the input editing requirements. (The order-processing program may be a subprogram in a hierarchy chart that depicts the structure of a complex on-line system.)

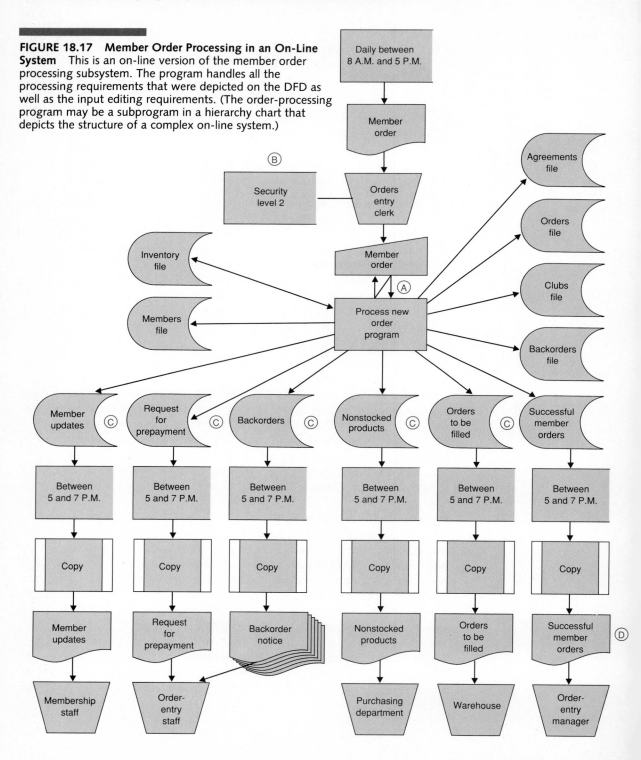

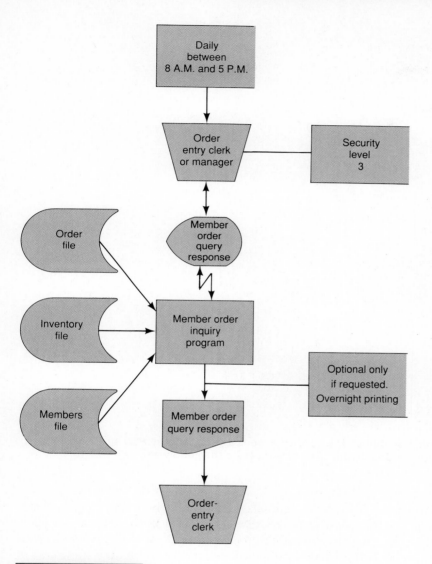

FIGURE 18.18   Systems Flowchart for an On-line Inquiry   The systems flowchart for an inquiry is relatively simple, as you can see.

(for instance, MEMBER and INVENTORY). These programs are similar to other transaction-processing programs except they only create, delete, and modify records in the master file. These requirements are normally excluded from DFDs, which tend to document the net input and output activities. However, they cannot be excluded from systems flowcharts. If the file isn't created and maintained, it can't be used and updated.

Earlier in this chapter, you learned that one of the most important tasks in documenting the methods and procedures for a system is the inclusion of

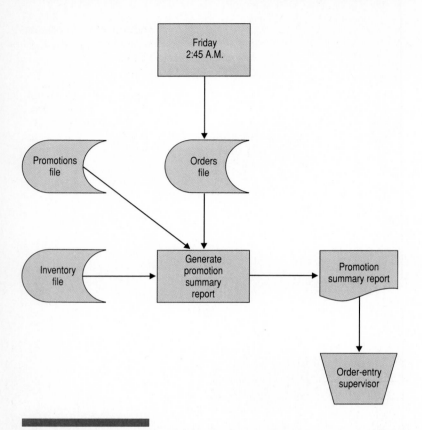

FIGURE 18.19 Scheduled Report Procedures Management reports are usually generated on a scheduled basis in *both* batch and on-line systems. The flowcharts are straightforward.

backup and recovery procedures. Let's think back. In documenting the batch and on-line versions of PROCESS NEW ORDER PROGRAM we made sure audit files for archiving the successfully processed member orders were generated. Suppose we lose a master or transaction file. What will we do?

A systems flowchart to back up and, if necessary, recover the MEMBER master file is shown in Figure 18.20. To recover the lost file, we recreate the MEMBER FILE from the last backup copy. Because this backup occurs weekly, to fully restore this file, we only need to reprocess those transactions that have taken place since the last backup. Similar backup and recovery procedures must be designed for every master and transaction file in the system.

It should be noted that in a database environment, the DBMS usually logs transactions automatically and that backup and recovery procedures are external to the systems analyst's responsibility.

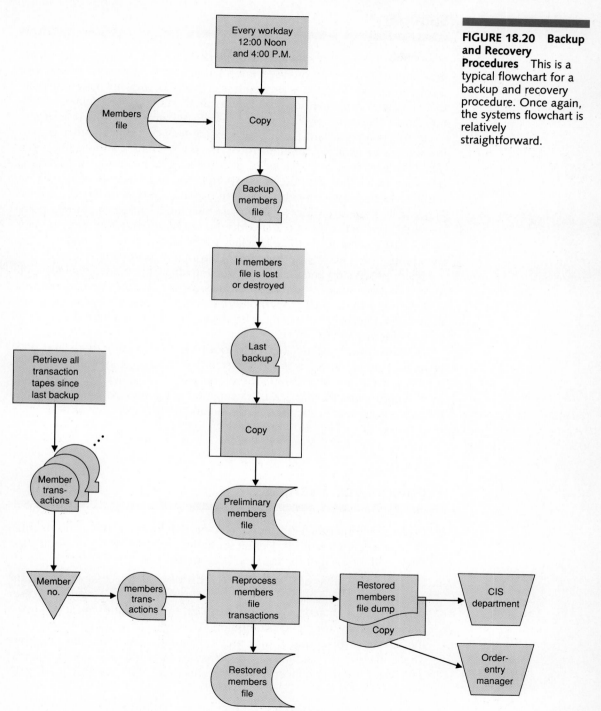

FIGURE 18.20   Backup and Recovery Procedures   This is a typical flowchart for a backup and recovery procedure. Once again, the systems flowchart is relatively straightforward.

Summary

Methods and procedures define the sequence of events that produce outputs from inputs in a computer-based information system. A method is a way of doing something. Procedures are step-by-step plans for implementing a method. Information systems methods can be classified on two scales: degree of centralized or distributed processing and degree of batch or on-line processing.

In the not-too-distant past, all processing was centralized—that is, performed on the central computer. Today, the analyst must consider the possibility of distributed processing. In distributed processing, the workload is offloaded to satellite computer systems. But, because data may be needed in multiple locations, we consider linking the satellite computers into some form of network. There are various network architectures, including point-to-point networks, bus networks, star networks, hierarchical networks, and ring networks. The analyst needs to be familiar with available networks and with the possibilities each network presents for distributed processing.

Batch processing and on-line processing alternatives are generally well known to most students and professionals. However, these two alternatives present the systems analyst with different requirements for internal control in systems. Internal controls must be documented in the methods and procedures for any system.

Systems flowcharts are a graphic tool used to depict methods and procedures. Although there is a national standard (ANSI) for systems flowchart symbols, it is somewhat dated. Therefore, like us, you need to adopt your own departmental standards for drawing and using systems flowcharts. This chapter has presented suggestions for expanding the ANSI standard.

Problems and Exercises

1. Explain the difference between batch, on-line, remote batch, and distributed processing methods. Define an input, and conceive a situation that might call for each of the four methods to be used.

2. Describe five different strategies for distributed processing. Make an appointment to visit a local computer facility (alternative—examine a case study in a distributed processing, network, or data communications textbook). Are they using microcomputers? If so, what networking strategy are they using or planning? If they aren't planning a network, find out why. There may be very good reasons. What data communications problems and issues are they concerned with at this time? What do they think will happen?

3. Prepare a systems flowchart to describe the backup and recovery procedures for restoring a lost or damaged master file (alternatively: a database).

4. An on-line computer program is to input transactions, update two master files, and print five reports, including an external document. What are the implications of printing so many reports? What alternative procedures could be followed?

5. Draw a systems flowchart for each alternative procedure you described in Exercise 4.

6. How many computer files, including basic input and output files, does your computer system allow to be open at one time? What are the implications for design of methods and procedures? Draw general systems flowcharts to show how you would deal with a program that requires more files than can be open at one time.

7. What symbols can be connected to the following symbols on a systems flowchart:
   a. Computer program (rectangle)
   b. Sort (diamond)
   c. Source document or printed report
   d. On-line input or keyboard
   e. On-line output or screen
   f. Data communications or upload or download arrow
   g. Manual operation (trapezoid)
   h. Disk file
   i. Tape file
   j. Extract (triangle)
   k. A database subschema

8. A data flow diagram indicates the computerized portion for the system. The DFD and the decisions made during output, file, and input design together aid in the construction of systems flowcharts. However, it is quite common during fact-finding in the study phase to find systems flowcharts of an existing computerized system. The analyst may wish to convert the systems flowcharts into data flow diagrams to aid in verifying the system with information systems professionals and, more importantly, end-users. Explain how systems flowcharts could easily be converted to data flow diagrams.

Projects and Minicases

1. Draw a systems flowchart to depict the following methods and procedures:
   a. Each morning, Monday through Friday, at 7:45, a program is executed to read the inventory file to identify products that are low in stock. A report is produced and delivered to the inventory purchasing clerk.
   b. Customers visiting a small computer sales store frequently request price quotes at a computer terminal. A salesperson at the store will execute a quote recording program. The program accepts data concerning the quote and checks an inventory file for availability, current prices, and

other product data needed to respond to the quote. The quote is then calculated and placed in a quote file (ISAM). A hard copy of the quote is generated for the customer and salesperson.

c. The Accounts Receivable Department fields phone calls from customers concerning disputes over invoices. Accounts Receivable uses a terminal to obtain information needed to resolve these disputes. An accounts receivable clerk may request information concerning a customer's account or an invoice.

You may expand on these general descriptions as you desire.

2. A magazine publishing company receives magazine subscriptions through the mail. Subscriptions are routed to the Data-Entry Department for batch input. The resulting input tape (sorted by subscription number) is read by a program that updates a customer and subscription file. The program also produces a daily subscription register. Totals appearing on the register are compared with totals on a batch control slip. If the totals don't match, the discrepancies are resolved. Otherwise, the register is delivered to the Order-Entry Department supervisor. Prepare a systems flowchart that describes the methods and procedures previously outlined. Be sure the batch input procedures allow for complete editing and correction of subscriptions.

Annotated References and Suggested Readings

Brill, Alan E. *Building Controls into Structured Systems.* New York: Yourdon Press, 1983. This book provides a comprehensive look at internal controls design, especially when using structured tools, such as data flow diagrams. The material covered is equally applicable when not using structured tools. This book is short and relatively easy to comprehend.

Fitzgerald, Jerry. *Internal Controls for Computerized Information Systems.* Redwood City, Calif.: Jerry Fitzgerald and Associates, 1978. This is our reference standard on the subject of internal controls. Fitzgerald teaches a powerful matrix approach to designing internal controls.

Page-Jones, Meiler. *The Practical Guide to Structured Systems Design.* New York: Yourdon Press, 1980. The primary emphasis of this book is program design, not systems flowcharting. However, for those of you familiar with the Structured Systems Design methodology, Page-Jones shows that systems flowcharts and structure charts are not mutually exclusive. Structured Systems Design is introduced in the next chapter of this textbook.

CHAPTER NINETEEN

Designing Structured Programs

Tower Lawn and Garden, Inc.

George Amana, a programmer/analyst for Tower Lawn and Garden, Inc.—a distribution center for lawn and garden equipment in northern Louisiana—had just sat down to lunch in the company cafeteria. Pete Wilcox, senior partner in the firm, joined him.

"Hi, George, why don't you join me for lunch," offered Pete as he gestured for George to be seated next to him. "Hey, you look pretty frustrated. What's the problem?"

George responded with a sigh, "I had to take over the sales information systems project that Judy left behind when she quit. It's total chaos. I was told that it was all but finished. Come to find out, she didn't finish several of the programs."

"But she did do a good job of specifying all the required inputs and outputs. What's so tough about the programs?" asked Pete. "Judy always preached about the benefits of structured programming. In fact, she taught me how to do it. Don't tell me she doesn't practice what she preaches!"

George replied, "No, her code is very well structured. And her documentation is adequate. It's just that the programs seem so poorly designed. Some of her subroutines are so long and complex that it's difficult to get a grasp on small enough pieces to test them for correctness. It seems like an all-or-nothing proposition. If I encounter a bug, I have to test large sections of code to zero in on the problem. Sometimes the bug turns out to be in an entirely different subroutine!"

"Why didn't Judy break the system into smaller pieces?" asked Pete.

"Oh, she did!" exclaimed George. "The subroutines are evidence of that. But it almost seems like she generated the subroutines on the fly—as if to say, 'Well, this piece of code is getting complex. I'd better put in a subroutine to finish it.' She left me a rough draft of a structure chart, but I just don't understand the reasons she factored the system the way she did."

"That's the way I write programs. I start by trying to draw a flowchart on a single page—sort of the high-level flowchart. Then I factor the more complex processes into more detailed processes that I implement as subroutines. It sounds like that may be what Judy did."

George responded, "Maybe she did. But that strategy causes the lower-level subroutines to be very dependent on other routines. I frequently encounter bugs that get traced back to other seemingly unrelated routines. I'm just getting further behind schedule. I may just have to write the programs from scratch."

Pete thought for a moment and then suggested, "Why don't you get some help? Barbara just finished her project. Maybe she can help you. You could divide up the work and get it done faster."

George grimaced, "Divide up the work? I don't see how. Judy's program specifications are just one big document. I'm not completely sure which file and report specifications to match up to which modules. For that matter, I'm not sure the programs themselves are fully documented."

"Well, I'm sorry George. I don't know what to tell you."

Discussion

1. If design specifications are thorough and complete and program code is well structured, how can the system still be difficult to construct?

2. How should subroutines in a program be conceived? How does Judy seem to have created them? What is the potential problem with creating subroutines "on the fly" (either during coding or during flowcharting)?

3. What effect does program and subroutine size have on testing?

4. What would Barbara require in order to take on responsibility for some of the programs that haven't been written? What does any programmer need to be able to write a new program? How would you organize the necessary documentation of program requirements?

What Will You Learn in This Chapter?

In this chapter, you will learn how to design good programs. You will also learn how to package program design specifications. You will know that you understand how to design programs and package design specifications into a format suitable for programmers when you can

1. Factor a program into manageable program modules around which complete specifications can be organized.

2. Associate the design documentation presented in Part Three of this book with the input, process, or output specifications of a computer program.

3. Determine which programs on a systems flowchart require detailed packaging.

4. Use structure charts and Structured English to package a computer program and its specifications.

We're nearly finished with systems design. You've designed the inputs, outputs, files, terminal dialogue, methods, procedures, and controls. You've selected appropriate computer equipment and packaged software (which has hopefully been delivered and installed during systems design). What's left? As the title of this chapter suggests, structured program design.

Let's make sure you understand what we mean by program design. From your programming courses, you may think of program design as *algorithm* or *logic* design. That is *not* the subject of this chapter. We don't intend to reteach you how to draw structured program flowcharts, to prepare pseudocode, or to construct box charts (sometimes called Nassi-Schneidermann charts). In our minds, that is clearly a subject for a programming textbook. Instead, we are concerned with how the programming specifications are presented to the computer programmer for implementation. To this end, we view program design as consisting of two components:

- *Modular design*—the decomposition of a program into manageable pieces.
- *Packaging*—the assembly of input, output, file, terminal dialogue, and processing specifications for each module.

Some readers are likely to interpret the material covered in this chapter as an invasion of the programmer's turf. It really varies from one computer information systems shop to another. Some shops insist that the analyst prepare detailed modular designs and program specifications (at a level close to pseudocode). Other shops believe that the analyst's job ends with general programming specification, leaving modular design to the programmer. Depending on your opinion, you may want to omit this chapter. However, we recommend the chapter for the following reasons:

- Your career may take you to organizations or management that prefer both of the approaches.
- The chapter helps tie the design specifications prepared in Chapters 13 through 18 to the program specifications that normally initiate systems implementation (discussed in detail in Chapter 20).
- In the absence of a company standard, you may want to consider a rigorous, personal standard for presenting specifications. Why? In Chapter 20, you will learn that the analyst is frequently engaged in a large number of activities during systems implementation. The more thorough and complete your programming specifications are, the less time you'll have to spend clarifying those specifications for the programmer.

Modular Design of Computer Programs

For those of you not familiar with modular design from your programming courses, we'll briefly review the concept. Large projects are more easily man-

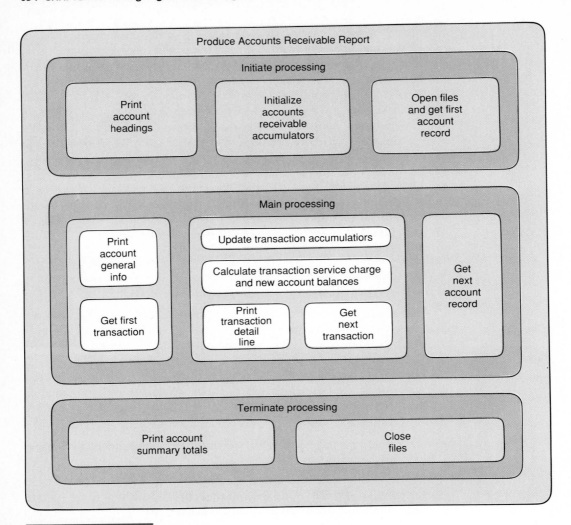

FIGURE 19.1 Modular Design This amoeba-like diagram is a useful way to depict the modular design approach, otherwise known as *divide and conquer*.

aged if they are broken into smaller pieces. You've seen us apply this concept with data flow diagrams. Computer programs can be similarly decomposed, as depicted in Figure 19.1. What we did here was to recursively factor a large program into smaller and smaller pieces called modules. We will now study how this is accomplished.

Modular Decomposition of Programs

What is a module? It could be a subroutine or subprogram. And it could be a main program. On the other hand, it could be a unit of measure smaller than any of those. For instance, a module could be a paragraph in a COBOL program. So, what is a module? We will define a **module** as a group of executable instructions with a single point of entry and a single point of exit. Some modules exist to perform single functions. These include READ A RECORD, EDIT A RECORD, CALCULATE PAY, and ADD A RECORD TO A FILE, to name a few. Other modules exist to *supervise* or *drive* the function modules.

The length of a module is important. Evidence suggests that modules should consist of a relatively limited number of lines of executable instructions. Most experts suggest a number between 24 lines (the most typical display screen size) and 60 (the number of lines printed on an average report). Consequently, this guideline would suggest that any program that cannot be written in fewer than 60 lines of code should be decomposed into modules.

Tools for Modular Design   There are two popular tools for depicting the modular design of programs. We'll use the simple example provided in Figure 19.1 to introduce these tools. The first tool we'll introduce is the Warnier/Orr bracket (see Figure 19.2). A **Warnier/Orr diagram** is nothing more than a hierarchy chart laid on its side. For this discussion, we want to focus on the tool itself, deferring our discussion of the Warnier/Orr methodology until the next section.

Brackets decompose modules into other lower-level modules that we'll call *submodules*. The Warnier/Orr diagram implies a sequence of execution that is read from top to bottom and left to right. A number in parentheses below a module indicates how many times that module executes (a *looping* concept). A plus sign between modules indicates that execution of those modules is mutually exclusive. In other words, each single execution of the calling module may call one or the other submodule—never both. Although we haven't seen it, a notation could easily be defined, say with an asterisk, to indicate that a module can call either or both submodules.

An alternative and somewhat more familiar tool is the structure chart. A **structure chart** (see Figure 19.3) is a treelike diagram. Structure charts, by whatever other name you might know them, may be familiar to you from your introductory programming course.

Structure chart modules are depicted by named rectangles. Modules are factored, from the top down, into submodules. Structure chart modules are presumed to execute in a top-to-bottom, left-to-right sequence. Named arrows are used to represent data (depicted by an arrow with a small circle on one end) or control flows (depicted by an arrow with a darkened circle on the one end) between modules.

We find Warnier/Orr diagrams easier to construct than structure charts, since there is no need for a template. But structure charts are more familiar

FIGURE 19.2 Warnier/Orr Notation
Warnier/Orr brackets are a popular and simple-to-use modular design tool. Compare this notation with Figure 19.1.

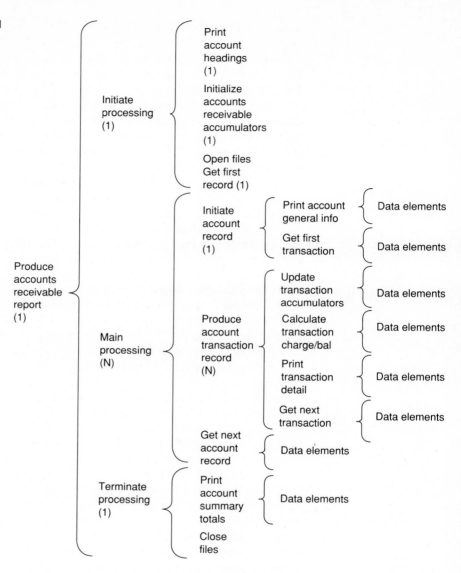

in the literature. Because it is more likely that you've encountered structure charts in your introductory programming course, we will use that notation throughout the remainder of this chapter. You can replace any of the structure charts we draw with an equivalent Warnier/Orr diagram. Let's move on to the strategies used to decompose a program (system) into modules.

Strategies for Modular Design Recall that structured design is a popular term for the decomposition of a program into modules. There are three popular strategies for structured design. We will review them briefly before we present an integrated strategy that we've found useful.

IBM's Hierarchy plus Input-Process-Output, called HIPO, was one of the earliest strategies for structured design. Using HIPO, the designer factors a program into logical functions, depicted as a structure chart (which they call a *vertical table of contents*). Each module is eventually documented with an IPO chart, which forces you to detail the inputs (these include inputs, file accesses, and subroutine parameter passing), processing requirements (narrative, pseudocode, or flowchart), and outputs (including reports, displays, file updates, and parameter passing). Although HIPO forces the decomposition of programs into modules, it doesn't really offer a strategy for doing so. Therefore, HIPO might be better thought of as a documentation tool than as a strategy. We'll show you how to take advantage of the HIPO documentation tool later in this chapter.

Ed Yourdon and Larry Constantine (Page-Jones, 1980) have developed what has become a popular strategy for determining an optimal structure chart for programs. Their technique is called *Structured Design* (a term we have already introduced), and it is based on the use of data flow diagrams. Essentially, you document programs with detailed logical data flow diagrams, study those diagrams, and convert the DFDs into *structure charts* (their term). They suggest two substrategies for developing the structure charts:

- *Transform Analysis.* To make a long story short, **transform analysis** is an examination of the DFD to divide the processes into those that perform input and editing, those that do processing (such as calculations), and those that do output. Although we have greatly simplified the strategy, it is based on the IPO concept about which you have learned throughout this book.

- *Transaction Analysis.* **Transaction analysis** is the examination of the DFD to identify processes that are distinct and therefore transaction centers. The resulting structure chart is factored into these transaction center modules, which may then be factored into IPO modules using transform analysis.

By using their strategy to divide a program into modules, you are able to end up with modules that are said to be *loosely coupled* and *highly cohesive*. Loosely coupled modules are less likely to be dependent on one another (remember the problem George had in the Tower Lawn and Garden minicase?). Highly cohesive modules contain instructions that collectively work together to solve a specific task. The data and control flow symbols depicted on a structure chart can serve as aids in determining the degree of coupling and cohesion of modules.

Another approach to developing an optimal modular structure has been suggested by Jean-Dominique Warnier and Ken Orr (Orr, 1977). This approach develops a program structure by working backward from the desired output data structure. This technique is called *Logical Design of Programs*. The out-

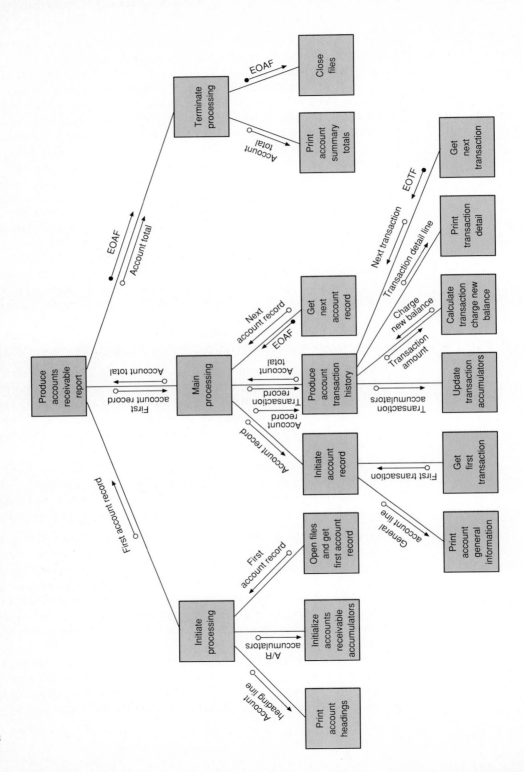

◄ **FIGURE 19.3   Structure Chart Notation**
Structure charts are another popular tool for modular design. Compare this diagram with Figures 19.1 and 19.2.

put data structure is first defined using the Warnier/Orr notation. Then the input, file, and/or database structure is defined using a similar notation. Finally, a program structure is defined from these structures. The resulting program structure usually reflects the input-process-output characteristics (or begin-process-terminate).

Both the Yourdon/Constantine and Warnier/Orr strategies have their die-hard advocates. We think the two strategies have much in common. Through-out this book, we have tried not to endorse any analysis or design method-ology. This chapter will be no exception. Instead, we'd like to present a simplified strategy for modular design. The strategy is based on common sense and the fundamental principles that underlie both the Yourdon/Con-stantine and the Warnier/Orr approaches.

How to Do Modular Design (a Simplified Approach)

Programs have been identified on both bounded data flow diagrams and systems flowcharts (Chapter 18). Given these programs, we want to break them into manageable modules around which program specifications will be written. Programmers can then build and test each module independently. Then modules can be integrated according to the structure chart and tested as a whole program. Now that you know what we're trying to accomplish, let's do it. We'll pay particular attention to how structure charts can be used to integrate large on-line programs, because the trend appears to be toward such systems.

Step 1: Define the High-Level Structure   Virtually all applications programs can initially be broken down into three main functions: INITIATE PROCESS-ING, MAIN PROCESSING (the body of the program), and TERMINATE PROCESSING. Normally, the initiation and termination functions are per-formed once. The main processing function normally executes several times. Figure 19.4 can be used as a starting point for any normal program structure. These three essential functions are loosely based on Yourdon/Constantine's transform analysis strategy.

Adams, Wagner, and Boyer (1983) have cataloged a number of common functional modules that may be included in the INITIATE and TERMINATE PROCESSING functions. This is consistent with the concept of defining highly cohesive structures. These start-up and close-down functions are described in Figure 19.5. Notice in Figure 19.4 that the GET FIRST INPUT or RECORD function is almost always the last submodule of the INITIATE PROCESSING function. For an on-line system, this module might be labeled DISPLAY MAIN MENU. Along the same lines, the STOP PROCESSING function is usually the last submodule of the TERMINATE PROCESSING.

What purpose does the MAIN PROGRAM module serve? It acts as a type of traffic cop, directing the execution of its subordinate modules. It executes

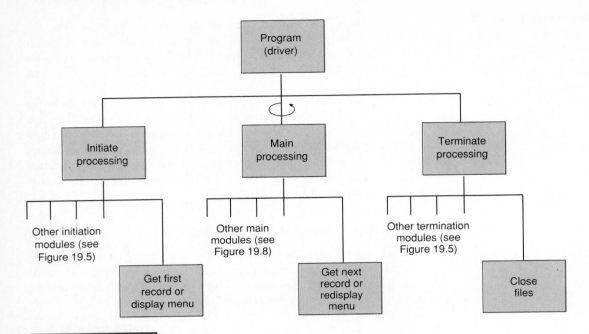

FIGURE 19.4   A De Facto Standard Structure Chart for Most Programs   Most programs can be initially factored into an INITIATE PROCESSING, MAIN PROCESSING, TERMINATE PROCESSING structure. Those modules can be further factored using a number of popular strategies.

the INITIATE PROCESSING module and begins executing the MAIN PROCESS-ING MODULE. When processing has been completely finished, it executes the TERMINATE PROCESSING module. That's it! For on-line systems, process-ing usually doesn't terminate—the main processing module is available for execution until the system is no longer needed. At that time, the TERMINATE PROCESSING module is executed.

The remainder of our strategy will focus on how to factor the MAIN PROCESSING module.

Step 2: Identify Transaction Centers   As a preface to factoring the MAIN PROCESSING module, the first question we like to ask is, "Does this program have transaction centers?" This question is based on Yourdon/Constantine's transaction analysis strategy. What we are really asking is, "Does this program support multiple transactions?" If so, we will factor the MAIN PROCESSING module according to those transactions. The following are examples that would lend themselves to this strategy:

- A file maintenance program typically supports at least three transactions: ADD A NEW RECORD, DELETE A RECORD, and MODIFY A RECORD. Each

FIGURE 19.5 Primitive Functions Performed by INITIATE and TERMINATE PROCESSING Modules This table, adapted from Adams, Wagner, and Boyer (1983), suggests highly cohesive primitive modules that are typically controlled by an INITIATE or TERMINATE processing module. Any of these functions may be factored into primitive subsets to further improve cohesion and possibility for reuse.

INITIATE PROCESSING FUNCTIONS:

BUILDING AND LOADING TABLES: Creating arrays to store tables, such as tax tables, actuary tables, and the like, and loading the data into those tables.

DEFINING CONSTANTS AND ACCUMULATORS: Constants are set in a dedicated module so those constants can be easily located if they need to be changed (for instance, SALES TAX PERCENT). Accumulators are used to count records and control totals during main processing.

OPENING FILES: Files must be opened before they can be read from or written to. It should be noted that some systems limit the number of files that can be open at any one time. If more files are needed than can be opened, then the program must be rewritten as multiple programs that pass intermediate results through temporary (scratch) files (which count as one open file).

FILE MERGING OR SORTING: This must be done before main processing can be done.

PRINTING REPORT HEADINGS: Why relegate report headings to a separate module? So they can be easily located if report headings need to be modified.

DISPLAYING (MAIN) MENU: For on-line systems, displaying the first menu and accepting the first choice from that menu are usually an initiation function.

GET FIRST INPUT RECORD: Read the first input record or file record to be processed.

TERMINATE PROCESSING FUNCTIONS:

CALCULATING CONTROL TOTALS: Performing arithmetic and statistical operations on totals accumulated during main processing functions.

PRINTING CONTROL TOTALS: Printing the accumulators and control totals maintained and calculated during main processing.

CLOSING FILES: The reverse of opening files. Disconnects the file from the program, thereby allowing other programs, which may have been locked out, to use those files.

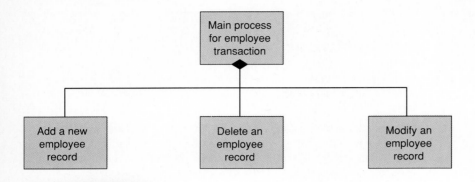

FIGURE 19.6 Transaction Centers for a Simple Program Generally, it is useful to factor a module into distinct submodules that act on single transactions. These transaction centers are then said to be loosely coupled, that is, less likely to impact one another.

transaction deserves its own module. Each transaction will cause the execution of one and only one of the transaction modules. For such a program, we would use the structure illustrated in Figure 19.6.

- An on-line system typically supports multiple levels of transactions. For instance, the main menu may offer three choices: EMPLOYEE FILE MAINTENANCE, PERSONNEL TRANSACTION, and EMPLOYEE INQUIRY. Each of these subfunctions consists of multiple transactions. EMPLOYEE FILE MAINTENANCE could be factored as described in the preceding example. PERSONNEL TRANSACTION could be factored into SICK LEAVE PROCESSING, TIME CARD PROCESSING, VACATION PROCESSING, and so on. The resulting hierarchy chart might look something like Figure 19.7.

Although data flow diagrams may help you identify transaction centers, it depends on how detailed the analyst drew those data flow diagrams (for

FIGURE 19.7 On-Line Transaction Centers On-line systems are particularly ▶ suited to transaction analysis since their capabilities are called on demand and integrated through a high-level control program.

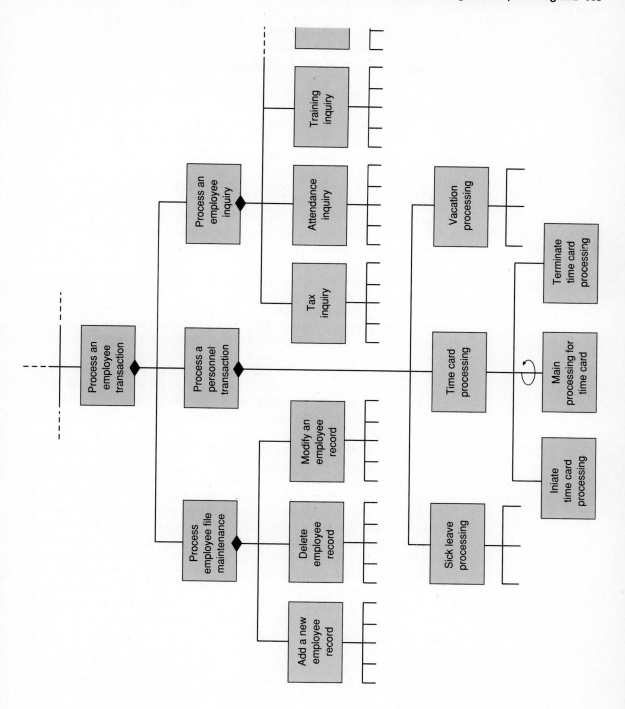

instance, many analysts won't factor the DFD process, MAINTAIN EMPLOYEE FILE, into three separate processes).

Finally, most transaction centers need to be further factored into their own INITIATE, PROCESS, and TERMINATE modules. For example, in Figure 19.7 we factored the PROCESS SICK LEAVE module into an initiate-process-terminate trio of submodules.

If the program you are trying to design cannot be factored into transactions (for instance, it supports a single transaction or only generates a single report), this step can be skipped.

Step 3: Factor the Initiate, Process, and Terminate Functions into Primitive Functions   At this point, we have factored our program into one or more iterations of INITIATE, PROCESS, and TERMINATE modules (*note:* we would have only one iteration if there were no multiple transaction centers). Now we can factor the INITIATE, PROCESS, and TERMINATE modules into their primitive functions. The typical primitive modules for INITIATE and TERMINATE modules were listed in Figure 19.5. There are two strategies for factoring the PROCESS modules into primitives.

Most PROCESS modules can be factored into the primitive functions described in Figure 19.8. Once again, these primitive functions are generally considered to be highly cohesive. We should be able to write the logic of these modules with 50 or fewer statements of code. Along those lines, it may be appropriate to factor one of the simple functions from Figure 19.8 into submodules. For example, an EDIT ORDER module may be factored into EDIT GENERAL ORDER DATA, EDIT ORDERED PARTS, EDIT CUSTOMER DATA, and so on. The first strategy is demonstrated by the structure chart that appears in Figure 19.9.

The second strategy is based on the Warnier/Orr data structure approach. If the output consists of a natural hierarchy, we like to use the Warnier/Orr data structure approach to factor the process. This frequently happens with multiple control break outputs (a concept that should be familiar to students of programming). For instance, suppose our process module is intended to produce a PART SALES SUMMARY REPORT. This report should contain detailed unit and dollar sales information for each part, each product line (consisting of multiple parts), and each warehouse zone (consisting of multiple product lines). We can factor the process module for this report into modules that correspond to the control breaks. This structure is illustrated in Figure 19.10. Notice that we added an INITIATE (to set accumulators) and TERMINATE (to format and print control totals) module to each control break processing module. This structure chart makes it possible to implement the program with a single flag, "end of file" (thus loosely coupling the hierarchy of modules), and with much less code. The final process module, PROCESS PART DETAIL LINE, is factored into primitive tasks by using the first strategy, the primitive functions introduced in Figure 19.8.

FIGURE 19.8 **Primitive Process** This list, again adapted from Adams, Wagner, and Boyer (1983), suggests primitive cohesive, MAIN PROCESSING functions. Again, these functions may have to be further factored to define small and reusable modules.

MAIN PROCESSING FUNCTIONS:

EDITING INPUT RECORDS: Performing picture, range, and completeness checks to make sure that data being input to the system for the first time is correct (this module will normally write to (or display) an errors report or file).

GETTING A SECONDARY RECORD: Reading an input or file record from a secondary source. For instance, if you are processing input ORDERS, you may have to retrieve a CUSTOMER RECORD for a credit check or retrieve PART records for an inventory check, all during main processing.

PERFORM CALCULATIONS: Performing arithmetic operations on data.

MAKING DECISIONS: Executing business policy decisions, such as credit checking, part availability, and discounting.

ACCUMULATING TOTALS: Where possible, totals should be accumulated in their own modules so those accumulators can be easily located and changed.

WRITING A DETAIL LINE: Recording a single detail line or transaction to a file or report that will contain many such detail lines.

WRITING A COMPLETE RECORD: Writing an entire record (as opposed to a detail line) to a report or file. For example, printing a paycheck, or updating a record in a master file.

GETTING THE NEXT RECORD: Retrieve or read the next record in the loop that drives the main processing routine.

REDISPLAY A MENU: Of options that are available in an on-line system.

And that, in an abbreviated form, is one possible strategy for program module design. We strongly urge you to study the full Warnier/Orr and Yourdon/Constantine strategies as part of your continuing education.

After performing modular design on all programs, we are ready to package programming specifications around the modular design.

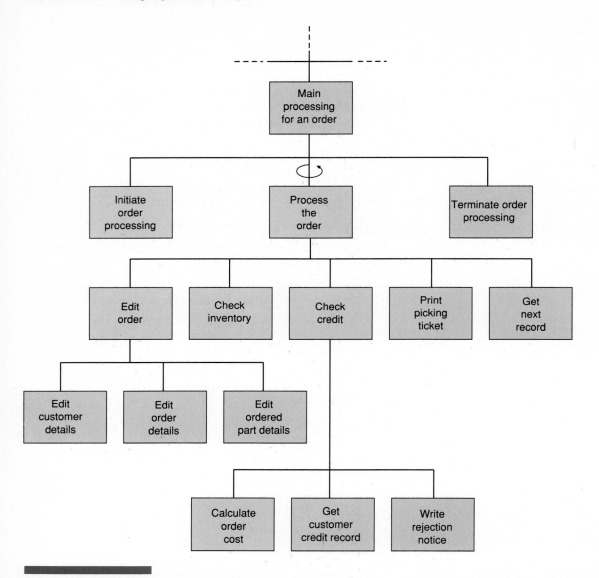

FIGURE 19.9 Factoring MAIN PROCESSING into Cohesive Primitives MAIN PROCESSING must eventually be factored down to loosely coupled, highly cohesive primitive modules.

Computer-Assisted Systems Engineering (CASE) for Program Design

Computer-Assisted Systems Engineering (CASE) was introduced in Chapter 5 as an enabling technology for systems analysis and design. Virtually all CASE

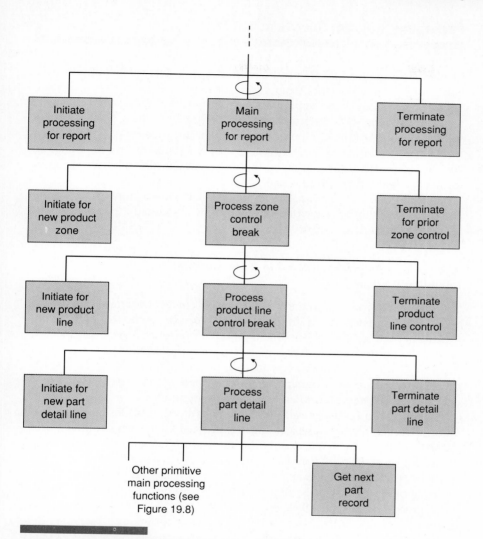

FIGURE 19.10 Data Structure Factoring For programs that produce outputs whose data structure is hierarchical, the MAIN PROCESSING can be factored according to that hierarchy (control breaks). Each of these data structure–oriented modules is flanked by modules that initiate and terminate the processing for that control break.

products include graphics capability for developing structure charts and a dictionary that supports the specifications of program module processing requirements. The structure chart and Structured English specification figures in the next section were created using the CASE product Excelerator.

Packaging Program Specifications

Using the design techniques presented in this unit, you have accumulated a good number of design specifications for the new system—perhaps you have separate stacks of documentation for the system outputs, files, database, inputs, terminal dialogue, systems flowcharts, and program modules. Now, put yourself in the shoes of the computer programmer. Are those specifications in a format that will help you write the programs? Not really. As a systems analyst, you are responsible for packaging that set of design documentation into a format suitable for the programmer.

As a direct result of modular design, you have a structure chart for the computer programs to be written. We can now package the program specifications around that structure chart.

What Does a Programmer Need in Order to Write Computer Programs?

You can't expect programmers to implement a correct program if they don't receive the necessary program specifications. You can avoid this problem by looking at packaging as salespeople look at their product. A good salesperson knows the product and knows how to sell it. In this section you're going to learn to look at a computer program as if it were a new product. Let's study the components of the program specifications package and see how the product can best be presented to end-users and programmers.

The program specifications package is a collection of design documentation that clearly communicates the requirements for each computer program in the system. What exactly are the requirements associated with a computer program? All programs perform three types of tasks, including input or reading of data, manipulation of input data, and output of data or information. In other words, all program tasks can be classified according to input, process, and output (IPO) requirements. This model will help us address the requirements for implementing a computer program.

Input Specifications   As a systems analyst, you are responsible for providing complete specifications of all sources of input data for each program. To the computer programmer, the term *input* has a broad meaning. Although batch and on-line inputs are still important, *input* also refers to file and database access. The specifications for all these program inputs have been documented as follows:

1. Master, transaction, and scratch file specifications (Chapter 13):
      Expanded data dictionary entries
      Record layout charts

2. Database specifications (Chapter 14):
    Expanded data dictionary entries
    Subschema
    Record layouts

3. Batch input file specifications (Chapter 16):
    Source document layout
    Expanded data dictionary entries
    Input record layout charts

4. On-line input specifications (Chapters 16 and 17):
    Source document layout
    Expanded data dictionary entries
    Display layout charts
    Terminal dialogue charts
    Prototype input screen

Processing Specifications   All programs execute processing tasks (such as sorting, summarizing, and calculating) on input data to produce outputs and information. These processing tasks are performed according to business policies and procedures. It is essential that the policies and procedures governing the processing tasks of a program be clearly explained to the programmer. A programmer can't implement a program that checks credit, for example, if the credit policies are not clear, accurate, and complete.

How complete should the processing requirements be? How close should the analyst come to specifying code? Generally, procedure specifications should represent a more general explanation of how the tasks of a program are to be accomplished. Program logic is intended to be much more detailed. For example, a procedure specification instruction might state

Sort the DAILY ORDERS FILE in ascending order by CUSTOMER ORDER NUMBER.

Alternatively, we could provide the pseudocode logic for an internal sort (Figure 19.11). Some programmers might like this detailed specification very much. Others might be offended by such a precise description. Where should the systems analyst draw the line? Many organizations have adopted standards that dictate exactly what the systems analyst must provide the programmer. In the absence of standards, perhaps analysts will simply document critical business formulas and decision rules. On the other hand, most systems analysts are required to simply provide the programmer with a clear and *concise* statement of the program processing requirements. The specifications for computer processing requirements may be given by:

Decision tables (covered in Chapter 11)

Structured English (covered in Chapter 11)

FIGURE 19.11
Pseudocode This is an example of pseudocode for a sorting requirement. The analyst should avoid this level of detail unless systems design standards call for it. This is the level of detail that the programmer would use to design logic.

Initialize the ORDER SORT array subscript X to 1.

For each record in the DAILY ORDER FILE, do the following:
Store DAILY ORDER FILE reccord in ORDER SORT array at subscript X location.

Add 1 to subscript X.

Initialize the IS SORT COMPLETE FLAG to "NO".

Initialize the RECORDS TO SORT variable equal to the subscript X.

Repeat the following steps until the SORT COUNTER variable equals X—1 or the SORT COMPLETE FLAG equals "YES":

Initialize SORT COMPLETE FLAG to "YES".

Subtract 1 from the RECORDS TO SORT variable.

Initialize the COMPARISON COUNTER to 1.

Repeat the following steps until the COMPARISON COUNTER is greater than the RECORDS TO SORT variable:

Calculate COMPARISON SUBSCRIPT using the following formula:

COMPARISON COUNTER + 1

If the CUSTOMER ORDER NUMBER for ORDER SORT array record at COMPARISON COUNTER location is less than the CUSTOMER ORDER NUMBER for ORDER SORT array record at COMPARISON SUBSCRIPT location, then:

Store ORDER SORT array record at location COMPARISON COUNTER in TEMPORARY STORAGE variable.

Store the ORDER SORT array record at location TEMPORARY STORAGE in ORDER SORT array at the COMPARISON COUNTER location.

Store the TEMPORARY STORAGE record in the ORDER SORT array at location COMPARISON COUNTER.

Set the IS SORT COMPLETED flag equal to "NO".

Add 1 to COMPARISON COUNTER variable.

Add 1 to SORT COUNTER variable.

Initialize the SORTED ORDER FILE COUNTER to 0.

For each record in the SORT ORDER array, do the following:
Store the SORT ORDER array record at the SORTED ORDER FILE COUNTER location in the SORTED ORDER FILE.

THE NEXT GENERATION

Automatic Code Generators—Will the Programmer Become Obsolete?

A continuing theme in these "Next Generation" features is the emergence of productivity tools for systems development. One of the great frustrations of the systems analyst occurs when carefully prepared program specifications don't get translated into the computer programs that were originally envisioned. This chapter has focused on tools and techniques that reorganize design specifications in a fashion suitable for programmers. Although it is true that improperly packaged and incomplete specifications are the cause of many program inadequacies, we don't intend to imply that the tools and techniques discussed in this chapter will eliminate the problem, although they will help. Will a future generation of tools and techniques promise to eliminate the problem? Perhaps!

It has long been suggested that programming, being a logical process, could be automated. In other words, we may be able to write programs that input (and insist on!) complete specifications and generate and test computer programs—all in a fraction of the time required by human programmers. And there are a few products that support this concept.

In this book, we have frequently referred to fourth-generation languages. You might consider them the answer to our problem. But although these end-user languages are good and getting better, they are not suitable for all information systems. They may be limited in *what* they can do. And they are frequently inefficient when compared with their third-generation language counterparts (such as COBOL, using conventional file organization techniques). Why? Prototypes are based on small files. When fully loaded files are installed and multiple end-users start accessing the system, the throughput and response time becomes unacceptable. Therefore, many prototype systems developed using fourth-generation languages are rewritten in languages such as COBOL after the prototype has been approved by end-users. Thus we want to address a different question: Is there any way to improve productivity when using such languages as COBOL, BASIC, FORTRAN, PL/1, and Pascal?

There is in the class we speak of. Higher Order Software (HOS), Inc., sells a product called USE.IT, which automates program design. The package forces you to recursively factor a program into binary (two) functions and subfunctions. The resulting structure can be translated by USE.IT into executable program code that HOS claims can be mathematically proven to be bug-free. Other products use Structured Design techniques to develop a program from a general idea by utilizing techniques such as Warnier/Orr. These programs can also generate usable code. We encountered a simpler approach at one *Fortune* 500 company. They have studied programs in their environment and developed programs that generate skeleton PL/1 code for functions they know to be needed in multiple applications.

In any case, we can expect to see more products of this type. They will become more sophisticated and will generate even more efficient program code. The impact on programmers is clear. We'll still need programmers to maintain operational systems and to enhance and customize the code from these program generators, but we will clearly need fewer programmers. That's no problem, though—we'll need more and better analysts because these program generators will be dependent on clear and complete program specifications.

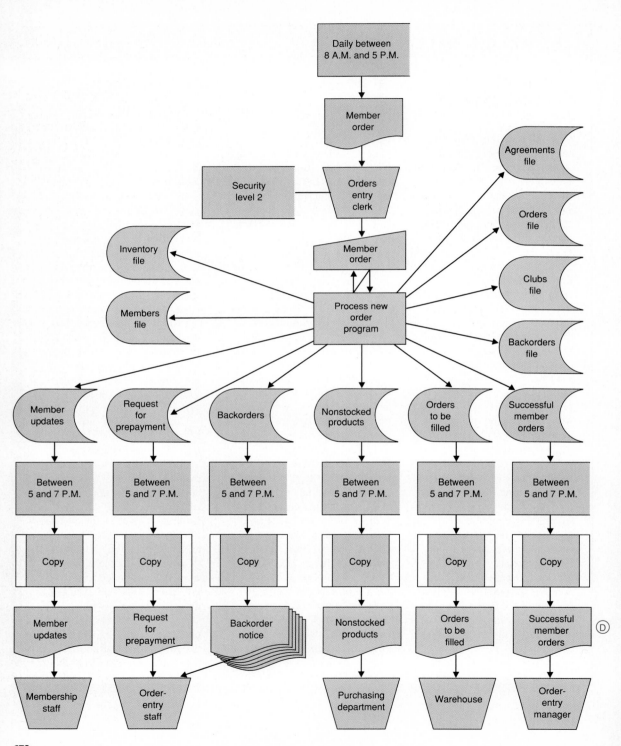

◄ **FIGURE 19.12**
Systems Flowchart A systems flowchart serves two purposes for program design. First, it identifies the programs to be designed. Second, it identifies the net inputs and outputs for that program. Notice the library programs. These programs don't normally have to be designed since they already exist.

In Figure 19.12, we've reproduced a systems flowchart from Chapter 18. The program PROCESS NEW ORDER is actually a subprogram in an on-line system. The partial structure chart for the on-line system is illustrated in Figure 19.13. This structure will be used to organize our program specifications.

Using Structured English, you would specify each module in the structure chart. Figure 19.14 represents the Structured English specifications for a module of the PROCESS NEW ORDER program. While we chose to use Structured English (and decision tables when appropriate) to present processing requirements, you could use alternative procedures or logic tools, such as program flowcharts.

Due to size constraints on the book, we cannot provide you with all of these detailed design specifications for each module on the structure chart appearing in Figure 19.13.

Output Specifications   Along the same lines as *inputs*, the term *outputs* means more to the programmer than we have suggested in this unit. In addition to printouts, forms, and displays, outputs include updates to files and databases. Output requirements include

1. Printed output specifications (Chapter 15):
   Expanded data dictionary entries
   Printer spacing charts
   Prototype reports

2. On-line output specifications (Chapter 15):
   Expanded data dictionary entries
   Display layout charts
   Terminal dialogue charts
   Prototype output screens

3. Master and transaction files updated specifications (Chapter 13):
   Expanded data dictionary entries
   Record layout charts

4. Database specifications (Chapter 14):
   Expanded data dictionary entries
   Subschema
   Record layouts

This concludes our discussion of program design and packaging. Try packaging a sample computer program.

Summary

The systems analyst's role in computer program design includes module design and packaging of design specifications. A module is defined as single entry–single exit group of instructions that performs a single function. To

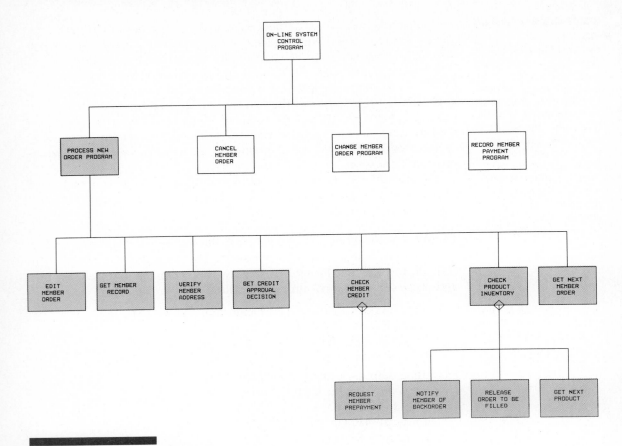

FIGURE 19.13 Structure Chart

deal with complexity of logic, programmers tend to break programs into modules. A better strategy is to break programs into code to deal with *functions*.

There are two popular tools for documenting modular structure, Warnier/Orr bracket charts and structure charts. Although often sold as distinct tools, they are actually quite similar. There are also two strategies for modular

```
TYPE Process                        NAME EDIT MEMBER ORDER

    Label EDIT              EXPLODES TO ONE OF:
          MEMBER           Data Flow Diagram
          ORDER            Structure Chart
                           Structure Diagram


    Location               ORDER = ENTRY STAFF

    Process Category       ON-LINE

    Duration Value         500
    Duration Type          DAY

    Manual or Computer C


    Description
FOR ALL INVALID MEMBER ORDER DATA FROM THE END = USER, DO THE FOLLOWING:
    SELECT THE APPROPRIATE CASE:

    CASE 1:  MEMBER NUMBER IS INVALID
             IF THE MEMBER NUMBER IS NOT EQUIVALENT TO A MEMBER NUMBER
             OF AN EXISTING MEMBER THEN:
                 ERROR MESSAGE = "MEMBER DOES NOT EXIST, PLEASE REENTER"
    CASE 2:  ORDER NUMBER IS INVALID
             IF THE ORDER NUMBER = ORDER NUMBER OF A PREVIOUS ORDER THEN:
                 ERROR MESSAGE = "ORDER NUMBER WAS ASSIGNED TO PREVIOUSLY
                                  ENTERED MEMBER ORDER, PLEASE REENTER"
    CASE 3:  ORDER DATE IS INVALID
             SELECT APPROPRIATE CASE:
                 CASE 2.1 ORDER DATE CONTAINS NO VALUES, THEN:
                     ERROR MESSAGE = "ORDER DATE MUST BE PROVIDED"
                 CASE 2.2 ORDER DATE CONTAINS INVALID MM/DD/YY VALUES,
                          THEN:
                     ERROR MESSAGE = "THE ORDER DATE IS INVALID,
                                      PLEASE REENTER"
    CASE 4:  PREPAID AMOUNT > TOTAL MEMBER ORDER COST, THEN:
             ERROR MESSAGE = "PREPAID AMOUNT EXCEEDS TOTAL COST OF MEMBER
                              ORDER, PLEASE REENTER"
    CASE 5:  PRODUCT NUMBER NOT VALID, THEN:
             ERROR MESSAGE = "INCORRECT PRODUCT NUMBER (DOES NOT EXIST),
                              PLEASE REENTER"
    CASE 6:  MEDIA CODE IS INVALID, THEN:
             ERROR MESSAGE = "MEDIA CODE IS INVALID, PLEASE RE-ENTER OR
                              PRESS F2 KEY FOR VALID CODES AND THEIR
                              MEANINGS"
    CASE 7:  QUANTITY ORDERED IS NOT > ZERO, THEN:
             ERROR MESSAGE = QUANTITY ORDERED MUST BE GREATER THAN ZERO,
                             PLEASE REENTER"
    CASE 8:  MEMBER RESPONSE STATUS IS NOT Y OR N, THEN:
             ERROR MESSAGE = "ENTER Y (YES) OR N (NO) IS REGARD TO
                              MEMBER'S ACCEPTANCE OF SELECTION OF
```

FIGURE 19.14   Structured English   Each module appearing on the structure chart in Figure 19.13 would be documented to communicate processing requirements.

design. The Yourdon/Constantine approach suggests that modules be defined by studying the flow of data between primitive functions. The Warnier/Orr approach suggests that modules be defined by studying the data structure of the outputs and inputs. This chapter presented a hybrid approach based on the two strategies.

Structure charts and Warnier/Orr charts adequately factor each program into manageable modules that can be assigned to programmers. IBM's Hierarchy plus Input-Process-Output (HIPO) provides a documentation tool for packaging detailed input, processing, and output details around the structure chart. In addition to the structure chart and input-process-output charts, the analyst must assemble the various data dictionaries, spacing charts, layouts, and systems flowcharts needed to present a complete program specification for the programmer.

Problems and Exercises

1. Obtain a copy of the documentation for a completed programming assignment. Study the program's source code to identify all referenced modules. Using a Warnier/Orr diagram and a structure chart, document the existing modular structure implemented by the program.

2. Study the processing requirements for the program you used for Exercise 1. Using the modular design approach suggested in this chapter, develop a new structure chart for the program. Compare the structure chart with the one derived in Exercise 1. Which would you prefer to work from as a programmer? Why?

3. Some typical initiate, processing, and terminate functions were presented in Figures 19.5 and 19.8. Can you identify other processing functions that might be included in the lists? How would you classify them—as initiate, processing, or terminate functions? Explain why.

4. Prepare a systems flowchart for the program in Exercise 1. Use the structure chart (from Exercise 2) and the systems flowchart to prepare an input-process-output chart or charts to package the program.

5. What value would an existing structure chart and an input-process-output chart of an existing program be to a systems analyst during the study phase?

6. What value would an existing structure chart and an input-process-output chart for a program be to a programmer who has to maintain the program?

7. What programs appearing on a systems flowchart does the systems analyst need to package? What correlations can be drawn between the systems flowchart and the input-process-output chart?

8. What are the transaction centers in the following program?

An on-line program allows an end-user to perform inquiries to obtain information concerning customer accounts, orders, invoices, and products. The end-user is allowed to obtain general information concerning an order or information about specific orders that have been placed on back order. The end-user who wishes to obtain information concerning orders placed on backorder may request information describing orders that have been backordered for less than one week, backordered for more than one week but less than two weeks, or backordered for more than a two-week period. The end-user may also perform inquiries to retrieve general information about a specific part and information concerning backordered parts.

Projects and Minicases

1. The Computer Information Systems Department at Northern Fence, Inc., is currently evaluating program design methods. Specifically, Northern is evaluating the structured design (Yourdon/Constantine) and the data structures (Warnier/Orr) methods. The CIS Department is planning to apply the two program design methods to a number of programming projects in order to more accurately evaluate each method. Once the projects are completed, the CIS Department believes it will be able to determine which method should be adopted as a company standard. Do you feel the CIS Department should view the methods as an either/or issue? If not, explain why not. How can the two methods be integrated?

Annotated References and Suggested Readings

Adams, David R., Gerald E. Wagner, and Terrence J. Boyer. *Computer Information Systems: An Introduction.* Cincinnati, Ohio: South-Western Publishing, 1983. We are indebted to this book's suggestion that most programs can be factored into initiate, main process, and terminate functions (Chapter 8). We adapted this approach to provide a high-level framework into which the various modular design strategies can be integrated.

Boehm, Barry. "Software Engineering." *IEEE Transactions on Computers,* vol. C-25, December 1976. This paper, a classic, described the logarithmic relationship between time and the cost to correct an error in the systems specifications.

Orr, Kenneth T. *Structured Systems Development.* New York: Yourdon Press, 1977. Although somewhat dated, this is still our favorite book on the Warnier/Orr method of modular design. It is short, easy to read, and contains numerous examples. It also contains an easy-to-read, albeit somewhat negative, discussion of the HIPO documentation technique.

Page-Jones, Meiler. *A Practical Guide to Structured Systems Design.* (2nd ed.) New York: Yourdon Press, 1988. This is our favorite book on the Yourdon/Constantine method of modular design, a methodology called *Structured Systems Design.* It is easy to read and contains numerous examples of both transform and transaction analysis. The concepts of coupling and cohesion are also covered in greater detail. This book is must reading for both analysts and programmers.

CHAPTER TWENTY

How to Implement and

Evaluate a

New Information System

Beck Electronic Supply

Tim Stallard is a systems analyst at Beck Electronic Supply. He has only been a systems analyst for 6 months. Unusual personnel turnover had thrust him into the position after only 18 months as a programmer. Now it's time for his semiannual job performance review. He enters the office of Ken Delphi, Assistant Director of MIS.

"Another six months!" exclaimed Ken. "It hardly seems that long since your last job performance review."

"I personally feel very good about my progress over the last six months," Tim replied. "This new position has been an eye-opener. I didn't realize that analysts do so much writing. I enrolled in some continuing education writing classes at the junior college. The courses are helping . . . I think."

"I wondered what you did. It shows in everything from your memos to your reports!" Ken commented. "More than any technical skills, your ability to communicate will determine your long-term career growth here at Beck. Now, let's look at your progress in other areas. Yes, you've been supervising the Materials Requirements Planning (MRP) project implementation for the last few months. This is your first real experience with the entire implementation process, right?"

"Yes. You know, I was a programmer for 18 months. I thought I knew everything there was to know about systems implementation. But this project has taught me differently."

"How's that?" Ken asked.

"The computer programming tasks have gone smoothly. In fact, we finished testing the entire system of programs six weeks ahead of schedule."

Ken interrupted, "I don't mean to interrupt, but I just want to reaffirm the role your design specifications played in accelerating the computer programming tasks. Bob has told me repeatedly that he had never seen such thorough and complete design specifications. The programmers seem to know exactly what to do."

"Thanks, Ken. That really makes me feel good. It takes a lot of time to prepare design specifications like that, but I think that it really pays off during implementation. Now, what was I going to say? Oh yes. Even though the programming and testing were completed ahead of schedule, the system still hasn't been placed into operation; it's two weeks late."

Ken replied, "That means you lost the six-week buffer plus another two weeks. What happened?"

"Well," answered Tim, "I'm to blame. I just didn't know enough about the nonprogramming activities of systems implementation. First, I underestimated the difficulties of training. My first-draft training manual made too many assumptions about computer familiarity. My end-users didn't understand the instructions, and I had to rewrite the manual. I also decided to conduct some training classes for the end-users. My instructional delivery was terrible, to put it mildly. I guess I never really considered the possibility that, as a systems analyst, I'd have to be a teacher. I think I owe a few apologies to some of my former instructors. Do you have any idea how much time goes into preparing for a class?"

Ken smiled, "Yes, especially when you're technically oriented and your audience is not."

Tim continued, "Anyway, that cost me more time than I had anticipated. But there are still other implementation problems that have to be solved. And I didn't budget time for them!"

"Like what?" asked Ken.

"Like getting data into the new files. We have to enter several thousand new records. And to top it off, management is insisting that we operate the new system in parallel with the old system for at least two months. Then, and only then, will they be willing to allow the old system to be discarded."

"Well, Tim, I think you're learning a lot. Obviously, we threw you to the wolves on this project. But I needed Bob's [Tim's mentor] experience and attention elsewhere. I knew when I pulled Bob off the project that it could introduce delays—I call it the *rookie factor.* Under normal circumstances, I would never have let you work on this alone. But you're doing a good job and you're learning. We have to take the circumstances into consideration. You'll obviously feel some heat from your end-users because the implementation is behind schedule, and I want you to deal with that on your own. I think you can handle it. But don't hesitate to call on Bob or me for advice. Let's talk about some training and job goals for the next six months. What do you think . . ."

Discussion

1. Above and beyond programming, what tasks do you think make up systems implementation? Can you think of any tasks that weren't described in this minicase?

2. Why is training so difficult? How do you feel about the prospects of becoming a "teacher"? How long do you think it takes to prepare for one hour of classroom instruction? What tasks do you think would be involved in preparing a lesson plan?

3. A 3,000-record master file must be created for a new system. Each record consists of 15 fields. The record length is 200 bytes. How long do you suppose it would take to create that file? If necessary, use your own typing speed as a performance gauge. What factors would affect how long it may take to get the file up and running?

4. What assumption did Tim make about transition from the old system to the new system? Why was it wrong? Can you think of any circumstances under which it would be correct?

What Will You Learn in This Chapter?

In this chapter you will learn about two systems implementation phases: (1) the construction of the new information system and (2) the delivery of the new information system. You will also learn about postimplementation review of the system. You will know that you understand the systems implementation process when you can

1. Define *systems implementation* and relate the term to the construction and delivery phases of the life cycle.
2. Describe the construction and delivery phases of the life cycle in terms of
   a. Purpose and objectives
   b. Tasks and activities that must or may be performed
   c. Sequence or overlap between tasks and activities
   d. Techniques used
   e. Skills you must master to perform the phase properly
3. Explain how the time spent on systems implementation can be managed.

Systems implementation is the fulfillment or carrying out of the design specifications to put a new information system into operation. In this chapter, we will study systems implementation. We will examine two systems implementation phases: the construction of the new information system and the delivery of the new information system.

Figure 20.1 illustrates these phases in terms of the systems development life cycle. For each of these two phases, we will study the *purpose* and *objectives,* specific *tasks* and *activities* that should be performed, and important *skills* to be mastered. As we did in Chapters 6 and 12, we will carefully build on the concepts and models that you learned in Part One.

You might be wondering why a chapter on systems implementation is included in a systems analysis and design book. First and foremost, implementation is part of the systems development life cycle you have been studying throughout this book. Second, as a systems analyst, you can expect to be directly involved in many, if not all, of the implementation activities. Finally, we want you to understand the close working relationship between the systems analyst and the computer programmer.

How to Construct the New Information System

Construction of the new information system is a phase that is familiar to most of you. This is the phase in which most of the computer programming occurs. Why not just call it *programming*? Because there's more to construction than programming! Although that may or may not surprise you, you'll soon understand the full implications of the construction phase.

Purpose and Objectives of the Construction Phase

The purpose of the **construction phase** in the life cycle is to build a working information system from the design specifications prepared during the design phase. To achieve the purpose of the construction phase, we must accomplish the following objectives.

Construct or Install Information Systems Components   In this phase we construct or install the hardware, software, and data storage components of the new system. Actually, *construct* may not be the best term for what happens during this phase. *Construct* implies building from scratch. Although many programs are built from scratch, others may actually be modified to fulfill new requirements. And still other software is installed. This is particularly true of purchased software. Those packages often must be customized, but it beats writing the programs from scratch.

Furthermore, hardware components are not constructed. New hardware was chosen during the acquisition phase of our life cycle. Now, during the construction phase, this hardware is installed so that the programmers can utilize it to construct the system. Note that the programmers may have to learn how to use new hardware and software packages before they can construct or modify the system.

Still, the most familiar and time-consuming activity is computer programming and maintenance. But do any other elements in the IPO components model have to be constructed or installed? Absolutely! Preprinted forms conceived during the design phase have to be physically printed. This is usually accomplished through a specialized forms manufacturer. Along similar lines,

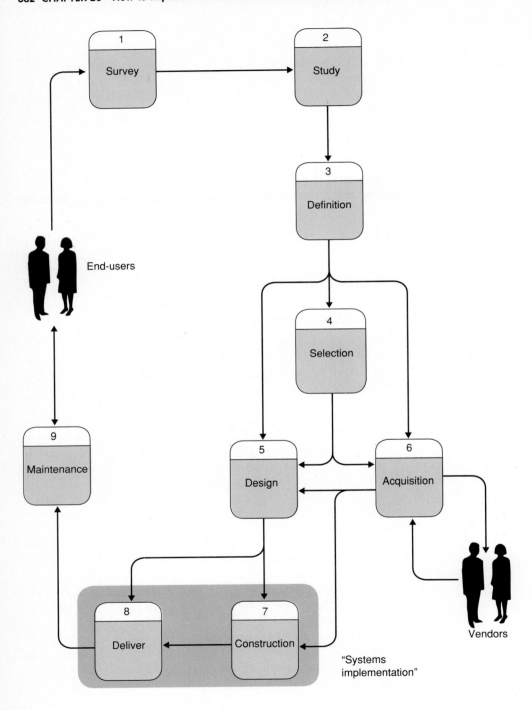

◄ **FIGURE 20.1 The Systems Implementation Process** Systems implementation is defined as the construction and delivery phases of your systems development life cycle.

stocks of supplies—for instance, diskettes, tapes, logbooks, and furniture—may have to be built up.

Implement Designed Information Systems Functions   It is very important that the programmer not stray from the design specifications without the analyst's approval. Why? Because the design specifications were prepared to fulfill end-user requirements. A change in these specifications could throw the system out of synchronization with the end-user's requirements. The computer programmer often has little or no direct contact with those end-users. Therefore, during the construction phase, the analyst is primarily concerned that the information system fulfills the transaction-processing, management reporting, and decision support capabilities specified during the definition and design phases.

Involve End-Users in Pertinent Construction Activities   You have already learned the importance of end-user involvement during the study, definition, selection, acquisition, and design phases of our life cycle. What about the construction phase? For most systems development, computer programmers will work alone to develop, debug, and test computer programs to fulfill design specifications. But, as we've said on several occasions in this book, a new class of friendly fourth-generation programming languages are generally considered to be more user friendly. The net result will be twofold:

1. Systems analysts will replace programmers during the construction phase of many projects. The analyst's familiarity with the project combined with the accelerated program development commonly experienced with 4GLs will enable systems to be constructed with greater efficiency.

2. End-users will become directly involved in the construction phase. Although analysts and programmers will implement the more difficult transaction and file maintenance programs, the end-users will be able to construct the management reporting and decision support programs. All 4GLs offer easy-to-learn and easy-to-use report generators and query facilities that can be directly applied by end-users.

How to Complete the Construction Phase

In this section, we want to identify and discuss the specific tasks, documentation, and skills required for the construction phase. Figure 20.2 illustrates the construction phase tasks in greater detail. As usual, note the overlap between tasks. As we further examine each task and its key documents, we will identify those skills you need to develop to perform the task.

Some of these tasks are specific to computer programming. Although we will identify those tasks and their associated skills, we will direct you to your programming curriculum for training in these subjects. We want only to place computer programming into the context of the systems development life cycle, of which it is a part.

Task: Install Hardware and Software (If Necessary)   Recall from Chapter 12 that the new information system may be built around new or upgraded technology that was ordered during the acquisition phase. A net output of the acquisition phase, the hardware/software configuration for the new system, is a net input in this task. Installation is frequently a two-step process.

The first step may be site preparation. New equipment may require a special environment—for instance, air-conditioning, furniture, printing rooms. While the hardware/software order is being filled, the analyst can work with computer center management or end-user management to prepare the site for installation of the technology. An example should drive the point home.

A new on-line system may specify that there will be an IBM 3179G terminal on each clerk's desk. Now is the time to plan for that delivery. The 3179G is a rather large terminal, and each clerk must make space for it. It is possible that now is the time to consider specialized furniture that can ergonomically handle the terminal and its cable connections. Telephone lines may need to be installed if the computer is at a remote site. In the computer center itself, a new controller may have to be installed to handle the new terminals. This controller will take up space. Where should it be located? These are issues to be decided before the equipment is delivered.

The second step is to install the delivered hardware and software. Hardware is normally installed and tested by the vendor (even microcomputer stores frequently offer this service). Software may or may not be installed by the vendor. In any case, specialists called *systems programmers* or *technical support staff* often become involved in the installation, testing, and modification of such software as operating systems, database management systems, word processors, spreadsheets, telecommunications software, and other general-purpose software packages.

Although site preparation and installation are not typically performed by systems analysts, they are included here as a construction phase task because construction cannot proceed without the required technology. On the other hand, the installation of new hardware and software usually goes hand in hand with the need to train programmers, operators, and end-users about how to use the new technology.

Task: Plan for Programming   Recall that an implementation plan was generated as the final task of the design phase. The implementation plan, included in the design specifications, specified systems test data and a schedule for the construction and delivery phases. This plan is rarely detailed enough to begin constructing the new system. Therefore, our first task is to plan the programming effort. The refined plan should include the following:

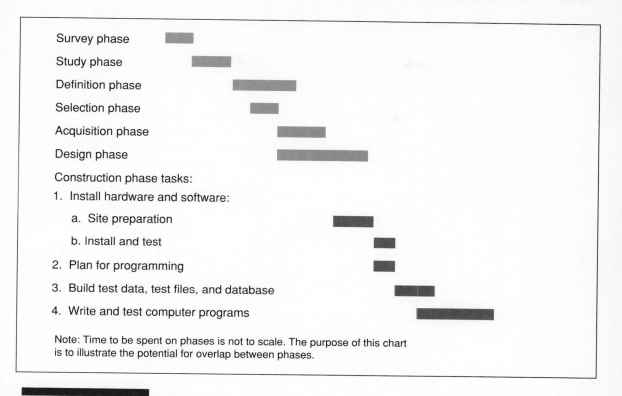

Survey phase

Study phase

Definition phase

Selection phase

Acquisition phase

Design phase

Construction phase tasks:

1. Install hardware and software:

    a. Site preparation

    b. Install and test

2. Plan for programming

3. Build test data, test files, and database

4. Write and test computer programs

Note: Time to be spent on phases is not to scale. The purpose of this chart is to illustrate the potential for overlap between phases.

FIGURE 20.2   Construction Phase Tasks   The construction phase can be broken down into the tasks suggested on this chart.

- *Review of the Design Specifications.* One major controversy that should be addressed is the freezing of the design specification. By **freezing,** we mean that changes to the design specifications are discouraged or prohibited. Why? With no discouragement against changes, end-users will continually be permitted to identify something they forgot or some new need or idea, and the system may never be constructed and delivered.

On the other hand, some experts dispute the idea of freezing the specifications. They argue that such an action is artificial and is not consistent with our goal to serve the end-user.

Both sides are right! We suggest that you tentatively freeze the document. If changes are proposed, ask yourself a simple question: "Is this a critical change that will make or break the system, or is it an enhancement that could be added later?" Critical changes require the specifications document to be modified. Do it! If the change isn't critical, log the change in an enhancements file. We'll discuss this file later.

- *Organization of the Programming Team.* Most large programming projects require a team effort. One popular organization strategy is the use of **chief programmer teams**. The team is managed by the chief programmer, a highly proficient and experienced programmer who assumes overall responsibility for the program design strategy, standards, and construction. The chief programmer oversees all coding and testing activities and helps out with the most difficult aspects of the programs.

  Other team members include a backup chief programmer, program librarian, programmers, and specialists. The backup programmer is able to assume the chief programmer's role as well as to perform normal programming activities. The program librarian maintains the program documentation and program library. The programmers, often selected because of specialized programming skills relevant to the project, code and test the programs. Specialists offer unique skills pertinent to the project (such as database techniques, telecommunications background).

  Where does the analyst fit in? That depends on the organization. Sometimes the analyst is the project manager to whom the chief programmer reports. At other times, the analyst is a consultant to the chief programming team, possibly as one of the specialists reporting to the chief programmer. Chief programming teams are formed and disbanded with each successive project.

- *Development of a Detailed Construction Phase Plan.* You don't just start programming. Most design specifications include numerous programs. Which programs should be written first? Many systems are built in **versions**. The first version implements the most critical aspects of the system, so that a version can be placed into operation before the system has been completely constructed.

  Another appropriate approach is to construct transaction-processing programs first. Implement these programs in the same sequence as that in which they would have to be run (NEW ORDER PROCESSING before ORDER CANCELATION before BILLING and so on). Then implement management reporting and decision support programs according to their relative importance. General file maintenance and backup and recovery programs are written last.

Task: Build Test Data and Test Files and Databases   Building test files and databases is a task unfamiliar to many students, who are accustomed to having an instructor provide them with the test data and files. This task must immediately precede other programming activities because files and databases are the resource shared by the computer programs to be written.

Test files and databases are constructed quickly using valid test data that conforms to project dictionary values specified during analysis and design. The files should be large enough to test various transactions and report capabilities but small enough so they can be quickly created and tested. These are not final files or databases, and the data should be representative but not necessarily real. Test files and databases are usually constructed by using editors and software utilities that are available on most computers.

To learn to appreciate this skill, commit yourself to initially preparing your own test data and test files when you receive your next programming assignment.

Task: Write and Test Computer Programs The major activity of the construction phase is the writing and testing of computer programs. And this is the activity with which many of you have the most experience. This book is not intended for a programming course. However, we'd like to summarize at least one appropriate computer program development cycle for comparison with our systems development life cycle. An appropriate program development life cycle is illustrated in Figure 20.3.

This particular program development life cycle begins with a review of program structure. By program structure, we mean the top-down, modular factoring of the program. This was covered in the last chapter. Some information systems shops insist that top-down, modular design is the programmer's responsibility. When that is the case, this first step could be changed from *review* to *design*.

The program development life cycle depicted in Figure 20.3 suggests a top-down implementation—that is, modules are designed, coded, and tested beginning with the top module. The upper-level modules *drive* the lower-level modules. Each module goes through three stages of development: algorithm design, coding, and testing. As modules are completed, they are integrated and tested. Eventually, the entire program is completed and tested as a whole. Each program for the new system goes through this cycle. After all of the programs have been individually coded and tested, the entire set of all programs that have been developed for the system are integrated and tested as a system.

Testing is an interesting skill that is often overlooked in academic courses on computer programming. If modules are coded top-down, they should be tested and debugged top-down *and as they are written*. Testing is not an activity to be deferred until after the entire program has been completely written! There are three levels of testing to be performed:

1. *Stub Testing.* Stub testing is the test performed on individual modules, whether they be main program, subroutine, subprogram, block, or paragraph. How can you test a higher-level module before coding its lower-level modules? Easy! You simulate the lower-level modules. These lower-level modules are often called *stubs*. Stub modules are subroutines, paragraphs, and the like that contain no logic. Perhaps all they do is print that they have been correctly called, and then control goes back to the parent module.

2. *Unit or Program Testing.* All of the modules that have been coded and stub tested are tested as an integrated unit. Eventually, all modules will have

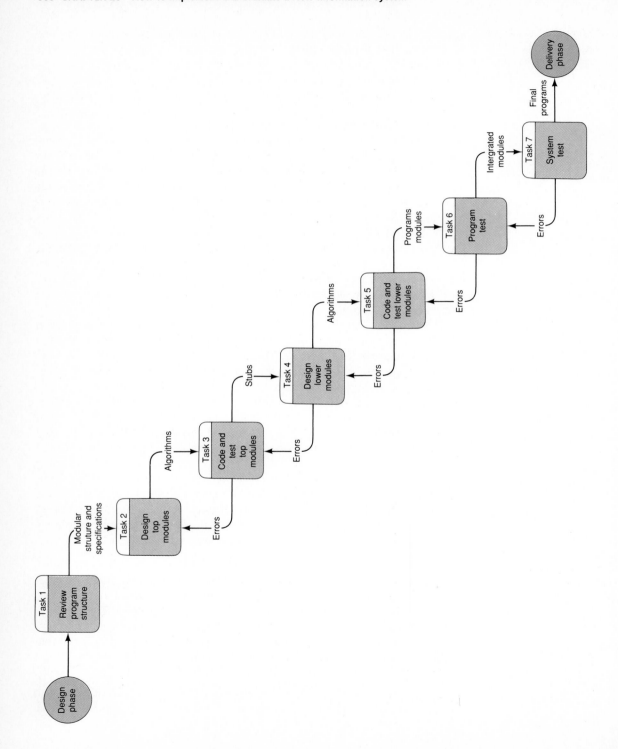

◀ **FIGURE 20.3 A Program Development Life Cycle** Like the systems development life cycle, there are many versions of a program development life cycle. This is one example. This approach suggests a top-down strategy. The high-level modules are built first.

been implemented, and that unit equals the program itself. Unit testing uses the test data that was created during the design phase.

3. *Systems Testing.* Just because each program works properly doesn't mean that the programs work *together* properly. The integrated set of programs should be run through a systems test to make sure that one program properly accepts, as input, the output of other programs.

We'll talk about additional tests when we discuss the delivery phase. Computer programming activities are frequently governed by information systems standards. These standards dictate program design, coding, testing, and documentation rules that are intended to promote a consistent style within all information systems. These standards are often subject to design and code walkthroughs that check the program for conformity to standards, as well as for logic and design errors.

Some information systems shops have a Quality Assurance Department staffed by specialists who check program documentation—and systems analysis and design documentation—for conformity to standards.

Programming skills and structured programming are beyond the scope of this book. As we conclude this section, we'd like to leave you with two opinions. You cannot be an effective systems analyst without computer programming experience—that experience helps you appreciate the importance of thorough systems design. Finally, the use of structured programming techniques results in programs that are easier to write, read, and (especially!) maintain. Learn structured programming, and learn how to prepare design specifications that are thorough. This will allow the programmers time to apply structured programming techniques!

Finally, we remind you that the construction of computer programs may be replaced or supplemented by the installation of purchased software. These packages were chosen during the selection phase. Now they may have to be modified or customized to fulfill your end-user and design requirements (specified in earlier phases). This modification to the construction phase is illustrated in Figure 20.4. Task 5 is the installation of the applications software package and task 6 is the modification of the software package.

How to Deliver and Evaluate the New System

Now we come to the last development phase in our life cycle: Deliver the new information system. The analyst is the principal figure in the delivery phase, regardless of his or her role in the construction phase. Let's set the stage for this phase.

From the definition and selection phases, we know what parts of the new system are manual and what parts are computerized. From the design phase, we know how all inputs, outputs, and procedures are implemented. And from

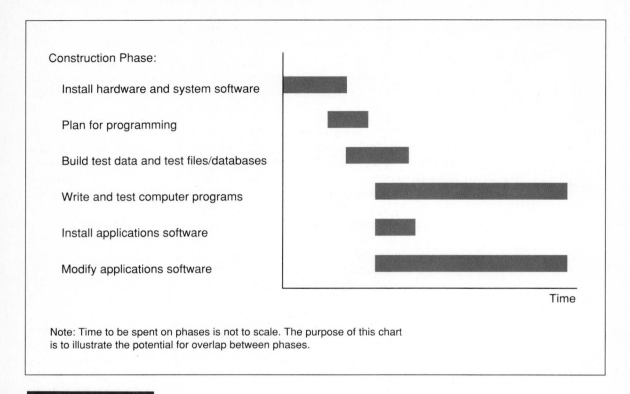

Construction Phase:

Install hardware and system software

Plan for programming

Build test data and test files/databases

Write and test computer programs

Install applications software

Modify applications software

Time

Note: Time to be spent on phases is not to scale. The purpose of this chart
is to illustrate the potential for overlap between phases.

FIGURE 20.4 A Modified Construction Phase This version of the construction
phase tasks reflects the installation and modification of software packages—as
opposed to writing programs in-house.

the construction phase, we will eventually have working hardware and software to support the new system. Now we must put the new system into
operation.

Purpose and Objectives of the Delivery Phase

The purpose of the delivery phase is to smoothly convert from the old system
to the new system. To achieve the purpose of the delivery phase, we must
accomplish the following objectives.

Train and Support End-Users The system is *for* the end-user! End-user
involvement is important in the delivery phase because the end-users will
inherit your successes and failures from this phase. Fortunately, end-user
involvement during this phase is rarely overlooked. The most important aspect
of their involvement in the delivery phase is training and advising of the end-

users. They must be trained to use equipment and to follow the procedures required of the new system. But no matter how good the training is, your end-users will become confused at times. Or perhaps they will find mistakes or limitations (an inevitable product despite the best of analysis, design, and implementation techniques). The analyst will help the end-users through the learning period until they become more familiar and comfortable with the new system.

Evaluate the Project and System   At the end of the delivery phase, the analyst should grade the system. How should it be graded? A number of criteria are possible. Did the system cost more or less than its budgeted development cost? Although this may affect the payback period for the system, the costs cannot be recovered. Did the system come in on schedule? This is good for a pat on the back, but if you didn't meet the deadline, you cannot get schedule overruns back. What's done is done!

You can only learn from your mistakes. The best criterion was, and still is, does the system fulfill the business mission? The system's grade relative to this criterion will determine the lifetime costs and usefulness of the system. If the system is successful, you will spend less time and money fixing it!

Make Smooth Transition to New Methods and Procedures Possible   The ways in which people perform their jobs are changing. The conversion from the old system to the new system must be carefully addressed. This applies not only to information systems methods and procedures but also to manual methods and procedures.

The key point is that the new system represents *change,* possibly dramatic. You should expect apprehension on the part of some of the end-users involved. No matter how involved and enthusiastic they have been, people are naturally apprehensive about change. The current system, however flawed it may be, represents something that the end-users understand—and familiarity breeds content. If you've done a good job of analysis, design, and implementation, end-user apprehension will not last long, and your reputation will be secure. Just remember, when you do something right, people don't always remember it, at least not for long. But when you do something wrong, people tend *not* to forget. Teachers and analysts know this better than most people.

How to Complete the Delivery Phase

Now we can discuss the specific tasks, documentation, and skills that make up the delivery phase. The basic activities of the delivery phase are illustrated in Figure 20.5. Let's look at each activity in greater detail.

Task: Install Files and Databases   During the construction phase, you built test files and test databases. But to place the system into operation, you need

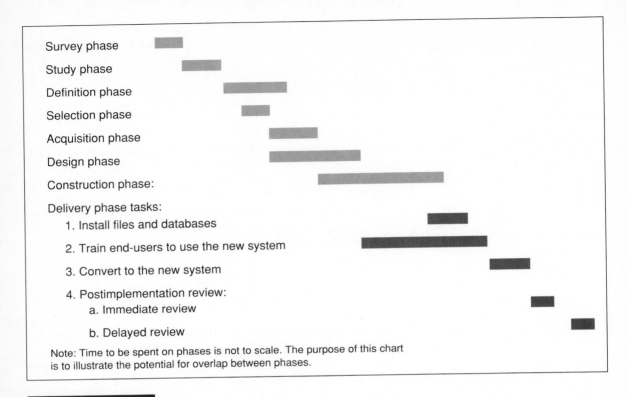

FIGURE 20.5   **Delivery Phase Tasks**   The delivery phase can be broken down into the tasks suggested on this chart.

fully loaded files and databases. Therefore, the first task we'll survey is installation of files and databases. At first, this activity may seem trivial. But consider the implications of loading a typical file, say, CUSTOMER ACCOUNT. Tens or hundreds of thousands of records may have to be loaded. Each must be input, edited, and confirmed before the file is ready to be placed into operation.

For computer files that are being converted to computer file structures or databases, the basic method is still complex. A special program must be written to read the old file and write to the new file, using the new structure. Then, if new fields were added to any of the records, additional programs must be written to initialize those fields, although you may be able to use the file maintenance programs mentioned in the paragraph. File conversion is an activity performed by data-entry personnel and computer programmers, not systems analysts.

As a systems analyst, you should calculate file and database sizes and estimate the time required to perform the task of installing them. The task itself is usually performed by data-entry personnel because end-users cannot release themselves for enough time to complete this task. Sometimes, temporary help must be hired for this one-time installation effort.

Task: Train End-Users to Use the New System A task more typically performed by systems analysts is to train end-users to use the new system. There are at least two fundamental training requirements: (1) training manuals and (2) the training itself. Many organizations hire special systems analysts who do nothing but write user documentation and training guides. If you have a skill for writing clearly, the demand for your services is out there! Figure 20.6 is a typical outline for a training manual. The golden rule should apply to user manual writing: "Write unto others as you would have them write unto you." You are not a business expert. Don't expect the reader to be a technical expert. Every possible situation and its proper procedure must be documented.

The actual training is built around the user manuals. Training can be performed one-on-one. However, group training is generally preferred. It is a better use of your time and it encourages group learning possibilities. Think about your education for a moment. Isn't it true that you really learn more from your fellow students and colleagues than from your instructors? Instructors facilitate learning and instruction, but you master specific skills through practice with large groups because, with them, common problems and issues can be addressed more effectively. Take advantage of the ripple effect of education. The first group of trainees can then train several other groups.

Related to user manuals and training is the complementary preparation of computer operations manuals and training. These manuals instruct computer operators on how to carry out information systems procedures as documented in systems flowcharts.

Once again, written and oral communications skills are critical. Familiarity with organizational behavior and psychology may also prove helpful. The delivery phase represents *change*, and people have a natural tendency to resist change or to look for fault in change. There is comfort in the status quo—even if the current system is fraught with problems.

Task: Convert to the New System Conversion to the new system from the old system is a significant milestone. After conversion, the ownership of the system *officially* transfers from the analysts and programmers to the end-users. Four conversion or installation strategies are commonly used:

1. *Abrupt Cutover.* On a specific date, usually a date that coincides with an official business period such as month, quarter, or fiscal year, the old system is terminated and the new system is placed into operation.

FIGURE 20.6   An Outline for a Training Manual   A good training manual or procedures manual can prevent many problems during the lifetime of a system. This is one outline for such a manual.

Training Manual End-Users Guide Outline

I. Introduction

II. Manual

    A. The manual system (a detailed explanation of peoples' jobs and standard operating procedures for the new system)

    B. The computer system (how it fits into the overall workflow)
        1. Terminal/keyboard familiarization
        2. First-time end-users
            a. Getting started
            b. Lessons

    C. Reference manual (for nonbeginners)

III. Appendixes

    A. Error messages

This is a high-risk approach because there may still be major problems that won't be uncovered until the system has been in operation for at least one business period. On the other hand, there are no transition costs. Abrupt cutover may be necessary if, for instance, a government mandate or business policy becomes effective on a specific date and the system couldn't be implemented prior to that date.

2. *Parallel Conversion.* Under this approach, both the old and new systems are operated for some period of time. This is done to ensure that all major problems in the new system have been solved before the old system is discarded. The final cutover may be either abrupt (usually at the end of one business period) or gradual, as portions of the new system are deemed adequate.

Obviously, this strategy minimizes the risk of major flaws in the new system causing irreparable harm to the business. But it also means that the cost of running two systems over some period of time must be incurred. Because running two editions of the same system on the computer could place an unreasonable demand on computing resources, this may only be possible if the old system is largely manual.

3. *Location Conversion.* When the same system will be used in numerous geographical locations, it is usually converted at one location (using either

abrupt or parallel conversion). As soon as that site has approved the system, it can be farmed to the other sites. Other sites can be cut over abruptly because major errors have been fixed. Furthermore, other sites benefit from the learning experiences of the first test site. Incidentally, the first production test site is often called a *beta test site*.

4. *Staged Conversion.* Like location conversion, staged conversion is a variation on the abrupt and parallel conversions. A staged conversion is based on the version concept introduced earlier. Each successive version of the new system is converted as it is developed. Each version may be converted using the abrupt, parallel, or location strategies.

What happens during the systems conversion? Training may occur, but we factored training out as a separate task that should begin well before systems conversion. The major activity is the **systems acceptance test**. At this point, we should differentiate between what we called a *systems test* earlier in the chapter and the *systems acceptance test* performed here.

The systems test was performed by programmers using test data. The systems acceptance test is performed by end-users using real data over an extended period of time. The systems acceptance test should be extensive. There are three levels of acceptance testing:

1. *Verification testing* runs the system in a simulated environment using simulated data. This simulated test is sometimes called *alpha testing*. The simulated test is primarily looking for errors and omissions regarding end-user and design specifications that were specified in the earlier phases but not fulfilled during construction.

2. *Validation testing* runs the system in a live environment using real data. This is sometimes called *beta testing*. During this validation, we are testing a number of items, including

    • *Systems Performance.* Is the throughput and response time for processing adequate to meet a normal processing workload? If not, some programs may have to be rewritten to improve efficiency, or processing hardware may have to be replaced or upgraded to handle the additional workload.

    • *Peak Workload Processing Performance.* Can the system handle the workload during peak processing periods? If not, we may have to improve hardware and/or software to increase efficiency or rethink our scheduling of processing—that is, consider doing some of the less critical processing during nonpeak periods.

    • *Human Engineering Test.* Is the system as easy to learn and use as anticipated? If not, is it adequate? Can enhancements to human engineering be deferred until after the system has been placed into operation?

    • *Methods and Procedures Test.* During conversion, the methods and procedures for the new system will be put to their first real test. Methods and procedures may have to be modified if they prove to be awkward and inefficient from the end-users' standpoint.

- *Backup and Recovery Testing.* Now that we have full-sized computer files and databases with real data, we should test all backup and recovery procedures. Simulate a data loss disaster, and test the time required to recover from that disaster. Also, do a before-and-after comparison of the data to ensure that data was properly recovered. It is crucial to test these procedures—don't wait until the first disaster to find an error in the recovery procedures.

3. *Audit testing* certifies that the system is free of errors and is ready to be placed into operation. Not all organizations require an audit. But many firms have an independent Audit or Quality Assurance staff that must certify a system's acceptability and documentation before that system is placed into final operation. There are independent companies that perform systems and software certification for end-users' organizations.

The systems acceptance test is the final opportunity for end-users, management, and information systems operations management to accept or reject the system. Hopefully, the analysts and programmers are well aware of the criteria for acceptance before this stage. Well-established systems objectives, systems requirements, and information systems and EDP audit policies are important if rework is to be minimized or eliminated.

As end-users and management uncover errors, the programs and procedures may have to be slightly modified. It might be useful to have regular support or maintenance programmers perform these modifications. This will enable us to test the quality of the program documentation while the programs are fresh in the minds of the original programmers.

As the system conversion progresses, the end-users will undoubtedly suggest enhancements. These enhancements should be added to the enhancements file we introduced earlier in the chapter.

Task: Postimplementation Review   The final task of the delivery phase is the postimplementation review. This task is sometimes called the *systems audit.* The review is intended to accomplish two goals:

1. Evaluate the operational information system that was developed.

2. Evaluate the systems development procedures to determine how the project could have been improved.

This is the easiest phase to skip. And that would be a major mistake. True, there are other projects waiting to be started or finished. But you have to learn to look at the long-term benefits of the task. How will you ever do a better job of systems analysis and design if you don't evaluate your current performance?

Evaluation of the new information system should not be done until some period of time after the conversion has been completed. The following elements should be reviewed:

- Does the new information system fulfill the goals and objectives identified and refined early in the project?
- Does the system adequately support the transaction-processing, management reporting, and decision support requirements of the business?
- Are the projected benefits being realized?
- How do the end-users feel about the new system? How can end-user relations be improved for future projects?
- Should any of the proposed enhancements to the system be addressed immediately? Enhancements should be prioritized.
- Are the internal controls adequate?

Ideally, these questions should be addressed at least twice. The first review should occur as soon as possible after the system has been placed into operation, since everything is fresh in everybody's mind at this point. The second review should occur after some reasonable operating period has elapsed—say, six months after delivery. By this time, we are out of the implementation phases and into the support and maintenance phases of the life cycle. Answers to the same questions may be very different after the newness of the system has worn off. Isn't that true of cars, houses, and other durable goods? Some organizations require annual or periodic reviews of all operational systems.

Evaluation of the systems development procedures should address the following:

- *Did the system come in under budget?* If not, why not? How could estimating procedures be improved? What factors should be taken into consideration when making future estimates?
- *Did the system come in on time?* If not, why not? How can estimating procedures be improved? How can procedures be improved to accelerate systems development efforts without sacrificing quality?
- *How did each person perform during the project?* What skills could be improved? What skills will be improved on the next project?
- *How did the project leader perform?* What leadership traits and skills could be improved? What improvement will the project leader address on the next project?

As you can see, the review process is a question-and-answer session intended to benefit future projects.

That concludes our survey of the construction and delivery phases. Figure 20.7 places the tasks into better perspective using a data flow diagram. The processes are tasks. The data flows are documentation passed from task to task.

THE NEXT GENERATION

The Role of CASE in Systems Operation

During the course of systems analysis, design, and implementation, we assemble a great deal of documentation that needs to be maintained for the benefit of those who will maintain, enhance, and eventually replace the new system. Contrary to popular belief, prototyping will not signal the end for all systems development documentation—it will merely reduce that documentation. Getting a grip on the volume of documentation generated during the project can be very difficult and frustrating. But help is on the way.

Throughout this book we've discussed the role of computer-assisted systems engineering (CASE) technology in systems development. But its potential extends beyond development and beyond implementation. In this box feature, we'd like to review and expand on that concept.

By way of review, a CASE workstation is a microcomputer-based set of software tools intended to enhance analyst productivity and integrate the myriad documentation associated with a project. Examples include Index Technology's Excelerator and Nastec's Design 2000, both of which run on IBM Personal Computers. These products include CAD/CAM graphics systems for drawing

data flow diagrams, systems flowcharts, program flowcharts, program structure charts, Warnier/Orr charts, database models such as entity relationship diagrams, and the like. That, in and of itself, is a productivity aid because these diagrams are subject to constant modification and redrawing. The workstations make changing the diagram no more difficult than using a word processor to change a document.

But the real power of the workstation is the integration of all such tools (including the graphics) with a central data dictionary to organize project details. The dictionary records both the requirements and the implementation decisions. It can also be used to record very detailed specifications for the working system. And therein lies additional value.

The CASE workstation can help maintenance programmers keep track of enhancement requests and changes made in response to those requests. As programming staffs change, there will be a point of reference for replacements. Implications of changes may be more carefully evaluated in light of their impact on the total system.

CASE will be ultimately linked with the final programs generated for the system. When this happens, programmers will

be more inclined to maintain specifications as changes are made. Ideally, we would see a day when the study phase of our life cycle is obsolete because a well-maintained specification of the existing system has been properly updated over time.

How about some real "blue-sky" projections? CASE researchers are currently investigating object-oriented programming techniques. Programs may be generated directly from object-oriented graphs and dictionaries. The implications are awesome. If programs can be automatically generated from systems models, then systems could be maintained through changes to those models. Think about it—systems maintained by their own documentation!

How to Manage the Time Spent on Implementation

We'll conclude this survey of the systems implementation process with a brief discussion of how much time should be spent on systems implementation. The amount of time that should be spent on implementation is a function of the following factors:

- The size of the project (as was the case with the analysis and design phases).
- The quality and completeness of the design specifications and the requirements statement.
- The techniques used to write computer programs—for instance, prototyping or Structured Design and programming.
- The type of system conversion used—for example, abrupt, parallel.

The second factor—quality and completeness of the design specifications and the requirements statement—is particularly important. Considerable empirical evidence suggests that the cost of fixing errors grows exponentially during the phases of system development (Boehm, 1976). For instance, an error that would cost $8 to fix in the study phase could cost $60 to fix in the design phase, $300 to fix in the construction phase, and $9,000 to fix after construction. Given the inflationary trend in analyst and programmer salaries since Boehm's research was published, these costs are not overstated! Thus, the quality of the work done in each phase dramatically affects all subsequent phases and the lifetime costs of the system.

At one time, it was not unusual for systems implementation to consume 75 percent or more of the total project schedule. But improvements in programming methods and technology—for example, Structured Programming and Structured Design)—have reduced these requirements. Programmers have still found themselves spending 60 percent or more of the total time for implementation, though. Why? Because systems analysis and design methods are often less rigorous in practice than they are in courses and books. Programmers have constantly found themselves backtracking to fulfill new requirements and design specifications not documented by the analysts. But you are now equipped with knowledge and skills to change that—and establish your own reputation for quality.

What effect will prototyping have on the time spent? As the prototyping trend continues, we see a reversal of the trend toward less time on the implementation phases. This is because prototyping is construction intensive. But the increased time spent on construction is often offset by a dramatically reduced time requirement for the analysis and design processes—particularly design, which is consolidated with construction when prototyping! The net result is much faster systems development and better systems.

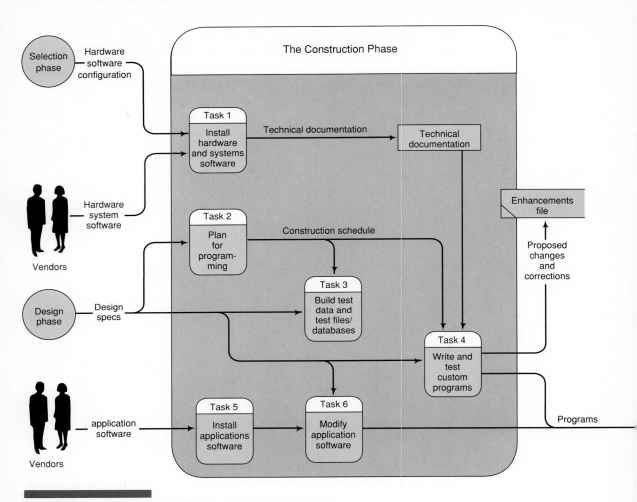

FIGURE 20.7 Data Flow View of Systems Implementation

FIGURE 20.7 (continued)

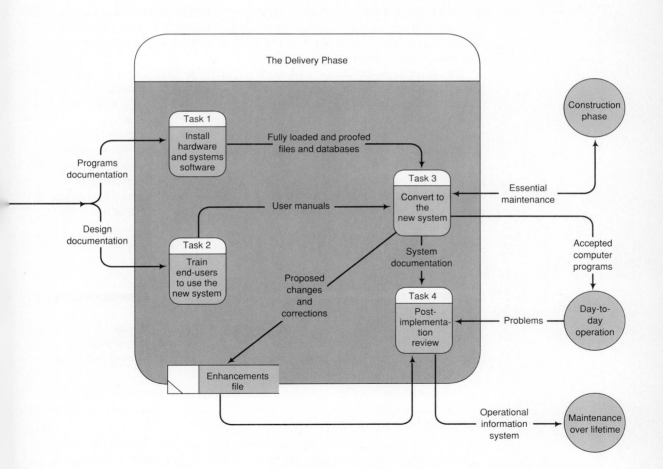

Summary

Systems implementation is the process whereby a new information system is placed into operation. Systems implementation consists of two phases: construction and delivery.

The purpose of the construction phase is to create the hardware and software elements for the new system. For the hardware elements, this may involve site preparation and installation. Site preparation includes both planning space for equipment and creating a favorable environment, involving factors such as temperature, humidity, and noise control. Installation includes taking delivery of new equipment, placing the equipment into operation, and making sure the equipment is functioning properly. For software elements, the construction phase includes computer programming. Not all programs are created from scratch. Some programs are modified. And those programs that were purchased—instead of being built in-house—are installed, tested, and, if necessary, modified.

The purpose of the delivery phase is to place the new system into operation. Delivery includes end-user training, computer staff training, file conversion and loading, and systems conversion. Of these, training is the most important; systems conversion is the most time consuming. Systems conversion is the process whereby a smooth transition is made from the current information system to the newly constructed information system. Strategies for systems conversion include *abrupt cutover, parallel conversion, site conversion,* and *staged conversion.* After the new system has been placed into operation, both the system and the project should be reviewed. Postimplementation reviews identify improvements—to both the system and the systems development process.

Problems and Exercises

1. How can a successful and thorough systems analysis and design be ruined by a poor systems implementation? How can poor systems analysis or design ruin a smooth implementation? For both questions, list some implementation consequences.

2. What skills are important during systems implementation? Create an itemized list. Identify computer, business, and general education courses that would help you develop or improve those skills.

3. How does your information systems pyramid model aid in systems implementation? Examine each face of the pyramid, and address issues and relevance to the systems implementation phases: construction and delivery.

4. What products of the systems design phases are used in the systems implementation phases? Why are they important? How are they used? What would happen if they were incomplete or inaccurate?

5. What are the end products of the construction and delivery phases? Explain the purpose and content of each of those products.

6. Explain how time spent on systems implementation can best be managed. Take a computer program or system that you constructed in a programming course. Put together a complete schedule to deliver that program or system. Write your schedule in the format of a memo to management. (Remember, management has a limited knowledge of computer vocabulary.)

7. What implications does the use of fourth-generation languages have for the construction phase?

8. How would the construction phase differ if the computer software for supporting the system was purchased?

9. Why should a systems analyst perform a postimplementation review? What types of benefits can be derived? Why do you really need two reviews?

10. What are four strategies commonly used to convert over to a new system? Describe situations for which each approach would be preferred and required.

Annotated References and Suggested Readings

Boehm, Barry. "Software Engineering." *IEEE Transactions on Computers,* vol. C-25, December 1976. This classic paper demonstrated the importance of catching errors and omissions before programming begins.

Metzger, Philip W. *Managing a Programming Project.* 2nd ed. Englewood Cliffs, N.J.: Prentice-Hall, 1981. This excellent book covers, in considerable detail, what we refer to as the construction and delivery phases. It is one of the few books to place emphasis solely on the implementation process.

PART FOUR

Skills That

Overlap Systems

Analysis and Design

Project Management Tools and Techniques

Fact-Finding Techniques

Communications Skills for the Systems Analyst

The modules of Part Four are not appendixes! We want to state that fact right away so that there is no confusion. There are a number of skills and issues that are important to all three major systems development phases: analysis, design, and implementation. We feel strongly that these comprehensive skills are more important than any specific tool or technique you learn from this book. In fact, they may very well be the ultimate critical success factor for all systems work. So why are they at the end of the book?

If we had placed these modules—a name chosen to distinguish them from the chapters you've been reading—in either the analysis unit (Part Two) or the design and implementation unit (Part Three), it would have understated their value relative to the unit that didn't contain that material. On the other hand, these modules do require some prerequisite knowledge—in most cases, at least Chapters 1 through 5.

By locating these modules at the end of the book, we give you and your instructor the flexibility of introducing them at your own preferred locations, after specific chapters. Each module begins by describing prerequisite chapters.

We cannot overstate the value of these modules. The material presented will have greater impact on your success as a systems analyst than any other material in the book. The modules teach skills that make the concepts, tools, and techniques in Parts One through Three easier to apply.

Module A introduces **project management tools and techniques**. All projects are dependent on the planning, control, and leadership principles that are surveyed. The module also presents two popular modeling techniques for project management: PERT and Gantt. These tools help you schedule activities, evaluate progress, and modify schedules.

Module B surveys **fact-finding techniques**. These techniques are used to solicit factual information, opinions, and requirements from end-users. They are a crucial prerequisite to all modeling techniques—for example, data flow diagrams, entity relationship data models, and so forth. Techniques surveyed include form sampling, research, observation, questionnaires, interviews, and group work sessions.

Feasibility and Cost-Benefit Analysis Techniques

Module C presents **communications skills for the systems analyst**. After you've collected facts and modeled systems, you must be able to present your findings and recommendations. In this module, you will learn how to plan and run meetings, conduct brainstorming sessions, conduct walkthroughs of documentation, make oral presentations, and write reports.

Finally, Module D introduces **feasibility and cost-benefit analysis techniques**. For any potential solution that you evaluate and possibly recommend to management, you must be prepared to defend its operational, technical, and economic feasibility. Especially important is economic feasibility, since most organizations are either profit-oriented, cost reduction-oriented, or both. Your ability to estimate costs and benefits and then analyze those numbers for cost-effectiveness is critical.

MODULE A

Project Management Tools and Techniques

When Should You Read This Module?

This module will prove most valuable if read after *any* of the following chapters:

- Chapter 4, "A Systems Development Life Cycle"—The life cycle itself is a project management tool.
- Chapter 6, "How to Analyze an Information System: A Problem-Solving Approach"—Even if you've already read this module, a review may help you put the systems analysis tasks in perspective.
- Chapter 12, "How to Design an Information System: Traditional and Prototyping Approaches"—Again, even if you've already read this module, a review will put the content of Chapter 12 in perspective.
- Chapter 20, "How to Implement and Evaluate a New Information System"—A review would again be appropriate.

Most of you are familiar with Murphy's law, which suggests that if anything can go wrong, it will. Murphy's law has motivated numerous pearls of wit and wisdom about projects, machines, and people, and why things go wrong. Unfortunately, many amusing laws, postulates, and theorems have been conceived from our failures as project and people managers. And although it's fun to laugh at our mistakes and shortcomings, we should never take them so lightly that we accept them as facts of life. Why? Because, in addition to basic analysis and design responsibilities, the systems analyst frequently assumes a project management role. The project manager is usually a senior systems analyst who must plan, staff, and control the project's many tasks.

The purpose of this module is to introduce you to project management. Because project management is applied during systems analysis, systems design, *and* systems implementation, we decided to place it in this supplementary module. You will learn about the importance of project management as well as guidelines, tools, and techniques for managing projects.

What Is Project Management?

For the systems project, **project management** is the process of directing the development of an acceptable system at a minimum cost within a specified time frame. Although the tools and techniques of systems analysis and design play a critical role in achieving successful systems, these methods are not sufficient on their own. Project mismanagement can deter or render ineffective the best of analysis and design methods.

What goes wrong in systems projects? What role does mismanagement play in these failures? How can the project manager avoid such failures? We'll answer these questions in this section.

What Goes Wrong in Systems Projects?

We can develop an appreciation for the importance of project management by studying the mistakes of other project managers. Failures and limited successes far outnumber very successful information systems. Why is that? True, many systems analysts and data processors are unfamiliar with or undisciplined in the tools and techniques of systems analysis and design. But that only partially explains the shortcomings of systems projects. Many projects suffer from poor leadership and management. Witness the following case.

The Anatomy of an Unsuccessful Project   Let's listen in on a postimplementation review for an information systems project:

JAN: Steven, I want to discuss the registration system project your team completed last month. Now that the system has been operational for a few weeks, we need to evaluate the performance of you and your team. Frankly, Steven, I'm a little disappointed.

STEVEN: Me too! I don't know what happened! We used the standard methodology and tools, but we still have problems.

JAN: Well, I've talked to several of the analysts, programmers, and end-users on the project, and I've drawn a few conclusions. Obviously, the end-users are less than satisfied with the system. You took some shortcuts in the methodology, didn't you?

STEVEN: We had to! We got behind schedule. We didn't have time to follow the methodology to the letter.

JAN: But now we have to do major parts of the system over. If you didn't have time to do it right, where will you find time to do it over? You see, Steven, systems development is more than tools, techniques, and methodologies. It's also a management process. In addition to your missing the boat on end-user requirements, I note two other problems. And both of them are management problems. The system was over budget and late. The projected budget of $35,000 was exceeded by 42 percent. The project was delivered 13 weeks behind schedule. Most of the delays and cost overruns occurred during programming. The programmers tell me that the delays were caused by rework of analysis and design specifications. Is this true?

STEVEN: Yes, for the most part.

JAN: Once again, those delays were probably caused by the shortcuts taken earlier. The shortcuts you took during analysis and design were intended to get you back on schedule. Instead, they got you further behind schedule when you got into the programming phase.

STEVEN: I see that now. But the project grew. How would you have dealt with the schedule slippage during analysis?

JAN: If I were you, I would have reevaluated the scope of the project when I first saw it changing. In this case, either project scope should have been reduced or project resources—schedule and budget— should have been increased. . . . (pause) . . . Don't be so glum! We all make mistakes. I had this very conversation with my boss seven years ago. You're going to be a good project manager. That's why I've decided to send you to this project management course and workshop . . .

This case highlighted the three common results of mismanaged projects:

- Unfulfilled or unidentified requirements
- Cost overruns
- Late delivery

These problems are not always caused by project mismanagement, but mismanagement certainly plays a role in such problems.

The Project Management Causes of Failed Projects What project management failures cause unfulfilled requirements and needs, cost overruns, and late deliveries? Before we get into project mismanagement causes, let's recognize that one possible cause of these problems could be inadequate systems analysis and design tools and techniques. But for purposes of this module, we will focus on causes that can be traced to project management.

In Steven's registration project, end-user requirements and needs were not fulfilled because shortcuts had been taken during the project. The systems development life cycle (discussed in Chapters 4, 5, 6, 12, and 20) provides a basic plan for a systems project. Each phase and activity is an important part of that plan. For all parts to work together, the life cycle must be monitored and managed!

Another common cause of unfulfilled requirements is lack of or imprecise targets. The problem in systems projects is that during the early phases, the scope of the project is rarely precise. And for many projects, the scope is never precisely defined. If the project leader fails to recognize this problem, the project team is frequently forced to make late changes to specifications and programs. As a result, the system doesn't meet requirements. And to further compound the problem, unless you know where you are going, you can't estimate how long it will take to get there and how much it will cost. That brings us to budget overruns.

What causes projects to exceed their budgets? One of the major problems with cost overruns is that many methodologies or project plans call for an unreasonably precise estimate of costs before the project begins. These estimates are made after a quick preliminary study or feasibility study. Think about it! Can you accurately estimate project costs before making a detailed study of the current system or defining end-user requirements? Can you estimate the costs of computer programming before a detailed systems design has been completed? It's not likely. The cost estimates of a project will change as you get further into the systems development process.

Poor estimating techniques are another cause of cost overruns. We suspect that many systems analysts estimate by making the best calculated estimate (*guesstimate?*) and then double that number. This is hardly a scientific approach. There are better approaches available; some useful techniques are discussed in Module D.

And finally, cost overruns are often caused by schedule delays. What causes the delays? Once again, we can point to premature estimates as a problem. These early estimates are based on the initial scope of the project. Because systems analysts—and information systems professionals in general—are eternal optimists, they often quote optimistic schedules and fail to modify those schedules as the true scope of the project becomes apparent.

Because many managers and analysts are often poor time managers, project schedules slip slowly but steadily. "So we've lost a day or two! It's no big deal. We can make it up later." This may be true, but then again, it might not. They fail to recognize the fact that in the systems development life cycle certain tasks are dependent on other tasks. Because of these dependencies, a one-day slip can set the whole schedule back. And when those one-day delays pile up, we inevitably find ourselves working 15-hour days at the end of the project.

Another cause of missed schedules is what Brooks (1975) has described as the *mythical man-month*. As the project gets behind schedule, the project leaders frequently try to solve the problem by assigning more people to the project team. It just doesn't work! There is no linear relationship between time and number of personnel. The addition of personnel creates more communications and political interfaces. The result? The project gets even further behind schedule.

You've probably noticed that the causes of failed projects are related. For instance, missed requirements may cause schedule slippages that, in turn, cause cost overruns. You might ask why somebody isn't able to recognize these problems and correct them. Somebody should. And that person is supposed to be the project manager or leader. Which brings us to a major cause of project failure: lack of management and leadership. Good computer programmers don't always go on to become good analysts. Similarly, good analysts don't automatically perform well as managers and leaders. To be a good project manager, the analyst must possess or develop skills in the basic functions of management.

The Basic Functions of the Project Manager

The project manager is *not* just a senior analyst who happens to be in charge. As Steven found out, a project manager must apply a set of skills different from those applied by the analyst. What skills must the project manager possess or learn? The basic functions of a manager or leader have been studied and refined by management theorists for many years. These functions include planning, staffing, organizing, scheduling, directing, and controlling.

Planning Project Tasks and Staffing the Project Team A good manager always has a plan. The manager estimates resource requirements and formulates a plan to deliver the target system. This is based on the manager's understanding of the requirements of the target system at that point in its development. A basic plan for developing an information system is provided by the systems development life cycle. Many firms have their own standard life cycles, and some firms have standards for the methods and tools to be used.

Each task required to complete the project must be planned. How much time will be required? How many people will be needed? How much will the task cost? What tasks must be completed before other tasks are started? Can some of the tasks overlap? These are all planning issues. Some of these issues can be resolved with a tool called the PERT chart, which is discussed later in this module.

Project managers frequently select analysts and programmers for the project team. The project manager should carefully consider the business and technical expertise that may be needed to successfully finish the project. The key is to match the personnel to the required tasks that have been identified as part of project planning.

Organizing and Scheduling the Project Effort Given the project plan and the project team, the project manager is responsible for organizing and scheduling the project. Members of the project team should understand their own individual roles and responsibilities as well as their reporting relationship to the project manager.

The project schedule should be developed with an understanding of task time requirements, personnel assignments, and intertask dependencies. Many projects present a deadline or requested delivery date. The project manager must determine whether a workable schedule can be built around such deadlines. If not, the deadlines must be delayed or the project scope must be trimmed. We will soon introduce a project scheduling tool called the Gantt chart.

Directing and Controlling the Project   Once the project has begun, the project manager becomes a leader. As a leader, the manager directs the team's activities and evaluates progress. Therefore, every project manager must demonstrate such people management skills as motivating, rewarding, advising, coordinating, delegating, and appraising team members. Additionally, the manager must frequently report progress to superiors.

Perhaps the manager's most difficult and important function is controlling the project. Few plans will be executed without problems and delays. We've already discussed the causes and effects of unsuccessful projects. The project manager's job is to monitor tasks, schedules, and costs in order to control those elements. If the project scope is increasing, the project manager is faced with a decision: Should the scope be reduced so the original schedule and budget will be met, or should the schedule and budget be revised? The project manager must be able to present the alternatives and their implications for the budget and schedule to the steering committee or management.

Space doesn't permit us to fully discuss these project management functions. But we will leave you with a reference to one small, inexpensive gem— Blanchard and Johnson's *One Minute Manager* (1982), one of the shortest but most powerful books ever written on how to direct and control subordinates. See the references for a full citation.

Next, we will introduce some of the more popular project management tools that are used to assist project managers.

Project Management Tools and Techniques

Two popular tools used by project managers are PERT charts and Gantt charts. PERT charts are most useful for project planning and modification; Gantt charts, for project scheduling and progress reporting.

Today, computer software such as Harvard's Project Manager, ABT's Project Management Workbench, and Microsoft's Project are being used to support project managers. Also, some CASE products provide useful interfaces between project dictionaries and project management software. For example, Excelerator and the Project Management Workstation can exchange common data.

PERT Chart: A Planning and Control Tool

PERT, which stands for *Project Evaluation and Control Technique,* was developed in the late 1950s to plan and control large weapons development projects for the U.S. Navy. It was developed to make clear the interdependence of project tasks when projects are being scheduled. Essentially, PERT is a graphic networking technique. Let's take a closer look at PERT charts, what they are, how to draw them, and how to use them.

You should be made aware that different PERT discussions and computer products use different terminology.

PERT Definitions and Symbols   On PERT charts, projects can be organized in terms of **events** and **tasks.** An event—also called a milestone—represents a point in time, such as the start or completion of a task or tasks.

A variety of symbols—circles, squares, and the like—have been used to depict events on PERT charts. For our discussion, we will use small circles, often called *nodes* (see Figure A.1), to represent an event. Each node is divided into three sections. The left half of the node includes an event identification number. This number is usually keyed to a legend that explicitly defines the event. The upper and lower right-hand quarters of the node are used to record the earliest and latest completion times for the event. Instead of dates, time is counted from TIME = 0, where 0 corresponds to the date on which the project is started. Every PERT chart has one beginning node that represents the start of the project and one end node that represents the completion of the project.

On a PERT chart, a task (also called an activity) is depicted by an arrow between nodes. The task identification letter and the expected duration of the task are recorded on the arrow. Look at the arrow between event 1 and event 2 in Figure A.1. The direction of the arrow indicates that event 1 must be completed prior to event 2. The expected duration of the task resulting in the completion of event 2 is four days. A dashed arrow is special. It represents a dependency between events. However, because there is no activity to be performed, there is no duration between the events. This is called a *dummy task.*

Estimating Project Time Requirements and Deriving the PERT Chart

Before drawing a PERT chart, you must estimate the time needed for each project task. The PERT chart will be used to indicate the estimated earliest finish and latest finish time for each event and the expected duration of each project task. Although these times are often expressed in terms of *person-days,* this approach is not recommended. Why? Because there is no proven linear relationship between project completion time and the number of

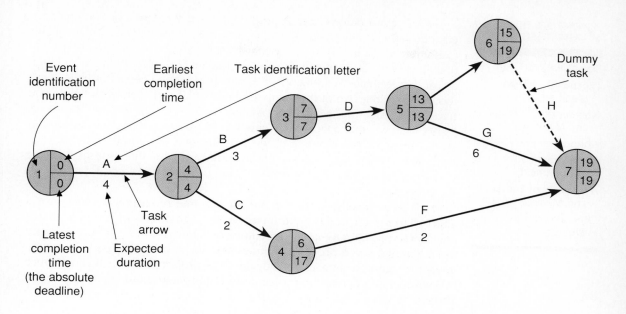

Event identification number

Earliest completion time

Task identification letter

Dummy task

Latest completion time (the absolute deadline)

Task arrow

Expected duration

slack time for event = latest completion time - earliest completion time

FIGURE A.1   PERT Notation   A PERT chart depicts events and tasks. Nodes represent events. Arrows represent tasks. Each event node includes an ID number, earliest completion time, and latest completion time. Each task arrow includes an ID letter and expected duration time estimate. A dashed arrow is a dummy task showing a dependency between tasks although no time is required.

people assigned to a project team. Many systems projects that were late have been further delayed by assigning additional personnel to them. Because two people can do a job in four days is no reason to assume that four people can do the same job in two days. We suggest that time be expressed in *calendar-days,* given the number of people assigned to the task.

Unfortunately, we cannot offer you a set of formulas to use to derive time requirements for any task. You must estimate these project time requirements. By *estimate* we don't mean that you simply make something up. A good systems analyst project manager draws on experience and data from previous projects. Some of the factors that might influence estimates are listed in the margin. There exist CASE products like SPQR/20 that help managers make better time estimates.

Other organizations have developed internal standards for deriving project time estimates in a more structured manner. These standards may involve examining a task in terms of its difficulty, skill requirements, and other identifiable factors. Alternatively, you could make an optimistic estimate and then adjust that estimate quantifiably by applying weighting factors to various criteria, such as the size of the team, the number of end-users with whom you have to interact, the availability of those end-users, and so on (Weinberg, 1979). Each weighting factor may either increase or decrease the estimate. As we go through this discussion, we'll try to give you some guidelines for estimating. When you begin to make estimates, seek the counsel of more experienced analysts (the more the better) until you are comfortable with the process.

Let's assume we're to derive the project time requirements and draw a PERT chart for a typical programming project. The project involves constructing a large program to update an employee master file. The project manager has identified seven program routines that are to be delegated to programmers for coding, testing, and debugging. The project planning table in Figure A.2 was used to derive the project time requirements and construct a PERT chart. Five steps are required:

1. *Make a list of all project tasks and events.* The first two columns in Figure A.3 provide an identification letter and a description for each task. The completion of a task is assigned an event identification number, which is entered into the third column.

2. *Determine intertask dependencies.* For each task, record the tasks that must be completed before and after the task in question is completed (columns 4 and 5).

3. *Estimate the duration of each task.* This can be determined as follows:
   a. Estimate the minimum amount of time it would take to perform the task. We'll call this the optimistic time (OT). The optimistic time estimate assumes that even the most likely interruptions or delays—such as occasional employee illnesses—will not happen.
   b. Estimate the maximum amount of time it would take to perform the task. We'll call this the pessimistic time (PT). The pessimistic time estimate assumes that anything that can go wrong will go wrong. All possible interruptions or delays—such as labor strikes, illnesses, training, inaccurate specification of requirements, equipment delivery delays, and underestimation of the systems complexity—are assumed to be inevitable.
   c. Estimate the most likely time that will be needed to perform the task (MLT). Don't just take the median of the optimistic and pessimistic times. Attempt to identify interruptions or delays that are likely to occur, such as occasional employee illnesses, inexperienced personnel, and occasional training.

Task ID	Task Description	Event ID Number	Preceding Event	Succeeding Event	Expected Duration	Earliest Finish	Latest Finish
A	CODE, TEST, AND DEBUG ROUTINE "A010 UPDATE MASTER FILE"	2	1	3	3	3	3
B	CODE, TEST, AND DEBUG ROUTINE "B010 INITIATE PROCESSING"	3	2	4	2	5	5
C	CODE, TEST, AND DEBUG ROUTINE "B020 PROCESS TRANSACTION"	4	3	5,6,7,8	2	7	7
D	CODE, TEST, AND DEBUG ROUTINE "C210 ADD EMPLOYEE RECORD"	5	4	8	7	14	14
E	CODE, TEST, AND DEBUG ROUTINE "C220 MODIFY EMPLOYEE RECORD"	6	4	8	6	13	14
F	CODE, TEST, AND DEBUG ROUTINE "C230 DELETE EMPLOYEE RECORD"	7	4	8	3	10	14
G	CODE, TEST, AND DEBUG ROUTINE "B030 TERMINATE PROCESSING"	8	4,5,6,7	9	2	14	14
H	COLLECTIVELY TEST AND DEBUG PROGRAM	9	8	NONE	5	19	19

FIGURE A.2 Project Planning Table A project planning table is used to prepare data for drawing a PERT chart. The table will also serve as a legend for the chart.

    d. Calculate the expected duration (ED) as follows:

$$ED = \frac{OT + (4 \times MLT) + PT}{6}$$

This commonly used formula provides a weighted average of the various estimates. The formula is based on experience and may be modified to reflect project history in any firm. Expected duration is recorded in column 6 of Figure A.2.

4. *Derive the earliest and latest completion times (ECT and LCT) for each task* as follows:

    a. The ECT for event n is equal to the largest ECT for the preceding events (column 4) plus the estimated duration time for the task culminating in event n. For the first event, the ECT is equal to zero.

    b. The LCT (also called the absolute deadline) for event n is equal to the smallest LCT for succeeding events minus the estimated duration time for the task culminating in event n. For the last event, the LCT equals the ECT.

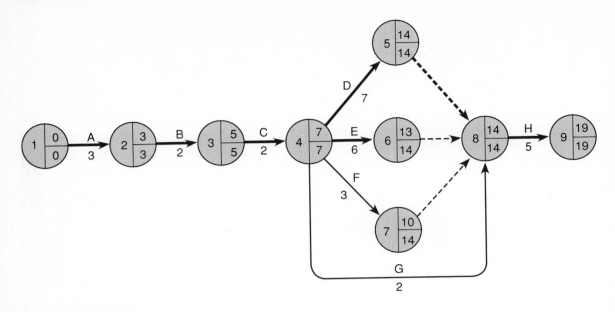

FIGURE A.3 Completed PERT Chart This is the PERT chart for the project planning table completed in Figure A.2. The bold arrows identify the critical path, those tasks for which there is no slack time.

The data in our table provides considerable planning and control assistance to the project manager.

5. Draw the PERT chart. The PERT chart includes sequencing and identification for all tasks and events along with their time estimates. Notice that our PERT chart (Figure A.3) contains three dummy tasks, represented by the dotted arrows. This means that events 5, 6, and 7 must occur before event 8. However, there is no associated time factor between the three events and event 8.

An alternative approach to deriving PERT charts is *backward scheduling*. The backward scheduling approach schedules activities starting with a proposed task or project completion date and working backward to schedule the tasks that must come before it. This approach is particularly useful when determining the feasibility of a proposed completion date. If all tasks preceding the prescribed date cannot be scheduled for completion prior to the current date, then the proposed completion date must be moved forward or the scope of the project must be reduced.

The Critical Path in a PERT Network Why are some of the solid arrows in Figure A.3 bolder than the others? This series of arrows represents the critical path. The **critical path** is a sequence of dependent project tasks that have the largest sum of estimated durations. It is the path that has no slack time built in. The slack time available for any task is equal to the difference between the earliest and latest completion times. If the earliest and latest completion times are equal, the task is on the critical path.

If any task on the critical path gets behind schedule, the whole project is thrown off schedule. Consequently, each task appearing on the critical path is referred to as a *critical task*. The critical path of PERT chart for the programming project in Figure A.3 consists of tasks A, B, C, D, and H. This path represents the expected completion time for the project. Critical tasks must be monitored closely by the project manager.

To find the critical path on a project's PERT chart, begin by identifying all the alternate paths or routes that exist from event 1 to the final event. For example, Figure A.3 contains four paths:

Path 1: A—B—C—D—dummy task—H

Path 2: A—B—C—E—dummy task—H

Path 3: A—B—C—F—dummy task—H

Path 4: A—B—C—G—H

After all paths have been identified, calculate the total expected duration time for each path. The total expected duration time for a path is equivalent to the sum of the expected duration times for each task in the path. For example,

Path 1: 3 + 2 + 2 + 7 + 0 + 5 = 19

Path 2: 3 + 2 + 2 + 6 + 0 + 5 = 18

Path 3: 3 + 2 + 2 + 3 + 0 + 5 = 15

Path 4: 3 + 2 + 2 + 2 + 5 = 14

You can now identify the critical path as the one having the largest total expected duration time. In our example, path 1 is the critical path and indicates the expected time for completing programming project is 19 days. But what if task G in path 4 had an expected duration time of 7 days? We would then have two critical paths containing tasks that the project manager would have to monitor closely!

Using PERT for Planning and Control Project managers find PERT charts particularly useful for communicating schedules of large systems projects to superiors. However, the primary uses and advantages of the PERT chart lie in its ability to assist in the planning and controlling of projects. In planning, the PERT chart aids in determining the estimated time required to complete a given project, in deriving actual project dates, and in allocating resources.

As a control tool, the PERT chart helps the manager identify current and potential problems. Particular attention should be paid to the critical path of a project. When a project manager identifies a critical task that is running behind schedule and that is in danger of upsetting the entire project schedule, alternative courses of action are examined. Corrective measures, such as the shuffling of human resources, might be taken. These resources are likely to be temporarily taken away from a noncritical task that is currently running smoothly. These noncritical tasks normally offer some *slack time* for the project.

Analysis of PERT   The classic approach to PERT presents one major problem when applied to information systems development. In the classic approach, a node represents the start *or* finish of a task or series of tasks. Consider the simple PERT chart presented in Figure A.4. It's a PERT chart for our high-level systems development life cycle (Chapter 2). Do you see the problem?

Classic PERT charting implies that the definition phase must be completed *before* either the selection phase or the design phase can be started. This, however, is not the case. Although neither the selection nor the design phase can be considered complete before the definition phase has been finished, both selection and design can begin *while* the definition phase is in progress. Classic PERT was developed to support projects that are often completed using an assembly-line approach. This does not include information systems! The tasks of systems development can overlap; it is only the *completion* of tasks that must occur in sequence. Don't assume that the next task cannot start until the prior task has been completed.

Gantt Chart: An Alternative Planning, Control, and Scheduling Tool

The **Gantt chart** is a simple time-charting tool that was developed by Henry L. Gantt in 1917. Gantt charts, which are still popular today, are effective for project scheduling and progress evaluation. Like PERT charts, Gantt charts involve a graphic approach. The popularity of Gantt charts stems from their simplicity: they are easy to learn, read, prepare, and use. Let's study this project management tool in more detail.

Gantt Chart Definitions and Symbols   The Gantt chart is a simple bar chart (see Figure A.5). Each bar represents a project task. *Task* has the same meaning in both PERT and Gantt charts. On a Gantt chart, the horizontal axis represents time. Because Gantt charts are used to schedule tasks, the horizontal axis should include dates. The tasks are listed vertically in the left-hand column. And that's all there is to the Gantt chart. We've been using these simple charts to illustrate the tasks of various life-cycle phases throughout this book (specifically, Chapters 6, 12, and 20).

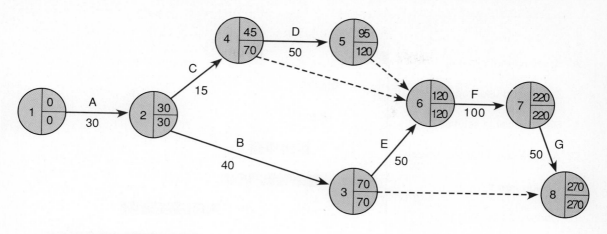

Event	Description	Task	Description
1	Project approved	N/A	N/A
2	Study phase report	A	Study the current system
3	Requirements statement	B	Define end-user requirements
4	New system proposal report	C	Select a feasible solution
5	Hardware software ordered	D	Acquire computer technology
6	Design specs report	E	Design the new system
7	Programs tested and debugged	F	Construct the new system
8	Postimplementation review	G	Deliver the system

FIGURE A.4 Sample PERT Chart for SDLC This simple PERT chart depicts interphase dependencies between tasks in the systems development life cycle. Each node, often called a milestone, represents the *completion* of a phase. PERT charts depict the sequence for completing tasks. However, PERT charts do not depict the potential for overlapping phases and tasks.

Notice that Gantt charts clearly depict the overlap of scheduled tasks. Because systems development tasks frequently overlap, this is a major advantage. But also notice that Gantt charts fail to clearly show the dependency of one task on another. That is a major strength of PERT charts. Let's briefly discuss the scheduling and evaluation uses of Gantt charts.

How to Use a Gantt Chart for Scheduling It's easy to use a Gantt chart to generate a schedule. First, identify the tasks that must be scheduled. (If you prepared a PERT chart first, this step would already have been completed.) Next, determine the duration of each task. You learned an appropriate time-estimating technique and formula in the preceding section. If you haven't already prepared a PERT chart, you should at least determine the interdepen-

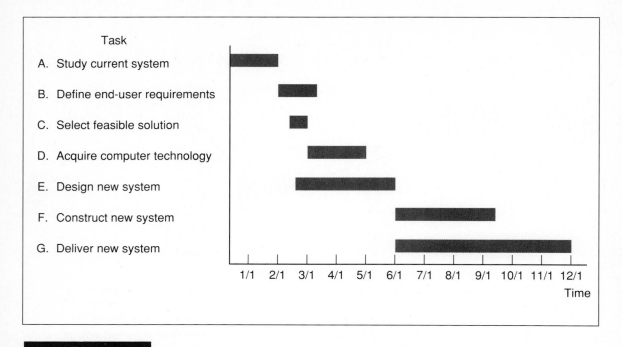

FIGURE A.5 Sample Gantt Chart for SDLC This is a sample Gantt chart for the systems development life cycle. Each bar represents a phase or task. The task is described in the left-hand column. The horizontal axis represents a calendar of dates.

dencies between tasks. Gantt charts cannot clearly show such dependencies, but it is imperative that the schedule recognize them. Now you are ready to schedule the tasks.

List each activity in the left-hand column of the Gantt chart. Record dates corresponding to the duration of the project on the horizontal axis of the chart. Determine starting and completion dates for each task. Careful! The start of any task may be dependent on at least the partial completion of a previous task. Additionally, the completion of a task is frequently dependent on the completion of a prior task. A simple example is in order.

Figure A.5 presents a Gantt chart for our high-level systems development life cycle. It offers some useful information that the PERT chart (Figure A.4) could not. Notice that the start of the definition phase appears to be dependent on the completion of the study phase. On the other hand, the selection phase is scheduled to begin after the definition phase has started but before that phase is completed. In fact, the definition phase overlaps with the entire selection phase.

If you compare the Gantt and PERT charts for the systems development life cycle, you will see that the sequence for *completing* phases (PERT) is maintained in the Gantt chart. Thus the preparation of a PERT chart (or, at the very least, understanding precedence between tasks) can significantly aid in the preparation of a Gantt chart.

Using Gantt Charts to Evaluate Progress One of the project manager's frequent responsibilities is to report project progress to superiors. Gantt charts frequently find their way into progress reports because they can conveniently compare the original schedule with actual performance. To report progress we must expand our Gantt charting conventions. If a task has been completed, completely shade in the bar corresponding to that task. If a task is partially completed, partially shade in the bar. The percentage of the bar that is shaded should correspond to the percentage of the task completed. Unshaded bars represent tasks that have not begun. Next, draw a bold vertical line that is perpendicular to the horizontal axis and that intersects the current date. You can now evaluate project progress.

Let's look at the sample Gantt chart in Figure A.6. The unshaded bar to the left of the current date (see ① on Figure A.6) is very much off schedule. This task is supposed to be completed but hasn't even been started. Completely shaded bars to the left of the current date (labeled ② on the figure) represent tasks that have been completed on schedule. Completely shaded bars to the right of the current date (labeled ③) represent tasks that have been completed ahead of schedule. Tasks that are currently in progress have to be evaluated relative to their shaded portions. If the shaded portion extends to or past the current date (see ④ on the figure), that task is on or ahead of schedule. Otherwise, the task is behind schedule (see ⑤). You should see that with a little practice, the Gantt chart can convey project progress at a glance.

A Comparison of the PERT and Gantt Charting Techniques

PERT and Gantt charting are frequently presented as mutually exclusive project management tools. PERT is usually recommended for larger projects with high intertask dependency. Gantt is recommended for simpler projects. But PERT and Gantt charting should not be considered as alternative project management approaches. All systems development projects have some intertask dependency. And all projects also offer opportunities for task overlapping. Therefore, PERT and Gantt charts can be used in a complementary manner to plan, schedule, evaluate, and control systems development projects.

Still, most information systems project managers seem to prefer Gantt charts because of their simplicity and ability to show the schedule of a project. Fortunately, project management software allows the best feature of PERT,

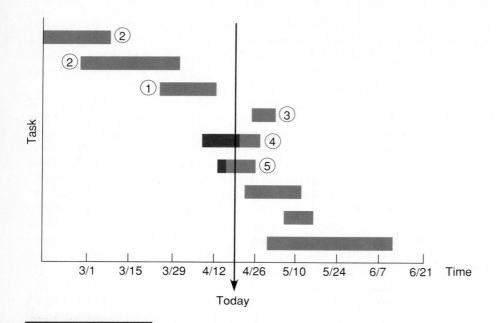

FIGURE A.6   Progress Reporting with Gantt Charts   Gantt charts can be annotated to clearly depict project progress. The bold vertical line represents today's date. Task bars are shaded in to reflect progress. Any unshaded bar to the left of today's date is behind schedule.

namely, the critical path analysis, to be incorporated into Gantt charts. As activities are entered, their duration and dependencies are entered. Gantt bars are scheduled to take into consideration the dependencies. Usually, the critical path is highlighted with boldfacing or color. Additionally, the amount of slack time in non–critical path activities is also highlighted. This can prove useful when deciding which activities to delay in order to get off-schedule activities back on track. Finally, most products allow Gantt charts to be temporarily displayed as PERT charts for a different perspective. Microsoft Project is an example of a fairly inexpensive PC tool that works this way.

Summary

In addition to systems analysis and design responsibilities, a systems analyst frequently assumes a project management role. Project mismanagement often leads to missed end-user requirements, cost overruns, and late delivery. The causes of these problems include shortcuts during systems development, imprecise targets, premature cost estimates, poor estimating techniques, poor time management, and lack of leadership. It is the project manager's responsibility to avoid these pitfalls and successfully complete the project on time and within budget. The project manager's basic functions include planning project tasks, staffing the project team, organizing and scheduling the project effort, directing the project team, and controlling project progress.

There are two major tools frequently used by analysts to plan, schedule, and control systems development projects. PERT charts graphically depict project tasks and events and show the dependency of tasks on one another. They also depict the time requirements for each task. Gantt charts graphically depict project tasks to show a project schedule. Gantt charts allow the project manager to show the overlapping of tasks in a project; they are also useful for depicting project progress. PERT and Gantt charts complement each other to provide the manager with an integrated planning, scheduling, evaluation, and control environment. This environment is being further enhanced by the availability of computer software to support the project management tools.

Problems and Exercises

1. What are some of the causes of mismanaged projects that result in missed requirements and needs, cost overruns, and late delivery? Explain how these problems that result from mismanaged projects are related.

2. Systems analysts have a tendency to assign additional people to a project that is running behind schedule. What are some of the potential problems with such an action?

3. What are the basic functions of a project manager?

4. Explain the advantages and disadvantages of a PERT chart and a Gantt chart. Explain how each tool can be best used by a project manager.

5. Why shouldn't estimated project time requirements be stated in terms of person-days?

6. Calculate the expected duration for the following tasks:

Task ID Letter	Optimistic Time (OT)	Pessimistic Time (PT)	Most Likely Time (MLT)
A	3	6	4
B	1	3	2
C	4	7	6
D	2	5	3

7. Derive the earliest and latest completion times (ECT and LCT) for each of the following:

Task ID Letter	Event ID Number	Preceding Event	Succeeding Event	Expected Duration
A	2	1	3	2
B	3	2	4	3
C	4	3	5,6	4
D	5	4	7	5
E	6	4	7	4
F	7	5,6	8,9	3
G	7	5,6	8,9	0
H	8	7	10	6
I	9	7	10	5
J	10	8	None	0
K	10	9	None	6

8. Draw the PERT chart described in Exercise 7. Be sure to include sequencing and identification for all tasks and events along with their time estimates. What is the critical path? What is the total expected duration time represented by the critical path?

9. Make a list of the tasks that you performed on your last programming assignment. Alternatively, make a list of the tasks required to complete your next programming assignment. Develop a PERT chart to depict the tasks and events and the dependency of tasks on one another. What is the critical path? How can the PERT chart aid in planning and scheduling the programming assignment?

10. Derive a Gantt chart to graphically depict the project schedule and overlapping of tasks for the programming assignment chosen in Exercise 9. How can the Gantt chart be used to evaluate the progress that is being, or has been, made?

11. Draw a PERT chart for the curriculum in which you are enrolled. Be sure to consider the prerequisites for all courses.

12. Draw a Gantt chart for your plan of study to get your degree. Annotate the graph to indicate your progress toward your degree or job objectives.

Annotated References and Suggested Readings

Blanchard, Kenneth, and Spencer Johnson. *The One Minute Manager.* 1982. A classic! Three simple rules to manage people!

Brooks, Fred. *The Mythical Man-Month.* Reading, Mass.: Addison-Wesley, 1975. A classic set of essays on software engineering, also known as systems analysis, design, and implementation. Emphasis is on managing complex projects.

Gildersleeve, Thomas. *Data Processing Project Management.* New York: Van Nostrand Reinhold, 1974. An entire book on project management! No coverage of PERT/CPM, Gantt, or any other tool, but an excellent book on personpower and resource management.

London, Keith. *The People Side of Systems.* New York: McGraw-Hill, 1976. Chapter 8, "Handling a Project Team," does an excellent job of teaching the people and leadership aspects of project management.

Senn, James A. *Analysis and Design of Information Systems.* New York: McGraw-Hill, 1984. Although his treatment of PERT/CPM and Gantt charting is of an introductory nature (like ours), we are indebted to Senn for his PERT/CPM symbol notation, which we emulated in this book.

Weinberg, Victor. *Structured Analysis.* New York: Yourdon Press, 1979. Chapter 11 contains some valuable guidelines and strategies for estimating. Although Weinberg uses his weighting strategy to adjust *costs* according to various factors, we have suggested that scheme would work equally well on *time* estimates.

Wiest, Jerome D., and Ferdinand K. Levy. *A Management Guide to PERT-CPM: With PERT-PDM, DCPM, and Other Networks.* 2d ed. Englewood Cliffs, N.J.: Prentice-Hall, 1977. A good source for more on PERT/CPM and other project planning and control networks.

MODULE B

Fact-Finding
Techniques

When Should You Read This Module?

You should study this module after reading *any* of the following chapters:

- Chapter 6, "How to Analyze an Information System: A Problem-Solving Approach"—All of the fact-finding techniques (sampling, observation, questionnaires, and interviews) are applicable to the survey, study, definition, and evaluation phases.
- Chapter 12, "How to Design an Information System: Traditional and Prototyping Approaches"—Fact-finding is important when designing report formats, terminal dialogue preferences, and data input methods.
- Chapter 20, "How to Implement and Evaluate a New Information System"—Fact-finding is particularly important during acceptance testing and post-implementation review.

Applying the tools and techniques of systems analysis and design in the classroom is easy. Applying those same tools and techniques in the real world may not work . . . that is, if they are not complemented by effective methods for collecting facts. Tools document facts, and conclusions are drawn from facts. If you can't collect the facts, you can't use the tools. Fact-finding skills must be learned and practiced.

As a systems analyst, you need an organized method of collecting facts. You especially need to develop a detective mentality to be able to discern relevant facts! The purpose of this module is to present fact-finding techniques, also known as information-gathering and data collection techniques. Although an entire textbook could be devoted to fact-finding techniques and strategies, no introductory systems course would be complete without the following survey.

Fundamentals of Fact-Finding

Before we leap headfirst into specific fact-finding methods, let's make sure we understand what we are trying to accomplish. The tools of systems analysis and design are used to *document* facts about an existing or proposed information system. These facts are in the domain of the business application and its end-users. Therefore, you must collect those facts in order to effectively apply the documentation tools and techniques. When should you use fact-finding techniques? What kinds of facts should you collect? And how will you collect the facts?

What Facts Does the Systems Analyst Need to Collect, and When?

There are many occasions for fact-finding during the systems development life cycle (discussed in Chapter 4). The most notable time is during each systems analysis phase (Chapter 6). During the *study phase,* the analyst must learn and understand the existing information system. During the *definition phase,* the analyst collects facts about end-user requirements and priorities. As the project proceeds into the *selection phase,* the analyst must solicit facts about alternative technical solutions and end-user preferences for computer solutions. And after systems design (Chapter 12) gets under way, the analyst must collect facts about report formats, terminal dialogue preferences, and input methods. Also, during the systems implementation (Chapter 20), fact-finding is particularly necessary in the *postimplementation review,* whereby the operational system and the project development process are evaluated. Thus we can see that fact-finding is an essential skill.

What types of facts must be collected? It would certainly be beneficial if we had a framework to help us determine what facts need to be collected, no matter what project we are working on. Fortunately, we have such a framework. Throughout the systems development process, we are looking at an existing or proposed information system. In Chapters 2 and 3, we saw that any information systems problem can be examined in four dimensions. Those four dimensions were depicted as the four faces of our pyramid model. As it turns out, the facts that describe any information system also correspond nicely with the dimensions of that pyramid model.

What Fact-Finding Methods Are Available?

Now that we have a framework for our fact-finding activities, we can introduce five common fact-finding techniques:

1. Sampling of existing documentation, forms, and files
2. Research and site visits
3. Observation of the work environment
4. Questionnaires
5. Interviews and group work sessions

An understanding of each of these techniques is essential to your success. An analyst usually applies several of these techniques during a single systems project. To be able to select the most suitable technique for use in any given situation, you will have to learn the advantages and disadvantages of each of the fact-finding techniques.

Sampling of Existing Documentation, Forms, and Files

Particularly when you are studying an existing system, you can develop a pretty good feel for the system by studying existing documentation, forms, and files. Because it would be impractical to study every occurrence of every form, analysts normally use sampling techniques to get a large enough cross section to determine what can happen in the system. A good analyst always gets facts first from existing documentation rather than from people.

Collecting Facts from Existing Documentation

What kind of documents can teach you about a system? The first document the analyst should seek out is the *organization chart*. What's the next step? One possibility is to trace the history that led to the project. To accomplish this, you may want to collect and review documents that describe the problem. These include

- Interoffice memoranda, studies, minutes, suggestion box notes, customer complaints, and reports that document the problem area
- Accounting records, performance reviews, work measurement reviews, and other scheduled operating reports
- Information systems project requests, past and present

In addition to documents that describe the problem, there are usually documents that describe the business function being studied or designed. These documents may include

- The company's mission statement and strategic plan
- Formal objectives for the organization subunits being studied
- Policy manuals that may place constraints on any proposed system
- Standard operating procedures (SOPs), job outlines, or task instructions for specific day-to-day operations
- Completed forms that represent actual transactions at various points in the processing cycle
- Manual and computerized files
- Manual and computerized reports

Don't forget to check for documentation of previous systems studies and designs performed by systems analysts and consultants. This documentation may include

- Various types of flowcharts and diagrams
- Data dictionaries
- Design documentation, such as inputs, outputs, and files
- Program documentation
- Computer operations manuals and training manuals

All documentation collected should be analyzed to determine how up to date it is. Don't discard outdated documentation. Just keep in mind that additional fact-finding will be needed to verify or update the facts collected. As you review existing documents, take notes, draw pictures, and use systems analysis and design tools to model what you are learning or proposing for the system.

Document and File Sampling Techniques

Sampling is the process of collecting sample documents, forms, and records. Do not sample blank forms—they tell you little about how the form is used, not used, or misused. When studying documents or records from a file, study enough samples to identify all the possible processing conditions and exceptions. How do you determine if the sample size is large enough to be representative? You use statistical sampling techniques.

How to Determine the Sample Size   The size of the sample depends on how representative you want the sample to be. There are many sampling issues and factors: a good reason to take an introductory statistics course. One simple and reliable formula for determining sample size is

sample size $= 0.25 \times$ (certainty factor/acceptable error)^2

The certainty factor depends on how certain you want to be that the data sampled will not include variations not in the sample. The certainty factor is calculated from tables (available in many industrial engineering texts); a partial example is given here.

Desired Certainty	Certainty Factor
95%	1.960
90%	1.645
80%	1.281

Suppose you want 90 percent certainty that a sample of invoices will contain no unsampled variations. Your sample size, SS, is calculated as follows:

$$SS = 0.25(1.645/0.10)^2 = 68$$

We need to sample 68 invoices to get the desired accuracy.

Selecting the Sample   How do we choose our 68 invoices? Two commonly used sampling techniques are randomization and stratification. In randomization, there is no predetermined pattern or plan for selecting sample data. Therefore, we just randomly choose 68 invoices. The stratification technique is a systematic approach for selecting sample data. This technique attempts to reduce the variance of the estimates by spreading out the sampling—for example, choosing documents or records by formula—and by avoiding very high or low estimates.

For computerized files, stratification sampling can be executed by writing a sample program. For instance, suppose our invoices were on a computer file that had a volume of approximately 250,000 invoices. Recall that our sample size needs to include 68 invoices. We will simply write a program that prints every 3,676th record (= 250,000/68). This is an example of stratification sampling on a computer file. For manual files and documents, we could execute a similar scheme.

Research and Site Visits

A second fact-finding technique is to thoroughly research the application and problem. Computer trade journals are a good source. Don't forget to study the trade journals typically read by your end-users. And you can learn how others have solved similar problems. You learn whether or not software packages exist to solve your problem.

A similar type of research involves visiting other companies or departments that have addressed similar problems. Memberships in professional societies can provide useful contacts.

Observation of the Work Environment

Observation is one of the most effective data collection techniques for obtaining an understanding of a system. In **observation**, the systems analyst either participates in or watches a person perform activities to obtain facts about the system. This technique is often used when the validity of data collected through other methods is in question or when the complexity of certain aspects of the system prevents a clear explanation by the end-users.

Collecting Facts by Observing People at Work

Even with a well-conceived observation plan, the systems analyst is not assured that fact-finding will be successful. You should become aware of the pros and cons of the technique of observation. Advantages and disadvantages include:

ADVANTAGES

1. Data gathered by observation can be highly reliable. Sometimes, observations are conducted to check the validity of data obtained directly from individuals.

2. The systems analyst is able to see exactly what is being done. Complex tasks are sometimes difficult to clearly explain in words. Through observation, the systems analyst can identify tasks that have been missed or inaccurately described by other fact-finding techniques. Also, the analyst can obtain data describing the physical environment of the task (for example, physical layout, traffic, lighting, noise level).

3. Observation is relatively inexpensive compared with other fact-finding techniques. Other techniques usually require substantially more employee release time and copying expenses.

4. Observation allows the systems analyst to do work measurements.

DISADVANTAGES

1. Because people usually feel uncomfortable when being watched, they may unwittingly perform differently when being observed. In fact, the famous Hawthorne Experiment proved that the act of observation can alter behavior.

2. The work being observed may not involve the level of difficulty or volume normally experienced during that time period.

3. Some systems activities may take place at odd times, causing a scheduling inconvenience for the systems analyst.

4. The tasks being observed are subject to various types of interruptions.

5. Some tasks may not always be performed in the manner in which they are observed by the systems analyst. For example, the systems analyst might have observed how a company filled several customer orders. However, the procedures the systems analyst observed may have been those steps used to fill a number of regular customer orders. If any of those orders had been special orders (such as an order for goods not normally kept in stock), the systems analyst would have observed a different set of procedures being executed.

6. If people have been performing tasks in a manner that violates standard operating procedures, they may temporarily perform their jobs correctly while you are observing them. In other words, people may let you see what they want you to see.

Guidelines for Observation

How does the systems analyst obtain facts through observation? Does one simply arrive at the observation site and begin recording everything that's viewed? Of course not. Much preparation should take place in advance. The analyst must determine how data will actually be captured. Will it be necessary to have special forms on which to quickly record data? Will the individual(s) being observed be bothered by having someone watch and record their actions? When are the low, normal, and peak periods of operations for the task to be observed? The systems analyst must identify the ideal time to observe a particular aspect of the system.

Observation should first be conducted when the workload is normal. Afterward, observations can be made during peak periods to gather information for measuring the effects caused by the increased volume. The systems analyst might also obtain samples of documents or forms that will be used by those being observed. As you can see, a great deal of planning and preparation must be done up front.

The sampling techniques discussed earlier are also useful for observation. In this case, the technique is called *work sampling*. **Work sampling** involves a large number of observations taken at random intervals. This technique is less threatening to the people being observed because the observation period is not continuous. When using work sampling, you need to predefine the operations of the job to be observed. Then calculate a sample size as you did for document and file sampling. Make that many random observations, being careful to observe activities at different times of the day. By counting the number of occurrences of each operation during the observations, you will get a feel for how employees spend their days.

With proper planning completed, the actual observation can be done. Effective observation is difficult to carry out. Experience is the best teacher; however, the following guidelines may help you develop your observation skills:

1. Determine the who, what, where, when, why, and how of the observation.
2. Obtain permission from appropriate supervisors or managers.
3. Inform those who will be observed of the purpose of the observation.
4. Keep a low profile.
5. Take notes during or immediately following the observation.
6. Review observation notes with appropriate individuals.

Don't:

1. Interrupt the individuals at work.
2. Focus heavily on trivial activities.
3. Make assumptions.

Questionnaires

The **questionnaire** is a special-purpose document that allows the analyst to collect information and opinions from respondents. The document can be mass-produced and distributed to respondents, who can then complete the questionnaire on their own time. Questionnaires allow the analyst to collect facts from a large number of people while maintaining uniform responses. When dealing with the large audience, no other fact-finding technique can tabulate the same facts as efficiently.

Collecting Facts by Using Questionnaires

The use of questionnaires has been heavily criticized and is often avoided by systems analysts. Why? Many systems analysts claim that the responses lack reliable and useful information. But questionnaires can be an effective method for fact gathering, and many of these criticisms can be attributed to the inappropriate use of the questionnaires by systems analysts. Before using questionnaires, you should first understand the pros and cons associated with their use.

ADVANTAGES

1. Most questionnaires can be answered quickly. People can complete and return questionnaires at their convenience.
2. Questionnaires provide a relatively inexpensive means for gathering data from a large number of individuals.
3. Questionnaires allow individuals to maintain anonymity. Therefore, individuals are more likely to provide the real facts, rather than telling you what they think their boss would want them to.
4. Responses can be tabulated and analyzed quickly.

DISADVANTAGES

1. The number of respondents is often low.
2. There's no guarantee that an individual will answer or expand on all of the questions.
3. Questionnaires tend to be inflexible. There's no opportunity for the systems analyst to obtain voluntary information from individuals or to reword questions that may have been misinterpreted.
4. It's not possible for the systems analyst to observe and analyze the respondent's body language.
5. There is no immediate opportunity to clarify a vague or incomplete answer to any question.
6. Good questionnaires are difficult to prepare.

Types of Questionnaires

There are two formats for questionnaires, free-format and fixed-format. A free-format questionnaire offers the respondent greater latitude in the answer. A question is asked, and the respondent records the answer in the space provided after the question. Examples of free-format questions are

1. What reports do you currently receive and how are they used?
2. Are there any problems with these reports (for example, are they inaccurate, is there insufficient information, or are they difficult to read and/or use)? If so, please explain.

Obviously, such responses may be difficult to tabulate. It is also possible that the respondents' answers may not match the questions asked. In order to ensure good responses in free-format questionnaires, the analyst should

- Phrase the questions in simple sentences and not use words—such as *good*—that can be interpreted differently by different respondents.
- Ask questions that can be answered with three or fewer sentences. (Otherwise, the questionnaire may take up more time than the respondent is willing to sacrifice.)

A fixed-format questionnaire contains questions that require specific responses from individuals. Given any question, the respondent must choose from the available answers. This makes the results much easier to tabulate. On the other hand, the respondent cannot provide additional information that might prove valuable. There are three types of fixed-format questions:

1. *Multiple choice,* in which the respondent is given a question and several answers. The respondent should be told if more than one answer may be selected. Some multiple-choice questions allow for very brief free-format responses when none of the standard answers apply. Examples of multiple-choice, fixed-format questions are

Do you feel that backorders occur too frequently?

[   ] YES   [   ] NO

Is the current accounts receivable report that you receive useful?

[   ] YES   [   ] NO   If no, please explain.

2. *Rating,* in which the respondent is given a statement and asked to use supplied responses to state an opinion. To prevent built-in bias, there should be an equal number of positive and negative ratings. The following is an example of a rating fixed-format question:

The implementation of quantity discounts would cause an increase in customer orders.

[   ] strongly agree
[   ] agree

[ ] no opinion
[ ] disagree
[ ] strongly disagree

3. *Ranking,* in which the respondent is given a question and several possible answers, which are to be ranked in order of preference or experience. An example of a ranking fixed-format question is

Rank the following transactions according to the amount of time you spend processing them:

———— % new customer orders
———— % order cancelations
———— % order modifications
———— % payments

Developing a Questionnaire

Good questionnaires are designed. If you write your questionnaires without designing them first, your chances of success are limited. The following procedure is effective:

1. Determine what facts and opinions must be collected and from whom you should get them. If the number of people is large, consider using a smaller, randomly selected group of respondents.

2. Based on the needed facts and opinions, determine whether free- or fixed-format questions will produce the best answers. A combination format that permits optional free-format clarification of fixed-format responses is often used.

3. Write the questions. Examine them for construction errors and possible misinterpretations. Make sure that the questions don't offer your personal bias or opinions. Edit the questions.

4. Test the questions on a small sample of respondents. If your respondents had problems with them or if the answers were not useful, edit the questions.

5. Duplicate and distribute the questionnaire.

✓ INTERVIEW
PURPOSES

Fact-Finding
Fact Verification
Clarifications
Generate Enthusiasm
Get End-User Involved
Identify Requirements
Solicit Ideas and
   Opinions

Interviews and Group Work Sessions

The personal interview is generally recognized as the most important and often used fact-finding technique. **Interviews** provide systems analysts the opportunity to collect information from individuals face to face. Interviewing can be used to achieve any of the goals listed in the margin. There are two roles assumed in an interview. The systems analyst is the **interviewer**, responsible for organizing and conducting the interview. The end-user, manager, or adviser is the interviewee, who is asked to respond to a series of questions. Unfortunately, many systems analysts are poor interviewers. In this section, you will learn how to conduct proper interviews.

Collecting Facts by Interviewing People

The most important element of an information system is people. And more than anything else, people want *to be in on things*. No other fact-finding technique places as much emphasis on people as interviews. But people have different values, priorities, opinions, motivations, and personalities. Therefore, to use the interviewing technique, you must possess good human relations skills for dealing effectively with different types of people. And like other fact-finding techniques, interviewing isn't the best method for all situations. Interviewing has its advantages and disadvantages, which should be weighed against those of other fact-finding techniques for every fact-finding situation.

ADVANTAGES

1. Interviews give the analyst an opportunity to motivate the interviewee to respond freely and openly to questions. By establishing rapport, the systems analyst is able to give the interviewee a feeling of actively contributing to the systems project.

2. Interviews allow the systems analyst to probe for more feedback from the interviewee.

3. Interviews permit the systems analyst to adapt or reword questions for each individual.

4. Interviews give the analyst an opportunity to observe the interviewee's nonverbal communication. A good systems analyst may be able to obtain information by observing the interviewee's body movements and facial expressions as well as by listening to verbal replies to questions.

DISADVANTAGES

1. Interviewing is a very time-consuming, and therefore costly, fact-finding approach.

2. Success of interviews is highly dependent on the systems analyst's human relations skills.

3. Interviewing may be impractical due to the location of interviewees.

Interview Types and Techniques

There are two types of interviews, unstructured and structured. In the *unstructured interview,* the interviewer conducts the interview with only a general goal or subject in mind and few, if any, specific questions. The interviewer counts on the interviewee to provide a framework and direct the conversation. This type of interview frequently gets off track, and the analyst must be prepared to redirect the interview back to the main goal or subject. For this reason, unstructured interviews don't usually work well for systems analysis and design.

In the *structured interview,* the interviewer has a specific set of questions to ask of the interviewee. Depending on the interviewee's responses, the interviewer will direct additional questions to obtain clarification or amplification. Some of these questions may be planned and others spontaneous. Questions may be either open-ended or closed-ended. *Open-ended questions* allow the interviewee to respond in any way that seems appropriate. An example of an open-ended question is "Why are you dissatisfied with the report of uncollectable accounts?" *Closed-ended questions* restrict answers to either specific choices or short, direct responses. An example of such a question might be "Are you receiving the report of uncollectable accounts on time?" or "Does the report of uncollectable accounts contain accurate information?" Realistically, most questions fall between the two extremes.

How to Conduct an Interview

Your success as a systems analyst is at least partially dependent on your ability to interview. A successful interview will involve selecting appropriate individuals to interview, preparing extensively for the interview, conducting the interview properly, and following up on the interview. Here we examine each of these aspects in more detail. Let's assume that you've identified the need for an interview and you have determined exactly what kinds of facts and opinions you need.

Select Interviewees Whom should you interview? You should interview the end-users of the information system you are studying. A formal organizational chart will help you identify these individuals and their responsibilities. You should attempt to learn as much as possible about each individual prior to the interview. Attempt to learn what their strengths, fears, biases, and motivations might be. The interview can then be geared to take the characteristics of the individual into account.

Always make an appointment with the interviewee. Never just drop in. Limit the appointment to somewhere between a half hour and an hour. The higher the management level of the interviewee, the less time you should schedule. If the interviewee is a clerical, service, or blue-collar worker, get their supervisor's permission before scheduling the interview. Be certain that the location you want for the interview will be available during the time the interview is scheduled. Never conduct an interview in the presence of your officemates or the interviewee's peers.

Prepare for the Interview Preparation is the key to a successful interview. An interviewee can easily detect an unprepared interviewer. In fact, the interviewee may very much resent the lack of preparation because it is a waste of valuable time. When the appointment is made, the interviewee should be notified about the subject of the interview. To ensure that all pertinent aspects of the subject are covered, the analyst should prepare an interview guide.

An **interview guide** is a checklist of specific questions the interviewer will ask the interviewee. It may also contain follow-up questions that will only be asked if the answers to other questions warrant the additional answers. A sample interview guide is presented in Figure B.1. Questions should be carefully chosen and phrased. Most questions begin with the standard *who, what, when, where, why,* and *how much* type of wording. Avoid the following types of questions:

- *Loaded questions,* such as "Do we have to have both of these columns on the report?" The question conveys the interviewee's personal opinion on the issue.

- *Leading questions,* such as "You're not going to use this OPERATOR CODE, are you?" The question leads the interviewee to respond "No, of course not," regardless of actual opinion.

- *Biased questions,* such as "How many codes do we need for FOOD-CLASSIFICATION in the INVENTORY FILE? I think 20 ought to cover it." Why bias the interviewee's answer with your own?

Additional guidelines for questions are provided in the margin. You should especially avoid threatening or critical questions. The purpose of the interview is to investigate, not to evaluate or criticize.

INTERVIEW QUESTION GUIDELINES

1. Use clear and concise language.
2. Don't include your opinion as part of questions.
3. Avoid long or complex questions.
4. Avoid threatening questions.
5. Don't use "you" when you mean a group of people.

Conduct the Interview The actual interview can be characterized as consisting of three phases: the opening, body, and conclusion. The opening is intended to influence or motivate the interviewee to participate and communicate by establishing an ideal environment. When establishing an environment of mutual trust and respect, you should identify the purpose and length of the interview and explain how the gathered data will be used. Possible ways to effectively begin an interview are to

1. Summarize the apparent problem, and explain how the problem was discovered.

2. Offer an incentive or reward for participation.

3. Ask the interviewee for advice or assistance.

The body of an interview represents the most time-consuming phase. During this phase, you obtain the interviewee's responses to your list of questions. Listen closely and observe the interviewee. Take notes concerning both verbal and nonverbal responses from the interviewee. It's very important

INTERVIEW AGENDA

INTERVIEWEE:	Jeff Bentley, Accounts Receivable Manager
DATE:	Tuesday, March 22, 1990
TIME:	1:30 P.M.
PLACE:	Room 223, Admin. Bldg.
SUBJECT:	Current Credit Checking Policy

1-2 min.   Open the interview.

Introduce ourselves.
Thank Mr. Bentley for the uses of his valuable time.
State the purpose of the interview--to obtain an
understanding of the existing credit checking policies.

5 min.   What conditions determine whether a customer's order
is approved for credit?

5 min.   What are the possible decisions or actions that might
be taken once these conditions have been evaluated?

3 min.   How are customers notified when credit is not approved
for their order.

1 min.   After a new order is approved for credit and placed in
the file containing orders that can be filled, a
customer might request a modification be made to the
order. Would the order have to go through credit
approval again if the new total order cost exceeds the
original cost?

1 min.   Who are the individuals that perform the credit checks?

1-3 min.   May I have permission to talk to those individuals to
learn specifically how they carry out the credit
checking process?

If so:

When would be an appropriate time to meet with each of
them?

1 min.   Conclude the interview:

Thank Mr. Bentley for his cooperation and assure him
that he will be receiving a copy of what transpired
during the interview.

---

21 minutes
+9 minutes for follow-up questions and redirection

30 minutes alloted for interview (1:30 P.M.-2:00 P.M.)

FIGURE B.1 Sample Interview Guide The sample interview guide represents an agenda that a systems analyst will use to obtain facts about a company's existing credit approval policy. Notice that the agenda is carefully laid out with specific time allocated to each question. Time should also be reserved for follow-up questions and redirecting the interview.

for you to keep the interview on track. Anticipate the need to adapt the interview to the interviewee. Often questions can be bypassed if they have been answered earlier in part of an answer to another question, or they can be deleted if determined to be irrelevant, based on what you've already learned during the interview. Finally, probe for more facts when necessary.

During the conclusion of an interview, you should express your appreciation and provide answers to any questions posed by the interviewee. The conclusion is very important for maintaining rapport and trust with the interviewee.

The importance of human relations skills in interviewing cannot be overemphasized. These skills must be exercised throughout the interview. In the margin, you will find a set of rules that should be followed during an interview.

Follow Up on the Interview   To help maintain good rapport and trust with interviewees, you should send them a memo that summarizes the interview. This memo should remind the interviewees of their contributions to the systems project and allow them the opportunity to clarify any misinterpretations that you may have derived during the interview. In addition, the interviewee should be given the opportunity to offer additional information they may have failed to bring out during the interview.

The Group Work Session   Separate interviews of end-users have always been the classic fact-finding technique. However, many analysts have discovered the great flaw of interviewing—separate interviews often lead to conflicting facts, opinions, and priorities. The end result is numerous follow-up interviews and/or group meetings. For this reason, many shops are using the group work session as a substitute for interviews.

Group work sessions are highly structured group meetings that get all the end-users, managers, and analysts in a single room for an extended period of time (four to eight hours). The goals are essentially the same as in an interview, except that you need a number of analysts to carry them out. One analyst serves as discussion leader and moderator. Another records facts, while another may record items that require further action or individual interviews. One example of the group work session approach is IBM's Joint Application Design (JAD). This and similar techniques generally require extensive training in order to work as intended. However, they can significantly decrease the time spent on fact-finding in one or more phases of the life cycle.

A Fact-Finding Strategy

At the beginning of this module, we suggested that an analyst needs an *organized* method for collecting facts. An inexperienced analyst will frequently

---

✓ **CONDUCTING THE INTERVIEW**

DO

1. Be courteous.
2. Listen carefully.
3. Maintain control.
4. Probe.
5. Observe mannerisms and nonverbal communication.
6. Be patient.
7. Keep interviewee at ease.
8. Maintain self-control.

AVOID

1. Continuing an interview unnecessarily.
2. Assuming an answer is finished or leading nowhere.
3. Revealing verbal and nonverbal cues.
4. Using jargon.
5. Revealing your personal biases.
6. Talking instead of listening.
7. Assuming anything about the topic and the interviewee.
8. Tape recording—a sign of poor listening skills.

jump right into interviews. "Go to the people. That's where the real facts are!" Wrong! This attitude fails to recognize an important fact of life: people must complete their day-to-day jobs! Your job is not their main responsibility. And your demand on their time is their money lost. Now you may be thinking, "But I thought you've been saying that the system is for people and that direct end-user involvement in systems development is essential! Aren't you contradicting yourselves?"

Not at all! Time is money. To waste your end-users' time is to waste your company's money. To make the most of the time that you spend with end-users, don't jump right into interviews. Instead, first collect all the facts you can by using methods other than asking people. Consider the following step-by-step strategy:

1. *Learn all you can from existing documents, forms, reports, and files.* You'll be surprised how much of the system becomes clear without any people contact.

2. *If appropriate, observe the system in action.* Agree not to ask questions. Just watch and take notes or draw pictures. Make sure that the workers know that you're not evaluating individuals. Otherwise, they may perform in a more efficient manner than normal.

3. *Given all the facts that you've already collected, design and distribute questionnaires to clear up things you don't fully understand.* This is also a good time to solicit opinions on problems and limitations. Questionnaires do require your end-users to give up some of their time. But *they* choose when to best make that sacrifice.

4. *Now you can conduct your interviews (or group work sessions).* Because you have already collected most of the pertinent facts by low-user-contact methods, you can use the interview to verify and clarify the most difficult issues and problems. You are unlikely to waste end-user time as you gather information you could have obtained in other ways.

5. Follow-up by using appropriate fact-finding techniques to verify facts (usually interviews or observation).

The strategy is not sacred. Although a fact-finding strategy should be developed for every pertinent phase of systems development, every project is unique. Sometimes observation and questionnaires may be inappropriate. But the idea should always be to collect as many facts as possible before using interviews.

Summary

Effective fact-finding techniques are crucial to the application of systems analysis and design methods during systems projects. Fact-finding is performed during the study, definition, evaluation, design, and implementation phases of the systems development life cycle. To support development activities, the analyst must collect facts about end-users, the business, data and information resources, and information systems components. There are five common fact-finding techniques: sampling, research, observation, questionnaires, and interviews.

The sampling of existing documents and files can provide many facts and details with little or no direct personal communication being necessary. The analyst should collect historical documents, business operations manuals and forms, and information systems documents. In order to ensure that an adequate number of documents have been studied, analysts often use sampling techniques. These techniques make it possible to collect a representative subset of the documents and minimize the chance of identifying exceptional events.

Research is an often overlooked technique based on the study of other similar applications. Site visits are a special form of research.

Observation is a fact-finding technique in which the analyst studies people doing their jobs. To minimize the chance that the observation time is not representative of normal workloads, the analyst can use work sampling to randomly collect observation data.

Questionnaires are used to collect similar facts from a larger number of individuals. Questionnaires can be either free-format or fixed-format.

Interviews are the most popular but time-consuming fact-finding technique. When interviewing, the analyst meets individually with people to gather information. Most systems analysis and design interviews are structured, meaning that the analyst has prepared a specific set of questions before the interview. After determining the need for an interview, the analyst arranges for appointments with the interviewee, carefully prepares the interview questions, conducts the interview, and summarizes the results. Group work sessions are many-on-many interviews that usually require special training. Because interviews are time consuming, the analyst should collect as many facts as possible by using the other fact-finding methods.

Problems and Exercises

1. Explain how the information systems pyramid model can serve as a framework in determining what facts need to be collected during systems development.

2. Explain how an organization chart can aid in planning for fact-finding. What are some of the potential drawbacks to using an existing organization chart?

3. A systems analyst wants to study documents stored in a large metal file cabinet. The cabinet contains several hundred records describing product warranty claims. The analyst wishes to study a sample of the records in the file and to be 95 percent certain (certainty factor = 1.960) that the data from which the sample is taken will not include variations not in the sample. How many sample records should the analyst retrieve to get this desired accuracy?

4. For the sample size in Exercise 3, explain two specific strategies for selecting the samples.

5. Describe how you would use form and/or file sampling in the following phases. If you think sampling would be inappropriate for any of these phases, explain why.
   a. Study of the current system
   b. Definition of end-user requirements
   c. Selection of a feasible solution
   d. Acquisition of computer equipment and software
   e. Design of the new information system
   f. Construction of the new information system
   g. Delivery of the new information system

6. Repeat Exercise 5 for the technique of observation.

7. Make a list of things that might affect your work performance when you are being observed performing your job. What could an observer do to eliminate these concerns or problems?

8. Repeat Exercise 5 for the questionnaire technique.

9. Give two examples of free-format questions and two examples of each of the following types of fixed-format questions:
   a. multiple choice
   b. rating
   c. ranking

10. Repeat Exercise 5 for the interviewing technique.

11. Explain the difference between a structured and an unstructured interview. When is each type of interview appropriately used?

12. Prepare a sample interview guide to use in obtaining from your academic adviser facts describing the course registration policies and procedures.

13. Mr. Art Pang is the Accounts Receivables Manager. You have been assigned to do a study of Mr. Pang's current billing system and need to solicit facts from his subordinates. Mr. Pang has expressed his concern that, although he wishes to support you in your fact-finding efforts, his people are extremely busy and must get their jobs done. Write a memo to Mr. Pang describing a fact-finding strategy that you could follow to maximize your fact-finding while minimizing the release time required for his subordinates.

Annotated References and Suggested Readings

Berdie, Douglas R., and John F. Anderson. *Questionnaires: Design and Use.* Metuchen, N.J.: Scarecrow Press, 1974. A practical guide to the construction of questionnaires. Particularly useful because of its short length and illustrative examples.

Davis, William S. *Systems Analysis and Design.* Reading, Mass.: Addison-Wesley, 1983. Provides useful pointers for preparing and conducting interviews.

Fitzgerald, Jerry, Ardra F. Fitzgerald, and Warren D. Stallings, Jr. *Fundamentals of Systems Analysis.* 2d ed. New York: Wiley, 1981. A useful survey text for the systems analyst. Chapter 6, "Understanding the Existing System," does a good job of presenting fact-finding techniques in the study phase.

Gildersleeve, Thomas R. *Successful Data Processing System Analysis.* Englewood Cliffs, N.J.: Prentice-Hall, 1978. Chapter 4, "Interviewing in Systems Work," provides a comprehensive look at interviewing specifically for the systems analyst. A thorough sample interview is scripted and analyzed in this chapter.

London, Keith R. *The People Side of Systems.* New York: McGraw-Hill, 1976. Chapter 5, "Investigation Versus Inquisition," provides a very good people-oriented look at fact-finding, with considerable emphasis on interviewing.

Lord, Kenniston W., Jr., and James B. Steiner. *CDP Review Manual: A Data Processing Handbook.* 2d ed. New York: Van Nostrand Reinhold, 1978. Chapter 8, "Systems Analysis and Design," provides a more comprehensive comparison of the merits and demerits of each fact-finding technique. This material is intended to prepare data processors for the Certificate in Data Processing examinations, one of which covers Systems Analysis and Design.

Salvendy, G., ed. *Handbook of Industrial Engineering.* New York: Wiley, 1974. A comprehensive handbook for industrial engineers; systems analysts are, in a way, a type of industrial engineer. Excellent coverage on sampling and work measurement.

Stewart, Charles J., and William B. Cash, Jr. *Interviewing: Principles and Practices.* 2d ed. Dubuque, Iowa: Brown, 1978. Popular college textbook that provides broad exposure to interviewing techniques, many of which are applicable to systems analysis and design.

Weinberg, Gerald M. *Rethinking Systems Analysis and Design.* Boston: Little, Brown, 1982. Chapter 3, "Observation," and Chapter 4, "Interviewing," are both excellent reading. Every analyst should be exposed to Weinberg's interesting style and thought process. The fables, stories, and anecdotes blend well into this insightful book.

MODULE C

Communications

Skills for the

Systems Analyst

When Should You Read This Module?

You should study this module after reading *any* of the following chapters:

- Chapter 6, "How to Analyze an Information System: A Problem-Solving Approach"—All of the communications skills discussed in this module (meetings, walkthroughs, presentations, and reports) are applicable to the survey, study, definition, and selection phases.

- Chapter 12, "How to Design an Information System: Traditional and Prototyping Approaches"—All of the communications skills in the module are applicable to the design and selection phase. During systems design, communication becomes more difficult because it is at this point that technical issues are introduced.

- Chapter 20, "How to Implement and Evaluate a New Information System"—All the communications skills discussed in this module are applicable to programmers and analysts who construct and deliver the information system.

Despite the availability of improved tools and methodologies, many information systems projects still fail due to a breakdown of communications. Information systems projects are frequently plagued by communications barriers that we erect between ourselves and the end-users. The business world has its own language to describe forms, methods, procedures, financial data, and the like. And the information systems industry has its own language of acronyms, terms, buzzwords, and procedures. As was the case in the biblical Tower of Babel project, a communications gap has developed between the end-user and the data processor.

The systems analyst is supposed to bridge this communications gap. A typical project enlists the participation of a diverse audience, both technical and nontechnical. The purpose of this module is to survey interpersonal communications skills, the cornerstone of successful systems development. Because communications skills are vital in all phases of systems development, we chose to locate this survey in a module rather than in any one section of the book.

Communicating with People

Because systems are built for people and by people, understanding people is an appropriate introduction to communications skills. With whom does the analyst communicate? What words influence these people?

Three Audiences for Interpersonal Communication During Projects

For years, English and communications scholars have told us that the secret of effective oral and written communications is to know your audience. Who is the communications audience during the systems development project? We can identify at least three distinct groups:

1. The *project team,* consisting of your colleagues—other analysts, programmers, and information systems specialists.
2. Your *end-users,* the people whose day-to-day jobs will be most affected, directly or indirectly, by the new system.
3. *Management,* who in addition to being an end-user, will approve systems expenditures and eventually own the new system.

You should recognize all of these audiences as end-users. Each audience has different levels of technical expertise, different perspectives on the system, and different expectations. End-users and management, in particular, present special problems. These people have day-to-day responsibilities and time constraints. Before communicating with any of them, ask yourself the following questions:

- What are the responsibilities of, and how might the new system affect, this person?
- What is the attitude of this person toward the existing or target system? enthusiasm? skepticism? hostility? apathy?
- What kind of information about the project does this person really need or want?
- How busy is this person? How much time and attention can I reasonably hope to get?

Use of Words: Turn-Ons and Turn-Offs

We communicate with words, oral and written. How important are words? Ask any politician. The wrong words at the wrong time, no matter what the intention, and the next election is history. But that's just politics, right? No. All businesses are political. And words are important, especially to the systems analyst who must carefully communicate with a diverse group of end-users, managers, and technicians. What words affect the attitudes and decisions of your audience?

On an upbeat note, let's first talk about words and phrases that appeal to end-users and managers. Leslie Matthies (1976), a noted author and consultant in the systems development field, has identified two categories of terms that influence managers: benefit terms and loss terms. Both can be used to sell ideas. Benefit terms sell themselves. Examples include *increase productivity, reduce inventory, improve customer relations, increase sales, increase profit margin,* and *reduce risk.* The list goes on and on. Loss terms can be used to sell proposed changes. Managers will accept ideas that eliminate loss terms. Examples of loss terms include *high costs, out-of-stock inventory, higher credit losses, increased processing errors, higher taxes, delays,* and *waste.* Again, the list is nearly endless. Benefit and loss terms can be used to sell ideas.

Are there other turn-on words or phrases? Yes! People like to feel they are part of the systems development effort. Avoid using the first-person pronoun *I.* People also like words of appreciation for their time and effort—systems development is your job, not theirs, and they are helping you. Make their names and department names a vivid part of any presentation. Most of all, people want respect. Words should be carefully chosen to show respect for people's feelings, knowledge, and skills.

Now, what about turn-off words or phrases? These can kill projects by changing the attitudes and opinions of management. Let's start with the oldest turn-off, the use of jargon. Jargon is important to the analyst and technician because it helps us communicate with the computing industry and our colleagues. But jargon has no place in the business end-user's world. Avoid terms such as JCL, EBCDIC, CPU, ROM, and DOS—leave your acronyms in the CIS offices! This includes the jargon you've learned in this book. Instead of saying "This is a *DFD* of your materials handling system," try saying "This is a *picture of the work and data flow* in your materials handling operation."

Other red-flag terms include those that attack people's performance or threaten their job. Before you candidly state that the current system is ridiculously inefficient and cumbersome, consider the possibility that an end-user who had much influence on its development and approval may be in your audience. Consider threats to job security when you get ready to propose the elimination of job steps. In other words, be diplomatic and tactful when you speak.

The remainder of this module will survey specific interpersonal communications techniques—specifically meetings, presentations, and walkthroughs—and written reports.

Meetings, Presentations, and Walkthroughs

During the course of a systems project, many team meetings are held. A **meeting** is intended to accomplish an objective as a result of discussion under leadership. Some possible objectives are listed under "Meeting Purposes" in the margin. The ability to coordinate or participate in a meeting is critical to the success of any project. In this section, we'll discuss how to run a meeting and then give extra attention to two special types of meetings: formal presentations and project walkthroughs.

How to Run a Group Meeting

Many meetings are poorly organized and conducted. Meetings are an expensive use of time because they consume time that could be spent on other productive work. The more individuals involved in a meeting, the more the meeting costs. Because meetings are essential, we must strive to offset the costs by maximizing benefits (in terms of project progress) realized during the meeting. It is not difficult to run a meeting if you are well organized. Without good organization, however, the meeting may prove chaotic or worthless to the participants. When planning and conducting meetings, try following the procedure outlined here.

MEETING PURPOSES

Presentation Definition
Brainstorming Ideas
Problem Solving
Conflict Resolution
Progress Analysis
Gather Facts
Merge Facts

Determine the Need for and Purpose of the Meeting   Why do you need a meeting? Every meeting should have a well-defined purpose that can be communicated to its participants. Meetings without a well-defined purpose are rarely productive. The purpose of every meeting should be attainable within 60 to 90 minutes, because longer meetings tend to become unproductive. However, when necessary, longer meetings are possible so long as they are divided into well-defined submeetings that are separated by breaks so people can catch up on their normal responsibilities. But it must be remembered that longer meetings are more likely to conflict with the participants' day-to-day responsibilities. The effect on the job can be as if everyone took a vacation on the same day.

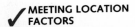

MEETING LOCATION FACTORS

Size of Room
Lighting
Noise From Outside Room
Seating Arrangement
Temperature
Audiovisual Needs

Schedule the Meeting and Arrange for Facilities   After deciding the purpose of the meeting, determine who should attend. (But take note: The larger the number of participants, the less the amount of work likely to be completed.) The participants are chosen to ensure that the purpose of the meeting will be attained. Given the number of participants, the meeting can now be scheduled. The date and time for the meeting will be subject to the availability of the meeting room and the prior commitments of the participants. Morning meetings are generally better than afternoon meetings because the participants are fresh and not yet caught up in the workday's problems. It is best to avoid scheduling meetings at the following times: late afternoon (people are too anxious to go home), before lunch, before holidays, and on the same day as other meetings for the same participants.

The meeting location is very important. The checklist in the margin identifies important factors to consider when selecting a meeting location. Seating arrangement is particularly important. If leader-group interaction is required, the group should face the leader but not necessarily other members of the group. If group-group interaction is needed, the team members, including the leader, should all face one another. Make sure that any necessary visual aids (flipchart, overhead projector, chalk, and so forth) are also scheduled for the room.

Write up an agenda for the meeting and distribute it well in advance. The agenda confirms the date, time, location, and duration of the meeting. It also states the meeting's purpose and offers a tentative timetable for discussion and questions. If participants should bring specific materials with them or review specific documents prior to the meeting, specify this in the agenda. Finally, the agenda may include any supplements—for example, reports, documentation, memoranda—that the participants will need to scan or study before or during the meeting.

Conduct the Meeting   Try to start on time, but do not start the meeting until everyone is present. If any important participant is more than 15 minutes late, then consider canceling the meeting. Once the meeting has started, try to discourage interruptions and delays, such as phone calls. Have enough copies of any handouts for all participants. Get off to a good start by listing or reviewing the agenda so that the discussion items become group property. Then, cover each item on the agenda according to the timetable developed when the meeting was scheduled. The group leader should ensure that no person dominates or is left out of the discussion. Decisions should be made by consensus opinion or majority vote. One rule is always in order: *Stay on the agenda and end on time!* If you do not finish discussing all items on the agenda, schedule another meeting.

Brainstorming can be an effective technique for generating ideas during group meetings. Contrary to what you might believe, brainstorming is a formal technique. It requires discipline. That's right! Here's how it works:

1. One person is appointed to record ideas. This person should use a flip-chart, chalkboard, or overhead projector that can be viewed by the entire group.

2. Within a specified time period, team members call out their ideas as quickly as they can think of them. *No idea is considered absurd or infeasible,* and the recorder or leader must insist that no idea be evaluated, analyzed, or otherwise critiqued at this point. Go for quantity of ideas.

   Alternatively, ideas can be generated by having each team member jot down a number of ideas. The ideas are then recorded after all members are finished.

3. After all ideas have been recorded, then and only then should they be analyzed.

Follow Up on the Meeting   As soon as possible after it is over, the minutes of the meeting should be published. The minutes are a brief, written summary of what happened during the meeting—items discussed, decisions made, and items for future consideration. The minutes are usually prepared by the *recording secretary,* a team member designated by the group leader.

How to Make a Formal, Oral Presentation

✓ PRESENTATION USES

Sell New System
Sell New Ideas
Sell Change
Head Off Criticism
Address Concerns
Verify Conclusions
Clarify Facts
Report Progress

Presentations are special meetings used to sell new ideas and gain approval for new systems. They may also be used for any of the purposes in the margin. In many cases, oral presentations set up or supplement a more detailed written presentation. Effective and successful presentations require three critical ingredients: preparation, preparation, and preparation. We're not trying to be funny. The time allotted to presentations is frequently brief; therefore, organization and format are critical issues. You cannot improvise and expect acceptance.

Presentations offer the advantage of impact through immediacy and spontaneity. The audience can respond to the presenter, who can use emphasis, timed pauses, and body language to convey messages not possible with the written word. Are there any disadvantages to the oral presentation? Yes—the material presented is easily forgotten because the words are spoken and the visual aids are transient. That's why presentations are often followed by a written report, either summary or detailed.

Preparing for the Oral Presentation   As is the case with most communication, it is particularly important to know your audience. This is especially true when your presentation is trying to sell new ideas and a new system. As Machiavelli wrote in his classic book *The Prince,*

> There is nothing more difficult to carry out, nor more dangerous to handle, than to initiate a new order of things. For the reformer has enemies in all who profit by the old order, and only lukewarm defenders in all those who would profit from the new order, this lukewarmness arising partly from fear of their adversaries—and partly from the incredulity of mankind, who do not believe in anything new until they have had actual experience of it. [From Machiavelli, Niccolo. *The Prince and Discourses,* trans. Luigi Ricci, 1940, 1950, Random House, Inc. Reprinted by permission of Oxford University Press.]

People tend to be opposed to change. There is comfort in the way things are today. To effectively present and sell change, you must be confident in your ideas and have the facts to back them up. Preparation is the key!

First define your expectations of the presentation—for instance, that you are seeking approval to continue the project, that you are trying to confirm facts, and so forth. A presentation is a summary of your ideas and proposals that is directed toward your expectations.

Executives are usually put off by excessive detail. To avoid this, your presentation should be carefully organized around the allotted time (usually 30 to 60 minutes). Although each presentation differs, you might try the following organization and time allocation:

I. Introduction (one-sixth of total time available)
   A. Problem statement
   B. Work completed to date

II. Part of the presentation (two-thirds of total time available)
   A. Summary of existing systems problems and limitations
   B. Summary description of the proposed system
   C. Feasibility analysis
   D. Proposed schedule to complete project

III. Questions and concerns from the audience (time here is not to be included in the time allotted for presentation and conclusion; it is determined by those asking the questions and voicing their concerns)

IV. Conclusion (one-sixth of total time available)
   A. Summary of proposal
   B. Call to action (request for whatever authority you require to continue systems development)

What else can you do to prepare for the presentation? Because of the limited time, use visual aids—predrawn flipcharts, overhead slides, and the like—to support your position. Just as a written paragraph does, each visual aid should convey a single idea. When preparing pictures or words, use the guidelines shown in Figure C.1. To hold your audience's attention, consider distributing photocopies of the visual aids at the start of the presentation. This way, the audience doesn't have to take as many notes.

Finally, practice the presentation in front of the most critical audience you can assemble. Play your own devil's advocate or, better yet, get somebody else to raise criticisms and objections. Practice your responses to these issues.

Conducting the Presentation   If you are well prepared, the presentation is 80 percent complete. There are a few additional guidelines that may benefit the actual presentation:

- *Dress professionally.* The way you dress influences people. John T. Malloy's books *Dress for Success* (now in its second edition) and *The Woman's Dress for Success Book* are excellent reading for both wardrobe advice and the results of studies regarding the effect of clothing on management.

- *Avoid the word I* when making the presentation. Use *you* and *we* to assign ownership of the proposed system to management, which is as it should be.

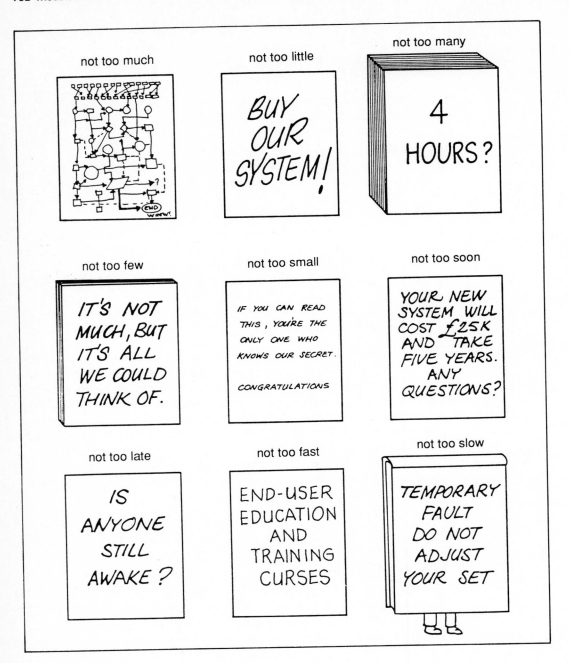

FIGURE C.1 Guidelines for Visual Aids Properly prepared visual aids both enhance and expedite a presentation. (Copyright Keith London. Reproduced by permission of Curtis Brown, Ltd.)

- *Maintain eye contact with the group and keep an air of confidence.* If you don't show management that you believe in your proposal, why should they believe in it?

- *Be aware of your own mannerisms.* Some of the most common mannerisms include using too many hand gestures, pacing, and repeatedly saying "you know" or "okay." Although mannerisms alone don't contradict the message, they can distract the audience.

Walkthroughs: A Peer Group Review Technique

✓ DOCUMENTATION
THAT CAN BE
VERIFIED THROUGH
WALKTHROUGHS

Data Flow Diagrams
Entity-Relationship Diagrams
Small Data Dictionaries
Policies and Procedures
Systems Flowcharts
Output Designs
Input Designs
File and Database Designs
Terminal Dialogue Designs
Program Structure Charts
Program Flowcharts
Program Test Data
Program Code
Program Documentation
User Manuals

The **walkthrough**, another type of meeting, is a peer group review of systems development documentation. Walkthroughs may be used to verify virtually any type of detailed documentation (some of these are listed for you in the margin). Why does peer group review tend to identify errors that go unnoticed by the analyst who prepared the documentation? Consider, if you will, the last paper or report you wrote. You probably gave that report to a colleague or teacher to review. And that colleague or teacher caught obvious errors that you didn't, right? Why didn't you catch them? Because like any author, you have mental blocks that prevent you from discovering errors in your own products.

Who Participates in the Walkthrough?   A walkthrough group should consist of seven or fewer participants. All parties in the walkthrough should be treated as equals. The analyst who prepared the documentation to be reviewed should present that documentation to the group during the walkthrough. Another analyst or key end-user is appointed as *coordinator* of the walkthrough. The coordinator schedules the walkthrough and ensures that each participant gets the documentation well before the meeting date. The coordinator also makes sure that the walkthrough is properly conducted and mediates disputes and problems that arise during the walkthrough. The coordinator has the authority to ask participants to stop a disagreement and move on. Finally, the coordinator designates a *recorder* to take notes during the walkthrough.

The remaining participants include end-users, analysts, or specialists who evaluate the documentation. These reviewers may also assume roles. For example, some reviewers may evaluate the accuracy of the documentation, while other reviewers comment on quality, standards, and technical issues. Participants must be willing to devote enough time to details. However, walkthroughs should never last more than 90 minutes. Our experience indicates that end-users particularly enjoy walkthroughs because the meetings encourage a sense of personal importance in the project.

How to Conduct a Walkthrough   All participants must agree to follow the same set of rules and procedures. Also, the participants must agree to review the *documentation;* this should not be done by the person who prepared the documentation. The basic purpose of the walkthrough is error *detection,* not error *correction.* The analyst who is presenting the documentation should seek only whatever clarification is needed to correct the errors. This maximizes the use of time! The analysts should not argue with the reviewers' comments. A defensive posture inhibits constructive criticism. The coordinator is responsible for seeing that these rules are properly outlined and followed. Reviewers should be encouraged to offer at least one positive and one negative comment in order to guarantee that the walkthrough is not superficial.

After the walkthrough, the coordinator asks the reviewers for a recommendation. There are three possible alternatives:

1. Accept the documentation in its present form.
2. Accept the documentation with the revisions noted.
3. Request another walkthrough because a large number of errors were found or because criticisms created controversy.

The walkthrough should be promptly followed by a report from the coordinator. The report contains a management summary that states what was reviewed, when the walkthrough occurred, who attended, and the final verdict. A sample set of forms used for walkthroughs in a real company are displayed in Figure C.2 (pages 805 and 806).

Written Reports

The business and technical report is the primary method used by analysts to communicate information about a systems development project. The purpose of the report is to either inform or persuade, possibly both. In these few pages, it is not possible to provide a comprehensive discussion of report writing. But because people make judgments about who we are and what we can accomplish based on our writing ability, we can offer some motivation for further study and some guidelines for current practice.

Business and Technical Reports for Systems Development

What types of formal reports are written by the systems analyst? Content outlines for several reports can be found in Chapters 6, 12, and 20, which place those reports in the context of the systems development life cycle phases. But an overview (or review) is appropriate here.

Systems Analysis Reports   The first major phase of the life cycle is the *study of the current system.* After completing this phase, the analyst normally prepares a report to verify with end-users their understanding of the current

WALKTHROUGH REPORT

Coordinator	Project

Segment for Review

Coordinator's checklist:

1. Confirm with developer that material is ready and stable _____

2. Issue invitations, assign responsibilities, distribute materials

   Date _____ Time _____ Duration _____

   Place _____

Responsibilities	Participants		Can attend	Received materials?
_____ _____	_____		_____	_____
_____ _____	_____		_____	_____
_____ _____	_____		_____	_____
_____ _____	_____		_____	_____
_____ _____	_____		_____	_____

Agenda:

_____ 1. All participants agree to follow the *SAME* set of rules.

_____ 2. New segment: walkthrough of material

_____   Old segment: item-by-item checkoff of previous action list

_____ 4. Group decision

_____ 5. Deliver copy of this form to project management.

Decision: _____   Accept product as is

              _____   Revise (no further walkthrough)

              _____   Revise and schedule another walkthrough

Signatures		

FIGURE C.2   Walkthrough Form   This standardized walkthrough form can be completed by the recorder, duplicated, and distributed to all meeting participants as a record of the walkthrough. (Courtesy of Cummins Diesel Engine, Columbus, IN)

WALKTHROUGH ACTION LIST— SCRIBE'S REPORT

Coordinator	Scribe	Date
Project	Segment	

=
fixed   Issues raised in review

FIGURE C.2   (Continued)

system and analyses of problems, limitations, and constraints in that system. This report might be titled a "Study and Analysis of the Current [insert name] System."

The second phase of the life cycle, *define end-user requirements,* results in a specification document called a *requirements statement.* This requirements statement, often large and complex, is rarely written up as a report to end-users and management. It is best reviewed in walkthroughs (in small pieces) with end-users and maintained as a reference for analysts and programmers.

The next formal report, the **systems proposal**, is generated after the *selection* phase has been completed. This report combines an outline of the end-user requirements from the definition phase with the detailed feasibility analysis of alternative solutions that fulfill those requirements. The report concludes with a recommended or proposed solution. This report is normally preceded or followed by an oral presentation to those managers and executives who will decide on the proposal.

Systems Design Reports The *design phase* results in detailed design specifications that are often organized into a technical design report. This report is quite detailed and is primarily intended for information systems professionals. It tends to be quite a large report because it contains numerous forms and charts.

The *acquisition phase* of systems development is only undertaken if the new system requires the purchase of new hardware or software. Several reports can be generated during this phase. The most important report—the **request for proposals**—is used to communicate requirements to prospective vendors who may respond with specific proposals. It was covered in Chapter 12. Especially when the selection decision involves significant expenditures, the analyst may have to write a report that defends the recommended proposal to management.

Systems Implementation Reports In a sense the most important report of all is written during the *construction* and *delivery* phases. Actually, it isn't a report; it's a manual—specifically, it's a user's manual and reference guide. This document explains how to use the computer system (such as what keys to push, how to react to certain messages, where to get help). How well this manual is written will frequently determine how many phone calls you'll get over the months that follow the conversion to the new system. In addition to computer manuals, the analyst may rewrite the standard operating procedures for the system. A standard operating procedure explains both the noncomputer and computer tasks and policies for the new system.

Length of a Written Report

Unfortunately, the written report is the most abused method used by analysts to communicate with end-users. We have a tendency to generate large, voluminous reports that look quite impressive. Sometimes such reports are necessary, but often they're not. If you lay a 300-page technical report on a manager's desk, you can expect that manager will skim it but not read it—and you can be certain it won't be studied carefully!

Report size is an interesting issue. After many bad experiences, we have learned to use the following guidelines to restrict report size:

To executive-level managers—one or two pages

To middle-level managers—three to five pages

To supervisory-level managers—less than ten pages

To clerk-level personnel—less than fifty pages

It is possible to organize a larger report to include subreports for managers who are at different levels. These subreports are usually included as early sections in the report and summarize the report, focusing on the bottom line: What's wrong; what do you suggest or want?

Organizing the Written Report

There is a general pattern to organizing any report. Every report consists of primary and secondary elements. **Primary elements** present the information. **Secondary elements** package the report so the reader can easily identify the report and its primary elements. Secondary elements also add professional polish to the report.

Primary Elements   As indicated in Figure C.3, the primary elements can be organized in one of two formats: factual and administrative. The *factual format* is very traditional and best suited to readers who are interested in facts and details as well as conclusions. This is the format we would use to specify detailed requirements and design specifications to end-users. On the other hand, the factual format is not appropriate for most managers and executives.

The *administrative format* is a modern, result-oriented format preferred by many managers and executives. This format is designed for readers who are interested in results, not facts. Notice that it presents conclusions or recommendations first. Any reader can read the report straight through, until the point at which the level of detail exceeds their interest.

Both formats include common elements. The *introduction* should include four components: purpose of the report, statement of the problem, scope of the project, and a narrative explanation of the contents of the report (sequential). The *methods and procedures* section should briefly explain how the information contained in the report was developed—for example, how the

Factual Format	Administrative Format
I. Introduction II. Methods and procedures III. Facts and details IV. Discussion and analysis of facts and details V. Recommendations VI. Conclusion	I. Introduction II. Conclusions and recommendations III. Summary and discussion of facts and details IV. Methods and procedures V. Final conclusion VI. Appendixes with facts and details

FIGURE C.3   Alternative Formats for Reports   The factual format is used to place emphasis on details. The administrative format is used to place emphasis on conclusions and recommendations.

study was performed or how the new system was designed. The bulk of the report will be in the *facts* section. This section should be named to describe the type of factual data to be presented (for example, "Existing Systems Description," "Analysis of Alternative Solutions," or "Design Specifications"). The *conclusion* should briefly summarize the report, verifying the problem statement, findings, and recommendations.

Secondary Elements   Figure C.4 shows the secondary, or packaging, elements of the report and their relationship to the primary elements. Many of

Letter of transmittal

Title page

Table of contents

List of figures, illustrations, and tables

Abstract or executive summary

(The primary elements—the body of the report, in either the factual or administrative format—are presented in this portion of the report.)

Appendixes

FIGURE C.4   Secondary Report Elements   Secondary elements are used to package the report. The same secondary elements are common to both the factual and administrative formats. Secondary elements exhibit organization and professionalism.

these elements are self-explanatory. We briefly discuss here those that may not be. No report should be distributed without a *letter of transmittal* to the recipient. This letter should be clearly visible, not inside the cover of the report. A letter of transmittal states what type of action is needed on the report. It can also call attention to any features of the project or report that deserve special attention. In addition, it is an appropriate place to acknowledge the help you've received from various people.

The *abstract* or *executive summary* is the one- or two-page capsule summary of the entire report. It helps the reader decide if the report contains information they need to know. It can also serve as the highest-level summary report. Virtually every manager reads these summaries. Most managers will read on, possibly skipping the detailed facts and appendixes.

Writing the Business or Technical Report

This is not a writing textbook. Do avail yourselves of every opportunity to improve your writing skills, through business and technical writing classes, books, audiovisual courses, and seminars. Writing can greatly influence career paths in any profession. Figure C.5 illustrates the proper procedure for writing a formal report. Some of the guidelines follow:

- *Paragraphs should convey a single idea.* They should flow nicely, one to the next. Poor paragraph structure can almost always be traced to outlining deficiencies.
- *Sentences should not be too complex.* The average sentence length should not exceed 20 words. Studies suggest that sentences longer than 20 words are more difficult to read and understand.
- *Write in the active voice.* The passive voice becomes wordy and boring when used consistently.
- *Eliminate jargon* (replace "DBMS" by "database management system software," and so on), *big words*, and *deadwood* (for example, substitute *so* for *accordingly*; try *useful* instead of *advantageous*; use *clearly* instead of *it is clear that*).

Get yourself a copy of *The Elements of Style* by William S. Strunk, Jr., and E. B. White. This classic little paperback may set an all-time record in value-cost ratio. Just barely bigger than a pocket-sized book, it is a virtual goldmine. Anything we might suggest about grammar and style can't be said any more clearly than in *The Elements of Style*.

Summary

Because the systems analyst is expected to bridge the language barrier between business end-users and information systems technicians, communications skills

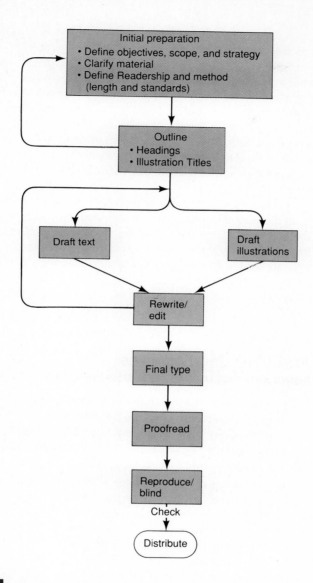

FIGURE C.5   How to Write a Report   (Copyright Keith London. Reproduced by permission of Curtis Brown, Ltd.)

have become an important part of the analyst's tool kit. Three distinct groups of people are part of the analyst's audience: colleagues, end-users, and management. Before directing any communication to any of these audiences, the analyst should profile the audience.

Systems analysts can expect to spend a considerable amount of time in meetings. To maximize the use of meeting time, the analyst should follow the following steps: determine the purpose of the meeting, schedule the meeting at an appropriate time and arrange for adequate facilities, conduct the meeting according to a preestablished agenda, and follow up on meeting results. If the meeting is intended to generate ideas, brainstorming is an effective technique. Oral presentations are a special type of meeting at which a person presents conclusions, ideas, or proposals to an interested audience. Preparation is the key to effective presentations. Walkthroughs are peer group evaluation meetings that seek to identify (but not correct) errors in systems development documentation.

Written reports are the most common communications vehicle used by analysts. Reports consist of primary and secondary elements. Primary elements contain factual information. Secondary elements package the report for ease of use. Reports may be organized in either the factual or administrative format. The factual format presents details before conclusions; the administrative format reverses that order. Managers like the administrative format because it is results-oriented and gets right to the bottom-line question, "What next?"

Problems and Exercises

1. The secret of effective oral and written communications is to know your audience. What are some of the things you would want to know about your audience prior to making a formal presentation to them? How could this knowledge be used to your advantage in formulating a presentation?

2. Identify the audience or audiences for each of the following systems development reports (chapter references are provided so you can skip those reports that you haven't covered):
   a. Feasibility survey (Chapter 6)
   b. Study and analysis of the current information system (Chapter 6)
   c. Requirements statement (Chapter 6)
   d. Systems proposal (Chapter 6)
   e. Design specifications for the new information system (Chapter 12)
   f. Request for proposals for hardware and software (Chapter 12)
   g. Analysis for proposals for hardware and software (Chapter 12)
   h. Program runbook (Chapter 20)
   i. End-user guide or training manual (Chapter 20)
   j. Postimplementation review (Chapter 20)
   k. Project progress report (Module A)
   l. Walkthrough report (Module B)
   m. Interview report (Module B)

3. Get permission to attend a board meeting or subcommittee meeting of a local organization (such as the Data Processing Management Association or the Association of Computing Machinery), school committee, or some other business meeting. Observe how the meeting is run by the leader. What was the purpose of the meeting? Did the purpose of the meeting appear to be understood by all the participants? Why or why not? Did the meeting start and end on time? If not, what caused the delay(s)? Was the meeting room reserved ahead of time? Did the room provide a comfortable atmosphere? Were there any problems with the meeting location? Was an agenda distributed prior to or during the meeting? Did the leader follow the agenda during the meeting? Were arrangements made for the minutes of the meeting to be published and distributed to appropriate individuals?

4. Analyze a team meeting for a team project to which you've been assigned. Use the criteria as in Exercise 3.

5. While attending the next lecture in each of your classes, observe the instructor's presentation of class material. Make a note of and learn from the techniques the professor uses to clearly deliver difficult material. If you feel comfortable about discussing the lecture with your professor, discuss your findings.

6. Arrange a formal walkthrough of one of your systems analysis and design assignments. Try to include your instructor and a few students. Prepare a walkthrough report.

7. In a current or future programming class, discuss the possibility of formal walkthroughs on one programming assignment. You'll need to secure permission so that your instructor will not consider the walkthrough cheating. Conduct walkthroughs as soon as you've completed the program design and immediately after coding (but before you compile or interpret the program). Analyze the impact the walkthroughs had on your productivity by comparing the number of compiles that you required to finish the assignment against the number of compiles that programmers who didn't use walkthroughs required.

8. Can word processing software packages aid your development of written communications skills? How?

9. Why do formal presentations usually accompany written reports?

10. Systems analysts have a tendency to generate written reports that are much too large for managers to read. How would you handle a size problem with a technical report?

11. What are some of the ways you might improve your written communications skills? Identify specific courses that help you improve your skills.

12. Try to obtain a systems development report outline or table of contents from an information systems shop. Was the report organized using the

factual format or the administrative format? Do you think everybody in the audience for the report read that report? Why or why not? Reorganize the outline or table of contents into an alternative format. Be sure to include secondary elements, even if they weren't included in the original report.

13. Take one of the report outlines in Chapters 6, 12, or 20 and prepare formal outlines that include primary and secondary elements for both the factual and administrative formats.

Annotated References and Suggested Readings

Gallagher, William J. *Writing the Business and Technical Report.* Boston, Mass.: CBI Publishing, 1978. A good professional development book for managers and professionals in any field.

Gildersleeve, Thomas R. *Successful Data Processing Systems Analysis.* Englewood Cliffs, N.J.: Prentice-Hall, 1978. Gildersleeve doesn't talk too much about tools in his books—that's why we like him! Chapter 5 discusses presentations, and Chapter 10 discusses interpersonal relations. They are worthwhile additional readings for any analyst.

London, Keith. *The People Side of Systems.* New York: McGraw-Hill, 1976. Chapters 9 through 12 provide an excellent treatment of presentations, reports, group meetings, and special illustrative techniques. Perhaps the best concentrated unit on communications skills in any single systems textbook we've seen.

Machiavelli, Niccolò. *The Prince and Discourses,* trans. Luigi Ricci. New York: Random House, 1950. Who says the classics aren't practical reading for students in management and business schools? Because it documents the power and political struggles that exist in both nations and organizations, this work is equally valuable to the systems analyst.

Malloy, John T. *The Woman's Dress for Success Book.* New York: Warner, 1975. The working woman's version of Malloy's successful book on how to dress for power and respect.

Malloy, John T. *Dress for Success,* 2nd ed. New York: Warner, 1987. Based on this best-selling book, John Malloy has been labeled "America's first wardrobe engineer." Like its sequel for women, this book teaches people how to dress for power and prestige. The guidelines are based on research conducted by Malloy.

Matthies, Leslie H. *The Management System: Systems Are for People.* New York: Wiley, 1976. Chapter 10 in Matthies' book is all about how to present and sell a new system to management. Some concepts we use were initially presented in this book.

Smith, Randi Sigmund. *Written Communication for Data Processing.* New York: Van Nostrand Reinhold, 1976. An excellent book on written communications for DP professionals—not just reports, but memos and letters too!

Strunk, William, Jr., and E. B. White. *The Elements of Style.* New York: Macmillan, 1979. A short handbook that summarizes everything most of us have forgotten from elementary school, high school, and college grammar classes. Many colleges require their students to have a copy. And we personally recommend that you keep that copy, at least until the English language becomes extinct.

Stuart, Ann. *Writing and Analyzing Effective Computer System Documentation.* New York: Holt, Rinehart & Winston, 1984. At last! A book for students about writing in the information systems environment. And a good book at that. Must reading!

MODULE D

Feasibility and

Cost-Benefit

Analysis Techniques

When Should You Read This Module?

This module will be most valuable if studied after *any* of the following chapters have been studied:

- Chapter 6, "How to Analyze an Information System: A Problem-Solving Approach"—Feasibility analysis is an integral part of three systems analysis phases: survey, study, and evaluation.
- Chapter 10, "How to Design an Information System: Traditional and Prototyping Approaches"—Feasibility is reassessed during the design and selection phases.

The Importance of Feasibility and Cost-Benefit Analyses

The best way to develop an appreciation for this module is to read the following scenario, one too often experienced by systems analysts.

"As you can see, we've designed a system that can fulfill all the needs you specified earlier in the project. Furthermore, this on-line system is very easy to learn and friendly to use. That concludes my presentation," Frank said. "Are there any questions?"

Benjamin Pierce, senior manager, responded, "That was quite a presentation, Frank. And the system is every bit as good as you claim. But how much is this going to cost me?"

"To develop the system, we estimate that it will cost $45,000. And the system will cost approximately $4,500 per year to operate. That includes an annual allowance for minor maintenance."

"And how much have we spent up to this point?" asked Ben.

"About $12,000 for analysis and design," Frank answered. "But that money can't be recovered, so we shouldn't consider it in our decision on whether to proceed."

Ben's tone turned more serious. "I agree that the project shouldn't be continued or canceled solely on the basis of the money spent so far. However, I disagree that the money is irrelevant. If the project is continued, I should think that the new system would eventually pay for itself. By my calculation, development costs total $57,000. And if the system will incur another $4,500 expenses per year, how long until it pays for itself?"

Frank answered, "You'll start receiving benefits immediately."

Ben countered, "But how many months or years will pass before the lifetime benefits surpass the lifetime costs?" There was a moderate silence after which Ben continued, "Look, I have the money you need to develop and operate this system. But I also have managers who are asking for new equipment and for facilities upgrades. Why should I give you the money and deny it to them?"

Frank seemed confused, "I don't understand. This system fulfills the needs that *you* and your staff specified, Ben."

Ben replied, "Frank, you are a marvelous computer professional. You understand my business work flow and needs. And you can design some of the best computer solutions to those needs that I've ever seen. But you've got a lot to learn about business."

This scenario demonstrates that analysts must learn to think in business terms. Computer applications are expanding at a record pace, and now more than ever, management expects information systems to pay for themselves. Information systems are a capital expenditure that must be justified, just as marketing must justify a new product and manufacturing must justify a new production plant. Will an investment return more money than it will cost? Are there other investments that will return even more money? In our opening scenario, Frank didn't pay attention to the fact that business is driven by economics! Dollars and cents! Frank *knows* this, and so do you; but we all easily forget that point when it comes to our fascination with the computer and the potential it holds for information systems in business.

This module deals with cost-benefit analysis and other feasibility issues of interest to the systems analyst. Few topics are more important! Feasibility analysis isn't really *systems analysis,* and it isn't really *systems design* either. Indeed, we consider feasibility analysis to be an ongoing activity of the systems development life cycle.

What Is Feasibility and When Do You Evaluate It?

Feasibility is a measure of how beneficial the development of an information system would be to an organization. **Feasibility analysis** is the process by which feasibility is measured. Feasibility analysis should be performed throughout the systems development life cycle. We call this a *creeping commitment approach* to feasibility. The scope and complexity of an apparently feasible project can change after the current problems are fully understood or after the end-users' needs have been defined in detail or after technical requirements are established. Thus, a project that is feasible at one stage of systems development may become less feasible or infeasible at a later checkpoint.

Feasibility Checkpoints in a Project

If you study your company's project standards or SDLC, you'll probably see a feasibility study, if it exists, but not the ongoing feasibility analysis. But look more closely! You'll probably see *go/no go* checkpoints or *management review* checkpoints. These checkpoints identify specific times during systems development at which feasibility is reevaluated. A project can be canceled, or resource estimates can be changed at such a time. Where are these checkpoints in a typical project?

Feasibility checkpoints can be installed into any systems development life cycle that you are using. Figure D.1 shows checkpoints added to our life cycle. The checkpoints are represented by diamonds. The diamonds indicate management decisions to be made after a phase is complete. A project may be canceled or revised at any checkpoint, despite whatever resources have been spent thus far.

This idea may bother you at first. Your natural inclination would be to justify continuing the project on the basis that you've already spent time and money. But the resources you have already spent cannot be recovered. They're *sunk*! A fundamental principle of management is never to throw good money after bad money. If the project is now infeasible, minimize your losses— cancel or revise the project scope! Let's briefly survey these checkpoints because each has a different perspective on feasibility.

Survey Phase Checkpoint   The first feasibility analysis is conducted during the survey phase. At this early stage of the project, feasibility is rarely more than a measure of the *urgency of the problem* and *first-cut estimate of development costs.* Do these problems or opportunities warrant the cost of a detailed study of the current system? Realistically, feasibility can't be accurately measured until the problems (study phase) and requirements (definition phase) are better understood.

Study Phase Checkpoint   The next checkpoint occurs after a study of the current system. Because the problem(s) are better understood, the analysts can make *better* estimates of development costs and the value of solving the problems. The minimum value of solving a problem is equivalent to the cost of that problem. For example, if inventory carrying costs are $35,000 over

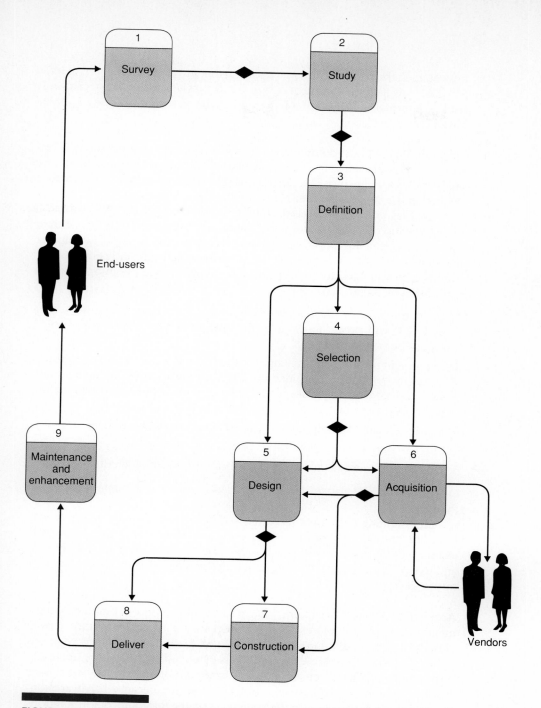

FIGURE D.1 Feasibility Checkpoints in the Systems Development Life Cycle This is our familiar SDLC with checkpoints, indicated by diamonds, for feasibility analysis. At any checkpoint, analysts and management reevaluate feasibility and determine whether to cancel, revise, or continue the project. The decision is part of the charter. This approach is called creeping commitment.

budget, then the potential benefit of solving that inventory problem is at least $35,000. However, development costs are still *guesstimates* since specific end-user requirements (definition phase) and technical solutions (design phase) haven't been determined.

Selection Phase Checkpoint   In some SDLCs, our selection phase *is* the feasibility study. Certainly, the selection phase is the most detailed feasibility analysis checkpoint since it is during this phase that alternative solutions are defined and a target computer-based solution is selected.

Problems and general requirements are defined prior to the selection phase. During this phase, alternative solutions are defined in terms of their input and output methods, storage requirements and methods, computer hardware and software requirements, processing methods, and people implications. The following solutions present the typical range of options that can be defined by the analyst:

1. Do nothing! Leave the current system alone. To ensure that any other option(s) will prove superior to the current system, this option should always be considered.

2. Implement noncomputer changes to the manual methods and procedures currently being used. This includes streamlining activities, reducing duplicate and unnecessary tasks, reorganizing office layouts, and eliminating redundant and unnecessary forms, among others.

3. Modify, enhance, or supplement an existing computer-based system.

4. Purchase a packaged applications software system.

5. Design and construct a customer applications system, including software.

Of course, an alternative may be a variation on or combination of these themes.

After defining these options (discussed in Chapter 6), each option is analyzed for operational, technical, and economic feasibility. This chapter will examine these three classes of feasibility criteria. One alternative is chosen, and that solution is passed along to the design and/or acquisition phases of our project.

Acquisition Phase Checkpoint   Because the selection of computer equipment and software involves economic decisions that usually require sizable outlays of cash, it shouldn't surprise you that feasibility analysis is required before a hardware or software contract is extended to the vendor. It should be pointed out that the selection phase (covered more extensively in Chapter 10) may be consolidated into the evaluation phase, because hardware/software selection may have a significant impact on the feasibility of the solutions being considered.

Design Phase Checkpoint   A final checkpoint is installed after the new system is designed. The detailed design specifications have been completed. Therefore, the complexity of the solution is well understood. Because computer programming (Construct the New System) and implementation (Deliver the New System) are the most time-consuming and expensive phases, the checkpoint after design gives us one last chance to update our project estimates and reevaluate feasibility.

Three Tests for Project Feasibility

So far, we've defined feasibility and determined when to perform a feasibility analysis. Most analysts have agreed that there are three categories of feasibility tests:

1. **Operational feasibility** is a measure of how well the solution of problems or a specific alternative solution will work in the organization. It is also a measure of how people feel about the system.
2. **Technical feasibility** is a measure of the practicality of a specific technical solution and of the availability of technical resources. It is also a measure of schedule feasibility, if deadlines are established.
3. **Economic feasibility** is a measure of the cost-effectiveness of a project or solution (this is often called a cost-benefit analysis).

In the remainder of this module we will focus on each of these feasibility criteria.

Measuring Project Feasibility

Recall the scenario that introduced this module. Frank's proposed system was very likely operationally and technically feasible. The people seemed to like the system because it filled their requirements. And Frank was unlikely to propose a technical solution he couldn't implement. Operational and technical feasibility criteria measure the worthiness of a problem or solution. Operational feasibility is people-oriented. Technical feasibility is computer-oriented.

But economic feasibility was questioned by the manager who had to make the financial decisions. Economic feasibility deals with the costs and benefits of the information system. The proposed project may very well have been economically feasible, but we can't tell because Frank didn't analyze costs and benefits. Actually, few systems are infeasible. Instead, different options tend to be *more* or *less* feasible than others. Let's take a closer look at the three feasibility criteria.

Operational Feasibility Criteria

Operational feasibility criteria measure the urgency of the problem (survey and study phases) or the acceptability of a solution (selection, acquisition, and design phases). How do you measure operational feasibility? There are two aspects of operational feasibility to be considered:

1. Is the problem worth solving or will the solution to the problem work?
2. How do the end-users and management feel about the problem (solution)?

Is the Problem Worth Solving or Will the Solution to the Problem Work?

Do you recall the PIECES framework for identifying problems that was discussed in Chapters 4 and 6? PIECES can be used as the basis for analyzing the urgency of a problem or the effectiveness of a solution. The following is a list of the questions that address these issues:

P *Performance.* Does the system provide adequate throughput and response time? *(Note:* the term *system,* used throughout this discussion, may refer either to the existing system or a proposed system solution, depending on which phase you're currently working in.)

I *Information.* Does the system provide end-users and managers with timely, pertinent, accurate, and usefully formatted information?

E *Economy.* No, we are not prematurely jumping into economic feasibility! The question here is, "Does the system offer adequate service level and capacity to reduce the costs of the business or increase the profits of the business?"

C *Control.* Does the system offer adequate controls to protect against fraud and embezzlement and to guarantee the accuracy and security of data and information?

E *Efficiency.* Does the system make maximum use of available resources including people, time, flow of forms, minimum processing delays, and the like?

S *Services.* Does the system provide desirable and reliable service to those who need it? Is the system flexible and expandable?

How Do the End-Users and Managers Feel About the Problem (Solution)?

It's not only important to evaluate whether a system *can* work. We must also evaluate whether a system *will* work. A workable solution might fail because of end-user or management resistance. The following questions address this concern:

- Does management support the system?

- How do the end-users feel about their role in the new system?
- What end-users or managers may resist or not use the system? People tend to resist change. Can this problem be overcome? If so, how?
- How will the working environment of the end-users change? Can or will end-users and management adapt to the change?

Technical Feasibility Criteria

Technical feasibility can only be evaluated after those phases during which technical issues are resolved, namely, after the evaluation and design phases of our life cycle have been completed. Today, very little is technically impossible. Consequently, technical feasibility looks at what is *practical* and *reasonable*. Technical feasibility addresses three major issues:

1. Is the proposed technology or solution practical?
2. Do we currently possess the necessary technology?
3. Do we possess the necessary technical expertise, and is the schedule reasonable?

Is the Proposed Technology or Solution Practical?   The technology for any defined solution is normally available. The question is whether that technology is mature enough that it can be easily applied to our problems. Some firms like to use state-of-the-art technology. But most firms prefer to use mature and proven technology. A mature technology has a larger customer base for obtaining advice concerning problems and improvements.

Do We Currently Possess the Necessary Technology?   Assuming the solution's required technology is practical, we must next ask ourselves, "Is the technology available in our information systems shop?" If the technology is available, we must ask if we have the capacity. For instance, "Will our current printer be able to handle the new reports and forms required of a new system?"

If the answer to either of these questions is no, then we must ask ourselves, "Can we get this technology?" The technology may be practical and available, and yes, we need it. But we simply may not be able to afford it at this time. Although this argument borders on economic feasibility, it is truly technical feasibility. If we can't afford the technology, then the alternative that requires the technology is not practical and is technically infeasible!

Do We Possess the Necessary Technical Expertise and Is the Schedule Reasonable?   This consideration of technical feasibility is often forgotten during feasibility analysis. We may have the technology, but that doesn't mean we have the skills required to properly apply that technology. For instance, we may have a database management system. However, the analysts and programmers available for the project may not know that DBMS well enough to properly apply it. True, all information systems professionals can learn new technologies. But that learning curve will impact the technical feasibility of the project; specifically, it will impact the schedule.

Given our technical expertise, are the project deadlines reasonable? Some projects are initiated with specific deadlines. You need to determine whether the deadlines are mandatory or desirable. For instance, a project to develop a system to meet new government reporting regulations may have a deadline that coincides with when the new reports must be initiated. Penalties associated with missing such a deadline may make meeting it mandatory. If the deadlines are desirable rather than mandatory, the analyst can propose alternative schedules.

It is preferable (unless the deadline is absolutely mandatory) to deliver a properly functioning information system two months late than to deliver an error-prone, useless information system on time! Missed schedules are bad. Inadequate systems are worse! It's a choice between the lesser of two evils.

Economic Feasibility Criteria

The bottom line in many projects is economic feasibility. During the early phases of the project, economic feasibility analysis amounts to little more than judging whether the possible benefits of solving the problem are worthwhile. Costs are practically impossible to estimate at that stage because the end-user's requirements and alternative technical solutions have not been identified. However, as soon as specific requirements and solutions have been identified, the analyst can weigh the costs and benefits of each alternative. This is called a cost-benefit analysis. Cost-benefit analysis is discussed in the last major section of this module.

The Bottom Line on Operational, Technical, and Economic Feasibility

You have learned that any alternative solution can be evaluated according to three criteria: operational, technical, and economic feasibility. How do you pick the best solution? It's not always easy. Operational and economic issues often conflict. For example, the solution that provides the best operational impact for the end-users may also be the most expensive and, therefore, the least economically feasible. The final decision can only be made by sitting down with end-users, reviewing the data, and choosing the best overall alternative.

Cost-Benefit Analysis Techniques

Economic feasibility has been defined as a cost-benefit analysis. How do you estimate costs and benefits? And how do you compare those costs and benefits to determine economic feasibility? Most schools offer complete courses on these subjects. They are usually called *Financial Management, Financial Decision Analysis,* or *Engineering Economics and Analysis.* Such a course should be included in your plan of study. This section presents an overview of the techniques.

How Much Will the System Cost?

Costs fall into two categories. There are costs associated with developing the system. And there are costs associated with operating a system. The former can be estimated from the outset of a project and should be refined at the end of each phase of the project. The latter can only be estimated once specific computer-based solutions have been defined (during the selection phase or later). Let's take a closer look at the costs of information systems.

The costs of developing an information system can be classified according to the phase in which they occur. Systems development costs are usually one-time costs that will not recur after the project has been completed. Many organizations have standard cost categories that must be evaluated. In the absence of such categories, the following lists should help:

- *Personnel costs*—the salaries of systems analysts, programmers, consultants, data-entry personnel, computer operators, secretaries, and the like who work on the project. Because many of these individuals spend time on many projects, their salaries should be prorated to reflect the time spent on the projects being estimated.

- *Computer usage*—Computer time will be used for one or more of the following activities: programming, testing, conversion, word processing, maintaining a project dictionary, prototyping, loading new data files, and the like. If a computing center charges for usage of computer resources, the cost should be estimated.

- *Training*—If computer personnel or end-users have to be trained, the training courses may incur expenses. Packaged training courses may be charged out on a flat fee per site, a student fee (such as $395 per student), or an hourly fee (such as $75 per class hour).

- *Supply, duplication, and equipment costs.*

- *Cost of any new computer equipment and software.*

Sample development costs for a typical solution are displayed in Figure D.2.

Estimated Costs for the On-Line System Alternative

DEVELOPMENT COSTS:

Personnel:

2 Systems analysts (400 hours/ea @ $35.00/hr)	$28,000
4 Programmers (250 hours/ea @ $25.00/hr)	25,000
1 Operator (50 hours @ $10.00/hr)	500
1 Secretary (75 hours @ $6.00/hr)	450
3 Data entry clerks (during file conversion—40 hours/ea @ $5.00/hr)	600

Computer usage:

500 hours @ $25.00	12,500

Supplies and expense:

Training (database—3 persons @ $395/person	1,185
Training users (150 hours @ $10.00/hr)	1,500
Duplication	300

New equipment:

2 Personal computers configured to emulate a terminal—also include printers	14,000
5 CRT terminals	2,500
7 New desks for office personnel	1,400

ANNUAL OPERATING COSTS (not incurred in existing system)

Personnel:

Systems analysts (maintenance—80 hours/year @ $35.00/hr)	2,800
Programmers (maintenance—200 hours/year @ $25.00/hr)	5,000
1 additional office clerk—2,000 hours/year @ $6.00/hr	12,000

Computer usage:

2,000 hours/year @ $45.00/hr—includes overhead	90,000

Supplies and expenses:

Prorated renewal of database software license	1,000
Preprinted forms (15,000/year @ .22/form)	3,300

FIGURE D.2   Costs for a Proposed Systems Solution   The costs for a proposed information system should be itemized into development costs and operating costs.

Almost nobody forgets systems development budgets when itemizing costs. On the other hand, it is easy to forget that a system will incur costs after it has been placed into operation. The lifetime benefits must recover both the developmental and operating costs. Unlike systems development costs, operating costs tend to recur throughout the lifetime of the system. The costs of operating a system over its useful lifetime can be classified as fixed and variable.

Fixed costs occur at regular intervals but at relatively fixed rates. Examples of fixed operating costs include

- Lease payments and software license payments
- Prorated salaries of information systems operators and support personnel (although salaries tend to rise, the rise is gradual and tends not to change dramatically from month to month)

Variable costs occur in proportion to some usage factor. Examples include

- Costs of computer usage (CPU time used, terminal connect time used, storage used) vary with the workload.
- Supplies (preprinted forms, printer paper used, punched cards, floppy disks, magnetic tapes, and other expendables) vary with the workload.
- Prorated overhead costs, such as utilities, maintenance, and telephone service. Overhead expenses can be allocated throughout the lifetime of the system using standard techniques of cost accounting.

Sample operating cost estimates for a solution are displayed in Figure D.2. After determining the costs and benefits for a possible solution, you can perform the cost-benefit analysis. Cost-benefit analysis techniques will be discussed shortly.

What Benefits Will the System Provide?

Because benefits or potential benefits become known prior to costs, we'll discuss benefits first. Benefits normally increase profits or decrease costs, both highly desirable characteristics of a new information system.

To as great a degree as possible, benefits should be quantified in dollars and cents. Benefits that can be easily quantified are called tangible benefits. **Tangible benefits** are usually measured in terms of monthly or annual savings or of profit to the firm. For example, consider the following scenario:

During the course of processing student housing applications, we discover that considerable data are being redundantly typed and filed. An analysis reveals that the same data is typed seven times, requiring an average of 44 additional minutes of clerical work per application. The office processes 1,500 applications per year. That means a total of 66,000 minutes or 1,100 hours of redundant work per year. If the average salary of a secretary is $6 per hour, the cost of this problem and the benefit of solving the problem is $6,600 per year.

✓ **TANGIBLE BENEFITS**

Fewer processing errors

Increased throughput

Decreased response time

Elimination of job steps

Reduced expenses

Increased sales

Faster turnaround

Better credit

Reduced credit losses

Alternatively, tangible benefits might be measured in terms of unit cost savings or profit. For instance, an alternative inventory valuation scheme may reduce inventory carrying cost by $0.32 per unit of inventory. Some examples of tangible benefits are listed in the margin.

Other benefits are intangible. **Intangible benefits** are those benefits believed to be difficult or impossible to quantify. Unless these benefits are at least identified, it is entirely possible that many projects would not be feasible. Examples of intangible benefits are listed in the margin.

Unfortunately, if a benefit cannot be quantified, it is difficult to accept the validity of an associated cost-benefit analysis. Why? Because that analysis is based on incomplete data. Some analysts dispute the existence of intangible benefits. They argue that all benefits are quantifiable; some are just more difficult than others. Suppose, for example, that improved customer goodwill is listed as a possible intangible benefit. Can we quantify goodwill? You might try the following analysis:

1. What is the result of customer "badwill"? The answer: the customer will submit fewer (or no) orders.

✓ **INTANGIBLE BENEFITS**

Improved customer goodwill

Improved employee morale

Improved employee job satisfaction

Better service to community

Better decision making

2. To what degree will a customer reduce orders? Your user may find it difficult to specifically quantify this impact. But you could try to have the end-user estimate the possibilities (or invent an estimate to which the end-user can react). For instance,

- There is a 50 percent (.50) chance that the regular customer would send a few orders—fewer than 10 percent of all their orders—to competitors to test their performance.
- There is a 20 percent (.20) chance that the regular customer would send as many as half their orders (.50) to competitors, particularly those orders we are historically slow to fulfill.
- There is a 10 percent (.10) chance that a regular customer would send us an order only as a last resort. That would reduce that customer's normal business with us to 10 percent of their current volume (90% or .90 loss).
- There is a 5 percent (.05) chance that a regular customer would choose not to do business with us at all (100% or 1.00 loss).

3. We can calculate an estimated business loss as follows:

$$
\begin{aligned}
loss = &\ .50 \times (.10 \text{ loss of business}) \\
&+ .20 \times (.50 \text{ loss of business}) \\
&+ .10 \times (.90 \text{ loss of business}) \\
&+ .50 \times (1.00 \text{ loss of business}) \\
= &\ .29 = 29\% \textit{ statistically estimated } \text{loss of business}
\end{aligned}
$$

4. If the average customer does $40,000/year of business, then we can expect to lose 29 percent or $11,600 of that business. If we have 500 customers, this can be expected to amount to a total of $5,800,000.

5. Present this analysis to management, and use it as a *starting point* for quantifying the benefit.

Is the Proposed System Economically Feasible (or Cost-Effective)?

There are three popular techniques to assess economic feasibility, also called *cost-effectiveness*. These three techniques are

- payback analysis
- return on investment
- net present value

The choice of techniques should take into consideration the audiences that will use them. Virtually all managers who have come through business schools are familiar with all three techniques. One concept that should be applied to each technique is the adjustment of cost and benefits to reflect the *time value of money*.

The Time Value of Money A concept shared by all three techniques is a reality of the financial world, the *time value of money*. A dollar today is worth more than a dollar one year from now. Why? Because you could invest that dollar today and, through accrued interest, have more than one dollar a year from now. Thus, you'd rather have that dollar today than in one year. That's why your creditors want you to pay your bills promptly—they can't invest what they don't have. The same principle can be applied to costs and benefits *before* a cost-benefit analysis is performed.

Some of the costs of a system will be accrued after implementation. Additionally, all benefits of the new system will be accrued in the future. Before cost-benefit analysis, these costs should be brought back to *current dollars*. An example should clarify the concept.

Suppose we are going to realize a benefit of $20,000 two years from now. What is the current dollar value of that $20,000 benefit? The current value of the benefit is the amount of money we would need to invest today to have $20,000 two years from now. If the current return on investments is running about 10 percent, an investment of $16,528 today would give us our $20,000 in two years (we'll show how to calculate this later). Therefore, the current value of the estimated benefit is $16,528—that is, we'd rather have $16,528 today than promise of $20,000 two years from now.

The same adjustment could be made on costs that are projected into the future. For example, suppose we are projecting a cost of $20,000 two years from now. What is the current dollar value of that $20,000 cost? The current value of the cost is the amount of money we would need to invest today to have $20,000 to pay the cost two years from now. Again, if we assume a 10 percent return on current investments, an investment of $16,528 today would give us the needed $20,000 in two years. Therefore, the current value of the estimated cost is $16,528—that is, we can fulfill our cost obligation of $20,000 in two years by investing $16,528 today.

Why go to all this trouble? Because projects are often compared against other projects that have different lifetimes. Time value analysis techniques have become the preferred cost-benefit methods for most managers. By time-adjusting costs and benefits, you can improve all of the cost-benefit techniques to be discussed in the remainder of this section.

Now, let's examine three cost-benefit analysis techniques.

Payback Analysis   The **payback analysis** technique is a simple and popular method for determining if and when an investment will pay for itself. Because systems development costs are incurred long before benefits begin to accrue, it will take some period of time for the benefits to overtake the costs. And after implementation, you will incur additional operating expenses that must be recovered. Payback analysis determines how much time will lapse before accrued benefits overtake accrued and continuing costs. This period of time is called the *payback period.*

In Figure D.3, we see an information system that will be developed at a cost of $100,000. The estimated net operating costs for each of the next 6 years are also recorded in the table. The estimated net benefits over the same 6 operating years are also shown. What is the payback period?

First, we need to adjust the costs and benefits for the time value of money (that is, adjust them to current dollar values). Here's how! The present value of a dollar in year n depends on something typically called a *discount rate.* The discount rate is a percentage similar to interest rates that you earn on your savings account. In most cases the discount rate for a business is the *opportunity cost* of being able to invest money in other projects—including the possibility of investing in the stock market, money market funds, bonds, and the like. Alternatively, a discount rate could represent what the company considers an acceptable return on its investments. This number can be learned by asking any financial manager, officer, or comptroller. Let's say that the discount rate for our sample company is 12 percent.

The *current value,* actually called the *present value,* of a dollar at any time in the future can be calculated using the following formula:

$$PV_n = \frac{1}{(1 + i)^n}$$

where PV_n is the present value of $1.00 n years from now and i is the discount rate.

	A	B	C	D	E	F	G	H
1	Net Present Value Analysis of On-Line Conversion							
2	Alternative for Member Service System							
3								
4								Year 6
5	Cash flow description	Year 0	Year 1	Year 2	Year 3	Year 4	Year 5	
6								
7	Analysis, design, and implementation cost	$100,000						
8	Operation and maintenance cost		$4,000	$4,500	$5,000	6,000	$7,000	$8,000
9								
10	Discount factors for 12%	1.00	0.89	0.80	0.71	0.64	0.57	0.51
11	Time-adjusted costs (current or present values)	100,000	3,572	3,587	3,560	3,816	3,969	4,056
12	Cumulative lefetime time-adjusted costs	100,000	103,572	107,159	110,719	114,535	118,504	122,560
13								
14	Benefits derived from operation of new system	0	25,000	30,000	35,000	50,000	60,000	70,000
15								
16	Discount factor of 12%	1.00	0.89	0.80	0.71	0.64	0.57	0.51
17	Time-adjusted benefits (current or present value)	0	22,325	23,910	24,920	31,800	34,020	35,490
18	Cumulative lifetime benefits	0	22,325	46,235	71,155	102,955	136,975	172,465
19								
20				– – – – – – – – – (payback period) – – – – – – – –				

FIGURE D.3   Payback Analysis for a Project   In payback analysis, you determine how much time will pass before the lifetime benefits exceed the lifetime costs. In this example, benefits will pass costs between years 4 and 5.

Therefore, the present value of a dollar two years from now is

$$PV_2 = \frac{1}{(1 + .12)^2} = 0.797$$

Does that bother you? Earlier we stated that a dollar today is worth more than a dollar a year from now. But it looks as if it is worth less, no? This is an illusion. The present value is interpreted as follows. If you have 79.7 cents today, it is better than having 79.7 cents two years from now. How much better? Exactly 20.3 cents better since that 79.7 cents would grow into one dollar in two years (assuming our 12% discount rate).

To determine the present value of any cost or benefit in year 2, you simply multiply 0.797 times the estimated cost or benefit. For example, the estimated operating expense in year 2 is $4,500. The present value of this expense is $4500 × 0.797 or $3,587 (rounded up). Fortunately, you don't have to calculate discount factors. There are tables similar to the partial one shown in Figure D.4 that show the present value of a dollar for different time periods and discount rates. Simply multiply this number times the estimated cost or benefit to get the present value of that cost or benefit.

Better still, most spreadsheets include built-in functions for calculating the present value of any cash flow, be it cost or benefit. All of the examples in this module were done with Microsoft Excel. The same tables can be

PRESENT VALUE OF A DOLLAR

Periods . . .	8%	10%	12%	14% . . . .
1	0.926	0.909	0.893	0.877
2	0.857	0.826	0.797	0.769
3	0.794	0.751	0.712	0.675
4	0.735	0.683	0.636	0.592
5	0.681	0.621	0.567	0.519
6	0.630	0.564	0.507	0.456
7	0.583	0.513	0.452	0.400
8	0.540	0.467	0.404	0.351
.				
.				
.				

FIGURE D.4   Partial Table for Present Value of a Dollar   This partial table is used to discount a dollar back to present value from the indicated years, using the indicated discount rates. More detailed versions of this table can be found in many accounting, finance, and economics books.

prepared with Lotus 1-2-3 or Borland's Quattro. The beauty of a spreadsheet is that, once the rows, columns, and functions have been set up, you simply enter the costs and benefits and let the spreadsheet discount the numbers to present value. (In fact, you can also program the spreadsheet to perform the cost-benefit analysis.)

Returning to Figure D.3, we have brought all costs and benefits for our example back to present value. Notice that the discount rate for year 0 is 1.000. Why? The present value of a dollar in year 0 is exactly $1. It makes sense. If you hold a dollar today, it is worth exactly $1!

Now that we've discounted the costs and benefits, we can complete our payback analysis. Look at the shaded rows, cumulative lifetime costs and benefits. The lifetime costs are gradually increasing over the 6-year period because operating costs are being incurred. But also notice that the lifetime benefits are accruing at a much faster pace. Lifetime benefits will overtake the lifetime costs between years 4 and 5. By extrapolating, we can estimate that the break-even will occur approximately 4.5 years after the system has been placed into operation. Is this information system a good or bad investment?

It depends! Many companies have a payback period guideline for all investments. In the absence of such a guideline, you need to determine a reasonable guideline before you determine the payback period. Suppose that the guideline states that all investments must have a payback period less than or equal to 5 years. Because our example has a payback period of 4.5 years, it is a good investment. If the payback period for the system were greater than 5 years, the information system would be a bad investment.

It should be noted that you can perform payback analysis without time-adjusting the costs and benefits. The result, however, would show a 3.9-year payback that looks more attractive than the 4.5-year payback that we calculated. Thus, non-time-adjusted paybacks tend to be overoptimistic and misleading.

Return on Investment Analysis The **return-on-investment (ROI) analysis** technique compares the lifetime profitability of alternative solutions or projects. The ROI for a solution or project is a percentage rate that measures the relationship between the amount the business gets back from an investment and the amount invested. The ROI for a potential solution or project is calculated as follows:

$$ROI = \frac{\text{estimated lifetime benefits} - \text{estimated lifetime costs}}{\text{estimated lifetime costs}}$$

Let's calculate the ROI for the same systems solution we used in our discussion of payback analysis. Once again, all costs and benefits should be time-adjusted. The time-adjusted costs and benefits were presented in rows 11 and 17 of Figure D.3. The estimated lifetime benefits minus estimated lifetime costs equals $49,815 (= $172,465 − $122,650). Therefore, the ROI is

$$ROI = 49,815/\$122,650 = .406 = 41\%$$

This is a lifetime ROI, *not* an annual ROI. Simple division by the lifetime of the system yields an average ROI of 6.7% per year. This solution can be compared with alternative solutions. The solution offering the highest ROI is the best alternative. However, as was the case with payback analysis, the business may set a minimum acceptable ROI for all investments. If none of the alternative solutions meets or exceeds that minimum standard, then none of the alternatives is economically feasible.

Once again, spreadsheets can greatly simplify ROI analysis through their built-in financial analysis functions.

We could have calculated the ROI without time-adjusting the costs and benefits. This would, however, result in a misleading 100.74% lifetime or a 16.8% annual ROI. Consequently, we recommend time-adjusting all costs and benefits to current dollars.

Net Present Value   The **net present value** of an investment alternative is considered the preferred cost-benefit technique by many managers, especially those who have substantial business schooling. Once again, you initially determine the costs and benefits for each year of the system's lifetime. And once again, we need to adjust all of the costs and benefits back to present dollar values.

Figure D.5 illustrates the net present value technique. We have brought all costs and benefits for our example back to present value. Notice again that the discount rate for year 0 (used to accumulate all development costs) is 1.000 because the present value of a dollar in year 0 is exactly $1.

After discounting all costs and benefits, subtract the sum of the discounted costs from the sum of the discounted benefits. This is called the *net present value* of the solution. If it is positive, the investment is good. If negative, the investment is bad. When comparing multiple solutions or projects, the one with the highest positive net present value is the best investment (*Note:* This even works if the alternatives have different lifetimes!) In our example, the solution being evaluated yields a net present value of $46,906. This means that if we invest $46,906 at 12 percent for 6 years, we will make the same profit that we'd make by implementing this information systems solution. This is a good investment provided no other alternative has a net present value greater than $46,906.

Once again, spreadsheets can greatly simplify net present value analysis through their built-in financial analysis functions.

Summary

Feasibility is a measure of how beneficial the development of an information system would be to an organization. Feasibility analysis is the process by which we measure feasibility. It is an ongoing evaluation of feasibility at various checkpoints in the life cycle. At any of these checkpoints, the project may be canceled, revised, or continued. This is called a creeping commitment approach to feasibility. There are three feasibility tests: operational, technical, and economic.

Operational feasibility is a measure of problem urgency or solution acceptability. It includes a measure of how the end-users and managers feel about the problems or solutions. Technical feasibility is a measure of how practical solutions are and whether the technology is already available within the organization. If the technology is not available to the firm, technical feasibility also looks at whether it can be acquired. Economic feasibility is a measure of whether a solution will pay for itself or how profitable a solution will be. For management, economic feasibility is the most important of our three measures.

	A	B	C	D	E	F	G	H	I
1	Net Present Value Analysis of On-Line Conversion								
2	Alternative for Member Service System								
3									
4									
5	Cash flow description	Year 0	Year 1	Year 2	Year 3	Year 4	Year 5	Year 6	Total
6									
7	Analysis, design, and implementation cost	-100,000							
8	Operation and maintenance cost		-4,000	-4,500	-5,000	-6,000	-7,000	-8,000	
9	Discount factors for 12%	1.00	0.89	0.80	0.71	0.64	0.57	0.51	
10	Present value of costs	-100,000	-3572	-3,587	-3,816	-3,816	-3,969	-4,056	
11	Total present value of cost								-122,560
12									
13	Benefits derived from operation of new system	0	25,000	30,000	35,000	50,000	60,000	70,000	
14	Discount factor of 12%	1.00	0.89	0.80	0.71	0.64	0.57	0.51	
15	Present value of benefits	0	22.325	23,910	24,920	31,800	34,020	35,490	
16	Total present value of benefits								172,465
17									
18	NET PRESENT VALUE								$49,906

To analyze economic feasibility, you itemize benefits and costs. Benefits are either tangible (easy to measure) or intangible (hard to measure). To properly analyze economic feasibility, try to estimate the value of all benefits. Costs fall into two categories: development and operating. Development costs are one-time costs associated with analysis, design, and implementation of the system. Operating costs may be fixed over time or variable with respect to system usage. Given the costs and benefits, economic feasibility is evaluated by the techniques of cost-benefit analysis.

Cost-benefit analysis determines if a project or solution will be cost-effective—if lifetime benefits will exceed lifetime costs. There are three popular ways to measure cost-effectiveness: payback analysis, return-on-investment analysis, and net present value analysis. Payback analysis defines how long it will take for a system to pay for itself. Return-on-investment and net present value analysis determine the profitability of a system. Net present value analysis is preferred because it can compare alternatives with different lifetimes.

Problems and Exercises

1. What is the difference between feasibility and feasibility analysis?
2. Explain what is meant by the creeping commitment approach to feasibility. What feasibility checkpoints can be built into a systems development life cycle?

3. Visit a local information systems shop. Try to obtain documentation of their systems development life cycle standards or guidelines. What feasibility checkpoints have they installed? What feasibility checkpoints do you think they should install? (*Note:* Don't be misled into believing that only during phases labeled *feasibility* is feasibility analyzed. There may be other points in the life cycle where this also happens.)

4. What are the three tests for project feasibility? How is each test for feasibility measured?

5. What feasibility criteria does the information systems shop you visited for Exercise 3 use to evaluate projects? How do their criteria compare against the criteria in this book? Have we omitted any tests that they feel are important? Have they omitted any tests we use?

6. Can you think of any technological trends or products that may be technically infeasible for the small- to medium-sized business at the current time? Defend your reasoning.

7. Whether or not you have information systems experience, you have experience with people who use computers (including friends, relatives, acquaintances, teachers, and fellow employees). Taking into consideration their biases for and against computers, identify issues that may make a proposed system operationally infeasible or unacceptable to those individuals.

8. What is the difference between a tangible and an intangible benefit? Give several examples of each. How would you quantify each in terms of dollars and cents—a measure that management can understand? Note that tangible benefits should be easy. Intangible benefits are harder, but pretend that management insists that you quantify the benefits.

9. What is the difference between fixed and variable operating costs? Give several examples of each.

10. What are some of the advantages and disadvantages of the payback analysis, return-on-investment analysis, and present value analysis cost-benefit techniques?

11. A new production scheduling information system for XYZ Corporation could be developed at a cost of $125,000. The estimated net operating costs and estimated net benefits over seven years of operation would be

	Estimated Net Operating Costs	Estimated Net Benefits
Year 0	$125,000	0
Year 1	$3,500	$26,000
Year 2	$4,700	$34,000
Year 3	$5,500	$41,000
Year 4	$6,300	$55,000
Year 5	$7,000	$66,000

Assuming a 12 percent discount rate, what would be the payback period for this investment? Would this be a good or bad investment? Why?

12. What is the ROI (return on investment) for the project in Exercise 11?

13. What is the net present value of the investment in Exercise 11 if the current discount rate is 12 percent?

Annotated References and Suggested Readings

Gildersleeve, Thomas R. *Successful Data Processing Systems Analysis.* 2nd ed. Englewood Cliffs, N.J.: Prentice-Hall, 1985. This book provides an excellent chapter on cost-benefit analysis techniques. We are indebted to Gildersleeve for the *creeping commitment* concept.

Gore, Marvin, and John Stubbe. *Elements of Systems Analysis.* 4th ed. Dubuque, Iowa: Brown, 1988. The feasibility analysis chapter suggests an interesting matrix approach to identifying, cataloging, and analyzing the feasibility of alternative solutions for a system.

Wetherbe, James. *Systems Analysis and Design: Traditional, Structured, and Advanced Concepts and Techniques.* 2nd ed. St. Paul, Minn.: West, 1984. Wetherbe pioneered the PIECES framework for problem classification. In this module, we extended that framework to analyze operational feasibility of solutions.

INDEX